T0329419

Food Waste Recovery

Processing Technologies and Industrial Techniques

Food Waste Recovery
Processing Technologies and Industrial Techniques

Charis M. Galanakis
Galanakis Laboratories, Chania, Greece

ELSEVIER

AMSTERDAM • BOSTON • HEIDELBERG • LONDON • NEW YORK
OXFORD • PARIS • SAN DIEGO • SAN FRANCISCO • SINGAPORE
SYDNEY • TOKYO

Academic Press is an Imprint of Elsevier

Academic Press is an imprint of Elsevier
125, London Wall, EC2Y 5AS, UK
525 B Street, Suite 1800, San Diego, CA 92101-4495, USA
225 Wyman Street, Waltham, MA 02451, USA
The Boulevard, Langford Lane, Kidlington, Oxford OX5 1GB, UK

Notices

Knowledge and best practice in this field are constantly changing. As new research and experience broaden our understanding, changes in research methods, professional practices, or medical treatment may become necessary.

Practitioners and researchers must always rely on their own experience and knowledge in evaluating and using any information, methods, compounds, or experiments described herein. In using such information or methods they should be mindful of their own safety and the safety of others, including parties for whom they have a professional responsibility.

To the fullest extent of the law, neither the publisher nor the authors, contributors, or editors, assume any liability for any injury and/or damage to persons or property as a matter of products liability, negligence or otherwise, or from any use or operation of any methods, products, instructions, or ideas contained in the material herein.

British Library Cataloguing-in-Publication Data
A catalogue record for this book is available from the British Library

Library of Congress Cataloging-in-Publication Data
A catalog record for this book is available from the Library of Congress

ISBN: 978-0-12-800351-0

For information on all Academic Press publications
visit our website at http://store.elsevier.com/

Publisher: Nikki Levy
Acquisition Editor: Patricia Osborn
Editorial Project Manager: Jaclyn Truesdell
Production Project Manager: Caroline Johnson
Designer: Victoria Pearson

Typeset by Thomson Digital

Printed and bound in United States of America

Working together
to grow libraries in
developing countries

www.elsevier.com • www.bookaid.org

Contents

SECTION II CONVENTIONAL TECHNIQUES

SECTION III EMERGING TECHNOLOGIES

List of Contributors

Carlos Alvarez
Food Chemistry and Technology Department, Teagasc Food Research Centre, Ashtown, Dublin, Ireland

Violaine Athès-Dutour
Génie et Microbiologie des Procédés Alimentaires, INRA/AgroParisTech, Paris, France

Francisco J. Barba
Department of Nutrition and Food Chemistry, Faculty of Pharmacy, Universitat de València, Valencia, Spain

Lorenzo Bertin
Department of Civil, Chemical, Environmental and Materials Engineering (DICAM), University of Bologna, Bologna, Italy

Chiranjib Bhattacharjee
Department of Chemical Engineering, Jadavpur University, Kolkata, West Bengal, India

Sangita Bhattacharjee
Department of Chemical Engineering, Heritage Institute of Technology, Kolkata, West Bengal, India

Silvia Alvarez Blanco
Department of Chemical and Nuclear Engineering, Universitat Politècnica de València, Valencia, Spain

Mladen Brnčić
Department of Process Engineering, Faculty of Food Technology and Biotechnology, University of Zagreb, Zagreb, Croatia

Camelia Bucatariu
Rural Infrastructure and Agro-Industry Division (AGS), Food and Agriculture Organization of the United Nations (FAO), Italy

Alfredo Cassano
Institute on Membrane Technology-Consiglio Nazionale delle Ricerche, University of Calabria, Rende Cosenza, Italy

Sudip Chakraborty
Institute on Membrane Technology, University of Calabria, ITM-CNR, Rende, Italy

Smain Chemat
Division Santé Centre de Recherches Scientifique et Technique en Analyses Physico-Chimiques (C.R.A.P.C.), Bon-Ismail, Algeria

Patrick J. Cullen
School of Food Science and Environmental Health, Dublin Institute of Technology, Dublin, Ireland; University of New South Wales, Sydney, Australia

Isabel C.N. Debien
LASEFI/DEA (Department of Food Engineering)/FEA (School of Food Engineering)/UNICAMP (University of Campinas), Campinas, São Paulo, Brazil

Maria Dolores del Castillo
Food Bioscience Group, Institute of Food Science Research (CSIC-UAM), Madrid, Spain

Qian Deng
Milne Fruit Products Inc., Prosser, Washington, USA

Stella Despoudi
School of Business and Economics, Loughborough University, Loughborough, UK

Dimitar Angelov Dimitrov
LADEC Ltd., Plovdiv, Bulgaria

Herminia Domínguez
Department of Chemical Engineering, University of Vigo, Pontevedra, Spain

Elena Falqué
Department of Analytical Chemistry, University of Vigo, Pontevedra, Spain

Milad Fathi
Department of Food Science and Technology, Faculty of Agriculture, Isfahan University of Technology, Isfahan, Iran

Dario Frascari
Department of Civil, Chemical, Environmental and Materials Engineering (DICAM), University of Bologna, Bologna, Italy

Charis M. Galanakis
Department of Research & Innovation, Galanakis Laboratories, Chania, Greece

Lia Noemi Gerschenson
Industry Department, Natural and Exact Sciences School (FCEN), Buenos Aires University (UBA), Buenos Aires, Argentina; National Scientific and Technical Research Council of Argentina (CONICET), Buenos Aires, Argentina

Adem Gharsallaoui
BioDyMIA (Bioingénierie et Dynamique Microbienne aux Interfaces Alimentaires), Bourg en Bresse, France

Tamara Dapčević Hadnađev
Institute of Food Technology, University of Novi Sad, Novi Sad, Vojvodina, Serbia

Ching Lik Hii
Department of Chemical and Environmental Engineering, University of Nottingham, Malaysia Campus, Malaysia

Henry Jaeger
Department of Food Science and Technology, University of Natural Resources and Life Sciences (BOKU), Vienna, Austria

Seid Mahdi Jafari
Department of Food Materials and Process Design Engineering, Faculty of Food Science and Technology, University of Agricultural Sciences and Natural Resources, Gorgan, Iran

Paula Jauregi
Department of Food and Nutritional Sciences, University of Reading, Reading, UK

Canan Kartal
Department of Food Engineering, Faculty of Engineering, Ege University, Bornova, Izmir, Turkey

Attila Kovács
Faculty of Agricultural and Food Sciences, Institute of Biosystems Engineering, University of West Hungary, Hungary

Dimitris P. Makris
School of Environment, University of the Aegean, Lemnos, Greece

Ioanna Mandala
Department of Food Science & Human Nutrition, Agricultural University of Athens, Food Process Engineering Laboratory, Athens, Greece

Nuria Martinez-Saez
Food Bioscience Group, Institute of Food Science Research (CSIC-UAM), Madrid, Spain

Maria Angela de Almeida Meireles
LASEFI/DEA (Department of Food Engineering)/FEA (School of Food Engineering)/UNICAMP (University of Campinas), Campinas, São Paulo, Brazil

N.N. Misra
School of Food Science and Environmental Health, Dublin Institute of Technology, Dublin, Ireland

Vassiliki S. Mitropoulou
Business Consultant, Athens, Greece

Marwen Moussa
Génie et Microbiologie des Procédés Alimentaires, INRA/AgroParisTech, Paris, France

Anne Maria Mullen
Food Chemistry and Technology Department, Teagasc Food Research Centre, Ashtown, Dublin, Ireland

Arijit Nath
Department of Chemical Engineering, Jadavpur University, Kolkata, West Bengal, India

Semih Otles
Department of Food Engineering, Faculty of Engineering, Ege University, Bornova, Izmir, Turkey

Ivan Nedelchev Panchev
Department of Physics, University of Food Technologies, Plovdiv, Bulgaria

Maria Papageorgiou
Department of Food Technology, Alexander Technological Educational Institute of Thessaloniki (ATEITh), Kentriki Makedonia, Greece

Paola Pittia
Faculty of Bioscience and Technology for Food, Agriculture and Environment, University of Teramo, Teramo, Italy

Milica Pojić
Institute of Food Technology, University of Novi Sad, Novi Sad, Vojvodina, Serbia

Juliana M. Prado
Centro de Ciências da Natureza (CCN), UFSCar (Federal University of São Carlos), Buri, Brazil

Krishnamurthy Nagendra Prasad
Chemical Engineering Discipline, School of Engineering, Monash University of Malaysia, Malaysia

Eduardo Puértolas
Food Research Division, AZTI-Tecnalia, Bizkaia, Spain

Francisco Amador Riera Rodriguez
Chemical Engineering and Environmental Technology Department, University of Oviedo, Oviedo, Spain

Julia Schmidt
Department of Food Science and Technology, University of Natural Resources and Life Sciences (BOKU), Vienna, Austria

Isabelle Souchon
Génie et Microbiologie des Procédés Alimentaires, INRA/AgroParisTech, Paris, France

Halagur Bogegowda Sowbhagya
Department of Plantation Products, Spices and Flavour Technology, CSIR – Central Food Technological Research Institute, Mysore, Karnataka, India

Giorgia Spigno
Institute of Oenology and Agro-Food Engineering, Università Cattolica del Sacro Cuore, Piacenza, Italy

Reza Tahergorabi
Department of Family and Consumer Sciences, Food and Nutritional Sciences, North Carolina Agricultural & Technical State University, Greensboro, North Carolina, USA

Renata Vardanega
LASEFI/DEA (Department of Food Engineering)/FEA (School of Food Engineering)/UNICAMP (University of Campinas), Campinas, São Paulo, Brazil

Ooi Chien Wei
Discipline of Chemical Engineering, School of Engineering, Monash University, Malaysia

Hiroshi Yoshida
Department of Materials Chemistry and Bioengineering, Oyama National College of Technology, Oyama, Tochigi, Japan

Preface

As long as food processing exists, nonconsumed materials will be considered a substrate of treatment, minimization, or prevention. On the other hand, the prospect of recovering high added-value compounds from these materials is a scenario that started few decades ago. The first successful efforts dealt with the recovery of oil from olive kernel; the production of essential oils; flavonoids, sugars, and pectin from citrus peel, as well as the recapture of protein concentrates and lactose from cheese whey. These commercially available applications inspired the scientific community to intensify its efforts towards the valorization of all kinds of food by-products for recovery purposes. Besides, the perpetual disposal of highly nutritional proteins, antioxidants, and dietary fibers in the environment is a practice that could not be continued for a long time within the sustainability and bioeconomy framework of the modern food industry. Indeed, the depletion of food sources, the fast-growing population, and the increasing need for nutritionally correct diets do not allow other alternatives to be considered. Nowadays, many relevant projects are in progress around the world and across different disciplines, whereas the existence of numerous scientific patents, articles, congresses, and commercialization efforts has brought about a wealth of literature in the field. Despite this plethora of information and the developed technologies that promise the recovery, recycling, and sustainability of valuable compounds inside the food chain, the respective shelf products remain rather limited. This is happening because the industrial implementation of the recovery processes is a complex approach, necessitating careful consideration of different aspects. A commercially feasible product can be manufactured only if a certain degree of flexibility and alternative choices can be adapted in the developing methodology. This book is envisaged to investigate the real full-scale applications and fill the gap between academia and industry within the particular topics. The main aim is to emphasize the advantages and disadvantages of processing technologies and techniques, as well as to provide a holistic approach for the recovery of valuable components from food wastes. This is conducted by adapting the different applied technologies in a recovery strategy, which could be implemented regardless of the nature of the food waste and the characteristics of the target compound in each case.

The book consists of 4 major sections and 16 chapters. Section I (Introduction) includes three chapters. Chapter 1 covers aspects of food waste management, valorization, and sustainability in the food industry. In Chapter 2, emphasis is given on the classification of food waste sources, the identification of the target compounds, and potential applications in each case. Chapter 3 focuses on the development of the *"Universal Recovery Strategy"*, which takes into account all the necessary aspects (e.g. substrate collection and deterioration, yield optimization, preservation of target compound functionality during processing, etc.) needed for the development of a recovery process. Moreover, it describes the five-stages recovery approach (macroscopic pretreatment, macro- and micromolecules separation, extraction, purification and isolation, and product formation). The technologies used for the recovery of valuable compounds from food wastes are presented in detail in Sections II and III. In particular, Chapters 4, 5, 6, 7, and 8 (Section II) describe the different conventional technologies implemented in each of the aforementioned five stages. Similarly, Chapters 9, 10, 11, 12, and 13 (Section III) explore the different emerging technologies applied in the respective stages. Section IV consists of three chapters that investigate implementation aspects and potential applications of recovered materials. Thereby, safety and cost issues of emerging versus conventional technologies are debated in Chapter 14. Patented methodologies, real-market products, and commercialized applications as adapted to the 5-Stage

Universal Recovery Process are discussed in Chapter 15. Finally, Chapter 16 explores the recovery and applications of enzymes from food wastes. The above chapters are authored by 50 experts from several countries in order to display the different perspectives and cover as many developments in the field as possible.

Conclusively, the ultimate goal of the book is to provide a handbook for anyone who wants to develop a food waste recovery application. It is intended to support researchers, scientists, food technologists, engineers, professionals, and students working or studying in the areas of food, by-products, and the environment. The most important feature of the book is that it covers recovery issues from an integral point of view, i.e. by investigating each stage separately and keeping a balance between the characteristics of the current conventional techniques and emerging technologies. Likewise, some key chapters (e.g. 14 and 15) provide information on how to develop an economic and safe recovery methodology, and at the same time give details about commercial products and industrial applications. These issues allow the reader to come closer and understand the success stories in the field.

I would like to take this opportunity to thank all the contributors of this book for their fruitful collaboration, by incorporating different topics and technologies in one integral and comprehensive text. I consider myself fortunate to have had the opportunity to collaborate with so many knowledgeable colleagues from Greece, Spain, Italy, Malaysia, India, Argentina, Algeria, Australia, Hungary, Japan, Ireland, Austria, Turkey, Brazil, Croatia, Bulgaria, France, Iran, Serbia, the United Kingdom, and the USA who have prepared the chapters based on their personal research experience. Their acceptance of editorial guidelines and book concept is highly appreciated. In addition, I gratefully acknowledge the support of Special Interest Group 5 (Food Waste Recovery) of ISEKI Food Association, which is the most relevant and fastest growing group worldwide in this particular field. I would also like to thank the acquisition editor Patricia M. Osborn for her honorary invitation to lead this project, and the production team of Elsevier, particularly Carrie Bolger and Jacklyn Truesdell, for their assistance during the editing process, as well as Caroline Johnson during production.

Last but not least, a message for you, the reader. In a collaborative project of this size it is impossible for it not to contain errors. Therefore, if you find errors or have any objections, please do not hesitate to contact me.

Charis M. Galanakis
Research & Innovation Department
Galanakis Laboratories
Chania, Greece
e-mail: cgalanakis@chemlab.gr

INTRODUCTION

I

FOOD WASTE MANAGEMENT, VALORIZATION, AND SUSTAINABILITY IN THE FOOD INDUSTRY

Semih Otles*, Stella Despoudi**, Camelia Bucatariu†, Canan Kartal*

*Department of Food Engineering, Faculty of Engineering, Ege University, Bornova, Izmir, Turkey;
**School of Business and Economics, Loughborough University, Loughborough, UK;
†Rural Infrastructure and Agro-Industry Division (AGS), Food and Agriculture Organization
of the United Nations (FAO), Italy

1.1 INTRODUCTION

Nutrition systems and food security were always among the first priorities of the Food and Agriculture Organization (FAO)'s efforts. Unfortunately, the food produced worldwide is not accessible to all people since the majority are living in developing regions (FAO, 2009b). According to the global estimates of the State of Food Insecurity in the World (FAO, IFAD, and WFP, 2014), about 805 million people were chronically undernourished between 2012 and 2014. Hunger reduction requires an integrated approach, including: (i) public and private investments to raise agricultural productivity; (ii) better access to inputs, land, services, technologies, and markets; (iii) measures to promote rural development; (iv) social protection of the most vulnerable population, including strengthening resilience to conflicts and natural disasters; and (v) specific nutrition programs to address micronutrient deficiencies in mothers and children under 5 years of age.

Today, food security is continuously stressed due to the scarcity of natural resources, population growth, fluctuating food prices, moderations in consumer habits, climate change, and food loss and waste (FAO, 2011a). For instance, the world population is estimated to reach 9 billion by 2050, requiring a 60–70% increase in food production (Foresight, 2011). Developing countries will play a fundamental role in population growth along with developed countries. Urbanization will continue, reaching up to 70% of the world's population (FAO, 2009a). Since 2007, the global population has been predominantly urban, whereas the United Nations World Urbanization Prospects to 2025 estimated a further expansion (United Nations, 2012). These shifts will generate a great demand for shelter, livelihood, and food supply. On the other hand, climate change and scarcity of natural resources restrict agriculture growth and food production. This means that a 70% increase in food production to feed 9 billion people will be very hard to achieve (Hodges et al., 2010).

Signs of degradation of natural resources worldwide (e.g. decline in land, water, and biodiversity) create critical concerns about meeting the future demands at a global level. In the next 50 years, not only the increase in population but also increasing urbanization and rising incomes will bring a rapid growth in food processing industries and alter food supply chains all over the world. Available but limited resources to feed the world population, pressures in natural resources, and environmental issues will thus gain much attention in the years to come. In particular, keeping a balance between future demands and sustainable supply will be the next challenge. The latter can only be met by (Foresight, 2011; FAO, 2012, 2014; HLPE, 2014):

1. using knowledge optimally,
2. introducing innovative science and technology,
3. reducing and preventing food loss and waste,
4. improving governance of the food system,
5. establishing sustainable diets.

1.2 DEFINITIONS OF "FOOD WASTE" AND "FOOD LOSS"

Food supply chains begin from the primary agricultural phase, proceed with manufacturing and retail, and end with household consumption. During this life cycle, food is lost or wasted because of technological, economic, and/or societal reasons. The definitions of "food waste" and "food loss" within the supply chain have been a subject of disagreement among related scientists. According to the European Union (EU) Commission Council Directive 2008/98/EC, "waste" is defined as "any substance or object, which the holder discards or intends or is required to discard". According to the Foresight Project report prepared by the Government Office for Science (Foresight Project, 2011), food waste is defined as "edible material intended for human consumption that is discarded, lost, degraded, or consumed by pests as food travels from harvest to consumer". Food supply chain and postharvest systems are two other terms under debate in the literature, whereas "postharvest loss" is generally referred as "food loss" and "spoilage" as well. The term "postharvest loss" contains agricultural systems and the onward supply of produce to the markets (Grolleaud, 2002; Parfitt et al., 2010). "Food loss" refers to quantitative and qualitative reductions in the amount and value of food (Premanandh, 2011). Qualitative loss means loss of caloric and nutritive value, loss of quality, and loss of edibility (Kader, 2005). Qualitative loss refers to the decrease in edible food mass throughout the part of the supply chain that specifically leads to edible food for human consumption.

In July 2014, the European Commission (European Commission, in press a and b) has announced its targets for the circular economy and waste management, and provided a "food waste" definition as "food (including inedible parts) lost from the food supply chain, not including food diverted to material uses such as bio-based products, animal feed, or sent for redistribution" (e.g. food donation). Concurrently, all member states of the European Union shall establish frameworks to collect and report levels of food waste across all sectors in a comparable way. The latest data are requested to develop national food waste prevention plans, aimed to reach the objective to reduce food waste by at least 30% between January 1, 2017 and December 31, 2025. To enable the process, the Commission shall adopt implementing acts by December 31, 2017 in order to establish uniform conditions for monitoring the implementation of food waste prevention measures taken by member states of the EU. However, on 16 December 2014, the EC announced the withdrawal of the Circular Economy package (proposed in July 2014) from the European Commission's new work programme. The package addressed policy and regulatory areas such as waste (including food waste), recycling, incineration, and landfill. A new proposal is targeted by end of 2015 (Euractiv, 2014).

The FAO (2014) (FAO SAVE FOOD, 2014) global voluntary definitional framework defined "food loss" as the decrease in quantity or quality of food, caused mainly by food production and supply system functioning or its institutional and legal framework. Thereby, "food loss" occurs throughout the food supply chain. Moreover, the FAO distinguishes "food waste" as an important part of "food loss", which refers to the removal of food fit for consumption from the supply chain by choice or food that has been left to spoil or expire as a result of negligence (predominantly but not exclusively) by the final consumer at the household level.

1.3 QUANTITIES OF LOST AND WASTED FOOD AND IMPACT ON FOOD AND NUTRITION SECURITY

Food waste has an important impact on food and nutrition security, food quality and safety, natural resources, and environmental protection. It impacts food systems' sustainability and economic development. For these reasons, food loss, food waste, co- and by-products management have already drawn the attention of food scientists and industry over the last decades. Indeed, there are an increasing number of scientific literature and reports relevant to food waste and treating methods. The latter includes the reduction of waste production, the valorization of co- and by-products, and improvement of waste management. In 2011, the FAO published a first report considering global food losses and food waste. According to the report, nearly one-third of worldwide food production for human consumption is lost or wasted. The amounts of food loss and waste along the food supply chains are, respectively, 54% of total loss and waste as upstream processes (including production and postharvest) and 46% of total loss and waste as downstream processes (including processing, distribution, and consumption) (Fig. 1.1) (FAO, 2011a).

Since 1974, in the United States, food waste has progressively increased by 50%, reaching more than 1400 kcal/person/day (Hall et al., 2009). Furthermore, not only is food wasted, but also large amounts of land, energy, water, and agricultural inputs are lost during the production of food. A European Commission technical report (published in 2010) indicated that around 90 million tonnes of food waste are

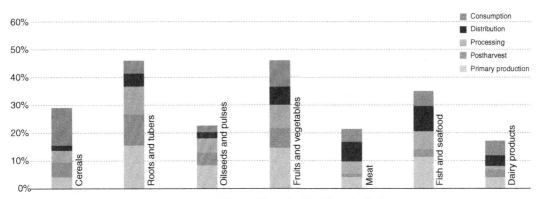

FIGURE 1.1 The Percentage of Food Loss and Waste Along the Food Supply Chains

FAO, 2011a

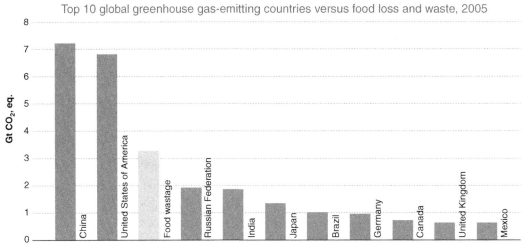

FIGURE 1.2 Global Greenhouse Gas-Emitting Countries Versus Food Loss and Waste

FAO, 2013a

generated within the European Union each year (European Commission, 2010). The percentage break-down of food waste according to this report is 39% manufacturing, 42% households, 14% food service/catering, and 5% retail/wholesale (2006 EUROSTAT data and various national sources provided by EU member states). Based on this study, it is expected that food waste would reach 126 Mt in 2020 (from about 89 Mt in 2006), without additional prevention policy or activities. From 2006 to 2020, food waste tonnages are expected to be 3.7 million in EU27 when the population increases by nearly 21 million. It can be estimated that at least 170 Mt of CO_2 eq. are emitted only because of food waste (1.9 t CO_2 eq./t of food wasted). For the manufacturing sector, 59 million tonnes of CO_2 eq. emitted per year and respective food waste amounts are responsible for ~35% of annual greenhouse gas emissions (Fig. 1.2) (FAO, 2013a). Without accounting for greenhouse gas emissions from land use change, the carbon footprint of uneaten food is estimated to be 3.3 Gt of CO_2 eq. (FAO, 2013a).

1.4 PROSPECTS

Until the end of the twentieth century, disposal of food, loss, and waste were not evaluated as a matter of concern. The prevalent policy was mainly to increase food production, without improving the efficiency of the food systems. This fact increased generation of food lost or wasted along supply chains. In the twenty-first century, escalating demands for processed foods have required identification of concrete opportunities to prevent depletion of natural resources, restrict energy demands, minimize economic costs as well as reduce food losses and wastes. According to the FAO (2014), a management strategy for resource optimization via waste reduction at source is producing the greatest benefit to the food processing industries and society. Moreover, recent changes in the legislative frameworks, environmental concerns, and increasing attention toward sustainability have stimulated industry to reconsider the concept of "recovery" as an opportunity.

Food processing industries generate a dramatic amount of waste (liquid and solid), consisting primarily of the organic residues of processed raw materials. Most of these materials, referred as "waste" by European legislations, could be utilized to yield value-added products. By-product is a common term in the food industry and represents a product formed during processing that may not count directly as a useful resource by its producer. However, by-products could still contain substances with a market value that can be turned into useful products (Chandrasekaran, 2013). The food industry used the term "waste" to characterize any kind of loss (e.g. raw materials, processed substances, energy, or even time) and lately the term "wasted by-products" is increasingly used. The possibility to recover by-products for food and feed needs varies from process to process because resources differ largely in each sector. For instance, vegetable by-products and plant residues may be used for the generation of innovative products such as dietary fibers, food flavors, food supplements, polyphenols, glucosinolates, protein concentrates, pectin, phytochemicals, and plant enzymes by upgrading them with value addition (Laufenberg et al., 2003; Tomás Barberán, 2007; Patsioura et al., 2011; Tsakona et al., 2012; Galanakis et al., 2013a, b, 2015b; Galanakis, 2011, 2015; Heng et al., 2015; Roselló-Soto et al., 2015; Deng et al., 2015). On the other hand, dairy potential "wastes" (that could also be classified as co- and by-products) contain active proteins, peptides, salts, fatty substances, and lactose (Kosseva, 2013b; Galanakis et al., 2014). Moreover, the meat industry by-products may constitute a considerable resource of protein and functional hydrolysates (Bhaskar et al., 2007).

1.5 ORIGIN OF FOOD WASTE AND FOOD LOSS

1.5.1 DISTRIBUTION IN THE DIFFERENT PRODUCTION STAGES

Food waste occurs predominantly, but not exclusively, at the final consumer stage of the food supply chain (e.g. household level). Food loss is generated during processing, distribution, retail, final consumption, and the postconsumption stages (FAO, 2011a, 2014; Parfitt et al., 2010). Figure 1.3 shows a generic model of the food supply chain and the different stages where food loss and food waste occur. Food losses at the first stage are usually due to mechanical damage and/or spillage during harvest

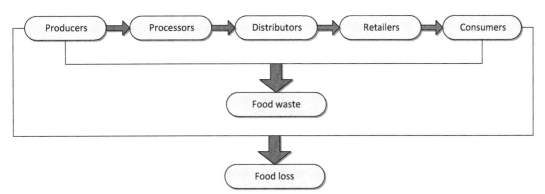

FIGURE 1.3 Food Supply Chain and Stages of Food Loss and Food Waste Occurrence

operations (e.g. threshing, fruit picking or crop sorting). Natural forces (e.g. temperature and weather conditions) and economic factors (e.g. regulations and public or private standards for quality and appearance) are the main causes of the above food losses (Kader, 2010). Food losses at the stages of postharvest, handling, storage, processing, and distribution are considered to be any decrease of edible food mass that was spoiled, spilled, or lost unintentionally. For example, food loss at the postharvest handling stage occurs due to spillage and degradation, lack of storage facilities, and transportation between farm and distribution (Kader, 2010; Akkerman and Van Donk, 2008). During storage, a considerable amount of food loss happens due to pests and microorganisms. However, food waste during industrial processing is due to spillages or degradation (e.g. juice production and canning). At this stage, food waste may also be generated during washing, peeling, slicing, boiling, process interruptions or by sorted-out crops. At the distribution stage, food waste is generated due to the lack of appropriate transportation methods, inappropriate packaging, time constraints, and supplier/buyer relationships along with weak infrastructure. At the retailer stage, food loss is usually referred as food waste since it is mainly generated due to a conscious decision to discharge food. This food is still safe and nutritious for human consumption. According to the Department for Environment Food & Rural Affairs (Defra, 2009), retailers' food waste is created due to poor demand forecasting, inventory mismanagement, temperature sensitivities, weather conditions during transportation, disposal of unsold food, inappropriate packaging, food policies, and their interpretation. Retailers throw away significant quantities of food that have reached dates with the following labels: "best before", "sell by", or "use by". Some retailers cooperate with charitable organizations and food recovery and redistribution entities (e.g. food banks) to distribute unsold food or advise consumers on how to use food that may be at risk of becoming waste (Kaye, 2011). Other retailers do not give away food; this avoids the risk of liability in case of food contamination. Besides, food waste at the consumer level arises due to individual shopping habits, lack of awareness, lack of knowledge on efficient food use, cultural issues, lack of appropriate shopping planning, packaging, and portion size issues (Defra, 2009). Governments around the world make efforts to reduce waste at this level, by diverting waste away from landfill through regulation, taxation, and public awareness (Mena et al., 2011). At the postconsumer stage, food waste can be classified as three types:

1. avoidable waste (food thrown away because it is no longer wanted or it has expired),
2. possibly avoidable waste (food that some people will eat and others will not, or that can be eaten when prepared in one way but not in another),
3. unavoidable waste (food waste from food preparation that was not edible under any circumstance; WRAP, 2009).

1.5.2 DISTRIBUTION IN TRANSITION AND INDUSTRIALIZED COUNTRIES

Food losses occurring from producer to distributor stage are estimated to be able to feed 1 billion people (Tomlinson, 2013). Food losses are also a waste of human effort, farm inputs, livelihoods, investments, and scarce natural resources such as water. In low-income countries, food losses and waste are, to date, mainly associated with the upstream supply chain (production to distribution), whereas the losses and waste in the industrialized world are more concerned with the downstream food supply chain (FAO, 2011a). As an example, reducing food losses in Africa is most important due to the structure of the food supply systems. At this region, losses come from wide-ranging technical and managerial limitations in harvesting techniques, storage, transportation, processing, cooling facilities, infrastructure, packaging, and marketing systems.

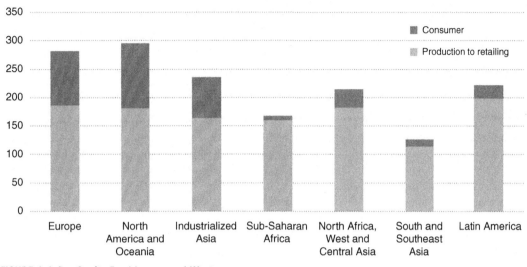

FIGURE 1.4 Per Capita Food Losses and Waste

FAO, 2011a

Some attempts have been made to quantify global food loss and waste. However, due to the variation of methodologies, appraisal levels and food products, it is difficult to estimate the actual loss figures (Premanandh, 2011). Figure 1.4 shows the per capita food loss and food waste for Europe, North America and Oceania, Industrialized Asia, Sub-Saharan Africa, North Africa, West and Central Asia, South and Southeast Asia, and Latin America (FAO, 2011a). As can be seen, the majority of food loss and waste is derived from production to retail stages, even in the industrialized countries of Europe. Food loss and waste at the consumer stage is mainly an issue for North America and Oceania, Europe, and Industrialized Asia. Sub-Saharan Africa and South and Southeast Asia seem to have minimum food waste at the consumer stage and the majority of the food is lost from production to retail stages.

In industrialized countries, food losses and waste are generated across supply chains and may be caused by managerial decisions, market signals, lack of access to appropriate technologies, and regulatory frameworks and their misinterpretation, along with social norms and inappropriate waste management strategies. A policy enabling environment, along with private sector targeted investments and civil society involvement, could facilitate rapid shifts toward a more sustainable food system, addressing concrete actions from primary production to the consumption level (FAO, 2014).

It is also important to note that food loss and waste figures differ from product to product. Food products can be classified into two groups (e.g. plant and animal) and seven subcategories (e.g. cereals, root and tubers, oil crops and pulses, fruit and vegetables, meat products, fish and seafood, and diary). Fruits, vegetables, roots, and tubers have the highest wastage levels. At the global level, 40–50% of root, fruits, and vegetables; 30% of cereals; 30% of fish; and 20% of oil seeds, meat, and dairy are lost or wasted (FAO, 2011a).

1.6 MANAGEMENT AND VALORIZATION STRATEGIES

The main trigger factors for sustainable waste management and valorization of food waste at the global level are as follows (Murugan et al., 2013):

1. renewed and stringent environmental legislations with increasing environmental concerns,
2. sustainable utilization of natural resources through technological developments,
3. waste disposal costs.

In particular, the food industry generates a high amount of biodegradable waste and discards large quantities of residues with a high biochemical oxygen demand and chemical oxygen demand contents. For this reason, worldwide legislative requirements for waste disposal have become increasingly restrictive over the last decade. Directives and regulations are enforcing handling and treatment of materials defined as wastes. Nevertheless, dealing with food wastes is difficult in many aspects. Inadequate biological stability and existence of pathogens can cause an increase in microbial activity. High water content (particularly for meat and vegetable wastes (FAO, 2006)) has an important effect on transport costs. Food wastes with high fat content are susceptive to oxidation and escalate spoilage due to continuous enzymatic activity (Russ and Meyer-Pittroff, 2004). General methods, such as incineration, anaerobic fermentation, composting, landfill, or using food residues for agricultural applications, such as animal feed or fertilizer, are the main strategies for waste minimization and valorization. Over the past few years, new management methods and treatments that focus on recovery and reutilization of valuable constituents from food wastes have generated more interest. Citrus by-products contain mainly pectin, flavonoids (Arvanitoyannis and Varzakas, 2008; Calvarano et al., 1996; El Nawawi, 1995), carotenoids (Chedea et al., 2008), fiber, and polyphenols (Larrauri et al., 1996). Huge quantities of water waste from olive oil production cause serious concerns on land and water environments, but could represent an alternative source of biologically active polyphenols (Mulinacci et al., 2001; Obied et al., 2005). Meat processing by-products could be a promising alternative source to recover functional ingredients such as proteins (Fonkwe and Singh, 1996; Jelen et al., 1979; Swingler and Lawrie, 1979).

1.6.1 POLICY IN THE EUROPEAN UNION

Waste prevention/minimization is the first priority in the Commission's Waste Management Strategies and Directive hierarchy (Fig. 1.5a). In 1989, the European Union published for the first time a strategy called "Community Strategy for Waste Management" (amended in 1996). This directive underlined that prevention of waste is the best option, followed by reuse, recycling, and energy recovery (the so-called "waste hierarchy"). Disposal (landfilling, incineration with low energy recovery) was defined as the worst environmental option. The Landfill Directive's (1999/31/EC) main purpose is to divert biodegradable waste (including food waste) out of landfills (Directive, 1999). At this time, the waste minimization strategy of the European Union included (Riemer and Kristoffersen, 1999):

1. waste prevention (more efficient production technologies),
2. internal recycling of production waste,
3. source-oriented improvement of waste quality,
4. reuse of products or parts of products, for the same purpose.

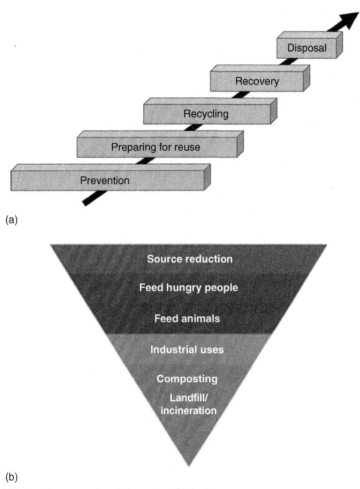

(a)

(b)

FIGURE 1.5 Food Waste Management and Recovery Hierarchy

(a) Waste management hierarchy according to EU Directive 2008/98/EC. (b) Food waste recovery hierarchy according to EPA, 2011.

Later, the Sixth Environmental Action Programme (6thEAP) provided the framework for a period of 10 years (2001–2010) and developed a vision integrating resource, product, and waste policies. The 6thEAP included four main issues: climate change, nature and biodiversity, health and quality of life, and natural resources and waste. Seven strategic policies including sustainable use of natural resources and recycling of waste were also mentioned. "Prevention" was later appended to the recycling of waste strategy (Derham, 2010; Kosseva, 2013a). In addition, the waste strategy of the European Union set out legal principles aimed at limitation of waste production as far as possible, taking responsibilities for waste by producer, precautionary behavior associated with waste problems and proximity principle for distance between production and disposal place (Sanders and Crosby, 2004).

Following this policy, the third strategy entitled "Taking Sustainable Use of Resources Forward – A Thematic Strategy on the Prevention and Recycling of Waste" was published in 2005. This strategy contains simplification, modernization, and full implementation of existing legislation "life-cycle thinking", promotion of waste prevention policies, and development of reference standards for recycling and elaboration of the EU's policy. Decrease in landfill, compost and energy recovery, better recycling, utilization of by-products, and prevention of waste were the potential impacts of the strategy (Derham, 2010).

In July 2014, the "Proposal for a Directive of the European Parliament and of the Council amending Directives 2008/98/EC on waste (...)" denoted that Member States of the European Union should set priorities based on the aforementioned waste management hierarchy (Directive, 2008a). The criteria, under which categories of potential food waste can be diverted toward human consumption and used as animal feed, were assessed carefully. Indeed, they provided a priority over other solutions such as composting, and creation of energy and landfill. This assessment should take into account the particular economic circumstances, health, and quality standards, but it should always be in line with the EU legislation regarding food safety and animal health. In addition, the biodegradable waste going to landfills should be reduced by up to 35% by 2016. Biodegradable food and kitchen waste from households, restaurants, caterers, and retail premises, and comparable waste from food processing plants, are covered by the EU legislation (Directive 2008/98/EC on Waste) definition of "bio-waste" (Directive, 2008b). The Green Paper on bio-waste (COM(2008) 811 final) highlights the risk of an increase in food waste generation (especially in EU12) and states that prevention of this food waste could save at least 15 Mt of CO_2 eq. emissions each year from disposal (Green Paper, 2008). Moreover, the stream of kitchen waste contains up to 80% water, making feed or energy recovery difficult. Besides, the economic and environmental costs of landfilling are high.

1.6.2 POLICY IN THE UNITED STATES

The United States Environmental Protection Agency (EPA) was created by Congress in 1970. The purpose was to consolidate a variety of federal research, monitoring, standard-setting, and enforcement activities in one agency with an ultimate goal to ensure environmental protection (www.epa.gov). The Food Recovery Hierarchy of EPA (Fig. 1.5b) starts with source reduction, which is the most preferred option, and ends with the least preferred landfill/incineration. In June 2013, EPA and the United States Department of Agriculture (USDA) announced a collaborative effort (the "US Food Waste Challenge") to raise awareness of the environmental, health, and nutrition issues created by food waste. The concept of the "US Food Waste Challenge" contains "reduce" for food loss and waste, "recover" for wholesome food for human consumption, and "recycle" for other uses including animal feed, composting, and energy generation (www.usda.gov). Extraction of the entire benefits from waste products and minimizing waste generation are the main goals of this hierarchy. The 3R (reduce/recover/recycle) application helps to minimize the amount of waste to disposal, to achieve more effective waste management, and finally to minimize associated health and environmental risks (Murugan and Ramasamy, 2013).

1.7 TREATMENT OF FOOD WASTE
1.7.1 VALORIZATION AS ANIMAL FEED

The management of food industry wastes can include numerous treatments (e.g. physical, chemical, thermal, and biological methods) with several advantages and disadvantages. For instance, valorization of food processing wastes as animal feed is one of the most traditional practices. Fat and protein-rich waste is suitable for omnivore animal feed, whereas substrates of high cellulose and hemicellulose content may be suitable for feeding ruminants (Russ and Schnappinger, 2006). However, the possible presence of toxic materials, which have an antinutritive effect and un-balanced nutrient compositions, may endanger both animals and humans (Murugan and Rama-samy, 2013). The transportation cost (due to the distance between waste production location and utilization location) often makes this feed source as costly as conventional animal feed (Russ and Schnappinger, 2006).

1.7.2 LANDFILLING

Landfilling is the common solid waste disposal method for many communities because it is one of the cheapest management options. It is defined as the disposal, compression, and embankment fill of waste at appropriate sites (Arvanitoyannis, 2008) and includes four common stages:

1. hydrolysis/aerobic degradation,
2. hydrolysis and fermentation,
3. acetogenesis, and
4. methanogenesis.

Oxidation takes place during the process and decomposition of the waste eventually leads to the production of methane (greenhouse gas) and groundwater pollution, due to the presence of organic compounds and heavy metals (Chen et al., 2006; Arvanitoyannis et al., 2008a, b).

Waste management strategies are focused on a number of policies in order to divert food waste from landfill and governments try to succeed in this goal through regulation, taxation, and public awareness. For example, the European Landfill Directive (1999/31/EC) and the Waste Framework Directive (2008/98/EC) contain a number of provisions, aimed at reducing landfilling. According to the waste policy of the EU, the quantity of food waste going to landfill was estimated as:

- 25% reduction in food waste going to landfill by 2010, compared with that produced in 2006 (based on Landfill Directive targets),
- 60% reduction in food waste going to landfill by 2013, compared with that produced in 2006 (based on Landfill Directive (50%) and Waste Framework Directive targets (10%)),
- 90% reduction in food waste going to landfill by 2020, compared with that produced in 2006 (based on Landfill Directive (65%), Waste Framework Directive (15%), and future bio-waste legislation following from the EC communication on future steps in bio-waste management in the European Union (10%)) (European Commission c, 2010).

Similarly, the waste strategy of the United Kingdom ("Waste Strategy 2000 for England and Wales") targets industrial waste to be landfilled by 85% of 1998 levels (Sanders and Crosby, 2004). In general, most policies target the diversion of waste handling from landfill to prevention, re-use, and recycling. Thereby, a balanced policy includes combined measures such as (i) landfill bans on food waste, (ii) landfill taxes that encourage diversion and make alternative treatments more attractive, (iii) development of composting or anaerobic digestion alternatives, (iv) development of the required infrastructure, and (v) establishment of a comprehensive treatment network (Kosseva, 2013a).

1.7.3 BIOFUEL CONVERSION METHODS

Food processing wastes contain a high amount of organic components that could be converted into energy and then recovered in the form of heat or electricity. Anaerobic digestion and thermochemical treatments (e.g. combustion, gasification, and pyrolysis) are the main biofuel conversion methods (Murugan et al., 2013). Wastes containing less than 50% moisture are suitable for thermochemical conversion, which converts energy-rich biomass into liquid or gaseous intermediate products. For instance, incineration is a thermal process that occurs by oxidizing the combustible material of the waste for heat production. Incineration is a viable option for food wastes with relatively low water content (<50% by mass) and an option for hazardous wastes. However, there are some increasing concerns about their emissions, adverse environmental impact, and high cost (Murugan et al., 2013). Anaerobic digestion is a widely used technology for the treatment of wastes with high (>50%) water content and organic value. During this process, a variety of microorganisms are used for the stabilization of food waste in the absence of oxygen. Organic substrates are degraded and the residual slurry could be used as fertilizer since it contains ammonia, phosphate, and various minerals (Nishio and Nakashimada, 2013). At the same time, biogas is produced. The latter is a mixture of methane, CO_2, and trace gases (water, hydrogen sulfide, or hydrogen). Biogas is used to generate electric power via thermal energy and is nowadays used to reduce the consumption of fossil energy and CO_2 emissions (Pesta, 2006).

1.7.4 COMPOSTING AND VERMICOMPOSTING

Composting is the aerobic degradation of organic materials into relatively stable products, by the action of a variety of microorganisms such as fungi, bacteria, and protozoa (Banks and Wang, 2006). Vermicomposting is the process that converts organic materials into a humus-like material by earthworms. Both processes produce fertilizers. Composting is producing a biomass that is able to improve the structural properties of the soil, increase its water capacity, increase its nutrients, support living soil organisms, and finally help organic materials return to the soil (Shilev et al., 2006). Temperature, pH, carbon/nitrogen (C/N) ratio, oxygen, and moisture content are the important conditions to optimize biological activity (Roupas et al., 2007).

1.7.5 RECOVERY AND VALORIZATION

The potential of food wastes to create new opportunities and markets has been underestimated until very recently. However, consumers' consciousness about environmental issues and legislative pressures increases the requirements of new methods for the recovery of food waste, rather than its disposal. Conventional methods such as animal feed or composting provide only partial utilization of food industry waste. As noted earlier, recovery of food wastes and their conversion to economically viable products could be an attractive option by implementing feasible strategies, supported at national and international level. Protection of natural resources such as water, energy, and land, preventing possible environmental impacts and sustainability in the food supply chain are global priorities, whereas the recovery of resources will play a vital role for the management strategies in the years to come. Among the different substrates described in Section 1.5.1, food processing by-products (generated at the third stage of Fig. 1.3) are the main materials that scientists utilize for recovery purposes. These substrates are of particular interest because of their low deterioration degree and existence in concentrated locations (Galanakis, 2012). Thereby, skins, husks, hulls, vegetable and fruit peels, seeds, animal meat, bones, or egg shells may be considered as wastes, but they contain considerable amounts of high value reusable materials (Chandrasekaran, 2013).

1.8 HOW FOOD WASTE RECOVERY IMPROVES SUSTAINABILITY OF FOOD SYSTEMS

Food waste recovery is a way of utilizing food wastes by recapturing their valuable compounds and/or developing new products with a market value (Galanakis, 2012, 2013; Galanakis and Schieber, 2014; Rahmanian et al., 2014). Utilizing food wastes could significantly reduce food waste levels and create new opportunities and benefits for everyone related to a food production system. Thus, reducing food waste through the recovery of its valuable components is an important way of increasing the sustainability of the food production systems. Organizations and in particularly food production systems can no longer ignore the need to act in a sustainable way.

According to the Brundtland Commission, "sustainability is the development that meets the needs of the present without compromising the ability of future generations to meet their own needs" (Brundtland, 1987). However, this definition is rather general and difficult for organizations to understand and apply. Corporate social responsibility (CSR) suggests that organizations have specific responsibilities to the society that go beyond their economic and legal obligations (McGuire, 1963). A dominant definition of CSR is Carroll's (1991) four-part definition that presents CSR as a multilayered concept that should embrace all business responsibilities; it is composed of four responsibilities: economic, legal, ethical, and philanthropic. The goal of CSR is to identify the negative impacts of an organization to society (as a whole), and try to reduce and eliminate them (Asbury and Ball, 2009). CSR is linked more to the business case of sustainability. The business case of sustainability refers to the tangible benefits that organizations have by engaging in CSR. Some of those benefits can be, for example, the reduction of costs and risks by being proactive in environmental issues.

According to the CSR concept, organizations should not only care about profits, but also about the impacts that the organization has on all of the involved stakeholders. The different stakeholders of an organization include shareholders, employees, customers, suppliers, and society. Based on the CSR elements, the triple bottom line (TBL) has been recently adopted (Fig. 1.6). The TBL considers organizational sustainability to include three components: (i) natural environment, (ii) society, and (iii) economic performance (Elkington, 1994). An organization's extended responsibilities means that people, planet, and profit should be considered as a whole system, needing balance. By balancing the social and environmental elements of sustainability, long-term profitability could be achieved. The TBL and the CSR concepts are used interchangeably by businesses. However, the TBL concept highlights the need to balance all three different sustainability elements.

A food system gathers all the elements (environment, people, inputs, processes, infrastructures, institutions, etc.), activities related to the production, processing, distribution, preparation, and consumption of food, and their socioeconomic and environmental outcomes (HLPE, 2014). A food system is defined as the sum of all the diverse elements and activities that lead to the production and consumption of food. A food system interfaces further with a wide range of other systems (energy, transport, etc.), and faces various constraints. Food system is a "descriptive" concept: its definition is not "normative" and does not preclude the performance of a food system, the generation of appropriate food security outcomes, as well as other socioeconomic and environmental outcomes (HLPE, 2014).

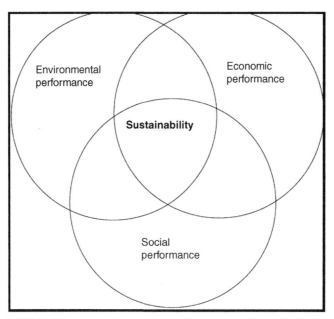

FIGURE 1.6 The TBL of Sustainability

Dao et al., 2011 adapted from Elkington, 1994

In a food system, sustainability could be illustrated through the product stewardship concept. Product stewardship can be defined as the shared responsibilities that all the participants in a product's life cycle have for minimizing its environmental and health impacts (Product Stewardship Institute, 2011). A product's responsibilities in a supply chain do not end when the product is delivered to customers and/or consumers. This means that product manufacturers, retailers, users, and disposers are responsible for the health, safety, and environmental impacts of their products across their life cycle (e.g. from raw material extraction to use and disposal). Thus, there is a need for balancing food product responsibilities (e.g. economic, social, and environmental) throughout the supply chain. Beer and Lemmer (2011) stated that environmental sustainability is not enough. In particular, food products must be politically, economically, environmentally, socially, and technologically sustainable. Achieving a sustainable food system means considering the political, economic, environmental, social, and technological impacts at each stage of the supply chain or else from production until the disposal and after-disposal stages of the food waste. In order to maximize the creation of shared value, organizations should adopt a long-term approach to sustainability such as exploring opportunities for developing innovative products that will enable societal wellbeing (European Commission, 2010).

The United Kingdom Sustainable Development Commission defined sustainable food systems as those that (Defra, 2006):

- Produce safe, healthy products in response to market demands, and ensure that all consumers have access to nutritious food, and to accurate information about food products.
- Support the viability and diversity of rural and urban economies and communities.

- Sustain the resource available for growing food and supplying other public benefits over time, except where alternative land uses are essential to meet other needs of society.
- Enable viable livelihoods to be made from sustainable land management, through both market and payments for public benefits.
- Achieve consistently high standards of environmental performance by reducing energy consumption, minimizing resource inputs, and using renewable energy wherever possible.
- Ensure a safe and hygienic working environment and high social welfare and training for all employees involved in the food chain.
- Sustain the resource available for growing food and supplying other public benefits over time, except where alternative land uses are essential to meet other needs of society.

The aforementioned definition of Defra (2006) should be applied to any food production system that aims to become sustainable.

As stated earlier, there is a need to decrease food loss and waste across the supply chain, but also to identify ways to best utilize discharged food mass (FAO, 2013b). Although waste arises at every stage of the food supply chain, the causes of waste vary depending on the stage of the supply chain. Effective food waste management will benefit all supply chain members. Reducing processing food wastes by recovering high-added value ingredients and developing new products can significantly improve the sustainability of the food production system, considering Elkington's (1994) definition and the following ways.

1.8.1 ECONOMIC SUSTAINABILITY IMPROVEMENTS

Effective waste management is critical to increase profitability levels of food chain members (Foresight, 2011). Reductions in energy and raw material usage can reduce costs significantly and simultaneously increase the environmental performance of the food system. This will be achieved through the efficient use of the materials and energy used for production. Efficient use of materials in the food waste recovery process has two meanings. First, utilizing the material that otherwise will have been thrown away, and second, using/processing that material in an efficient way. For example, potato peels and processing wastewater could be used for the extraction of phenols (Oreopoulou and Tzia, 2007). Through this process, potato peels intended to be thrown away were transformed into a new material with an economic value. Cheese processing whey is another example that could be used to recover monosaccharides and oligopeptides, prior to their implementation in nutritional supplements (Madureira, Tavares, Gomens, Pintado, and Malcata, 2010). Through the recovery of those valuable materials huge energy savings are achieved. If those materials were not produced, there would be a need to purchase new ones.

Recovering high added-value ingredients from food waste could also enable compliance with different types of food regulations, which are used to check regulatory compliance of other food products. For example, water insoluble fibers could improve intestinal regulation and consequently could be used to supplement food products or ready meals (Rodriguez et al., 2006). Thereafter, economic benefits could be achieved by these products without the need to buy any new ingredients or develop completely new products.

Another economic improvement that could be gained through the recovery of valuable components of food wastes is the reuse of wastewater. For example, olive mill wastewater could be valorized as a source of bioactive phenols and pectin (Galanakis et al., 2010a, b, c, d, e). Through this process, not only are the natural resources saved (e.g. wastewater), but new materials are also created. Following the appropriate food waste recovery stages, the yield of the target compounds could be maximized, and thus an economic value could be gained from them. Finally, the recovery of high added-value material could result in the development of innovative products and/or materials. Defra (2006) highlights the need for science-based innovation in order to develop a sustainable food system.

1.8.2 SOCIAL, AND ENVIRONMENTAL SUSTAINABILITY IMPROVEMENTS

The World Food Summit (1996) was the first to define food security as "availability at all times of adequate world food supplies of basic food stuffs to sustain a steady expansion of food consumption and to offset fluctuations in production and prices". In recent times, food and nutrition security has been defined as "a situation that exists when all people at all times have physical, social and economic access to sufficient, safe and nutritious food that meets their dietary needs and food preferences for an active and healthy life" (FAO, 2011b). The aforementioned definition considers the different preferences and needs of a diverse population. Specifically, food and nutrition security comprises four elements: food availability, food access, food utilization, and food stability (FAO, 2013c). Food availability is the consistent availability of a sufficient quantity of food. Improving food availability can increase food security (Yang and Hanson, 2009). Since it is important to increase production and optimize the resource used as food to feed an ever-increasing population, it is also vitally important to utilize the currently produced food effectively. This approach includes appropriate materials that are nowadays classified or could be at risk of becoming food waste. Thus, there is an absolute need for recovery, redistribution, and valorization methods, too. Using the appropriate food waste valorization methods means that high value-added ingredients are extracted and could be used to develop new food products or even extend their shelf-life to be consumed in a longer period of time (Oreopoulou and Tzia, 2007). The creation of new food products could steadily increase food availability. For example, phenols and carotenoids from fruit by-products could be used as natural food or beverage preservatives as they extend the shelf-life of the product and increase antioxidant activity (Galanakis, 2012; Galanakis et al., 2010d, 2015a). This could increase food availability by delaying a product's deterioration, having the product available on the shelf for longer and improving people's livelihoods as a whole. The recovery of food wastes' valuable components could also help in promoting the viability and diversity of rural and urban economies. Existing and new methods of successful food waste materials recovery could create new job opportunities. It is quite a new area and there is a lot of potential in creating innovative and sustainable solutions. Finally, the recovery of food wastes for optimal use enables the reutilization of a product's resources and the recapture of a product's energy nutrients. Through the recovery of food wastes, the environmental impact of the food industry is reduced significantly, as it prevents, among other things, the usage of additional primary resources (FAO, 2013a). This approach is translated to a more effective usage of natural resources and minimization of food waste going to landfill.

REFERENCES

Akkerman, R., Van Donk, D.P., 2008. Development and application of a decision support tool for reduction of product losses in the food-processing industry. J. Clean. Prod. 16 (3), 335–342.

Arvanitoyannis, I.S., 2008. Waste management in food packaging industries. In: Arvanitoyannis, I.S. (Ed.), Waste Management for the Food Industries. Elsevier Inc., Oxford, pp. 941–1045.

Arvanitoyannis, I.S., Kassaveti, A., Ladas, D., 2008a. Food waste treatment methodologies. In: Arvanitoyannis, I.S. (Ed.), Waste Management for the Food Industries. Elsevier Inc., Oxford, pp. 345–410.

Arvanitoyannis, I.S., Ladas, D., Mavromatis, A., 2008b. Wine waste management: treatment methods and potential uses of treated waste. In: Arvanitoyannis, I.S. (Ed.), Waste Management for the Food Industries. Elsevier Inc., Oxford, pp. 413–452.

Arvanitoyannis, I.S., Varzakas, T.H., 2008. Fruit/fruit juice waste management: treatment methods and potential uses of treated waste. In: Arvanitoyannis, I.S. (Ed.), Waste Management for the Food Industries. Elsevier Inc., Oxford, pp. 569–628.

Asbury, S., Ball, R., 2009. Do the Right Thing: The Practical, Jargon-Free Guide to Corporate Social Responsibility. IOSH Services Ltd, Wigston, United Kingdom.

Banks, C.J., Wang, Z., 2006. Treatment of meat wastes. In: Wang, L.K., Hung, Y., Lo, H.H., Yapijakis, C. (Eds.), Waste Treatment in the Food Processing Industry. Taylor & Francis Group, Florida, pp. 67–100.

Beer, S., Lemmer, C., 2011. A critical review of "green" procurement: life cycle analysis of food products within the supply chain. Worldwide Hosp. Tourism Themes 3 (3), 229–244.

Bhaskar, N., Modi, V.K., Govindaraju, K., Radha, C., Lalitha, R.G., 2007. Utilization of meat industry by products: protein hydrolysate from sheep visceral mass. Bioresour. Technol. 98, 388–394.

Brundtland, G.H., 1987. Our common future. Report of the World Commission on Environment and Development. Oxford University Press, Oxford.

Calvarano, M., Postorino, E., Gionfriddo, F., Calvarano, I., Bovalo, F., 1996. Naringin extraction from exhausted Bergamot peels. Perfumer Flavorist 21, 1–3.

Carroll, A.B., 1991. The pyramid of corporate social responsibility: towards the moral management of organizational stakeholders. Bus. Horizons, 39–48, Jul–Aug.

Chandrasekaran, M., 2013. Need for valorization of food processing by-products and wastes. In: Chandrasekaran, M. (Ed.), Valorization of Food Processing By-Products. Taylor & Francis Group, Florida, pp. 91–108.

Chedea, V.S., Kefalas, P., Socaciu, C., 2008. Patterns of carotenoid pigments extracted from two orange peelwastes (valencia and navel var.). J. Food Biochem. 34, 101–110.

Chen, J.P., Yang, L., Bai, R., Hung, Y.T., 2006. Bakery waste treatment. In: Wang, L.K., Hung, Y.T., Lo, H.H., Yapijakis, C. (Eds.), Waste Treatment in the Food Processing Industry. Taylor & Francis Group, Florida, pp. 271–290.

Dao, V., Langella, I., Carbo, J., 2011. From green to sustainability: information technology and integrated sustainability framework. J. Strategic Infor. Syst. 20, 63–79.

Defra, 2006. Food industry sustainability strategy. http://www.defra.gov.uk/publications/files/pb11649-fiss2006-060411.pdf (accessed 25.01.12.).

Defra, 2009. UK food security assessment: our approach. http://archive.defra.gov.uk/foodfarm/food/pdf/food-assess-approach-0908.pdf (accessed 25.01.12.).

Deng, Q., Zinoviadou, K.G., Galanakis, C.M., Orlien, V., Grimi, N., Vorobiev, E., Lebovka, N., Barba, F.J., 2015. The effects of conventional and non-conventional processing on glucosinolates and its derived forms, isothiocyanates: extraction, degradation and applications. Food Eng. Rev. in press.

Derham, J., 2010. EU policy and legislation for waste management, IGI Seminar on EU Directives and the Geosciences, EPA.

Directive 1999/31/EC of 26 April 1999 on the landfill of waste. Available from: http://eur-lex.europa.eu/legal-content/EN/TXT/?uri=CELEX:31999L0031.

Directive 2008/98/EC of the European Parliament and of the Council of 19 November 2008 on waste and repealing certain directives. Available from: http://eur-lex.europa.eu/legal-content/EN/TXT/?uri=CELEX:32008L0098.

El Nawawi, S.A., 1995. Extraction of citrus glucosides. Carbohydr. Polym. 27, 1–4.

Elkington, J., 1994. Towards the sustainable corporation. Calif. Manag. Rev. Winter, 90–100.

EPA, 2011. Generators of food waste. http://www.epa.gov/osw/conserve/materials/organics/food/fd-gener.htm#food-hier (accessed 18.02.12.).

European Commission, 2012. Guidance on the interpretation of key provisions of Directive 2008/98/EC on waste. Available from: http://ec.europa.eu/environment/waste/framework/pdf/guidance_doc.pdf.

European Commission c, 2010. Preparatory study on food waste across EU 27 – Technical report.

European Commission a. Available from: http://ec.europa.eu/environment/circular-economy/index_en.htm.

European Commission b. "Proposal for a iDirective of the European Parliament and of the Council amending Directives 2008/98/EC on waste (…)", 94/62/EC on packaging and packaging waste, 1999/31/EC on the landfill of waste, 2000/53/EC on end-of-life vehicles, 2006/66/EC on batteries and accumulators and waste batteries and accumulators, and 2012/19/EU on waste electrical and electronic equipment. Available from: http://eur-lex.europa.eu/legal-content/EN/TXT/?uri=CELEX:52014PC0397.

Euractiv, 2014. Circular economy package to be ditched and re-tabled (available at http://www.euractiv.com/sections/sustainable-dev/circular-economy-package-be-ditched-and-re-tabled-310866).

European Union. Communication from the Commission to the Council and the European Parliament on future steps in bio-waste management in the European Union, Brussels, 18.5.2010 COM(2010)235 final. Available from: http://eur-lex.europa.eu/legal-content/EN/TXT/?uri=CELEX:52010DC0235.

FAO SAVE FOOD: Global initiative on food loss and waste reduction. Available from: http://www.fao.org/save-food/en/.

FAO, 2006. Postharvest management of fruit and vegetables in the Asia-Pacific region. http://www.apo-tokyo.org/00e-books/AG-18_PostHarvest/AG-18_PostHarvest.pdf (accessed 09.01.12.).

FAO, 2009a. How to feed the world in 2050. Rome, 12–13 October.

FAO, 2009b. The state of food insecurity in the world. ftp://ftp.fao.org/docrep/fao/012/i0876e/i0876e.pdf (accessed 09.01.12.).

FAO, 2011a. Global food losses and food waste: extent, causes and prevention. Food and Agriculture Organization of the United Nations, Rome.

FAO, 2011b. How to feed the world in 2050. http://www.fao.org/fileadmin/templates/wsfs/docs/expert_paper/How_to_Feed_the_World_in_2050.pdf (accessed 09.01.12.).

FAO, 2012. Sustainable diets and biodiversity. Directions and solutions for policy, research and action, Rome. Available from: http://www.fao.org/docrep/016/i3004e/i3004e.pdf.

FAO, 2013a. Food wastage footprint: impacts on natural resources. Summary report, Rome.

FAO, 2013b. Toolkit: reducing the food wastage footprint. http://www.fao.org/docrep/018/i3342e/i3342e.pdf (accessed 02.03.14.).

FAO, 2013c. The State of Food Insecurity in the World, The multiple dimensions of food security http://www.fao.org/docrep/018/i3434e/i3434e.pdf (accessed 08.08.14).

FAO, 2014. Definitional framework of food loss. Available from: http://www.fao.org/fileadmin/user_upload/save-food/PDF/FLW_Definition_and_Scope_2014.pdf.

FAO, IFAD, and WFP. 2014. The state of food insecurity in the world 2014. Strengthening the enabling environment for food security and nutrition. FAO, Rome. Available from: http://www.fao.org/3/a-i4030e.pdf.

Fonkwe, L.G., Singh, R.K., 1996. Protein recovery from mechanically deboned turkey residue by enzymatic hydrolysis. Process Biochem. 31, 605–616.

Foresight Project, 2011. The Future of Food and Farming: Challenges and Choices for Global Sustainability. The Government Office for Science, London, United Kingdom.

Foresight, 2011. Foresight Project on Global Food and Farming Futures. Synthesis Report C7: Reducing Waste. The Government Office for Science, London, United Kingdom.

Galanakis, C.M., 2011. Olive fruit and dietary fibers: components, recovery and applications. Trends Food Sci. Technol. 22, 175–184.

Galanakis, C.M., 2012. Recovery of high added-value components from food wastes: conventional, emerging technologies and commercialized applications. Trends Food Sci. Technol. 26, 68–87.

Galanakis, C.M., 2013. Emerging technologies for the production of nutraceuticals from agricultural by-products: a viewpoint of opportunities and challenges. Food Bioprod. Process. 91, 575–579.

Galanakis, C.M., 2015. Separation of functional macromolecules and micromolecules: from ultrafiltration to the border of nanofiltration. Trends Food Sci. Technol. 42, 44–63.

Galanakis, C.M., Chasiotis, S., Botsaris, G., Gekas, V., 2014. Separation and recovery of proteins and sugars from Halloumi cheese whey. Food Res. Int. 65, 477–483.

Galanakis, C.M., Goulas, V., Tsakona, S., Manganaris, G.A., Gekas, V., 2013a. A knowledge base for the recovery of natural phenols with different solvents. Int. J. Food Prop. 16, 382–396.

Galanakis, C.M., Kotanidis, A., Dianellou, M., Gekas, V., 2015a. Phenolic content and antioxidant capacity of Cypriot wines. Czech J. Food Sci. 33, 126–136.

Galanakis, C.M., Markouli, E., Gekas, V., 2013b. Fractionation and recovery of different phenolic classes from winery sludge via membrane filtration. Sep. Purif. Technol. 107, 245–251.

Galanakis, C.M., Patsioura, A., Gekas, V., 2015b. Enzyme kinetics modeling as a tool to optimize food biotechnology applications: a pragmatic approach based on amylolytic enzymes. Crit. Rev. Food Sci. Technol. 55, 1758–1770.

Galanakis, C.M., Schieber, A., 2014. Editorial. Special issue on recovery and utilization of valuable compounds from food processing by-products. Food Res. Int. 65, 299–330.

Galanakis, C.M., Tornberg, E., Gekas, V., 2010a. A study of the recovery of the dietary fibres from olive mill wastewater and the gelling ability of the soluble fibre fraction. LWT-Food Sci. Technol. 43, 1009–1017.

Galanakis, C.M., Tornberg, E., Gekas, V., 2010b. Clarification of high-added value products from olive mill wastewater. J. Food Eng. 99, 190–197.

Galanakis, C.M., Tornberg, E., Gekas, V., 2010c. Dietary fiber suspensions from olive mill wastewater as potential fat replacements in meatballs. LWT-Food Sci. Technol. 43, 1018–1025.

Galanakis, C.M., Tornberg, E., Gekas, V., 2010d. Recovery and preservation of phenols from olive waste in ethanolic extracts. J.Chem. Technol. Biotechnol. 85, 1148–1155.

Galanakis, C.M., Tornberg, E., Gekas, V., 2010e. The effect of heat processing on the functional properties of pectin contained in olive mill wastewater. LWT-Food Sci. Technol. 43, 1001–1008.

Green Paper on the management of bio-waste in the European Union, Brussels, 3.12.2008 COM(2008) 811 final. Available from: http://eur-lex.europa.eu/LexUriServ/LexUriServ.do?uri=COM:2008:0811:FIN:EN:PDF.

Grolleaud, M., 2002. Post-harvest losses: discovering the full story. Overview of the Phenomenon of Losses During the Post-Harvest System. FAO, Agro Industries and Post-Harvest Management Service, Rome, Italy.

Hall, K.D., Guo, J., Dore, M., Chow, C.C., 2009. The progressive increase of food waste in America and its environmental impact. PLoS One 4 (11), 1–6.

Heng, W.W., Xiong, L.W., Ramanan, R.N., Hong, T.L., Kong, K.W., Galanakis, C.M., Prasad, K.N., 2015. Two level factorial design for the optimization of phenolics and flavonoids recovery from palm kernel by-product. Ind. Crop. Prod. 63, 238–248.

HLPE, 2014. Food losses and waste in the context of sustainable food systems. A report by the High Level Panel of Experts on Food Security and Nutrition of the Committee on World Food Security. Rome 2014. Available from: http://www.fao.org/fileadmin/user_upload/hlpe/hlpe_documents/HLPE_Reports/HLPE-Report-8_EN.pdf.

Hodges, R., Buzby, J.C., Benett, B., 2010. Postharvest losses and waste in developed and less developed countries: opportunities to improve resource use. J. Agric. Sci. 149(S1), 37–45.

Jelen, P., Earle, M., Edwardson, W., 1979. Recovery of meat protein from alkaline extracts of beef bones. J. Food Sci. 44, 327–331.

Kader, A.A., 2005. Increasing food availability by reducing postharvest losses of fresh produce. Proc. 5th Int. Postharvest Symposium. http://ucce.ucdavis.edu/files/datastore/234-528.pdf (accessed 15.01.12.).

Kader, A.A., 2010. Handling of horticultural perishables in developing vs. developed countries. Proc. 6th Int. Postharvest Symposium. http://ucce.ucdavis.edu/files/datastore/234-1875.pdf (accessed 15.01.12.).

Kaye, L., 2011. Food retailers must do more to reduce food waste. http://www.theguardian.com/sustainable-business/food-retailers-must-reduce-waste.

Kosseva, M.R., 2013a. Recent European legislation on management of wastes in the food industry. In: Kosseva, M.R., Webb, C. (Eds.), Food Industry Wastes Assessment and Recuperation of Commodities. Elsevier Inc., Oxford, pp. 3–16.

Kosseva, M.R., 2013b. Functional food and nutraceuticals derived from food industry wastes. In: Kosseva, M.R., Webb, C. (Eds.), Food Industry Wastes Assessment and Recuperation of Commodities. Elsevier Inc., Oxford, pp. 103–120.

Larrauri, J.A., Ruperez, R., Borroto, B., Saura-Calixto, F., 1996. Mango peels as a new tropical fibre: preparation and characterization. LWT 29, 729–733.

Laufenberg, G., Kunza, B., Nystroem, M., 2003. Transformation of vegetable waste into value added products: (A) the upgrading concept; (B) practical implementations. Bioresour. Technol. 87, 167–198.

Madureira, A.R., Tavares, T., Gomes, A.M.P., Pintado, M.E., Malcata, F.X., 2010. Invited review: Psychological properties of bioactive peptides obtained from whey proteins. J.Diary Sci. 93(2), 473–55.

McGuire, J.W., 1963. Business and Society. McGraw-Hill, New York.

Mena, C., Adenso-Diaz, B., Yurt, O., 2011. The causes of food waste in the supplier–retailer interface: evidences from the UK and Spain. Resour. Conserv. Recy. 55 (6), 48–658.

Mulinacci, N., Romani, A., Galardi, C., Pinelli, P., Giaccherini, C., Vincieri, F.F., 2001. Polyphenolic content in olive oil waste waters and related olive samples. J. Agric. Food Chem. 49 (8), 3509–3514.

Murugan, K., Chandrasekaran, V.S., Karthikeyan, P., Al-Sohaibani, S., 2013. Current state-of-the-art of food processing by-products. In: Chandrasekaran, M. (Ed.), Valorization of Food Processing By-Products. Taylor & Francis Group, Florida, pp. 35–62.

Murugan, K., Ramasamy, K., 2013. Environmental concerns and sustainable development. In: Chandrasekaran, M. (Ed.), Valorization of Food Processing By-Products. Taylor & Francis Group, Florida, pp. 739–756.

Nishio, N., Nakashimada, Y., 2013. Manufacture of biogas and fertilizer from solid food wastes by means of anaerobic digestion. In: Kosseva, M.R., Webb, C. (Eds.), Food Industry Wastes: Assessment and Recuperation of Commodities. Elsevier Inc., Oxford, pp. 121–136.

Obied, H.K., Allen, M.S., Bedgood, D.R., Prenzler, P.D., Robards, K., Stockmann, R., 2005. Bioactivity and analysis of biophenols recovered from olive mill waste. J. Agric. Food Chem. 53, 823–837.

Oreopoulou, V., Tzia, C., 2007. Utilization of plant by-products for the recovery of proteins, dietary fibers, antioxidants, and colorants. In: Oreopoulou, V., Russ, W. (Eds.), Utilization of By-Products and Treatment of Waste in the Food Industry. Springer Science+Business Media, New York, pp. 209–232.

Parfitt, J., Barthel, M., Macnaughton, S., 2010. Food waste within food supply chains: quantification and potential for change to 2050. Philos. Trans. R. Soc. B 365, 3065–3081.

Patsioura, A., Galanakis, C.M., Gekas, V., 2011. Ultrafiltration optimization for the recovery of β-glucan from oat mill waste. J. Membr. Sci. 373, 53–63.

Pesta, G., 2006. Anaerobic digestion of organic residues and wastes. In: Oreopoulou, V., Russ, W. (Eds.), Utilization of By-Products and Treatment of Waste in the Food Industry. Springer Science+Business Media, New York, pp. 53–72.

Premanandh, J., 2011. Factors affecting food security and contribution of modern technologies in food sustainability. J. Sci. Food Agric. 91 (15), 2707–2714.

Product Stewardship Institute, 2011. http://productstewardship.us/displaycommon.cfm?an=1&subarticlenbr=55 (accessed 17.04.11.).

Rahmanian, N., Jafari, S.M., Galanakis, C.M., 2014. Recovery and removal of phenolic compounds from olive mill wastewater. J. Am. Oil. Chem. Soc. 91, 1–18.

Riemer, J., Kristoffersen, M., 1999. Information on Waste Management Practices: A Proposed Electronic Framework. EEA, Copenhagen.

Rodriguez, R., Jimenez, A., Fernadez-Bolanos, J., Guillen, R., Heredia, A., 2006. Dietary fibre from vegetable products as source of functional ingredients. Trends Food Sci. Technol. 17, 3–15.

Roselló-Soto, E., Barba, F.J., Parniakov, O., Galanakis, C.M., Grimi, N., Lebovka, N., Vorobiev, E., 2015. High voltage electrical discharges, pulsed electric field and ultrasounds assisted extraction of protein and phenolic compounds from olive kernel. Food Bioprocess Technol. 8, 885–894.

Roupas, P., De Silva, K., Smithers, G., 2007. Waste management and co-product recovery in red and white meat processing. In: Waldron, K. (Ed.), Handbook of Waste Management and Co-Product Recovery in Food Processing. Woodhead Publishing Limited, Cambridge, pp. 305–331.

Russ, W., Meyer-Pittroff, R., 2004. Utilizing waste products from the food production and processing industries. Crit. Rev. Food Sci. Nutr. 44, 57–62.

Russ, W., Schnappinger, M., 2006. Waste related to the food industry: a challenge in material loops. In: Oreopoulou, V., Russ, W. (Eds.), Utilization of By-Products and Treatment of Waste in the Food Industry. Springer Science+Business Media, New York, pp. 1–14.

Sanders, B., Crosby, K.S., 2004. Waste legislation and its impact on the food industry. In: Waldron, K., Faulds, C., Smith, A. (Eds.), Total Food Exploiting Co-Products – Minimizing Waste. IFR, Norwich.

Shilev, S., Naydenov, M., Vancheva, V., Aladjadjiyan, A., 2006. Composting of food and agricultural wastes. Utilization of By-Products and Treatment of Waste in the Food Industry. Springer Science+Business Media, New York, pp. 283–302.

Swingler, G.R., Lawrie, R.A., 1979. Improved protein recovery from some meat industry by-products. Meat Sci. 3, 63–73.

Tomás Barberán, F.A., 2007. High-value co-products from plant foods: nutraceuticals, micronutrients and functional ingredients. In: Waldron, K. (Ed.), Handbook of Waste Management and Co-Product Recovery in Food Processing. Woodhead Publishing Limited, Cambridge, pp. 448–469.

Tomlinson, I., 2013. Doubling food production to feed the 9 billion: a critical perspective on a key discourse of food security in the UK. J. Rural Stud. 29, 81–90.

Tsakona, S., Galanakis, C.M., Gekas, V., 2012. Hydro-ethanolic mixtures for the recovery of phenols from Mediterranean plant materials. Food Bioprocess Technol. 5, 1384–1393.

United Nations, 2012. Department of Economic and Social Affairs, Population Division: World Urbanization Prospects, the 2011 Revision. United Nations, New York.

WRAP, 2009. Household food and drink waste in UK. http://www.wrap.org.uk/sites/files/wrap/Household_food_and_drink_waste_in_the_UK_-_report.pdf.

Yang, R., Hanson, P.M., 2009. Improved food availability for food security in Asia-Pacific region. Asia Pac. J. Clin. N. 18 (4), 633–637.

CLASSIFICATION AND TARGET COMPOUNDS

2

Anne Maria Mullen*, Carlos Álvarez*, Milica Pojić**, Tamara Dapčević Hadnađev**, Maria Papageorgiou[†]

*Food Chemistry and Technology Department, Teagasc Food Research Centre, Ashtown, Dublin, Ireland;
**Institute of Food Technology, University of Novi Sad, Novi Sad, Vojvodina, Serbia;
[†]Department of Food Technology, Alexander Technological Educational Institute of Thessaloniki (ATEITh),
Kentriki Makedonia, Greece

2.1 INTRODUCTION

According to the Food and Agriculture Organization (FAO), a third of the total weight of edible parts of food produced for human consumption is lost or wasted. Food by-products and wastes are composed of highly complex components many of which can command high value. These can be extracted, purified, and characterized by existing and emerging technologies, leading to the development of new commercial applications in food and nonfood (pharmaceutical, biomedical, cosmetic, etc.) sectors.

The wastes originating from various branches of the food industry can be divided in two main groups and seven subcategories (Galanakis, 2012):

1. Plant origin:
   a. Cereals
   b. Root and tubers
   c. Oil crops and pulses
   d. Fruit and vegetables
2. Animal origin:
   a. Meat products
   b. Fish and seafood
   c. Dairy products

Regardless of the branch of the food industry under investigation, by-products and/or waste streams are generated across various stages of the supply chain. Many research efforts are focused on the recovery of valuable compounds from these by-products or wastes generated during agricultural and food processing stages. These materials are abundant and often concentrated within the industry. Depending on the source (e.g. excluding meat/fish) these can be less susceptible to deterioration compared with the wastes produced at the end of the food supply chain. Waste produced at the end of the chain tends to be highly dispersed (e.g. in individual households), which complicates their valorization as sources for valuable components recovery due to the need for an additional collection step and reduction in biological stability due to microbial growth (Galanakis, 2012). In this chapter some of the high added-value biomolecules are identified in the different by-products generated by the key food sectors. The corresponding target compounds in each case as well as their potential applications in key sectors such as food, pharmaceutical, or biomedical are outlined. The reader is guided to a number of relevant published articles for more in-depth descriptions, which are beyond the scope of this chapter.

2.2 CEREALS

The term "cereals" refers to the members of the Gramineae family, comprising nine species: (i) wheat (*Triticum*), (ii) rye (*Secale*), (iii) barley (*Hordeum*), (iv) oat (*Avena*), (v) rice (*Oryza*), (vi) millet (*Pennisetum*), (vii) corn (*Zea*), (viii) sorghum (*Sorghum*), and (ix) triticale, which is a hybrid of wheat and rye. Cereals represent the most important source of food, both for direct human consumption and meat production since, for example, 660 Mt of cereals are used as livestock feed annually.

Cereals and respective derivatives are an important source of carbohydrates, proteins, lipids, vitamins, mainly of B-complex and vitamin E, inorganic and trace elements. They are known to process a broad range of phytochemicals not only with well-established beneficial effects to human health but also antinutritional factors like phytic acid (McKevith, 2004; Fardet, 2010). Epidemiological studies have shown that regular consumption of whole cereal grains is associated with reduced risk of developing chronic diseases such as cardiovascular disease (Anderson, 2003), some cancers (Chatenoud et al., 1998), and type II diabetes (Venn and Mann, 2004). The beneficial phytochemicals are mainly found in the outer seed coat (bran). Further processing to obtain refined products (e.g. white bread, pasta, and white rice) results in the removal of a substantial portion thereof. The bioactive compounds are unevenly distributed in different parts of the seed depending on the type of cereals (Liu, 2007).

The most common processing step that cereal grains undergo is dry milling (wheat and rye), pearling (e.g. rice, oat, barley), and malting. On milling, the bran and germ layers are separated from the starchy endosperm as by-products along with hulls and polish waste. Wet milling, on the other hand, aims towards the extraction of the maximum possible amount of native or undamaged starch granules, thus providing steep solids (rich in nutrients valuable for the pharmaceutical industry), germ (intended for the oil crushing industry), bran and gluten (vital and not vital) as by-products. Pearling is an abrasive technique that gradually removes the seed coat (testa and pericarp), aleurone, subaleurone layers, and the germ. This process results in polished grain and by-products with a high concentration of bioactive compounds. Malting of grains includes steeping, germination, and kilning. During this process, fermentable sugars and starch of the grain are consumed by the enzymes, leaving behind spent grain.

2.2.1 WHEAT STRAW

Wheat straw, rice straw, and corn stover are the most abundant lignocellulosic biomasses among agricultural residues in the world (Kim and Dale, 2004). Wheat straw consists mainly of cellulose (28–39%), hemicelluloses (23–24%), lignin (16–25%), and fewer contents of ash and protein (Carvalheiro et al., 2009). Among the several hemicelluloses, xylans have been found to contribute to lowering blood cholesterol, decreasing the postprandial glucose and insulin responses, as well as exhibiting antitumor effects (Asp et al., 1993). The xylan fragment of hemicellulose found in cereal wastes (straw, corn cob) can be further exploited for the production of xylooligosaccharides (XOs). XOs are value-added oligosaccharides with a functional role as prebiotics. They are indigestible and maintain gastrointestinal health (Kabel et al., 2002). XOs are generally produced by enzymatic hydrolysis

(Pellerin et al., 1991; Yoon et al., 2006) or by controlled acid hydrolysis (Akpinar et al., 2009) of the xylan fragments. Corn cob xylan and wheat straw xylan are also reported in the literature as enzymatic substrates (Pellerin et al., 1991; Yoon et al., 2006; Aachary and Prapulla, 2009).

2.2.2 WHEAT MILL FRACTIONS

The weight ratio of wheat bran to milled wheat is 25%. Therefore, the resulting biomass of wheat bran can be estimated to be 150 Mt (Prückler et al., 2014). Three different tissues comprise the wheat bran: the pericarp, representing 4–5% of the wheat grain, testa (about 1% of the wheat grain), and the aleurone layer (6–9% of the wheat grain) (Hemery et al., 2007). Wheat germ and parts of the endosperm top up this milling output. Generally, wheat bran comprises approximately 12% water, 13–18% protein, 3.5% fat, and 56% carbohydrates (Apprich et al., 2014). Wheat bran is rich in dietary fibers (44%), with arabinoxylan (AX), β-glucan, cellulose, and lignin as major components. AX is known to absorb large amounts of water and influence significantly the water balance, rheological properties of dough, and retrogradation of starch and bread quality. Wheat $(1\rightarrow3)$, $(1\rightarrow4)$-β-glucans are present only at about 2% as opposed to barley (3–11%) and oat (3–7%). EFSA's claim on β-glucan "contributing to maintenance of normal blood cholesterol" and "reduction of postprandial glycaemic responses" is confined to oat and barley β-glucans (EFSA, 2011).

The outer tissues of the pericarp are rich in insoluble dietary fiber and xylans with high arabinose to xylose ratios and substitution by ferulic acid (FA) dehydrodimers. The latter act as cross-linkers between the polymer chains. Pericarp and testa layers also contain significant amounts of lignin. On the other hand, the aleurone layer is composed of relatively linear AXs with low arabinose to xylose ratio (Saulnier et al., 2007). The AXs in the aleurone layer have appreciable amounts of FA, which is recognized for its antioxidant properties (Rhodes et al., 2002). The aleurone contains high levels of β-glucans compared with whole grain, which can explain its high content of soluble dietary fiber. It is rich in phenolic compounds, too. Among them, the most abundant ones belong to the chemical class of hydroxycinnamic acids (mainly FA). Derivatives of benzoic acid (e.g. syringic and vanillic acid), lignans, and lignins have also been found in the aleurone layer (Brouns et al., 2012). On the other hand, alkylresorcinols seem to be localized in the testa (220–400 mg/100 g). Components of milling by-products (wheat and rice bran, wheat germ, and corn fiber) with beneficial health effects include soluble dietary fibers, dietary fibers, alkylresorcinol, ferulic acid, β-glucans, arabinoxylan, lignans, and sterols (Jiang and Wang, 2005; Prückler et al., 2014).

Furthermore, rye and wheat brans contain 400–1000 ng folate/g dm (Kamal-Eldin et al., 2009), whereas wheat aleurone fractions contain over 1000 ng/g (Hemery et al., 2011).

2.2.3 RICE MILL FRACTIONS

Rice bran is a coproduct obtained from rice milling and constitutes about 10% of the weight of rough rice (Hu et al., 1996). Other by-products of rice milling are hull, germ, bran layers, and fine brokens. The objective of a rice milling system is to remove the husk and the bran layers from paddy rice in order to deliver whole white rice kernels ready for milling. An ideal milling process will result in different fractions: 20% husk, 8–12% bran, and 68–72% milled rice or white rice depending on the variety. Rice bran is used as an ingredient for animal feeds (e.g. ruminants and poultry), but it has a potential to be used for food applications, too, due to its hypocholesterolemic, anticancer, and antitumor activity (Han et al., 2001).

The constituents of rice bran are lipids (at a concentration of 15–22% that allows the economic viable extraction of bran oil after stabilization), carbohydrates, fiber, ash, moisture, and 10–16% highly nutritional protein (Juliano, 1985). Prior to food use, bran is stabilized to inactivate the antinutrients including lipases, trypsin inhibitors, hemagglutinin–lectins, and phytates (Khan et al., 2011). Rice bran has been incorporated as a protein supplement in bakery products and breakfast cereals, and as binder ingredient for meats and sausages.

Rice bran protein concentrates have been incorporated in bread, beverages, confections, and weaning foods. Rice bran is also a rich source of steryl ferulate esters, referred to as oryzanols (Seetharamaiah and Prabhakar, 1986). Butsat and Siriamornpun (2010) indicated that rice husk is a valuable source of antioxidants like phenolic acids. Additionally, rice bran is a potential source of tocopherols, tocotrienols, and phenolic compounds (Nicolosi et al., 1994; Butsat and Siriamornpun, 2010).

2.2.4 OAT MILL FRACTIONS

Oats are a good source of soluble dietary fiber, β-glucan, unsaturated fatty acids, and phytochemicals such as vitamins, phenolic acids and avenanthramides (Shewry et al., 2008; Welch et al., 2011). In general, positive effects of oats are associated with β-glucan due to its beneficial effect on serum cholesterol levels (EFSA, 2011). Oat bran is the edible outermost layer of oat kernel. It is produced by grinding clean oats or rolled oats after the separation of the flour. With this process, more than 50% of the starting material is obtained with a total β-glucan content of at least 5.5% (dry-weight basis) and a total dietary fiber content of at least 16.0% (dry-weight basis). Oat mill waste originated from rolled oat grains has been used for the successful recovery of high molecular weight β-glucan by employing an optimized ultrafiltration process (Patsioura et al., 2011; Galanakis, 2015). However, the researchers reported a small degree of separation between β-glucan and proteins. With regard to antioxidants, tocol concentration in oats ranges between 13.6 and 36.1 mg/kg (Peterson and Qureshi, 1993; Shewry et al., 2008), while it is distributed unevenly in the kernel.

2.2.5 BARLEY MILL FRACTIONS

Barley is a good source of tocols, which have well-known antioxidative properties and are important quenchers of reactive oxygen species and lipid radicals. During milling, the tocols content of whole grain is concentrated in certain milling fractions such as bran, germ, etc. The pearling by-products of commercial hulled barley stock have higher concentrations of tocopherols and tocotrienols. Panfili et al. (2008) reported a seven- and a fivefold increase for tocopherols and tocotrienols in these by-products, respectively. Therefore, it is a challenge for the manufacturers to incorporate pearling by-products into food. Barley β-glucan, either as a component of fractionated barley meal or as a concentrated extract, is known to have a range of beneficial physiological effects (Newman et al., 1989).

2.2.6 BARLEY MALT

Brewers' spent grain is the residual solid fraction of barley malt remaining after the production of wort from the brewing industry. Approximately 3.4 Mt of brewers' spent grain are produced every year in the European Union. Brewers' spent grain is used in animal feed due to its high protein and fiber concentration. However, it is also a rich source of bioactive ingredients including phenolic acids, such as hydroxycinnamic acids, FA, *p*-coumaric acid, sinapic acid, and caffeic acid (Bartolome et al., 2002; Szwajgier et al., 2010), as well as tocols (Peterson and Qureshi, 1993).

2.3 **ROOT AND TUBERS**

According to the FAO (1994) seven primary root and tuber crops can be distinguished: (i) potatoes (Irish potato), (ii) sweet potatoes, (iii) cassava (manioc, mandioca, yucca), (iv) yams, (v) edible aroids such as yautia (malanga, new cocoyam, ocumo, tannia), (vi) taro (dasheen, eddoe, old cocoyam) referred as cocoyams, and (vii) roots and tubers (e.g. arracacha, arrowroot, chufa, sago palm, oca and ullucu, yam bean, jicama, mashua, and Jerusalem artichoke). Cassava, potato, and sweet potato account for 93% of total root and tuber crops used for human consumption, while yams and aroids account for 7% (FAO, 1997; Tréche, 1996).

Approximately 55% of root and tuber production is consumed as food, while the remaining part is used as animal feed or to produce starch, distilled spirits, alcohol and other minor products (FAO, 1997). Roots and tubers have the highest wastage rates after fruits and vegetables (Gustavsson et al., 2011).

2.3.1 **POTATO PROCESSING WASTE**

Potato (*Solanum tuberosum*) is the fourth largest crop grown worldwide after rice, wheat, and maize (Leo et al., 2008). Trends in the potato sector are changing in a manner that consumption of potato is shifting from fresh potatoes to processed products such as frozen potatoes, mostly French fries ("chips" in the United Kingdom), potato crisp ("chips" in the United States), dehydrated potato flakes, potato flour, and potato starch (Schieber and Saldana, 2008; Charmley et al., 2006). This fact leads to an increased amount of waste generated in the potato industry. Depending on the obtained potato product, the following processing waste and by-products can be distinguished: (i) potato peels; (ii) cutting waste; (iii) wastewater from French fry, chip, and starch manufacturing plants, as well as potato slops/ stillage (distillery wastewater); (iv) potato pulp and liquor (fruit juice) after potato starch processing; (v) chip scraps (broken fried chips), etc. (Charmley et al., 2006).

Potato peel is a good source of carbohydrates, whereas other components are presented in lower amounts. The proximate chemical composition of potato peel is 11.2% moisture, 7.6% ash, 64.5% carbohydrates, 13.5% protein, and 3.4% sugars (Mabrouk and El Ahwany, 2008). However, this composition could vary with potato variety, growing conditions, and other parameters. At present, a great amount of this waste is used for feeding livestock (Al-Weshahy and Rao, 2012).

Recently, utilization of potato peel as a source of natural antioxidants has been investigated extensively due to the fact that it contains a 10-fold higher amount of phenolic compounds than potato flesh (Malmberg and Theander, 1984). Among them, chlorogenic and gallic acids are the most abundant (Sotillo et al., 1994; Singh et al., 2011). Other phenolic acids – caffeic, *p*-coumaric, ferulic, vanillic, and protocatechuic acids – are also present in lower amounts (Schieber and Saldana, 2008). Depending on the extraction technique and the polarity of the solvent (Galanakis et al., 2013a; Tsakona et al., 2012), the content of phenolic compounds in potato peel ranges from 25 to 125 mg/100 g (Singh and Saldaña, 2011). The phenolic content in potato peel depends on potato variety, where higher amounts of phenolic compounds can be found in red and purple colored potato peel compared with yellow and brown potato peel (Al-Weshahy and Rao, 2009; Im et al., 2008). In addition to phenolic compounds, potato peel is a good source of other functional ingredients such as dietary fiber (50% per weight) (Camire et al., 1995). The fiber content is influenced by the method of peeling. For instance, the abrasion method (used for potato chip manufacture) results in more starch and less dietary fiber (26.60% dry basis) than the steam peeling method (used for dehydrated potato production), which gives 56.5% (dry basis) total dietary fiber (Camire et al., 1997). Potato peel is also a source of glycoalkaloids, mostly α-chaconine and α-solanine, which are toxic for humans (Padmaja and Jyothi, 2012). However, it is possible to enzymatically convert solanaceous steroid glycoalkaloids into their aglycones and metabolites, in order to perform detoxification and produce raw materials for hormone synthesis (Wijbenga et al., 1999). Moreover, glycoalkaloids and their aglycones may also have anti-inflammatory, antihyperglycemic effects, and antibiotic activity against pathogenic bacteria, viruses, protozoa, and fungi (Friedman, 2006), as well as an anticarcinogenic effect against a series of human cancer cells in vitro (Friedman et al., 2005). Potato peel has been reported to be an excellent substrate for the production of enzymes such as endo-β-1-4-mannase (Mabrouk and El Ahwany, 2008), α-amylase, neutral and alkaline proteases, and polygalacturonate lyase (Mahmood et al., 1998), as well as for bioethanol production (Padmaja and Jyothi, 2012).

Potato pulp, the residue of rasped potatoes after starch extraction and fruit juice separation (decanting), is mainly valorized for feed, but it has also been used as a source of pectins. Under alkaline conditions, products containing a high content of pectic substances (25.8% estimated on the basis of galacturonic acid contents) were obtained (Turquois et al., 1999). Starch isolation and cellulose enzyme preparation can also be performed. Potato pulp may also be utilized for the replacement of wood fiber in paper making and as a substrate for yeast and B12 production (Kosseva, 2013). From fruit juice, protein can be recovered by coagulation and centrifugation or by ultrafiltration and used as an additive for cattle fodder or, after purification, as the ingredient of human food products (Lisinska, 1989).

Stillage (distillery wastewater) that is generated from the fermentation of a starch-based feedstock (e.g. potato) contains high nutritive value substances such as vitamins (mostly group B), proteins rich in exogenous amino acids, and mineral components and thus can be used as fodder, for direct soil fertilization or for the production of organic fertilizers (Krzywonos et al., 2009).

2.3.2 CASSAVA PROCESSING WASTE

Cassava (*Manihot esculenta*) is a major root crop cultivated in tropical developing countries. While Nigeria is the world's largest producer (FAO, 2013), Thailand is the largest exporting country of cassava. In Africa, cassava is used to prepare various food products from its submerged fermentation (e.g. garri and fufu), whereas in Brazil, Costa Rica, and Bolivia, it is mostly refined to farinha (flour) (Onilude, 1996). However, in several countries such as Thailand, India, Indonesia, and Vietnam, it is also used for the manufacture of starch and bioethanol (Padmaja and Jyothi, 2012).

There are five categories of wastes generated during the processing of cassava tubers (Aro et al., 2010): (i) starch residues (also called bagasse, pomace, pulp or fibrous residue), (ii) peels, (iii) stumps, (iv) whey, and (v) effluent.

Pomace is the fibrous residue (up to 17% of the tuber) that remains after the removal of starch. Peels represent from 5 to 15% of the root. They are obtained after washing and peeling the tubers. Cassava stumps are the ends trimmed off the tubers as they are prepared for washing and peeling. Cassava whey is the liquid pressed out of the tuber after it has been crushed. It can be mixed together with pomace to form an effluent (or slurry) (Aro et al., 2010; Heuzé et al., 2014).

Cassava pomace proximate composition includes 75–85% water as well as 1–4% protein, 15–30% starch, and more than 35% nonsoluble dietary fiber (dry matter basis) (Ubalua, 2007; Kosoom et al., 2009). Due to high starch content, pomace has found its application as a raw material for ethanol production (Ray et al., 2008). Moreover, it was found that pomace can be used as a substrate for the production of xanthan (Woiciechowski et al., 2004) and pullulan (Ray and Moorthy, 2007), which are microbial polysaccharides and hydrocolloids. The latest are widely used as stabilizing and thickening agents in food products.

Cassava peels are characterized with low protein content (<6% dry mater basis) and a high crude fiber content (10–30% dry mater basis). On the other hand, they contain natural toxic substances (e.g. phytates), which contribute to low phosphorous availability in nonruminants and lethal cyanogenic glycosides (Heuzé et al., 2014). In order to reduce cyanogenic content below the acceptable levels of 50 mg/kg (Nwokoro and Ekhosuehi, 2005), cassava peel has to be processed via drying, ensiling, soaking, and sun-drying operations (Salami and Odunsi, 2003; Tewe, 1992). Cassava peels can be used as a substrate for the production of commercially important enzymes, such as cellulase, α-amylase, glucoamylase, and xylanase (Ofuya and Nwajiuba, 1990; Onilude, 1996; Silva et al., 2009).

2.3.3 SWEET POTATO WASTE

China produces 80–85% of the world's sweet potato (*Ipomoea batatas*) tubers. Other sweet potato producers are Indonesia, Vietnam, India, the Philippines, and Japan in Asia; Brazil in South America, and the USA in North America; and Nigeria, Uganda, Tanzania, Rwanda, Burundi, Madagascar, Angola, and Mozambique in Africa (FAO, 2010). Sweet potatoes are typically processed in the cannery, distillery, and starch industries. Therefore, a variety of by-products (e.g. peels, vines and leaves, cannery wastes, distillery by-products and starch waste) is produced. Due to the similarity in type and composition of potato and sweet potato wastes, there are minor differences in the sorts of waste reutilization. Sweet potato peels contain phenolic compounds up to 6 mg/g (dry matter basis) (Anastácio and Carvalho, 2013). Truong et al. (2007) have found that sweet potato leaves have the highest phenolic acid content followed by the peel, whole root, and flesh tissues. Chlorogenic acid is the highest in root tissues, whereas 3,5-di-O-caffeoylquinic acid and/or 4,5-di-O-caffeoylquinic acid is predominant in the leaves. Waste powder from orange flesh sweet potato roots contains other antioxidants, too, such as β-carotene and α-tocopherol (Okuno et al., 2002). Besides, dietary fibers have been extracted from sweet potato residues after starch isolation, where the average yield and dietary fiber content of obtained product was 10 and 75%, respectively (Mei et al., 2010). Finally, sweet potato forage is a source of protein (15–30% dry matter basis, depending on the proportion of leaves and stems), where stems contain less protein than the leaves (An, 2004).

2.4 OILCROPS AND PULSES
2.4.1 PULSES PROCESSING WASTE

Pulses are the second most important foodstuff in the world after cereals as they are a rich source of proteins and other important nutrients such as carbohydrates, dietary fiber, vitamins and minerals (Oomah et al., 2011). Moreover, the presence of certain phytochemicals such as polyphenols, flavonoids, and phytosterols with pronounced health benefits has also been reported (Sreerama et al., 2010; Oomah et al., 2011). The FAO defines pulses as crops harvested solely from the dry seed of leguminous plants, excluding green beans and green peas (considered as vegetables), groundnut (*Arachis hypogaea*), and soybean (*Glycine max*) (considered as oilseeds). The most cultivated and consumed pulses include lentil, pea, chickpea, common bean, mung bean, pigeon pea, cowpea, and lupin (Tiwari et al., 2011). Due to their high protein content, they have already found applications in meat and pasta production, ready-to-eat breakfast cereals, baby food, snack food, texturized vegetable protein, pet foods, dried soups, and dry beverages (Tiwari et al., 2011; Kiosseoglou and Paraskevopoulou, 2011). Along with the food applications of pulse processing by-products, nonfood applications (e.g. animal feed, bioethanol production, biodegradable, and edible films) are also possible (Del Campo et al., 2006; Han and Gennadios, 2005), but less studied.

Pulses are typically processed in small-scale facilities or niche markets by dehulling, puffing, grinding, and splitting. The main products are flours and milling fractions (e.g. protein, starch and fiber), which can be utilized in the food industry either directly or pretreated (e.g. after extrusion cooking) to improve the nutritive value, functionality, and/or texture.

Dehulling and separation of cotyledons during milling involves the application of strong abrasive forces that lead to broken grains and powder products. The commercial pulses milling process yields approximately 75% of final products and 25% of by-products. The latter consists mainly of husk (up to 14%); powder (up to 12%); broken (up to 13%), shriveled, and unprocessed seeds.

The recovery of pulse by-products involves several stages such as grinding and fractionation (either by sieving or air classification) in order to obtain powdery/floury bulk material. In this regard, the first stage of pulse by-product recovery is sizing, fractionation of broken grains, coarse, fine fractions, and husk, which are removed from the process. Since husk contains a significant amount of tannins and insoluble dietary fibers (75–87%) (Dalgetty and Baik, 2003; Tiwari et al., 2011; Girish et al., 2012), an increase in protein digestibility of the remaining material is achieved. By further milling and sieving of the husk fraction, the husk can be reduced to a desirable particle size to be used as a dietary fiber source in food applications. The resultant material has a positive technological functionality reflected in control of hydration and rheological properties, fat/oil retention, surface area characteristics and porosity, particle size and bulk volume, and ion exchange capacity (Tosh and Yada, 2010). Coarse and fine fractions are sifted once again to obtain cotyledon material and husk. Separated cotyledon material together with broken seeds is subjected to powdering and subsequent sieving to obtain husk-enriched fraction and flour (Patras et al., 2011). Narasimha et al. (2004) reported the recovery of approximately 30–35% of the cotyledon material from pulse by-products and its utilization in production of traditional Indian snack products, which contained up to 50% of this material.

Sometimes the direct use of pulse flour for specific purposes (e.g. bread fortification) is limited due to its characteristic flavor caused by accompanying constituents and fatty acid oxidation products. On the other hand, by isolating bioactive compounds, it is possible to obtain products of neutral flavor, which can therefore be incorporated in high amounts without negative effects on the sensory properties (Fechner et al., 2011). Fechner et al. (2011) gave an overview of a multistage process for isolation of lupine dietary fiber and its application in bread baking. Further, pulse proteins can be isolated and fractionated in the form of protein concentrates and isolates using extraction techniques such as salt extraction (micellization), ultrafiltration, and water extraction.

However, the reutilization of pulse by-products is still insufficiently explored, requiring an interdisciplinary research by food technologists, food chemists, nutritionists, and toxicologists (Patras et al., 2011).

2.4.2 **OILSEED PROCESSING WASTE**

The most cultivated oilseed crops used for oil production and protein-rich residues are rapeseed (canola), sunflower, flax, peanut, and sesame. Rapeseed and sunflower have the highest significance (Lindhauer, 2004). Oilseeds such as rapeseed and sunflower are primarily grown for oil processing, however, crops such as soybeans are primarily grown for proteins, whereas cotton is primarily grown for fibers. The last two crops yield oil as a by-product. Moreover, peanuts are not only solely used for edible oil processing but also for direct human consumption, while linseed is used for the production of both edible and industrial oil (Day, 2004).

Oil content in oilseed crops accounts for 50% of sunflower and rapeseed, 56% of peanut, as well as 15–25% of soybean and cottonseed (Day, 2004). Oilseeds are processed by mechanical pressing and solvent extraction in order to remove oil (Daun, 2004). Both methods generate a waste consisting of peels, seeds, defatted oilseed meals, and oil sludge. Oilseed processing includes cleaning, dehulling, and cooking followed by oil extraction. Each of these steps produces a certain amount of by-products to be recovered. In the cleaning phase, separated material consists of stems, pods, leaves, broken grain, dirt, small stones, and extraneous seeds commonly found in the bulk oilseed material. These by-products could be reutilized as animal feed. By dehulling, a certain amount of hulls is obtained that could be utilized for extraction and recovery of residual functional compounds. The next step comprises seed grinding, followed by heating to enable complete destruction of the oil cells. The application of high temperatures in this phase has several effects: coagulation of proteins, insolubilization of the phospholipids, increased oil fluidity, destruction of microorganisms, and inactivation of enzymes. Preconditioned oilseeds are subjected to mechanical pressing, which squeezes the oil from the seed, or direct solvent extraction. In this way, crude oil and oilseed meal are obtained (O'Brien, 2009). Considering the fact that oil extraction is performed with solvents not always regarded as food-friendly (e.g. hexane or hydrocarbon mixtures rich in hexane), the recovery of oilseed meals from these processes is more complicated and less safe (Daun, 2004; Oreopoulou and Tzia, 2007; Galanakis, 2014).

Oilseed cakes are rich in proteins, dietary fibers, other bioactive compounds such as colorants, antioxidants, and other substances with positive health benefits, which make them suitable for valorization either as human food or feed (Oreopoulou and Tzia, 2007). Application of hemp seed oil press-cake in order to increase the nutritional profile of gluten-free products has already been shown (Dapčević Hadnađev et al., 2014; Radočaj et al., 2014). Taking into account the severity of protein malnutrition among world population and the renewability aspects of oilseed meals, oilseed proteins are considered a valuable alternative to animal proteins (Moure et al., 2006). However, their nutritional value is determined by their amino acid composition, noting that certain oilseed proteins are deficient in sulfur amino acids compared with proteins of animal origin. Amino acid composition of oilseed meals is often compared with soybean, which is considered a good source of amino acid for infants and children according to WHO (Tan et al., 2011; Rodrigues et al., 2012). However, the presence of certain antinutrients (condensed tannins, α-amylase inhibitors, trypsin inhibitors, phytic acid, glucosinolates, or saponins) may limit their conversion into edible-grade products and utilization in human nutrition. These compounds influence protein digestibility, organoleptic properties, and bioavailability of macro- and micro-elements. Pojić et al. (2014) demonstrated the possibility to fractionate hemp flour by sieving to obtain the fractions that were clearly differentiated in terms of crude protein, crude fiber, lipids, phenolic compounds, as well as antinutrient content. This pretreatment could be used to concentrate valuable target compounds and consequently facilitate their recovery.

In recent decades, an increasing trend of application of oilseed proteins in food, cosmetic and pharmaceutical industries has been observed (Moure et al., 2006; Oreopoulou and Tzia, 2007; Rodrigues et al., 2012). Valorization of this kind of product is based on the extraction, concentration and isolation of proteins from defatted meals, during which the reduction or total elimination of antinutrients is achieved. The obtained products are categorized on the basis of their protein content as flours (containing up to 50% of proteins), concentrates (containing up to 65–70% of proteins), and isolates (containing up to 90–95% of proteins) (Oreopoulou and Tzia, 2007). Protein isolates have good functional properties, such as emulsifying capacity, filmogenic properties, and water solubility. Moreover, protein isolates could be subjected to protein hydrolysis in order to obtain functional peptides and essential amino acids (Rodrigues et al., 2012).

2.4.3 RECOVERY OF PULSES AND OILSEED BY-PRODUCTS FOR NONFOOD APPLICATION

Recently, the utilization of biopolymers such as proteins, polysaccharides, lipids, or their combinations for the preparation of edible/biodegradable films has been demonstrated (Mariniello et al., 2003; Su et al., 2010). It was found that protein films possess superior properties over lipid and polysaccharide films in terms of gas barrier and mechanical properties (Jia et al., 2009; Gupta and Nayak, 2014). Since pulse and oilseed by-products are characterized by high protein content, they have been designated as suitable materials for the isolation of film forming materials.

Soy protein was found to be a superior material over other plant protein sources, as it yields more flexible, smoother, and clearer films (Gontard and Guilbert, 1994). However, films made from one type of natural polymer can express certain positive properties, but at the same time poor ones for other quality aspects. On the other hand, by combining film-forming polymers of different origin, it is possible to prepare composite edible films with improved properties (Jia et al., 2009). Beans (black and white) and chickpeas have also showed some potential to be used in the manufacture of compression-molded plastics, as explored by Salmoral et al., 2000a, b. In this case, bean protein isolates or defatted whole flours were blended with glycerol and the extracted starch was used as plasticizer. Apart from the application of biodegradable packaging materials, soy meal could be utilized for the production of adhesives. However, soy protein adhesives are characterized by low gluing strength and water resistance. For this reason, Huang and Sun (2000) proposed the modification of soy protein isolates by urea and guanidine hydrochloride solutions.

2.5 FRUIT AND VEGETABLES

According to the report by the FAO of the United Nations (FAO, 2014), 45% of fruit and vegetable wastes and by-products from the fruit and vegetable processing industry are generated around the world. These losses occur throughout the entire food supply chain. A large quantity of fruit and vegetable wastes is disposed in landfills or rivers, which represents a threat to the environment due to their high biodegradability, leachate, and methane emissions (Misi and Forster, 2002). However, these resources have a great potential to be used for the recovery of value-added products (Wadhwa and Bakshi, 2013).

Agricultural fruit and vegetable waste is the consequence of mechanical damage and/or spillage during harvest operation (e.g. threshing or fruit picking) or crops sorted out postharvest to meet quality standards (Gustavsson et al., 2011). Depending on fruit and vegetable processing technology (e.g. drying and dehydration, juice technology, fruit jams, canning, jellies, marmalade, paste production, vegetable pickles, and sauerkraut technology), solid (e.g. pomace, pulp, peels, cores, seeds, and stems), as well as liquid (e.g. juices, wash water, chilling water and cleaning chemicals) waste streams are produced. In general, fruit and vegetable wastes are mostly composed of water (80–90%) and hydrocarbons with relatively small amounts of proteins and fat (Mirabella et al., 2014). However, since these wastes also contain a significant amount of biologically active compounds, they can be used for the recovery of value-added products like polyphenols, glucosinolates, dietary fibers, essential oils, pigments, enzymes, organic acids, etc. (Galanakis, 2013; Galanakis et al., 2013b, 2015a, b; Galanakis and Schieber, 2014; Deng et al., 2015; Heng et al., 2015). Table 2.1 summarizes the target compounds from fruit and vegetable wastes and by-products that have been recovered.

Table 2.1 Fruit and Vegetable By-Products and Corresponding Functional Ingredients for Recovery

Fruits and Vegetables	By-Product	Target Ingredient	References
Mandarin	Peel	Flavanone glycoside – narirutin	Kim et al. (2004)
	Peel	Essential oil (limonene and limonene/γ-terpinene)	Lota et al. (2000)
	Leaf	Essential oil (sabinene/linalool, linalool/ γ-terpinene and methyl N-methylanthranilate)	Lota et al. (2000)
Orange	Peel	Flavanone glycoside – hesperidin	Di Mauro et al. (1999)
		Apocarotenoid	Chedea et al. (2010)
		Essential oil (limonene)	Farhat et al. (2011)
		Cellulose	Bicu and Mustata (2011)
Lemon	By-product	Pectin	Masmoudi et al. (2008)
Apple	Pomace	Pectin	Wang et al. (2007)
	Skin	Phenols	Schieber et al. (2001)
Peach	Pomace	Pectin	Pagan et al. (1999)
Apricot	Kernel	Protein	Sharma et al. (2010)
Grape	Pomace	Dietary fiber	Schieber et al. (2001)
	Skin	Phenols	Pinelo et al. (2006)
	Wine lees	Food preservative – calcium tartrate	Braga et al. (2002)
		Pigment enocyanin	Braga et al. (2002)
Black currant	Seed residue after oil extraction	Phenols	Bakowska-Barczak et al. (2009)
Banana	Bracts (leaves below calyx)	Anthocyanin pigments	Pazmiño-Durán et al. (2001)

(Continued)

Table 2.1 Fruit and Vegetable By-Products and Corresponding Functional Ingredients for Recovery *(cont.)*

Fruits and Vegetables	By-Product	Target Ingredient	References
Kiwi	Pomace	Soluble and insoluble dietary fiber	Martin-Cabrejas et al. (1995)
Pear	Pomace	Soluble and insoluble dietary fiber	Martin-Cabrejas et al. (1995)
Pineapple	Core, peel, crown and extended stem	Proteolytic enzyme – bromelain	Umesh Hebbar et al. (2008)
Mango	Seed kernels	Phenolic compounds	Abdalla et al. (2007)
	Peel	Polyphenols, carotenoids, vitamins, enzymes and dietary fibers	Ajila et al. (2010)
Carrot	Peel	β-Carotene	Chantaro et al. (2008)
		Phenols	Chantaro et al. (2008)
	Pomace	Carotenoids	Zhang and Hamauzu (2004)
Tomato	Pomace	Lycopene	Lavecchia and Zuorro (2008)
	Peel	Carotenoids (lycopene, lutein, β-carotene, and *cis*-β-carotene)	Knoblich et al. (2005)
	Seeds	Lycopene	Knoblich et al. (2005)
		Dietary fiber	Knoblich et al. (2005)
Cauliflower	Floret and curd	Pectin	Femenia et al. (1997)
Broccoli	Leaves or stalks	Glucosinolates, phenolic acids, flavonoids, vitamin C	Domínguez-Perles et al. (2010)

Apart from being a substrate for bioactive compound isolation, fruit and vegetable by-products can be used as functional additives in food. Vergara-Valencia et al. (2007) demonstrated that mango dietary fiber concentrate from unripe fruit could be applied as bakery product ingredient in order to increase its antioxidant capacity. Moreover, apple pomace, a by-product of the apple juice industry, could be used as a source of dietary fiber and polyphenols in cake production (Sudha et al., 2007), while raspberry (Górecka et al., 2010), white grape (Mildner-Szkudlarz et al., 2013), and blueberry (Mišan et al., 2014) pomace have been used for cookie enrichment. Carrot pomace could be used in bread (Osawa et al., 1994), cake, dressing, and pickles (Osawa et al., 1995) and onion pomace in snacks (Kee et al., 2000). Citrus by-products (lemon albedo and orange dietary fiber powder) have been added to cooked and dry-cured sausages to increase their dietary fiber content (Fernández-López et al., 2004), while orange juice fibers (peel, pulp, and seeds) have been used as a fat replacer in ice cream (Crizel de Moraes et al., 2013). Olive by-products have been used for the recovery of polyphenols and pectin, which have been proposed as antioxidants and fat substitutes in foodstuff, respectively (Galanakis et al., 2010a, b, c, d, e; Galanakis, 2011; Rahmanian et al., 2014; Roselló-Soto et al., 2015). Finally, utilization of fruit and vegetable waste for ethanol, biogas, single cell protein, beverages (cider, beer, and wine), and vinegar production has also been reported (Wadhwa and Bakshi, 2013).

2.6 **MEAT PRODUCTS**

2.6.1 **FIFTH QUARTER**

The term fifth quarter refers generally to all the nonmeat parts of a carcass. It includes all of the products, other than meat, which are harvested from the carcass in an abattoir. These parts could comprise offal, other edible coproducts (e.g. blood, fat, stomach, tendons, membranes, etc.), and animal by-products. In some countries, the industry has secured markets for many of these products; however, these are quite often low-value markets. Also some of the products hold either a neutral (no cost to dispose of) or negative (costs money to dispose of) value. Many of these materials are rich sources of valuable components such as protein, lipids, minerals, etc., which in their own right can command a higher value than the original source material (e.g. blood plasma proteins are higher value than blood). An important aspect of recovery of additional value from meat processing is adherence to the strict legislation associated with animal by-products, etc. In addition to many edible nonmeat parts, which are considered viable sources of high value components, a number of animal by-products are also generated. According to European legislation (Article 3 of Regulation (EC) 1069/2009), an animal by-product is "entire bodies or parts of animals, products of animal origin or other products obtained from animals that are not intended for human consumption". This legislation splits by-products into three categories on the basis of risk for human health. Categories 1 and 2 are considered as the highest risk material while materials in category 3 are considered as low risk. Materials of category 1 can only be incinerated, whereas materials of category 2 can be used as fertilizer, compost, or anaerobic digestion plants after being pressure-rendered. Category 3 materials include parts of animals that have been passed fit for human consumption, but are not suitable either because they are not normally eaten or for commercial reasons have not been processed in that way. Nevertheless, they can be used for pet food or animal feeding stuffs (under several restrictions according to Transmissible Spongiform Encephalopathies Regulations). More importantly, they can be used for the manufacture of derived products (e.g. medical devices), as specified by the Regulations. Recently, blood from young ruminants and ruminants passed a Transmissible Spongiform Encephalopathies test and were reclassified from category 2 to category 3. Unfortunately, the low economic value of these materials can sometimes mean that they are often mixed together and processed as a single batch. An overview of the breakdown of carcasses of different species into various components is presented in Fig. 2.1. Due to new animal by-products regulations and improved export opportunities, the market is changing significantly. The new regulations make provision for the introduction of new technologies or methods for the authorization of such operations.

There are many markets for a variety of fifth quarter type products, which are being capitalized upon by meat processors across the world. Relatively low cost and simple steps are taken to stabilize and further process these products: washing, mincing, rendering, smoking, slicing, or soaking. Nevertheless, the added value of the final product remains quite low and does not always reflect the value that can be generated from these materials. Recovering high value functional components (functional proteins, lipids, and other biomolecules such as proteoglycans) from these products can provide a viable route to innovation, growth, and increased market opportunities for the meat sector. However,

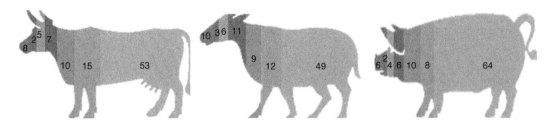

■ Fat and blood   ■ Red offal   ■ Gut content   ■ Hide   ■ Stomach/Intestines   ■ Feet/Head   ■ Carcase

FIGURE 2.1 Percentage of Fifth Quarter Tissue Weights Relative to Total Live Weight of Bovine, Ovine, and Porcine

Adapted from Stanley, 2009

as highlighted by Waldron (2007), any approach to further exploiting the value of these products needs to be guided as far as possible by the principle of total exploitation without generating more waste in the process. Ensuring that all the derived components are of commercial value is critical to the ultimate success of this approach.

Table 2.2 compares the direct use of fifth quarter products with their valorization as substrates for the recovery of low and high added-value compounds. Despite the numerous efforts in the field, it is still necessary to employ more effective and novel tools to analyze these products, for example, for nutritional properties and to develop new technological applications (Toldrá et al., 2012). The recovery of protein-rich functional coproducts from meat-processing streams represents an area of significant opportunity to enhance the economic performance and improve the environmental impact of the meat industry.

2.6.2 HIGHER VALUE PRODUCTS FROM MEAT PROCESSING SOURCES

Blood is one of the most problematic by-products of the meat industry due the large amount produced (3–4 L per pig, up to 18 L per bovine) and its high polluting capacity. However, blood has high content of good-quality proteins (around 15–18% in wet base). The current regulation encourages recovering the blood and finding new ways of using blood proteins (European Directive 2008/98/CE). The main obstacle for the full utilization of blood cellular fraction is the presence of hemoglobin, which provides a strong color, distinctive flavor, and influences oxidation. However, several industrial applications have been developed (Bah et al., 2013).Whole proteins from plasma (as albumin, fibrinogen, and globulins) can be easily separated by chemical precipitation (Moure et al., 2003) and then dried. Such proteins have relevant functional properties like gelation, foaming, emulsifying, or thickeners, which allow them to be utilized in the food industry as dietary supplements (Ramos-Clamont et al., 2003; Álvarez et al., 2009; Liu et al., 2010; Ofori and Hsieh, 2011). More recent applications of blood proteins include the development of products with enhanced functional properties and novel biological activities such as antioxidant (Chang et al., 2007; Sarmadi and Ismail, 2010; Sun et al., 2011; Alvarez et al., 2012), antimicrobial (Nedjar-Arroume et al., 2008; Catiau et al., 2011), mineral-binding (Lee and Song, 2009), antigenotoxic (Park and Hyun, 2002), opioid (Piot et al., 1992), or antihypertensive (Yu et al., 2006).

Table 2.2 Comparison Between Direct Uses of Fifth Quarter and High Added-Value Products Obtained with Novel Extraction Techniques

ABP	Direct Preparation[a]	Uses[a]	Revalorizing Techniques	High Added-Value Products	References
Liver	Frozen, fresh or refrigerated	Braised, broiled, fried, patty and sausage	Enzymatic hydrolysis	Antioxidant peptides	Di Bernardini et al. (2011b)
Heart	Whole or sliced	Braised, cooked, luncheon meat, patty, loaf	Isoelectric solubilization/ precipitation	High value protein with low ash, fat, and cholesterol	Dewitt et al. (2002)
	Frozen, fresh, or refrigerated		Phosphate buffer washing	Myofibrillar concentrate as texturizing	Ionescu et al. (2007)
Skin	Fresh, refrigerated	Gelatin	Collagen recovery	Barrier membrane, drug delivery, fibroblast scaffolds, bioengineered tissues	Kew et al. (2011)
			Enzymatic hydrolysis and chromatographic purification	Antioxidant peptides and liver protectors	Lee et al. (2012)
				Antioxidant activity, antimicrobial properties, antihypertensive, biomimetic tissue	Li et al. (2007); Di Bernardini et al. (2011a); Ichimura et al. (2009); Zeugolis et al. (2008b)
			Collagen hydrolysis		
Blood	Fresh or refrigerated	Black pudding, sausages, blood and barley loaf	Enzymatic hydrolysis	Antioxidant, antibacterial, antihypertensive or iron-binding peptides	Sun et al. (2011); Nedjar-Arroume et al. (2008); Lee and Song (2009);
			Chemical hydrolysis	Predigested peptides for animal and pet food	Alvarez et al. (2012)
			Ethanol precipitation	Purified protein as food ingredient	Álvarez et al. (2009)
			High pressure treatment	Peptides and biopreserved blood	Toldrà et al. (2011)
			Subcritical water hydrolysis	Amino acid and peptides production	Rogalinski et al. (2005)
Bone	Frozen, fresh, or refrigerated	Gelatin, soup, jellied products	Subcritical water Alkaline extraction	Hydroxyapatite and collagen New kind of sausages	Barakat et al. (2009) Boles et al. (2000)
Lung	Frozen, fresh, or refrigerated	Blood preparations, pet food	ISP[b] and membrane filtration	Protein concentrates with good functional properties	Darine et al. (2010)
Feathers and hair	Incineration, rendering	Feather or hair meal	Keratinolytic bacteria fermentation and enzymatic hydrolysis	Keratinolytic protease production, culture medium, soil assessment, separation membranes	Chaturvedi et al. (2014); Gousterova et al. (2012); Sueyoshi et al. (2011)

[a]Two columns adapted from Ockerman and Hansen (1999).
[b]ISP: isoelectric solubilization/precipitation.

Collagen is the major component of mammalian connective tissue, which accounts for 30% of total protein weight. Twenty-seven different types of collagen have been identified with type I being the most abundant and the most widely occurring in connective tissue. Collagen molecules are composed of three α-chains intertwined into a triple helix. This particular structure (mainly stabilized by intra- and interchain hydrogen bonding) is the product of an almost continuous repeating sequence of Gly-X-Y, where X and Y are typically proline and hydroxyproline, respectively (Asghar and Henrickson, 1982). The most abundant sources of gelatin are pig skin (46%), bovine hide (29%), and pork and cattle bones (23%).

Two types of gelatin are obtainable, depending on the pretreatment procedure, and are known commercially as type-A gelatin (isoelectric point at pH~8–9) and type-B gelatin (isoelectric point at pH~ 4–5) obtained under acid and alkaline pretreatment conditions, respectively (Gómez-Guillén et al., 2011). Industrial applications call for one or the other gelatin type, depending on the degree of collagen cross-linking in the raw material (Gómez-Guillén et al., 2011). Collagen shows a great versatility and, for example, can be used as a calcifiable matrix system for implantable biomaterials (Singla and Lee, 2002). Collagen can also form scaffolds for liposomes, which control the release rate of entrapped drugs by thermal triggering (López-Noriega et al., 2014). It can also be used in gene transferring, by subcutaneous collagenous pellets for somatic gene therapy (Sano et al., 2003) or for probiotic bacteria microencapsulation by spray-drying technology (Li et al., 2009). Zeugolis et al. have used collagen extracted from bovine Achilles tendons to create extruded fibers capable of forming scaffolds for tissue engineering and regeneration of tendons and ligaments (Zeugolis et al., 2008a; Kew et al., 2011). Collagen and gelatin can be used as a source of bioactive peptides by enzymatic methods with many studies focusing on pig and bovine skins. However, gelatin from poultry, ducks, and other minority sources has also been studied in recent years (Lee et al., 2012; Lasekan et al., 2013), particularly in Asiatic countries. Other proteins recovered from pork and beef sources have also demonstrated a strong heat induced gelation and emulsification ability (Darine et al., 2010). An ongoing project focused in recovery high added value functional proteins and bio-active peptides from beef and porcine offal is being carried out in Teagasc Food Research Centre (Mullen et al., 2015).

Feather and hair have also a high protein content, but their low content of certain amino acids (Lys, Met, His, and Trp) makes them less attractive for feed supplementation, as artificial addition of amino acids or blood complementation is required (Brotzge et al., 2014). A promising approach for feathers and hair is as culture media for screening keratinolytic bacteria aiming to characterize and recover new keratinase enzymes (Nam et al., 2002; Riffel et al., 2003; Chaturvedi et al., 2014). Such enzymes have a potential use in leather industry for the development of more ecofriendly hair removal processes.

2.7 FISHERIES BY-PRODUCTS

In 2012, according to FAO, almost 160 million tons of fish were produced between captures (58%) and aquaculture (42%). Although sustainability is essential to maintain a constant capture rate, aquaculture production is still growing by 4–5% per year. In general, at least 25–30% of the total fish or seafood weight is considered a waste. The remains of the fish are usually called by-products. According to EU regulation, they can be classified in category 3 if properly treated. By-products are typically considered for animal feeding, fertilizers or just discarded. On the other hand, fisheries wastes and by-products contain valuable compounds such as peptides, proteins, natural pigments, collagen, fatty acids, chitin, chitosan and calcium, which could be recovered and reutilized in different applications (Arvanitoyannis and Kassaveti, 2008; Galanakis, 2012).

2.7.1 APPLICATIONS

Protein and oil are by far the main compounds in fisheries waste and hence most of the research has been focused on developing and enhancing techniques to recover these valuable compounds. For instance, surimi wastewater contains myofibrillar protein obtained from fish flesh, which has been extensively washed with cold water. The protein content in wastewater can be as high as 2.3%, composed by enzymes, low molecular weight proteins, and sarcoplasmatic proteins. Oral administration of peptides recovered from sales of sea bream showed a decrease in blood pressure in laboratory rats. Similar results were obtained after hydrolyzing skate skin with α-chymotrypsin (Lee et al., 2011).

When assessing the quality of proteins used for food purposes, the essential amino acid content is very important. Protein recovered from fisheries wastes (regardless of method used) shows an amino acidic profile of advanced characteristics (Chen et al., 2007). Peptides have acquired more relevance over the past few decades, as according to recent studies, they show promising bioactivities, that is, antihypertensive, antioxidant, antidiabetic, antimicrobial, or mineral binding (Korhonen and Pihlanto, 2006; Tierney et al., 2010; Fitzgerald et al., 2012). Peptides from fish wastes have been employed to create new biofilms (Rocha et al., 2014; Santos et al., 2014). The latest materials can be used as edible food coating, able to enlarge the shelf-life of food products. Collagen extracted from fish skin has various industrial applications in cosmetics and medicine. For instance, it can be used as an alternative to mammalian collagen. Gelatin extracted from fish collagen has caught the interest of the scientific community, due to the huge number of species having very different intrinsic characteristics (Gómez-Guillén et al., 2011). Compared with mammalian gelatins, fish gelatins are characterized by low gelling strength and temperatures as well as good swelling capacity, which allow them to be used in capsules of controlled drug release (Zohuriaan-Mehr et al., 2009).

Oil and fatty acids from marine wastes are mainly used as sources of polyunsaturated fatty acids (PUFAs). Linolenic (ALA, 18:3ω-3), eicosapentaenoic (EPA, 20:5ω-3), and docosahexaenoic acids (DHA, 22:6ω-3) are the main ω-3 PUFAs, whereas linoleic (L, 18:2ω-6) and arachidonic acids (AN, 20:4ω6) are the main ω-6 PUFAs in fillets obtained from farm-raised trout (Chen et al., 2007). Key applications of oil with high PUFAs content include cosmetics or food supplements. Industrial applications for recovered oil as biodiesel have also been developed (Kato et al., 2004).

The color of many fish and shellfish is due to the presence of carotenoids. These pigments can be detected in the flesh (salmon, trout, or tuna) or integument (lobster, shrimp, and crab). Natural carotenoids can be used as pigments, supplements, or as a source of vitamin A. One of the most abundant carotenoids, astaxanthin, is a powerful antioxidant, with anticancer activity and a photoprotective effect (Guerin et al., 2003). Once the pigments of shells have been extracted, the resulting product is available to be used for chitin/chitosan extraction. Chitosan is the deacetylated form of chitin and is soluble in acidic solutions. Industrial applications of such polymers include antimicrobial agents, edible films, food supplements, emulsifiers, thickeners, purification of water, or chromatography. Extended reviews about their applications can be consulted (Shahidi et al., 1999; Rinaudo, 2006).

Calcium can be recovered from bones, frames, and crustaceous wastes, once protein, oil, and other compounds of interest have been extracted. Food additive is the main application for calcium recovered from fisheries wastes, that is, to handle calcium deficiencies in diets or animal feeding (Coward-Kelly et al., 2006). On the other hand, shells from mollusks are composed mainly of calcium carbonate (95% wt%) and low amounts of protein. Kuji et al. (2006) developed a new low-cost process that involved the use of high-pressure CO_2 solution for solubilization of shells. Other methods use high concentrated acetic acid solutions to dissolve calcium. Finally, integrated extraction systems have been developed for recovering a variety of components such as pigments, glucosamine, chitin, and chitosan from shrimp by-products (Cahú et al., 2012). Hydroxyapatite is another compound that can be extracted from bones; as well as serving in bone substitute applications it has also been used, for example, as an ion exchanger for cleaning wastewaters (Ferraro et al., 2013).

Fish wastes are also a source of enzymes, which can be extracted and then used for enzymatic assisted extraction techniques. Such enzymes are often used in autolysis processes. Autolysis of wastes has been successfully applied (Silva et al., 2014; Cao et al., 2009) for hydrolysates production, whereas the yield in terms of protein recovery was similar to those obtained by using commercial proteases.

2.8 DAIRY PRODUCTS

The dairy industry is considered as one of the largest processing industries in the food sector, and consequently is one of the largest generators of food waste and other industrial effluents (Demirel et al., 2005; Pesta et al., 2007). The demand for milk and its products is constantly increasing, being the highest in Europe and North America. This is a consequence of a variety of products produced by the dairy industry from fresh milk to yogurt, fermented milk products, butter, milk powders, cheese, ice cream, and novel functional dairy products. Thus, the production in the dairy industry is carried out in several separate and multifunctional processes, which could be roughly divided into fluid milk and processed milk processing, where each process itself generates a certain amount of waste (Sherman, 2007; Arvanitoyannis and Kassaveti, 2008).

Milk processing is generally conducted throughout the following processing steps:

- clarification or filtration,
- blending and mixing,
- pasteurization and homogenization,
- process manufacturing,
- packaging, and
- cleaning.

2.8.1 DAIRY PROCESSING WASTE

The first processing operation in the dairy industry (centrifugation of fresh milk) generates a certain amount of sludge (275 g of sediment/t of centrifuged fresh milk), which can be utilized in the animal feed industry due to its high nutritional value (Özbay and Demirer, 2007; Yadav and Garg, 2011). Dairy sludge could be mixed with other organic residues and bulking agents to improve the structure and nutrient content, and used prior to the production of vermicompost (Gratelly et al., 1996; Nogales et al., 1999; Yadav and Garg, 2011).

The generation of waste in milk processing is continued by rinsing and cleaning the tanks, transport lines and technical equipment wherein a certain amount of milk is collected (0.04% per mass of finished product). Cheese production generates waste consisting of whey and cheese residues (Russ and Schnappinger, 2007; Pesta et al., 2007). Waste streams from dairy processing contain high amounts of organic matter composed of carbohydrates, proteins, and fats, which are therefore characterized by high biological and chemical oxygen demand (Demirel et al., 2005; Pesta et al., 2007). A large number of dairies have already implemented systems for aerobic wastewater treatments, but aerobic process generates a high amount of sludge requiring its subsequent treatment. Moreover, it is characterized by relatively high energy consumption (Doble and Kumar, 2005). Anaerobic digestion is an alternative choice that enables the production of biogas as a renewable source of energy and a valuable effluent that can be used as soil fertilizer (Demirel et al., 2005; Pesta et al., 2007). Combined (anaerobic/aerobic) treatment systems for dairy wastewaters are also possible. Since it is difficult to degrade milk fat, the application of this process requires a suitable bioreactor design to avoid undesirable fat accumulation (Demirel et al., 2005).

2.8.2 WHEY AS THE MOST ABUNDANT DAIRY BY-PRODUCT

With the increasing popularity of cheese and related cheese products, a generation of an increasing amount of waste (whey and cheese residues) has posed the need to cope with the problem of cheese waste disposal and its further processing (Kosikowski, 1979; Audic et al., 2003). Moreover, in the cheese making process a certain amount of wastewater may arise as a result of cheese rinsing required for certain types of cheeses (Britz et al., 2006).

Although the generated amount of whey and cheese residues is dependent on the type of cheese produced, the cheese making process yields approximately 10% of cheese, while the remaining 90% of material is separated as whey (Russ and Schnappinger, 2007). These yields indicate that whey is characterized by a very high specific waste index, being the highest compared with other types of agricultural by-products (Russ and Meyer-Pittroff, 2004).

Whey can be utilized either directly as a component of different products formulation or indirectly, that is, when used as a substrate for fermentation, whey protein and lactose concentrates/isolates recovery, and biopolymers production (Patel and Singh, 2012; Galanakis et al., 2012a; Mandal et al., 2013).

2.8.3 DIRECT UTILIZATION OF WHEY

Direct application of whey implies its utilization either in liquid or dried form. In the first case, it is usually used as a water replacer without any changes in its composition (Patel et al., 2007). This is the case of soft whey-based beverage production, wherein native sweet, diluted, or acid whey is mixed with different additives. The latter includes fruits (tropical fruits, apples, pears, peaches, strawberries, or cranberries), cereals, and cereal-based products, isolates of vegetable proteins, chocolate, cocoa, vanilla extracts, and other flavorings (Beucler et al., 2005; Jeličić et al., 2008; Singh and Singh, 2012). In recent years, the development of functional whey beverages obtained by fermentation with probiotic bacteria has been reported (Magalhães et al., 2010; Shukla et al., 2013; Bulatović et al., 2014). The utilization of whey in dried form requires previous pretreatment by applying condensing techniques such as spray-drying technology, mechanical vapor recompression, ultrafiltration, and reverse osmosis (Jeličić et al., 2008). More specifically, sweet whey powder could be utilized by the dairy, baking, and confectionary industries, as well as by the baby food and meat industries (Kosikowski, 1979; Britz et al., 2006; Indrani and Rao, 2011). Direct application of acid whey powder, due to its astringent flavor is more limited than that of sweet whey powder. It can be used as an auxiliary raw material in cheese production as well as a component for the production of cheese powders, sauces, spread, margarines, bread, biscuits, crackers, and snack foods (Kosikowski, 1979).

2.8.4 **INDIRECT UTILIZATION OF WHEY**

The indirect utilization of whey implies reprocessing it for the recovery of valuable nutrients such as proteins and carbohydrates (Sherman, 2007). The most common techniques applied for the separation of whey to its main components (e.g. protein, lactose, and delactosed permeate) are membrane techniques such as ultrafiltration or reverse osmosis (Pesta et al., 2007; Galanakis, 2012; Galanakis et al., 2014).

Whey proteins are characterized by their exceptionally high nutritional quality, predominantly determined by amino acid composition and physiological functionality. They possess also unique physical functional properties such as gelling, foaming, water binding, and emulsifying capacity, which justify its inclusion as ingredients in the food and related industries. Continuous separation chromatographic technology as well as membrane processing technique with the application of membrane adsorbers (e.g. ion exchange) have been employed in order to obtain specific whey protein isolates (enriched in β-lactoglobulin and/or glycomacropeptide), lactoferrin, and bioactive factors from whey (Smithers, 2008).

Deproteinized whey permeates, whey, delactosed permeate, or their mixture could be used as raw material for ethanol fermentation in order to obtain bioethanol as a liquid fuel (Christensen et al., 2011; Wagner et al., 2014) or neutral spirits (Patel, 2010; Dragone et al., 2008). The production of whey beer and whey wine is also possible (Kosikowski, 1979; Jeličić et al., 2008). Ethanol fermentation requires enzymatic pretreatment to facilitate the conversion of lactose to ethanol, most commonly conducted with β-galactosidase (Pesta et al., 2007). In order to prevent the disturbance of the fermentation process, caused by high mineral content of deproteinized whey permeates, whey and delactosed permeates, a demineralization process should be performed by ion exchange technique, electrodialysis, gel filtration, or nanofiltration (Suárez et al., 2009; Pan et al., 2011; Galanakis et al., 2012b). Thus, specific milk salts are recovered and used as nutritional supplements and salt substitutes, especially for the production of sports and health beverages, feed and for the cosmetic industry. Deproteinized permeate in concentrated form is used for the recovery of lactose and lactose derivates by chemical and enzymatic modifications (Audic et al., 2003; Pesta et al., 2007; Gänzle et al., 2008).

Starting from the fermentation of deproteinized whey permeates, the production of organic acids (acetic, propionic, lactic, lactobionic, citric, gluconic), polysaccharides (xanthan gum, dextrans, phosphomannans, gellans, pullulans), certain vitamins, and amino acids is possible, too (Audic et al., 2003; Pesta et al., 2007).

REFERENCES

AACC (American Association of Cereal Chemists)1989. AACC Committee adopts oat bran definition. Cereal Foods World 34, 1033.

Aachary, A.A., Prapulla, S.G., 2009. Value addition to corncob: production and characterization of xylooligosaccharides from alkali pretreated lignin-saccharide complex using *Aspergillus oryzae* MTCC 5154. Bioresource Technol. 100 (2), 991–995.

Abdalla, A.E.M., Darwish, S.M., Ayad, E.H.E., El-Hamahmy, R.M., 2007. Egyptian mango by-product 1. Compositional quality of mango seed kernel. Food Chem. 103, 1134–1140.

Ajila, C.M., Aalami, M., Leelavathi, K., Rao, U.J.S.P., 2010. Mango peel powder: a potential source of antioxidant and dietary fiber in macaroni preparations. Innov. Food Sci. Emerg. Techn. 11, 219–224.

Akpinar, O., Erdogan, K., Bostanci, S., 2009. Production of xylooligosaccharides by controlled acid hydrolysis of lignocellulosic materials. Carbohydr. Res. 344 (5), 660–666.

Alvarez, C., Rendueles, M., Diaz, M., 2012. Production of porcine hemoglobin peptides at moderate temperature and medium pressure under a nitrogen stream. Functional and antioxidant properties. J. Agric. Food Chem. 60 (22), 5636–5643.

Alvarez, C., Rendueles, M., Diaz, M., 2012. The yield of peptides and amino acids following acid hydrolysis of haemoglobin from porcine blood. Anim. Prod. Sci. 52 (5), 313.

Álvarez, C., Bances, M., Rendueles, M., Díaz, M., 2009. Functional properties of isolated porcine blood proteins. Int. J. Food Sci. Tech. 44 (4), 807–814.

Al-Weshahy, A., Rao, V.A., 2009. Isolation and characterization of functional components from peel samples of six potatoes varieties growing in Ontario. Food Res. Int. 42, 1062–1066.

Al-Weshahy, A., Rao, A.V., 2012. Potato peel as a source of important phytochemical antioxidant nutraceuticals and their role in human health – a review. In: Rao, V. (Ed.), Phytochemicals as Nutraceuticals – Global Approaches to Their Role in Nutrition and Health. InTech, Rijeka, Croatia.

An, L.V., 2004. Sweet potato leaves for growing pigs: biomass yield, digestion and nutritive value. Doctoral thesis, Swedish University of Agricultural Sciences, Uppsala, Acta Universitatis Agriculturae Sueciae Agraria, p. 470.

Anastácio, A., Carvalho, I.S., 2013. Phenolics extraction from sweet potato peels: key factors screening through a Placket–Burman design. Ind. Crop. Prod. 43, 99–105.

Anderson, J.W., 2003. Whole grains protect against atherosclerotic cardiovascular disease. Proc. Nutr. Soc. 62 (1), 135–142.

Apprich, S., Tirpanalan, Ö., Hell, J., Reisinger, M., Böhmdorfer, S., Siebenhandl-Ehn, S., Kneifel, W., 2014. Wheat bran-based biorefinery 2: valorization of products. LWT-Food Sci. Technol. 56 (2), 222–231.

Aro, S.O., Aletor, V.A., Tewe, O.O., Agbede, J.O., 2010. Nutritional potentials of cassava tuber wastes: a case study of a cassava starch processing factory in south-western Nigeria. Livest. Res. Rural Dev. 22, 57–62.

Arvanitoyannis, I.S., Kassaveti, A., 2008. Dairy waste management: treatment methods and potential uses of treated waste. In: Arvanitoyannis, I.S. (Ed.), Waste Management for the Food Industries. Elsevier Academic Press, London, UK, pp. 801–860.

Asghar, A., Henrickson, R., 1982. Chemical, biochemical, functional, and nutritional characteristics of collagen in food systems. Adv. Food Res. 28, 231–372.

Asp, H.G., Bjorck, I., Nyman M., 1993. Physiological effect of cereal dietary fibre. Carbohydr. Polym. 21, 183–187.

Audic, J.-L., Chaufer, B., Daufin, G., 2003. Non-food applications of milk components and dairy co-products: a review. Lait 83, 417–438.

Bah, C.S.F., et al.,2013. Slaughterhouse blood: an emerging source of bioactive compounds. Compr. Rev. Food Sci. F. 12 (3), 314–331.

Bakowska-Barczak, A.M., Schieber, A., Kolodziejczyk, P., 2009. Characterization of Canadian black currant (*Ribes nigrum* L.) seed oils and residues. J. Agr. Food Chem. 57, 11528–11536.

Barakat, N.A.M., Khil, M.S., Omran, A.M., Sheikh, F.A., Kim, H.Y., 2009. Extraction of pure natural hydroxyapatite from the bovine bones bio waste by three different methods. J. Mater. Process. Tech. 209 (7), 3408–3415.

Bartolomé, B., Santos, M., Jiménez, J.J., Del Nozal, M.J., Gómez-Cordovés, C., 2002. Pentoses and hydroxycinnamic acids in brewer's spent grain. J. Cereal Sci. 36(1), 51–58.

Beucler, J., Drake, M., Foegeding, E., 2005. Design of a beverage from whey permeate. J. Food Sci. 70, S277–S285.

Bicu, I., Mustata, F., 2011. Cellulose extraction from orange peel using sulfite digestion reagents. Bioresource Technol. 102, 10013–10019.

Boles, J., Rathgeber, B., Shand, P., 2000. Recovery of proteins from beef bone and the functionality of these proteins in sausage batters. Meat Sci. 55 (2), 223–231.

Braga, F.G., Silva, F.A.L.E., Alves, A, 2002. Recovery of winery by-products in the Douro demarcated region: production of calcium tartate and grape pigments. Am. J. Enol. Vitic. 53, 41–45.

Britz, T.J., van Schalkwyk, C., Hung, Y.T., 2006. Treatment of dairy processing wastewaters. In: Wang, L.K., Hung, Y.T., LoH, H., Yapijakis, C. (Eds.), Waste Treatment in the Food Processing Industry. CRC Press, Taylor & Francis Group, Boca Raton, USA, pp. 8–28.

Brotzge, S.D., Chiba, L.I., Adhikari, C.K., Stein, H.H., Rodning, S.P., Welles, E.G., 2014. Complete replacement of soybean meal in pig diets with hydrolyzed feather meal with blood by amino acid supplementation based on standardized ileal amino acid digestibility. Livest. Sci. 163 (0), 85–93.

Brouns, F., Hemery, Y., Price, R., Anson, N.M., 2012. Wheat aleurone: separation, composition, health aspects, and potential food use. Crit. Rev. Food Sci. Nutr. 52 (6), 553–568.

Bulatović, M.L.J., Rakin, M.B., Mojović, L.J.V., Nikolić, S.B., VukašinovićSekulić, M.S., Đukić Vuković, A.P., 2014. Improvement of production performance of functional fermented whey-based beverage. Chem. Ind. Chem. Eng. Q. 20, 1–8.

Butsat, S., Siriamornpun, S., 2010. Antioxidant capacities and phenolic compounds of the husk, bran and endosperm of Thai rice. Food Chem. 119 (2), 606–613.

Cahú, T.B., Santos, S.D., Mendes, A., Córdula, C.R., Chavante, S.F., Carvalho, Jr., L.B., Bezerra, R.S., 2012. Recovery of protein, chitin, carotenoids and glycosaminoglycans from Pacific white shrimp (*Litopenaeus vannamei*) processing waste. Process Biochem. 47 (4), 570–577.

Camire, M.E., Zhao, J., Dougherty, M.P., Bushway, R.J., 1995. *In vitro* binding of benzo[a]pyrene by extruded potato peels. J. Agr. Food Chem. 43, 970–973.

Camire, M.E., Violette, D., Dougherty, M.P., McLaughlin, M.A., 1997. Potato peel dietary fiber composition: effects of peeling and extrusion cooking processes. J. Agr. Food Chem. 45, 1404–1408.

Cao, W., Zhang, C., Hong, P., Ji, H., Hao, J., Zhang, J., 2009. Autolysis of shrimp head by gradual temperature and nutritional quality of the resulting hydrolysate. LWT-Food Sci. Technol. 42 (1), 244–249.

Carvalheiro, F., Silva-Fernandes, T., Duarte, L.C., Gírio, F. M., 2009. Wheat straw autohydrolysis: process optimization and products characterization. Appl. Biochem. Biotech. 153(1–3), 84–93.

Catiau, L., Traisnel, J., Delval-Dubois, V., Chihib, N.E., Guillochon, D., Nedjar-Arroume, N., 2011. Minimal antimicrobial peptidic sequence from hemoglobin alpha-chain: KYR. Peptides 32 (4), 633–638.

Chang, C.-Y., Wu, K.-C., Chiang, S.-H., 2007. Antioxidant properties and protein compositions of porcine haemoglobin hydrolysates. Food Chem. 100 (4), 1537–1543.

Chantaro, P., Devahastin, S., Chiewchan, N., 2008. Production of antioxidant high dietary fiber powder from carrot peels. LWT-Food Sci. Technol. 41, 1987–1994.

Charmley, E., Nelson, D., Zvomuya, F., 2006. Nutrient cycling in the vegetable processing industry: utilization of potato by-products. Can. J. Soil Sci. 86, 621–629.

Chatenoud, L., Tavani, A., La Vecchia, C., Jacobs, D.R., Negri, E., Levi, F., Franceschi, S., 1998. Whole grain food intake and cancer risk. Int. J. Cancer 77 (1), 24–28.

Chaturvedi, V., Bhange, K., Bhatt, R., Verma, P., 2014. Production of kertinases using chicken feathers as substrate by a novel multifunctional strain of *Pseudomonas stutzeri* and its dehairing application. Biocatal. Agric. Biotechnol. 3 (2), 167–174.

Chedea, V.S., Kefalas, P., Socaciu, C., 2010. Patterns of carotenoid pigments extracted from two orange peel wastes (Valencia and Navel var.). J. Food Biochem. 34, 101–110.

Chen, Y.C., Tou, J.C., Jaczynski, J., 2007. Amino acid, fatty acid, and mineral profiles of materials recovered from rainbow trout (*Oncorhynchus mykiss*) processing by-products using isoelectric solubilization/precipitation. J. Food Sci. 72 (9), C527–C535.

Christensen, A.D., Kádár, Z., Oleskowicz-Popiel, P., Thomsen, M.H., 2011. Production of bioethanol from organic whey using *Kluyveromyces marxianus*. J. Ind. Microbiol. Biot. 38, 283–290.

Coward-Kelly, G., Agbogbo, F.K., Holtzapple, M.T., 2006. Lime treatment of keratinous materials for the generation of highly digestible animal feed: 2. animal hair. Bioresource Technol. 97 (11), 1344–1352.

Crizel de Moraes, T., Jablonski, Rios de Oliveira, A., Rech, R., Flôres Hickmann, S., 2013. Dietary fiber from orange byproducts as a potential fat replacer. LWT-Food Sci. Technol. 53, 9–14.

Dalgetty, D.D., Baik, B.-K., 2003. Isolation and characterization of cotyledon fibers from peas, lentils, and chickpeas. Cereal Chem. 80, 310–315.

Dapčević Hadnađev, T., Hadnađev, M., Pojić, M., 2014. OSA starches as improvers in gluten-free bread from hemp seed meal, Proceedings of II International Congress Food Technology, Quality and Safety, October 28th–30th, Novi Sad, Serbia, 166–170.

Darine, S., Christophe, V., Gholamreza, D., 2010. Production and functional properties of beef lung protein concentrates. Meat Sci. 84 (3), 315–322.

Daun, J.K., 2004. Processing/Canola, Encyclopedia of Grain Science. Elsevier Academic Press, Oxford, UK.

Day, L., 2004. In: Wrigley, C., Corke, H., Walker, C.E. (Eds.), Lipid chemistry. Encyclopedia of Grain Science. Elsevier Academic Press, Oxford, UK.

Del Campo, I., Alegría, I., Zazpe, M., Echeverría, M., Echeverría, I., 2006. Diluted acid hydrolysis pretreatment of agri-food wastes for bioethanol production. Ind. Crop. Prod. 24, 214–221.

Demirel, B., Yenigun, O., Onay, T.T., 2005. Anaerobic treatment of dairy waste waters: a review. Process Biochem. 40, 2583–2595.

Deng, Q., Zinoviadou, K.G., Galanakis, C.M., Orlien, V., Grimi, N., Vorobiev, E., Lebovka, N., Barba, F.J., 2015. The effects of conventional and non-conventional processing on glucosinolates and its derived forms, isothiocyanates: extraction, degradation and applications. Food Eng. Rev., in press.

Dewitt, C., Gomez, G., James, J., 2002. Protein extraction from beef heart using acid solubilization. J. Food Sci. 67 (9), 3335–3341.

Di Bernardini, R., Harnedy, P., Bolton, D., Kerry, J., O'Neill, E., Mullen, A.M., Hayes, M., 2011a. Antioxidant and antimicrobial peptidic hydrolysates from muscle protein sources and by-products. Food Chem. 124 (4), 1296–1307.

Di Bernardini, R., Rai, D.K., Bolton, D., Kerry, J., O'Neill, E., Mullen, A.M., Harnedy, P., Hayes, M., 2011b. Isolation, purification and characterization of antioxidant peptidic fractions from a bovine liver sarcoplasmic protein thermolysin hydrolyzate. Peptides 32 (2), 388–400.

Di Mauro, A., Fallico, B., Passerini, A., Rapisarda, P., Maccarone, E., 1999. Recovery of hesperidin from orange peel by concentration of extracts on styrene-divinylbenzene resin. J. Agr. Food Chem. 47, 4391–4397.

Doble, M., Kumar, A., 2005. Biotreatment of Industrial Effluents. Elsevier Inc, Oxford, UK.

Domínguez-Perles, R., Martínez-Ballesta, M.C., Carvajal, M., García-Viguera, C., Moreno, D., 2010. Broccoli-derived by-products – a promising source of bioactive ingredients. J. Food Sci. 75, 383–392.

Dragone, G., Mussatto, S.I., Oliveira, J.M., Teixeira, J.A., 2008. Alcoholic beverage from cheese whey: identification of volatile compounds. Bioenergy: challenges and opportunities, International Conference and Exhibition on Bioenergy: April 6th–9th. Universidade do Minho. Guimarães, Portugal.

EFSA, 2011. Scientific opinion on the substantiation of health claims related to beta-glucans and maintenance of normal blood cholesterol concentrations (ID 754, 755, 757, 801, 1465, 2934) and maintenance or achievement of a normal body weight (ID 820, 823). EFSA J. 7 (9), 1–18.

FAO, 1994. Definition and classification of commodities. Available from: http://www.fao.org/es/faodef/fdef02e.htm.

FAO, 1997. Report on the inter-centre review of root and tuber crops research in the CGIAR, Appendix 4 – global production and consumption of roots and tubers. Available from: http://www.fao.org/wairdocs/tac/x5791e/x5791e0q.htm.

FAO, 2010. FaoStat: Agriculture Data. HYPERLINK "http://faostat.fao.org/site/567/DesktopDefault.aspx?PageID=567" \l "ancor"http://faostat.fao.org/site/567/DesktopDefault.aspx?PageID=567#ancor (accessed 02.09.2010).

FAO, 2013. FAOSTAT, Food and agricultural commodities production. Available from: http://faostat.fao.org/site/339/default.aspx.

FAO, 2014. Save food – global initiative on food loss and waste reduction. Available from: http://www.fao.org/docrep/015/i2776e/i2776e00.pdf.

Fardet, A., 2010. New hypotheses for the health-protective mechanisms of whole-grain cereals: what is beyond fibre? Nutr. Res. Rev. 23 (1), 65–134.

Farhat, A., Fabiano-Tixier, A.-S., Maataoui, M.E., Maingonnat, J.-F., Romdhane, M., Chemat, F., 2011. Microwave steam diffusion for extraction of essential oil from orange peel: kinetic data, extract's global yield and mechanism. Food Chem. 125, 255–261.

Fechner, A., Schweiggert, U., Hasenkopf, K., Jahreis, G., 2011. Lupine kernel fiber: metabolic effects in human intervention studies and use as a supplement in wheat bread. In: Preedy, V.R., Watson, R.R., Patel, V.B. (Eds.), Flour and Breads and Their Fortification in Health and Disease Prevention. Elsevier Inc, Oxford, pp. 463–473.

Femenia, A., Lefebvre, A.-C., Thebaudin, J.-Y., Robertson, J.A., Bourgeois, C.-M., 1997. Physical and sensory properties of model foods supplemented with cauliflower fiber. J. Food Sci. 62, 635–639.

Fernández-López, J., Fernández-Ginés, J.M., Aleson-Carbonell, L., Sendra, E., Sayas-Barberá, E., Pérez-Alvarez, J.A., 2004. Application of functional citrus byproducts to meat products. Trends Food Sci. Tech. 15, 176–185.

Ferraro, V., Carvalho, A.P., Piccirillo, C., Santos, M.M.L., Castro, P.M., E Pintado, M., 2013. Extraction of high added value biological compounds from sardine, sardine-type fish and mackerel canning residues – a review. Mater. Sci. Eng. 33 (6), 3111–3120.

Fitzgerald, C., Mora-Soler, L., Gallagher, E., O'Connor, P., Prieto, J., Soler-Vila, A., Hayes, M., 2012. Isolation and characterization of bioactive pro-peptides with *in vitro* renin inhibitory activities from the macroalga *Palmaria palmata*. J. Agr. Food Chem. 60 (30), 7421–7427.

Friedman, M., 2006. Potato glycoalkaloids and metabolites: roles in the plant and in the diet. J. Agr. Food Chem. 45, 1523–1540.

Friedman, M., Lee, K.-R., Kim, H.-J., Lee, I.-S., Kozukue, N., 2005. Anticarcinogenic effects of glycoalkaloids from potatoes against human cervical, liver, lymphoma and stomach cancer cells. J. Agr. Food Chem. 53, 6162–6169.

Galanakis, C.M., 2011. Olive fruit and dietary fibers: components, recovery and applications. Trends Food Sci. Tech. 22, 175–184.

Galanakis, C.M., 2012. Recovery of high added-value components from food wastes: conventional, emerging technologies and commercialized applications. Trends Food Sci. Tech. 26, 68–87.

Galanakis, C.M., 2013. Emerging technologies for the production of nutraceuticals from agricultural by-products: a viewpoint of opportunities and challenges. Food Bioprod. Process. 91, 575–579.

Galanakis, C.M., 2015. Separation of functional macromolecules and micromolecules: from ultrafiltration to the border of nanofiltration. Trends Food Sci. Tech. 42(1), 44–63.

Galanakis, C.M., Schieber, A., 2014. Editorial of Special Issue on "Recovery and utilization of valuable compounds from food processing by-products". Food Res. Int. 65, 299–300.

Galanakis, C.M., Tornberg, E., Gekas, V., 2010a. A study of the recovery of the dietary fibres from olive mill wastewater and the gelling ability of the soluble fibre fraction. LWT-Food Sci. Technol. 43, 1009–1017.

Galanakis, C.M., Tornberg, E., Gekas, V., 2010b. Clarification of high-added value products from olive mill wastewater. J. Food Eng. 99, 190–197.

Galanakis, C.M., Tornberg, E., Gekas, V., 2010c. Dietary fiber suspensions from olive mill wastewater as potential fat replacements in meatballs. LWT-Food Sci. Technol. 43, 1018–1025.

Galanakis, C.M., Tornberg, E., Gekas, V., 2010d. Recovery and preservation of phenols from olive waste in ethanolic extracts. J. Chem. Technol. Biotechnol. 85, 1148–1155.

Galanakis, C.M., Tornberg, E., Gekas, V., 2010e. The effect of heat processing on the functional properties of pectin contained in olive mill wastewater. LWT-Food Sci. Technol. 43, 1001–1008.

Galanakis, C.M., Fountoulis, G., Gekas, V., 2012a. Nanofiltration of brackish groundwater by using a polypiperazine membrane. Desalination 286, 277–284.

Galanakis, C.M., Kanellaki, M., Koutinas, A.A., Bekatorou, A., Lycourghiotis, A., Kordoulis, C.H., 2012b. Effect of pressure and temperature on alcoholic fermentation by *Saccharomyces cerevisiae* immobilized on γ-alumina pellets. Bioresource Technol. 114, 492–498.

Galanakis, C.M., Goulas, V., Tsakona, S., Manganaris, G.A., Gekas, V., 2013a. A knowledge base for the recovery of natural phenols with different solvents. Int. J. Food Prop. 16, 382–396.

Galanakis, C.M., Markouli, E., Gekas, V., 2013b. Fractionation and recovery of different phenolic classes from winery sludge via membrane filtration. Sep. Purif. Technol. 107, 245–251.

Galanakis, C.M., Chasiotis, S., Botsaris, G., Gekas, V., 2014. Separation and recovery of proteins and sugars from Halloumi cheese whey. Food Res. Int., doi: 10.1016/j.foodres.2014.03.060.

Galanakis, C.M., Kotanidis, A., Dianellou, M., Gekas, V., 2015a. Phenolic content and antioxidant capacity of Cypriot Wines. Czech J. Food Sci. 33, 126–136.

Galanakis, C.M., Patsioura, A., Gekas, V., 2015b. Enzyme kinetics modeling as a tool to optimize food biotechnology applications: a pragmatic approach based on amylolytic enzymes. Crit. Rev. Food Sci. Technol., 55, 1758–1770.

Galanakis, C.M., Schieber, A., 2014. Editorial. Special Issue on Recovery and utilization of valuable compounds from food processing by-products, Food Res. Int. 65, 299–230.

Gänzle, M.G., Haase, G., Jelen, P., 2008. Lactose: crystallization, hydrolysis and value-added derivatives. Int. Dairy J. 18, 685–694.

Girish, T.K., Pratape, V.M., Prasada Rao, U.J.S., 2012. Nutrient distribution, phenolic acid composition, antioxidant and alpha-glucosidase inhibitory potentials of black gram (*Vigna mungo* L.) and its milled by-products. Food Res. Int. 46, 370–377.

Gómez-Guillén, M., Giménez, B., López-Caballero, M.A., Montero, M., 2011. Functional and bioactive properties of collagen and gelatin from alternative sources: a review. Food Hydrocolloid. 25 (8), 1813–1827.

Gontard, N., Guilbert, S., 1994. Bio-packaging: technology and properties of edible and/or biodegradable material of agricultural origin. In: Mathlouthi, M. (Ed.), Food Packaging and Preservation. Springer US Science+Business Media, pp. 159–181.

Górecka, D., Pachołek, B., Dziedzic, K., Górecka, M., 2010. Raspberry pomace as a potential fiber source for cookies enrichment. Acta Sci. Pol. Technol. Aliment. 9, 451–461.

Gousterova, A., Nustorova, M., Paskaleva, D., Naydenov, M., Neshev, G., Vasileva-Tonkova, E., 2012. Assessment of feather hydrolysate from thermophilic actinomycetes for soil amendment and biological control application. Int. J. Environ. Res. 6 (2), 467–474.

Gratelly, P., Benitez, E., Elvira, C., Polo, A., Nogales, R., 1996. Stabilization of sludges from a dairy processing plant using vermicomposting. In: Rodriguez-Barrueco, C. (Ed.), Fertilizers and Environment. Kluwer, The Netherlands, pp. 341–343.

Guerin, M., Huntley, M.E., Olaizola, M., 2003. *Haematococcus astaxanthin*: applications for human health and nutrition. Trends Biotechnol. 21 (5), 210–216.

Gupta, P., Nayak, K. K. 2014. Characteristics of protein-based biopolymer and its application. Polymer Eng. Sci.

Gupta, S., Chandi, G.K., Sogi, D.S., 2008. Effect of extraction temperature on functional properties of rice bran protein concentrates. Int. J. Food. Eng. 4, 2–19.

Gustavsson, J., Cederberg, C., Sonesson, U., van Otterdijk, R., Meybeck, A., 2011. Global food losses and food wastes: extent, causes and prevention, food and agriculture organization of the United Nations. (Rome) ISBN 978-92-5-107205-9.

Han, J.H., Gennadios, A., 2005. Edible films and coatings: a review. In: Han, J.H. (Ed.), Innovations in Food Packaging. Elsevier Ltd, California, USA, pp. 239–262.

Han, B.S., Park, C.B., Takasuka, N., Naito, A., Sekine, K., Nomura, E., Tsuda, H., 2001. A ferulic acid derivative, ethyl 3-(4′-geranyloxy-3-methoxyphenyl)-2-propenoate, as a new candidate chemopreventive agent for colon carcinogenesis in the rat. Cancer Sci. 92 (4), 404–409.

Hemery, Y., Rouau, X., Lullien-Pellerin, V., Barron, C., Abecassis, J., 2007. Dry processes to develop wheat fractions and products with enhanced nutritional quality. J. Cereal Sci. 46 (3), 327–347.

Hemery, Y., Chaurand, M., Holopainen, U., Lampi, A.M., Lehtinen, P., Piironen, V., Rouau, X., 2011. Potential of dry fractionation of wheat bran for the development of food ingredients, part I: influence of ultra-fine grinding. J. Cereal Sci. 53 (1), 1–8.

Heng, W.W., Xiong, L.W., Ramanan, R.N., Hong, T.L., Kong, K.W., Galanakis, C.M., Prasad, K.N., 2015. Two level factorial design for the optimization of phenolics and flavonoids recovery from palm kernel by-product. Ind. Crop. Prod. 63, 238–248.

Heuzé, V., Tran, G., Bastianelli, D., Archimède, H., Lebas, F., Régnier, C., 2014. Cassava peels, cassava pomace and other cassava by-products. Feedipedia.org. A programme by INRA, CIRAD, AFZ and FAO. Available from: http://www.feedipedia.org/node/526.

Hu, W., Wells, J.H., Shin, T.S., Godber, J.S., 1996. Comparison of isopropanol and hexane for extraction of vitamin E and oryzanols from stabilized rice bran. J. Am. Oil Chem. Soc. 73 (12), 1653–1656.

Huang, W., Sun, X., 2000. Adhesive properties of soy proteins modified by urea and guanidine hydrochloride. J. Am. Oil Chem. Soc. 77, 101–104.

Ichimura, T., Yamanaka, A., Otsuka, T., Yamashita, E., Maruyama, S., 2009. Antihypertensive effect of enzymatic hydrolysate of collagen and Gly-Pro in spontaneously hypertensive rats. Biosci., Biotechnol., Biochem. 73 (10), 2317–2319.

Im, H.W., Suh, B.S., Lee, S.U., Kozukue, N., Ohnisi-Kameyama, M., Levin, C.E., Friedman, M., 2008. Analysis of phenolic compounds by high-performance liquid chromatography and liquid chromatography/mass spectrometry in potato plant flowers, leaves, stems, and tubers and in home-processed potatoes. J. Agr. Food Chem. 56, 3341–3349.

Indrani, G., Rao, G.V., 2011. South Indian parotta: an unleavened flat bread. In: Preedy, V.R., Watson, R.R., Patel, V.B. (Eds.), Flour and Breads and their Fortification in Health and Disease Prevention. Elsevier Inc, London, UK, pp. 27–36.

Ionescu, A., Aprodu, I., Zara, M., Vasile, A., Porneala, L., 2007. Evaluating of some functional properties of the myofibrillar protein concentrate from the beef heart. Sci. Stud. Res. VIII(2), 155–170.

Jeličić, I., Božanić, R., Tratnik, L.J., 2008. Whey-based beverages – a new generation of dairy products. Mljekarstvo 58, 257–274.

Jia, D., Fang, Y., Yao, K., 2009. Water vapor barrier and mechanical properties of konjacglucomannan–chitosan–soy protein isolate edible films. Food Bioprod. Process. 87, 7–10.

Jiang, Y., Wang, T., 2005. Phytosterols in cereal by-products. J. Am. Oil Chem. Soc. 82, 439–444.

Juliano, B.O., 1985. Polysaccharides, proteins, and lipids of rice. In: Juliano, B.O. (Ed.), Rice Chemistry, Technology. American Association of Cereal Chemists, Inc.: St. Paul, Minnesota, 59–174.

Kabel, M.A., Schols, H.A., Voragen, A.G.V., 2002. Complex xylo-oligosaccharides identified from hydrothermally treated Eucalyptus wood and brewery's spent grain. Carbohydr. Polym. 50 (1), 191–200.

Kamal-Eldin, A., Lærke, H.N., Knudsen, K.E.B., Lampi, A.M., Piironen, V., Adlercreutz, H., Iman, P., 2009. Physical, microscopic and chemical characterisation of industrial rye and wheat brans from the Nordic countries. Food Nutr. Res. 53, 1–11.

Kato, S., Kunisawa, K., Kojima, T., Murakami, S., 2004. Evaluation of ozone treated fish waste oil as a fuel for transportation. J. Chem. Eng. Jpn. 37 (7), 863–870.

Kee, H.J., Ryu, G.H., Park, Y.K., 2000. Preparation and quality properties of extruded snack using onion pomace and onion. Korean J. Food Sci. Technol. 32, 578–583.

Kew, S.J., Gwynne, J.H., Enea, D., Abu-Rub, M., Pandit, A., Zeugolis, D., Brooks, R.A., Rushton, N., Best, S.M., Cameron, R.E., 2011. Regeneration and repair of tendon and ligament tissue using collagen fibre biomaterials. Acta Biomater. 7 (9), 3237–3247.

Khan, S.H., Butt, M.S., Sharif, M.K., 2011. Biological quality and safety assessment of rice bran protein isolates. Int. J. Food Sci. Technol. 46, 2366–2372.

Kim, S., Dale, B.E., 2004. Global potential bioethanol production from wasted crops and crop residues. Biomass and Bioenergy, 26(4), 361–375.

Kim, W.C., Lee, D.Y., Lee, C.H., Kim, C.W., 2004. Optimization of narirutin extraction during washing step of the pectin production from citrus peels. J. Food Eng. 63, 191–197.

Kiosseoglou, V., Paraskevopoulou, A., 2011. Functional and physicochemical properties of pulse proteins. In: Tiwari, B.K., Gowen, A., McKenna, B. (Eds.), Pulse Foods: Processing, Quality and Nutraceutical Applications. Elsevier Inc, London, UK, pp. 57–362.

Knoblich, M., Anderson, B., Latshaw, D., 2005. Analyses of tomato peel and seed byproducts and their use as a source of carotenoids. J. Sci. Food Agric. 85, 1166–1170.

Korhonen, H., Pihlanto, A., 2006. Bioactive peptides: production and functionality. Int. Dairy J. 16 (9), 945–960.

Kosikowski, F.V., 1979. Whey utilization and whey products. J. Dairy Sci. 62, 1149–1160.

Kosoom, W., Ruangpanit, Y., Rattanatabtimtong, S., Attamangkune, S., 2009. Effect of feeding cassava pulp on growth performance of nursery pigs. Proceeding of the 47th Kasetsart University Annual Conference, Kasetsart, March, 17–20 2009, 125–131.

Kosseva, M. R., 2013. Sources, characterization, and composition of food industry wastes. In: Kosseva, M.R., Webb, C., (Eds), Food Industry Wastes Assessment and Recuperation of Commodities. Elsevier Inc., London, UK, pp. 37–60.

Krzywonos, M., Cibis, E., Miśkiewicz, T., Ryznar-Luty, A., 2009. Utilization and biodegradation of starch stillage (distillery wastewater). Electron. J. Biotechnol., 12(1), 1–12.

Kuji, Y., Iizuka, A., Kitajima, Y., Yamasaki, A., Yanagisawa, Y., 2006. Development of a new recycling process of shell waste using high-pressure carbon dioxide solution. In: The 2006 Annual Meeting.

Lasekan, A., Abu Bakar, F., Hashim, D., 2013. Potential of chicken by-products as sources of useful biological resources. Waste Manage. 33 (3), 552–565.

Lavecchia, R., Zuorro, A. 2008. Process for the extraction of lycopene. World Intellectual Property Organization. WO/2008/055894.

Lee, S.H., Song, K.B., 2009. Purification of an iron-binding nona-peptide from hydrolysates of porcine blood plasma protein. Process Biochem. 44 (3), 378–381.

Lee, J.K., Jeon, J.K., Byun, H.G., 2011. Effect of angiotensin I converting enzyme inhibitory peptide purified from skate skin hydrolysate. Food Chem. 125 (2), 495–499.

Lee, S.J., Kim, Y.S., Hwang, J.W., Kim, E.K., Moon, S.H., Jeon, B.T., Jeon, Y.J., Kim, J.M., Park, P.J., 2012. Purification and characterization of a novel antioxidative peptide from duck skin by-products that protects liver against oxidative damage. Food Res. Int. 49 (1), 285–295.

Leo, L., Leone, A., Longo, C., Lombardi, D.A., Raimo, F., Zacheo, G., 2008. Antioxidant compounds and antioxidant activity in early potatoes. J. Agr. Food Chem. 56, 4154–4163.

Li, B.F., Chen, X., Wang, B.J., Wu, Y., 2007. Isolation and identification of antioxidative peptides from porcine collagen hydrolysate by consecutive chromatography and electrospray ionization–mass spectrometry. Food Chem. 102 (4), 1135–1143.

Li, X.Y., Chen, X.G., Cha, D.S., Park, H.J., Liu, C.S., 2009. Microencapsulation of a probiotic bacteria with alginate-gelatin and its properties. J. Microencapsulation 26 (4), 315–324.

Lindhauer, M.G., 2004. Grain production and consumption/Europe. In: Wrigley, C., Corke, H., Walker, C.E. (Eds.), Encyclopedia of Grain Sci. Elsevier Academic Press, California, USA.

Lisinska, G., 1989. Potato Science and Technology. Elsevier, NY, USA.

Liu, R.H., 2007. Whole grain phytochemicals and health. J. Cereal Sci. 46 (3), 207–219.

Liu, Q., Kong, B., Xiong, Y.L., Xia, X., 2010. Antioxidant activity and functional properties of porcine plasma protein hydrolysate as influenced by the degree of hydrolysis. Food Chem. 118 (2), 403–410.

López-Noriega, A., Ruiz-Hernández, E., Quinlan, E., Storm, G., Hennink, W.E., O'Brien, F.J., 2014. Thermally triggered release of a pro-osteogenic peptide from a functionalized collagen-based scaffold using thermosensitive liposomes. J. Control. Release 187 (0), 158–166.

Lota, M.L., de Rocca Serra, D., Tomi, F., Casanova, J., 2000. Chemical variability of peel and leaf essential oils of mandarins from Citrus reticulata Blanco. Biochem. Syst. Ecol. 28, 61–78.

Mabrouk, M.E.M., El Ahwany, A.M.D., 2008. Production of β-mannanase by Bacillus amyloliquefaciens 10A1 cultured on potato peels. Afr. J. Biotechnol. 7, 1123–1128.

Magalhães, K.T., Pereira, M.A., Nicolau, A., Dragone, G., Domingues, L., Teixeira, J.A., de Almeida Silva, J.B., Schwan, R.F., 2010. Production of fermented cheese whey-based beverage using kefir grains as starter culture: Evaluation of morphological and microbial variations. Bioresource Technol. 101, 8843–8850.

Mahmood, A.U., Greenman, J., Scragg, A.H., 1998. Orange and potato peel extracts: analysis and use as Bacillus substrates for the production of extracellular enzymes in continuous culture. Enzyme Microb. Tech. 22, 130–137.

Malmberg, A., Theander, O., 1984. Free and conjugated phenolic acids and aldehydes in potato tubers. Swed. J. Agric. Res. 14, 119–125.

Mandal, S., Puniya, M., Sangu, K.P.S., Dagar, S.S., Singh, R.A., Puniya, A.K., 2013. Dairy by-products: Wastes or resources? The shifting perception after valorization. In: Chandrasekaran, M. (Ed.), Valorization of Food Processing By-Products. CRC Press, Taylor & Francis Group, Boca Raton, pp. 617–648.

Mariniello, L., Di Pierro, P., Esposito, C., Sorrentino, A., Masi, P., Porta, R., 2003. Preparation and mechanical properties of edible pectin-soy flour films obtained in the absence or presence of transglutaminase. J. Biotechnol. 102, 191–198.

Martin-Cabrejas, M.A., Esteban, R.M., Lopez-Andreu, F.J., Waldron, K., Selvendran, R.R., 1995. Dietary fiber content of pear and kiwi pomaces. J. Agr. Food Chem. 43, 662–666.

Masmoudi, M., Besbes, S., Chaabouni, M., Robert, C., Paquot, M., Blecker, C., Attia, H., 2008. Optimization of pectin extraction from lemon by-product with acidified date juice using response surface methodology. Carbohydr. Polym. 74, 185–192.

McKevith, B., 2004. Nutritional aspects of cereals. Nutr. Bull. 29 (2), 111–142.

Mei, X., Mu, T.H., Han, J.J., 2010. Composition and physicochemical properties of dietary fiber extracted from residues of 10 varieties of sweet potato by a sieving method. J. Agr. Food Chem. 58 (12), 7305–7310.

Mildner-Szkudlarz, S., Bajerska, J., Zawirska-Wojtasiak, R., Górecka, D., 2013. White grape pomace as a source of dietary fibre and polyphenols and its effect on physical and nutraceutical characteristics of wheat biscuits. J. Sci. Food Agric. 93 (2), 389–395.

Mirabella, N., Castellani, V., Sala, S., 2014. Current options for the valorization of food manufacturing waste: a review. J. Clean. Prod. 65, 28–41.

Mišan, A., Šarić, B., Nedeljković, N., Pestorić, M., Jovanov, P., Pojić, M., Tomić, J., Filipčev, B., Hadnađev, M., Mandić, A., 2014. Gluten-free cookies enriched with blueberry pomace: optimization of baking process. World Acad. Sci. Eng. Technol., Int. J. Biol., Veter. Agric. Food Eng. 8, 330–333.

Misi, S.N., Forster, C.F., 2002. Semi-continuous anaerobic co-digestion of agrowastes. Environ. Technol. 23, 445–451.

Moure, F., Rendueles, M., Díaz, M., 2003. Coupling process for plasma protein fractionation using ethanol precipitation and ion exchange chromatography. Meat Sci. 64 (4), 391–398.

Moure, A., Sineiro, J., Domínguez, H., Parajó, J.C., 2006. Functionality of oilseed protein products: a review. Food Res. Int. 39, 945–963.

Mullen, A.M., Drummond, L., Lynch, S., Álvarez., 2015. Protein recovery from the beef and pork fifth quarter – a multidisciplinary approach. EuroFood Chem., Madrid, Spain.

Nam, G.W., Lee, D.W., Lee, H.S., Lee, N.J., Kim, B.C., Choe, E.A., Hwang, J.K., Suhartono, M.T., Pyun, Y.R., 2002. Native-feather degradation by Fervidobacterium islandicum AW-1, a newly isolated keratinase-producing thermophilic anaerobe. Arch. Microbiol. 178 (6), 538–547.

Narasimha, H.V., Ramakrishnaiah, N., Pratape, V.M., Sasikala, V.B., 2004. Value addition to by product from dhal milling industry in India. J. Food Sci. Techn. 41, 492–496.

Nedjar-Arroume, N., Dubois-Delval, V., Adje, E.Y., Traisnel, J., Krier, F., Mary, P., Kouach, M., Briand, G., Guillochon, D., 2008. Bovine hemoglobin: an attractive source of antibacterial peptides. Peptides 29 (6), 969–977.

Newman, R.K., Newman, C.W., Hofer, P.J., Goering, K.J., 1989. Effects of a barley beta-glucan fraction and barley oil on chick serum lipids. Cereal Food. World 34, 768.

Nicolosi, R.J., Rogers, E.J., Ausman, L.M., Orthoefer, F.T., 1994. Rice bran oil and its health benefits. In: Marshall, W., Wadsworth, J. (Eds.), Rice Science Technology. Marcel Dekker, Inc, New York.

Nogales, R., Elvira, C., Benitez, E., Thompson, R., Gomez, M., 1999. Feasibility of vermicomposting dairy biosolids using a modified system to avoid earthworm mortality. J. Environ. Sci. Heal. B 34, 151–169.

Nwokoro, S.O., Ekhosuehi, E.I., 2005. Effect of replacement of maize with cassava peel in cockerel diets on performance and carcass characteristics. Trop. Anim. Health. Pro. 37, 495–501.

O'Brien, R.D., 2009. Fats and oils processing. Fats and Oils: Formulating and Processing for Applications. CRC Press, Taylor, Francis Group, Boca Raton, pp. 73–196.

Ockerman, H.W., Hansen, C.L., 1999. Animal By-Product Processing & Utilization. CRC Press, Boca Raton, Florida, USA.

Ofori, J.A., Hsieh, Y.-H.P., 2011. Blood-derived products for human consumption. Revel. Sci. 1, 14–21.

Ofuya, C.O., Nwajiuba, C.J., 1990. Fermentation of cassava peels for the production of cellulolytic enzymes. J. Appl. Microbiol. 68, 171–177.

Okuno, S., Yoshinaga, M., Nakatani, M., Ishiguro, K., Yoshimoto, M., Morishita, T., Uehara, T., Kawano, M., 2002. Extraction of antioxidants in sweetpotato waste powder with supercritical carbon dioxide. Food Sci. Technol. Res. 8 (2), 154–157.

Onilude, A.A., 1996. Effect of cassava cultivar, age and pretreatment processes of cellulase and xylanase production from cassava waste by *Trichoderma harzianum*. J. Basic Microbiol. 36, 421–431.

Oomah, B.D., Patras, A., Rawson, A., Singh, N., Compos-Vega, R., 2011. Chemistry of pulses. In: Tiwari, B.K., Gowen, A., McKenna, B. (Eds.), Pulse Foods: Processing, Quality and Nutraceutical Applications. Elsevier Inc, London, UK, pp. 9–55.

Oreopoulou, V., Tzia, C., 2007. Utilization of plant by-products for the recovery of proteins, dietary fibers, antioxidants, and colorants. In: Oreopoulou, V., Russ, W. (Eds.), Utilization of By-Products and Treatment of Waste in the Food Industry. Springer Science+Business Media, LLC, New York, pp. 209–232.

Osawa, K., Chinen, C., Takanami, S., Kuribayashi, T., Kurokouchi, K. 1994. Studies on effective utilisation of carrot pomace. I. Effective utilisation to bread. Research Report of the Nagano State Laboratory of Food Technology 22, pp. 24–28.

Osawa, K., Chinen, C., Takanami, S., Kuribayashi, T. Kurokouchi, K. 1995. Studies on effective utilisation of carrot pomace. II. Effective utilisation to cake, dressings and pickles. Research Report of the Nagano State Laboratory of Food Technology 23, pp. 15–18.

Özbay, A., Demirer, G.N., 2007. Cleaner production opportunity assessment for a milk processing facility. J. Environ. Manage. 84, 484–493.

Padmaja, G., Jyothi, A.N.M., 2012. Roots and tubers. In: Chandrasekaran, M. (Ed.), Valorization of Food Processing By-Products. CRC Press, Boca Raton, Florida, USA, pp. 377–414.

Pagan, J., Ibarz, A., Llorca, M., Coll, L., 1999. Quality of industrial pectin extracted from peach pomace at different pH and temperature. J. Sci. Food Agric. 79, 1038–1042.

Pan, K., Song, Q., Wang, L., Cao, B., 2011. A study of demineralization of whey by nanofiltration membrane. Desalination 267, 217–221.

Panfili, G., Fratianni, A., Criscio, T.D., Marconi, E., 2008. Tocol and β-glucan levels in barley varieties and in pearling by-products. Food Chem. 107 (1), 84–91.

Park, K.-J., Hyun, C.-K., 2002. Antigenotoxic effects of the peptides derived from bovine blood plasma proteins. Enzyme Microb. Tech. 30 (5), 633–638.

Patel, N., 2010. The production of a potable alcoholic spirit from New Zealand dairy proteins, lactose and whey ethanol. Doctoral dissertation, Auckland University of Technology.

Patel, A.A., Singh, A.K., 2012. Trends in new product development with reference to functional food. In: Lecture Compendium of National Training Programme on Innovative Trends in Dairy and Food Products Formulation. National Dairy Research Institute, India, pp. 1–7.

Patel, A., Stamatakis, S., Young, S., Friedheim, J., 2007. Advances in inhibitive water-based drilling fluids – can they replace oil-based muds? In: International Symposium on Oilfield Chemistry. Society of Petroleum Engineers.

Patel, S.K., Singh, M., Kumar, P., Purohit, H.J., Kalia, V.C., 2012. Exploitation of defined bacterial cultures for production of hydrogen and polyhydroxybutyrate from pea-shells. Biomass Bioenerg. 36, 218–225.

Patras, A., Oomah, B.D., Gallagher, E., 2011. By-product utilization. In: Tiwari, B.K., Gowen, A., McKenna, B. (Eds.), Pulse Foods: Processing, Quality and Nutraceutical Applications. Elsevier Inc, London, UK, pp. 325–362.

Patsioura, A., Galanakis, C.M., Gekas, V., 2011. Ultrafiltration optimization for the recovery of b-glucan from oat mill waste. J. Membrane Sci. 373, 53–63.

Pazmiño-Durán, A.E., Giusti, M.M., Wrolstad, R.E., Glória, M.B.A., 2001. Anthocyanins from banana bracts (*Musa × paradisiaca*) as potential food colorants. Food Chem. 73, 327–332.

Pellerin, P., Gosselin, M., Lepoutre, J.P., Samain, E., Debeire, P., 1991. Enzymic production of oligosaccharides from corncob xylan. Enzyme Microb. Tech. 13 (8), 617–621.

Pesta, G., Meyer-Pittroff, R., Russ, W., 2007. Utilization of whey. In: Oreopoulou, V., Russ, W. (Eds.), Utilization of By-Products and Treatment of Waste in the Food Industry. Springer Science+Business Media LLC, New York, pp. 193–207.

Peterson, D.M., Qureshi, A.A., 1993. Genotype and environment effects on tocols of barley and oats. Cereal Chem. 70, 157–162, http://faostat.fao.org/site/567/desktopdefault.aspx#ancor.

Pinelo, M., Arnous, A., Meyer, A.S., 2006. Upgrading of grape skins: significance of plant cell-wall structural components and extraction techniques for phenol release. Trends Food Sci. Tech. 17, 579–590.

Piot, J.-M., Zhao, Q., Guillochon, D., Ricart, G., Thomas, D., 1992. Isolation and characterization of two opioid peptides from a bovine hemoglobin peptic hydrolysate. Biochem. Biophys. Res. Commun. 189 (1), 101–110.

Pojić, M., Mišan, A., Sakač, M., Dapčević Hadnađev, T., Šarić, B., Milovanović, I., Hadnađev, M., 2014. Characterization of by-products originating from hemp oil processing. J. Agr. Food Chem. 62, 12436–12442.

Prückler, M., Siebenhandl-Ehn, S., Apprich, S., Höltinger, S., Haas, C., Schmid, E., Kneifel, W., 2014. Wheat bran-based biorefinery 1: composition of wheat bran and strategies of functionalization. LWT-Food Sci. Technol. 56 (2), 211–221.

Radočaj, O., Dimić, E., Tsao, R., 2014. Effects of hemp (*Cannabis sativa* L.) seed oil press-cake and decaffeinated green tea leaves (*Camellia sinensis*) on functional characteristics of gluten-free crackers. J. Food Sci. 79 (3), C318–C325.

Rahmanian, N., Jafari, S.M., Galanakis, C.M., 2014. Recovery and removal of phenolic compounds from olive mill wastewater. J. Am. Oil Chem. Soc. 91, 1–18.

Ramos-Clamont, G., Fernández-Michel, S., Carrillo-Vargas, L., Martinez-Calderón, E., Vázquez-Moreno, L., 2003. Functional properties of protein fractions isolated from porcine blood. J. Food Sci. 68 (4), 1196–1200.

Ray, R.C., Moorthy, S.N., 2007. Exopolysaccharide (Pullulan) production from cassava starch residue by Aureobasidium pullulans strain MTCC 1991. J. Sci. Ind. Res. 66, 252–255.

Ray, R.C., Mohapatra, S., Panda, S., Kar, S., 2008. Solid substrate fermentation of cassava fibrous residue for production of α-amylase, lactic acid and ethanol. J. Environ. Biol. 29, 111–115.

Rhodes, D.I., Sadek, M., Stone, B.A., 2002. Hydroxycinnamic acids in walls of wheat aleurone cells. J. Cereal Sci. 36, 67–81.

Riffel, A., Lucas, F., Heeb, P., Brandelli, A., 2003. Characterization of a new keratinolytic bacterium that completely degrades native feather keratin. Arch. Microbiol. 179 (4), 258–265.

Rinaudo, M., 2006. Chitin and chitosan: Properties and applications. Prog. Polym. Sci. 31 (7), 603–632.

Rocha, M.D., Loiko, M.R., Tondo, E.C., Prentice, C., 2014. Physical, mechanical and antimicrobial properties of *Argentine anchovy* (*Engraulis anchoita*) protein films incorporated with organic acids. Food Hydrocolloid. 37, 213–220.

Rodrigues, I.M., Coelho, J.F.J., Carvalho, M.G.V.S., 2012. Isolation and valorisation of vegetable proteins from oilseed plants: methods, limitations and potential. J. Food Eng. 109, 337–346.

Rogalinski, T., Herrmann, S., Brunner, G., 2005. Production of amino acids from bovine serum albumin by continuous sub-critical water hydrolysis. J. Supercrit. Fluids 36 (1), 49–58.

Roselló-Soto, E., Barba, F.J., Parniakov, O., Galanakis, C.M., Grimi, N., Lebovka, N., Vorobiev, E., 2015. High voltage electrical discharges, pulsed electric field and ultrasounds assisted extraction of protein and phenolic compounds from olive kernel. Food Bioprocess Tech. 8, 885–894.

Russ, W., Meyer-Pittroff, R., 2004. Utilizing waste products from the food production and processing industries. Crit. Rev. Food Sci. 44, 57–62.

Russ, W., Schnappinger, M., 2007. Waste related to the food industry: a challenge in material loops. In: Oreopoulou, V., Russ, W. (Eds.), Utilization of By-Products and Treatment of Waste in the Food Industry. Springer Science+Business Media LLC, New York, pp. 1–13.

Salami, R.I., Odunsi, A.A., 2003. Evaluation of processed cassava peel meals as substitutes for maize in the diets of layers. Int. J. Poultry Sci. 2, 112–116.

Salmoral, E.M., Gonzalez, M.E., Mariscal, M.P., 2000a. Biodegradable plastic made from bean products. Ind. Crop. Prod. 11, 217–225.

Salmoral, E.M., Gonzalez, M.E., Mariscal, M.P., Medina, L.F., 2000b. Comparison of chick pea and soy protein isolate and whole flour as biodegradable plastics. Ind. Crop. Prod. 11, 227–236.

Sano, A., Maeda, M., Nagahara, S., Ochiya, T., Honma, K., Itoh, H., Miyata, T., Fujioka, K., 2003. Atelocollagen for protein and gene delivery. Adv. Drug Deliver. Rev. 55 (12), 1651–1677.

Santos, T.M., Souza Filho, M.D.S.M., Caceres, C.A., Rosa, M.F., Morais, J.P.S., Pinto, A., Azeredo, H., 2014. Fish gelatin films as affected by cellulose whiskers and sonication. Food Hydrocolloid. 41, 113–118.

Sarmadi, B.H., Ismail, A., 2010. Antioxidative peptides from food proteins: a review. Peptides 31 (10), 1949–1956.

Saulnier, L., Sado, P.E., Branlard, G., Charmet, G., Guillon, F., 2007. Wheat arabinoxylans: exploiting variation in amount and composition to develop enhanced varieties. J. Cereal Sci. 46 (3), 261–281.

Schieber, A., Saldana, M.D.A., 2008. Potato peels: a source of nutritionally and pharmacologically interesting compounds – a review. Food 3, 23–29.

Schieber, A., Stintzing, F.C., Carle, A., 2001. By-products of plant food processing as a source of functional compounds – recent developments. Trends Food Sci. Tech. 12, 401–413.

Seetharamaiah, G.S., Prabhakar, J.V., 1986. Oryzanol content of Indian rice bran oil and its extraction from soap stock. J. Food Sci. Techn. 23 (5), 270–273.

Shahidi, F., Arachchi, J.K.V., Jeon, Y.J., 1999. Food applications of chitin and chitosans. Trends Food Sci. Tech. 10 (2), 37–51.

Sharma, P.C., Tilakratne, B.M., Anil, G., 2010. Utilization of wild apricot kernel press cake for extraction of protein isolate. J. Food Sci. Technol. Mysore 47, 682–685.

Sherman, R., 2007. Chain management issues and good housekeeping procedures to minimize food processing waste. In: Waldron, K. (Ed.), Handbook of Waste Management and Co-Product Recovery in Food Processing, Volume 1, Woodhead Publishing Limited, Cambridge, England, pp. 39–58.

Shewry, P.R., Piironen, V., Lampi, A.M., Nyström, L., Li, L., Rakszegi, M., 2008. Phytochemicals in oat varieties in HEALTHGRAIN diversity screen. J. Agr. Food Chem. 56, 9777–9784.

Shukla, M., Jha, Y.K., Admassu, S., 2013. Development of probiotic beverage from whey and pineapple juice. J. Food Processing Technol. 4.

Silva, T.M., Alarcon, R.F., deLima Damasio, A.R., 2009. Use of cassava peel as carbon source for production of amylolytic enzymes by *Aspergillus niveus*. Int. J. Food. Eng. 5, 1–11.

Silva, J.F.X., Ribeiro, K., Silva, J.F., Cahú, T.B., Bezerra, R.S., 2014. Utilization of tilapia processing waste for the production of fish protein hydrolysate. Anim. Feed Sci. Tech. 196, 96–106.

Singh, P.P., Saldaña, M.D.A., 2011. Subcritical water extraction of phenolic compounds from potato peel. Food Res. Int. 44, 2452–2458.

Singh, A.K., Singh, K., 2012. Utilization of whey for the production of instant energy beverage by using response surface methodology. Adv. J. Food Sci. Tech. 4, 103–111.

Singh, A., Sabally, K., Kubow, S., Donnelly, D.J., Gariepy, Y., Orsat, V., Raghavan, G.S.V., 2011. Microwave-assisted extraction of phenolic antioxidants from potato peels. Molecules 16, 2218–2232.

Singla, A., Lee, C.H., 2002. Effect of elastin on the calcification rate of collagen-elastin matrix systems. J. Biomed. Mater. Res. 60 (3), 368–374.

Smithers, G.W., 2008. Whey and whey proteins – From "gutter-to-gold". Int. Dairy J. 18, 695–704.

Sotillo, R.D., Hadley, M., Holm, E.T., 1994. Potato peel waste: Stability and antioxidant activity of a freeze-dried extract. J. Food Sci. 59, 1031–1033.

Sreerama, Y.N., Neelam, D.A., Sashikala, V.B., Pratape, V.M., 2010. Distribution of nutrients and antinutrients in milled fractions of chickpea and horse gram: seed coat phenolics and their distinct modes of enzyme inhibition. J. Agr. Food Chem. 58, 4322–4330.

Stanley, B., 2009. Exploiting the potential of the fifth quarter. Spring Barrow Lodge Farm. Swannymote Road, Grace Dieu, Nr Coalville, Leicestershire, LE67 5UT.

Su, J.-F., Huan-g, Z., Yuan, X.-Y., Wang, X.-Y., Li, M., 2010. Structure and properties ofcarboxymethyl cellulose/soy protein isolate blend edible films crosslinked by Maillard reactions. Carbohydr. Polym. 79, 145–153.

Suárez, E., Lobo, A., Alvarez, S., Riera, F.A., Álvarez, R., 2009. Demineralization of whey and milk ultrafiltration permeate by means of nanofiltration. Desalination 241, 272–280.

Sudha, M.L., Baskaran, V., Leelavathi, K., 2007. Apple pomace as a source of dietary fiber and polyphenols and its effect on the rheological characteristics and cake making. Food Chem. 104, 686–692.

Sueyoshi, Y., Hashimoto, T., Yoshikawa, M., Watanabe, K., 2011. Transformation of intact chicken feathers into chiral separation membranes. Waste Biomass Valorization 2 (3), 303–307.

Sun, Q., Shen, H., Luo, Y., 2011. Antioxidant activity of hydrolysates and peptide fractions derived from porcine hemoglobin. J. Food Sci. Technol. 48 (1), 53–60.

Szwajgier, D., Borowiec, K., 2012. Phenolic acids from malt are efficient acetylcholinesterase and butyrylcholinesterase inhibitors. J. Inst. Brew. 118 (1), 40–48.

Szwajgier, D., Waśko, A., Targoński, Z., Niedźwiadek, M., Bancarzewska, M., 2010. The use of a novel ferulic acid esterase from Lactobacillus acidophilus K1 for the release of phenolic acids from brewer's spent grain. J. Inst. Brew. 116 (3), 293–303.

Tan, S.H., Mailer, R.J., Blanchard, C.L., Agboola, S.O., 2011. Canola proteins for human consumption: Extraction, profile, and functional properties. J. Food Sci. 76, R16–R28.

Tewe, O.O., 1992. Detoxification of cassava products and effects of residual toxins on consuming animals. In: Machin, D., Nyvold, S., (Eds.), Roots, Tubers, Plantains and Bananas in Animal Feeding, Proceedings of the FAO Expert Consultation held in CIAT, Cali, Colombia, January 21–25, 1991, FAO Animal Production and Health Paper – 95.

Tierney, M.S., Croft, A.K., Hayes, M., 2010. A review of antihypertensive and antioxidant activities in macroalgae. Bot. Mar. 53 (5), 387–408.

Tiwari, B.K., Gowen, A., McKenna, B., 2011. Introduction. In: Tiwari, B.K., Gowen, A., McKenna, B. (Eds.), Pulse Foods: Processing, Quality and Nutraceutical Applications. Elsevier Inc, London, UK, pp. 1–7.

Toldrá, F., Aristoy, M.C., Mora, L., Reig, M., 2012. Innovations in value-addition of edible meat by-products. Meat Sci. 92 (3), 290–296.

Toldrà, M., Parés, D., Saguer, E., Carretero, C., 2011. Hemoglobin hydrolysates from porcine blood obtained through enzymatic hydrolysis assisted by high hydrostatic pressure processing. Innov. Food Sci. Emerg. Technol. 12 (4), 435–442.

Tosh, S.M., Yada, S., 2010. Dietary fibers in pulse seeds and fractions: characterization, functional attributes, and applications. Food Res. Int. 43, 450–460.

Tréche, S., 1996. Tropical root and tuber crops as human staple food, Conférence présentée au I Congresso Latino Americano de Raizes Tropicais, Sao Pedro, Brésil.

Truong, V.D., McFeeters, R.F., Thompson, R.T., Dean, L.L., Shofran, B., 2007. Phenolic acid content and composition in leaves and roots of common commercial sweetpotato (*Ipomea batatas* L.) cultivars in the United States. J. Food Sci. 72, 343–349.

Tsakona, S., Galanakis, C.M., Gekas, V., 2012. Hydro-ethanolic mixtures for the recovery of phenols from Mediterranean plant materials. Food Bioprocess Tech. 5, 1384–1393.

Turquois, T., Rinaudo, M., Taravel, F.R., Heyraud, A., 1999. Extraction of highly gelling pectic substances from sugar beet pulp and potato pulp: influence of extrinsic parameters on their gelling properties. Food Hydrocolloid. 13, 255–262.

Ubalua, A.O., 2007. Cassava wastes: treatment options and value addition alternatives. Afr. J. Biotechnol. 6, 2065–2073.

Umesh Hebbar, H., Sumana, B., Raghavarao, K.S.M.S., 2008. Use of reverse micellar systems for the extraction and purification of bromelain from pineapple wastes. Bioresource Technol. 99, 4896–4902.

Venn, B.J., Mann, J.I., 2004. Cereal grains, legumes and diabetes. Eur. J. Clin. Nutr. 58 (11), 1443–1461.

Vergara-Valencia, N., Granados-Pérez, E., Agama-Acevedo, E., Tovar, J., Ruales, J., Bello-Pérez, L.A., 2007. Fibre concentrate from mango fruit: characterization, associated antioxidant capacity and application as a bakery product ingredient. LWT-Food Sci. Technol. 40, 722–729.

Wadhwa, M., Bakshi, M.P.S., 2013. Utilization of Fruit and Vegetable Wastes as Livestock Feed and as Substrates for Generation of Other Value-Added Products. FAO, Rome.

Wagner, C., Benecke, C., Buchholz, H., Beutel, S., 2014. Enhancing bioethanol production from delactosed whey permeate by upstream desalination techniques. Eng. Life Sci., DOI: 10.1002/elsc.201300138.

Waldron, K. 2007. Waste minimization, management and co-product recovery in food processing: an introduction. In: Handbook of Waste Management and Co-product Recovery in Food Processing, Woodhead Publishing Ltd, Cambridge, UK, vol. 1, pp. 3–20.

Wang, S., Chen, F., Wu, J., Wang, Z., Liao, X., Hu, X., 2007. Optimization of pectin extraction assisted by microwave from apple pomace using response surface methodology. J. Food Eng. 78, 693–700.

Welch, R.W., Webster, F.H., Wood, P.J., 2011. Nutrient composition and nutritional quality of oats and comparisons with other cereals. Oats: Chemistry and Technology, second ed. 95–107. Francis H. Webster, Francis Webster & Associates, Branson, Missouri, USA, Peter J. Wood, Guelph Food Research Centre, Guelph, Ontario, Canada.

Wijbenga, D.-J.W., Binnema, D.J., Veen, A., Bos, H.T.P., 1999. Conversion and removal of steroid glycoalkaloids. Patent Application. WO/1999/043837.

Woiciechowski, A.L., Soccol, C.R., Rocha, S.N., Pandey, A., 2004. Xanthan gum production from cassava bagasse hydrolysate with *Xanthomonas campestris* using alternative sources of nitrogen. Appl. Biochem. Biotech. 118, 305–312.

Yadav, A., Garg, V.K., 2011. Industrial wastes and sludges management by vermicomposting. Rev. Environ. Sci. Biotechnol. 10, 243–276.

Yoon, K.Y., Woodams, E.E., Hang, Y.D., 2006. Enzymatic production of pentoses from the hemicellulose fraction of corn residues. LWT-Food Sci. Technol. 39 (4), 388–392.

Yu, Y., Hu, J., Miyaguchi, Y., Bai, X., Du, Y., Lin, B., 2006. Isolation and characterization of angiotensin I-converting enzyme inhibitory peptides derived from porcine hemoglobin. Peptides 27 (11), 2950–2956.

Zeugolis, D.I., Paul, R.G., Attenburrow, G., 2008. Engineering extruded collagen fibers for biomedical applications. J. Appl. Polym. Sci. 108 (5), 2886–2894.

Zeugolis, D.I., Paul, R.G., Attenburrow, G., 2008. Factors influencing the properties of reconstituted collagen fibers prior to self-assembly: animal species and collagen extraction method. J. Biomed. Mater. Res. A 86A (4), 892–904.

Zhang, D., Hamauzu, Y., 2004. Phenolic compounds and their antioxidant properties in different tissues of carrots. J. Food Agric. Environ. 2, 95–100.

Zohuriaan-Mehr, M.J., Pourjavadi, A., Salimi, H., Kurdtabar, M., 2009. Protein- and homo poly (amino acid)-based hydrogels with super-swelling properties. Polym. Advan. Technol. 20 (8), 655–671.

THE UNIVERSAL RECOVERY STRATEGY

3

Charis M. Galanakis

Department of Research & Innovation, Galanakis Laboratories, Chania, Greece

3.1 INTRODUCTION

Food wastes are generated in different forms and compositions, following regional, seasonal, and processing characteristics in each case. Moreover, they generally contain lower concentrations of valuable compounds compared with the initial sources (i.e. fruits or vegetables). This fact results in higher processing cost, lower recovery yield, and ultimately lower revenues. Food wastes are already processed materials, susceptible to microbial growth and require both preservation and fast treatment. Thereby, collection at the source is crucial and extended transportation should be avoided. Following the above considerations, the development of an economically feasible, sustainable, and safe recovery of high added-value compounds from food wastes requires a holistic approach, taking into account parameters such as:

1. waste minimization prior to the recovery process,
2. the abundance and distribution of food wastes at the source of their production (typically food industries),
3. the proper collection and mixing of food wastes in order to minimize variations in the content of their components (this way variations in the final products are also avoided),
4. the development of a production line near but not inside the food industries in order to ensure minimum transportation and at the same time meet their Hazard Analysis and Critical Control Points requirements,
5. the development of a methodology that provides the highest recovery yield of different compounds and discharges minimum quantities of by-products in the environment,
6. the nondestructive separation of valuable compounds, their recapture in different streams, and their reutilization in different products,
7. the addition of food grade materials and the utilization of green solvents,
8. the proper management of the selected stages and technologies,
9. the preservation of the compounds' functional properties from source to final product,
10. the development of qualitative products with constant concentration of target compounds and stable sensory characteristics.

Figure 3.1 illustrates the *"Universal Recovery Strategy"* that is presented for the first time in this book. Initially, the different forms and compositions of wastes that exist for a particular source (i.e. a fruit or a vegetable) are identified. The next step is to collect all the necessary information concerning

Food Waste Recovery. http://dx.doi.org/10.1016/B978-0-12-800351-0.00003-1

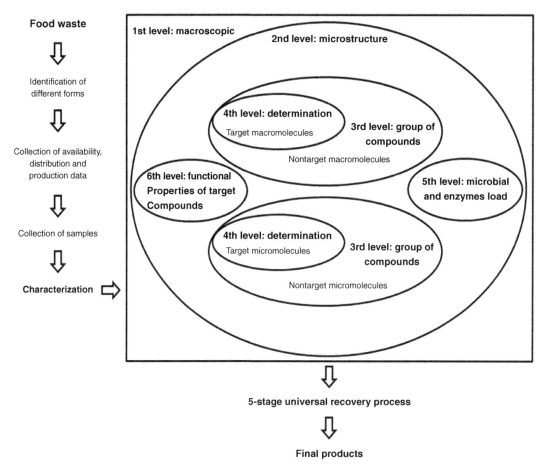

FIGURE 3.1 Development of the *Universal Recovery Strategy*

food waste availability, distribution in different locations, production frequency (i.e. seasonal or not), and finally generated quantities. Not all of these data may look of particular interest for the development of the recovery process; however, they are necessary in business plans that evaluate the overall potentiality and economic growth of the developed process. Thereafter, samples are collected and characterized at six levels.

The first level includes the determination of macroscopic characteristics, i.e. the different phases (water, oils, and solids) within waste mixtures. The adjustment of the phases' content could be an effective way to fit in the different compositions of the initial substrates. The second level is the determination of microstructure characteristics in order to get an overview of the waste matrix prior to designing the recovery process. The third level is the determination of compound groups such as total phenols, total sugars, carotenoids, dietary fibers, proteins, etc. This step allows one to have a clear view of the main macro- and micromolecules content prior to designing their separation. Also, these determinations (typically based on spectrophotometric techniques) allow the fast screening of numerous samples and optimization conditions. The fourth level of characterization is the identification of particular target macro- and micromolecules. The fifth level of characterization includes the determination of microbial and enzyme load. The latter causes deterioration of the substrate and in many cases can diminish functional properties of target compounds (sixth level of characterization). For example, enzymes like pectin methyl esterase, pectic lyase, and polygalacturonase of fruit by-products can affect pectin de-esterification and solubilization during processing. In addition, polypheoloxidases can diminish the antioxidant properties of polyphenols.

The final step of the recovery strategy is the so-called *"5-Stage Universal Recovery Process,"* which was briefly presented in a recent review (Galanakis, 2012) and herein is described in detail in paragraphs 3.4 and 3.5, Sections II and III.

3.2 CHARACTERISTICS OF TARGET COMPOUNDS
3.2.1 CALCULATIONS

The characterization of the target compounds according to their structural and other characteristics (i.e. polarity, molecular volume, and polar surface area (PSA)) is very important for the selection of the appropriate solvent (Section 3.5) and applied technologies (Section 3.6). Most of these parameters can be easily calculated online using the software provided by Molinspiration Cheminformatics (2005). Thereby, molecular polarity can be expressed using the octanol/water partition coefficient ($\log P$), which describes the hydrophobicity of each molecule. Calculation of $\log P$ can be conducted using the logistic tally miLogP2.2, based on chemical group contributions for each of the assayed compounds. Calculated and experimental $\log P$ have been fitted for a training set of more than 12,000 molecules (mostly drugs). In this way, hydrophobicity values of 35 smaller and 185 fragments have been obtained, whereas the intramolecular hydrogen bonding contribution to $\log P$ and charge interactions can be characterized.

Molecular volume is a very useful parameter for the prediction of molecules' transport properties. Respective calculation can be performed with various methods, i.e. those requiring generation of three-dimensional molecular geometries or those based on fragment contribution such as McGowan volume approximation. Calculation using Molinspiration software is conducted by fitting sum of fragment contributions to the apparent three-dimensional volume for a training set of 12,000 molecules, as above. Molecular geometries of these molecules were optimized by the semiempirical AM1 method, whereas calculated volume is expressed in cubic angstroms (Å^3). This robust method is practically able to estimate molecular volume of numerous organic and organometallic molecules.

Another parameter that characterizes the molecular transport properties is PSA. This parameter is defined as the sum of the polar atoms' surfaces (i.e. oxygen, nitrogen, and attached hydrogen) in a molecule. The calculation of PSA in a conventional way is time consuming because a reasonable three-dimensional molecular structure is needed prior to determination. On the other hand, the calculation of topological polar surface area (TPSA) is simpler and provides practically similar results. TPSA is defined as the sum of tabulated surface contributions of polar fragments, i.e. bonding pattern of atoms (Ertl et al., 2000). The contribution of these fragments can be least squares fitting to the single conformer three-dimensional PSA for 34,810 drugs from the World Drug Index.

3.2.2 STRUCTURAL CHARACTERISTICS OF TARGET MACROMOLECULES

The structural characteristics of common macromolecules found in food wastes are shown in Table 3.1. Macromolecules could be either soluble or insoluble in water, whereas they are generally insoluble in alcohols. Molecular weight (MW), intermolecular polarity, charge, and isoelectric point (pI) are the main aspects affecting their recovery with different technologies. For instance, pectin is an example of water soluble dietary fibers (Galanakis, 2011). Homo- and rhamnogalacturonans are the most common pectin types in sources like olive mill waste and winery sludge. The first polymer is compiled from a backbone of α-1,4-linked galacturonic acids (McNeil et al., 1984). Homogalacturonans have numerous hydroxyl groups around their molecules, forming hydrogen bonds. Likewise, they are negatively

Source	Potential Waste	Macromolecule	Molecular Weight (kDa)	Charge	References
Oat	Mill processing sludge	β-glucan	122	Neutral	Patsioura et al. (2011)
Grape and olive	Wine sludge and mill waste	Homogalacturonan	70–250	Negative depending on methylation degree	Galanakis et al. (2010e, 2013)
Grape and olive	Wine sludge and mill waste	Rhamnogalacturonan	70–250	Negative depending on methylation degree	Galanakis et al. (2010e, 2013)
Olive	Mill waste	Arabinan	8–10	Neutral	Galanakis et al. (2010e)
Grape and olive	Wine sludge and mill waste	Arabinogalactan	n.a.	Neutral	Galanakis et al. (2010e, 2013)
Cheese whey		Immunoglobulin	150–1000	Positive	Zydney (1998)
Cheese whey		Bovine serum albumin	69	Positive	Zydney (1998)
Cheese whey		α-Lactalbumin	14	Weakly negative	Zydney (1998)
Cheese whey		β-Lactoglobuin	18	Positive	Zydney (1998)
Oat	Mill processing sludge	Globulin	20–35	Positive	Klose and Arendt (2012)
Oat	Mill processing sludge	Albumin	14–17	Weakly positive	Klose and Arendt (2012)
Oat	Mill processing sludge	Prolamin	17–34	Positive	Klose and Arendt (2012)
Grape	Winery sludge	Pigments of tannins and anthocyanins	n.a.	Weakly positive	Remy et al. (2000); Mateus et al. (2002)

Table 3.1 Structural Characteristics of Macromolecules Found in Food Wastes

n.a., not available.
Modified from Galanakis (2015).

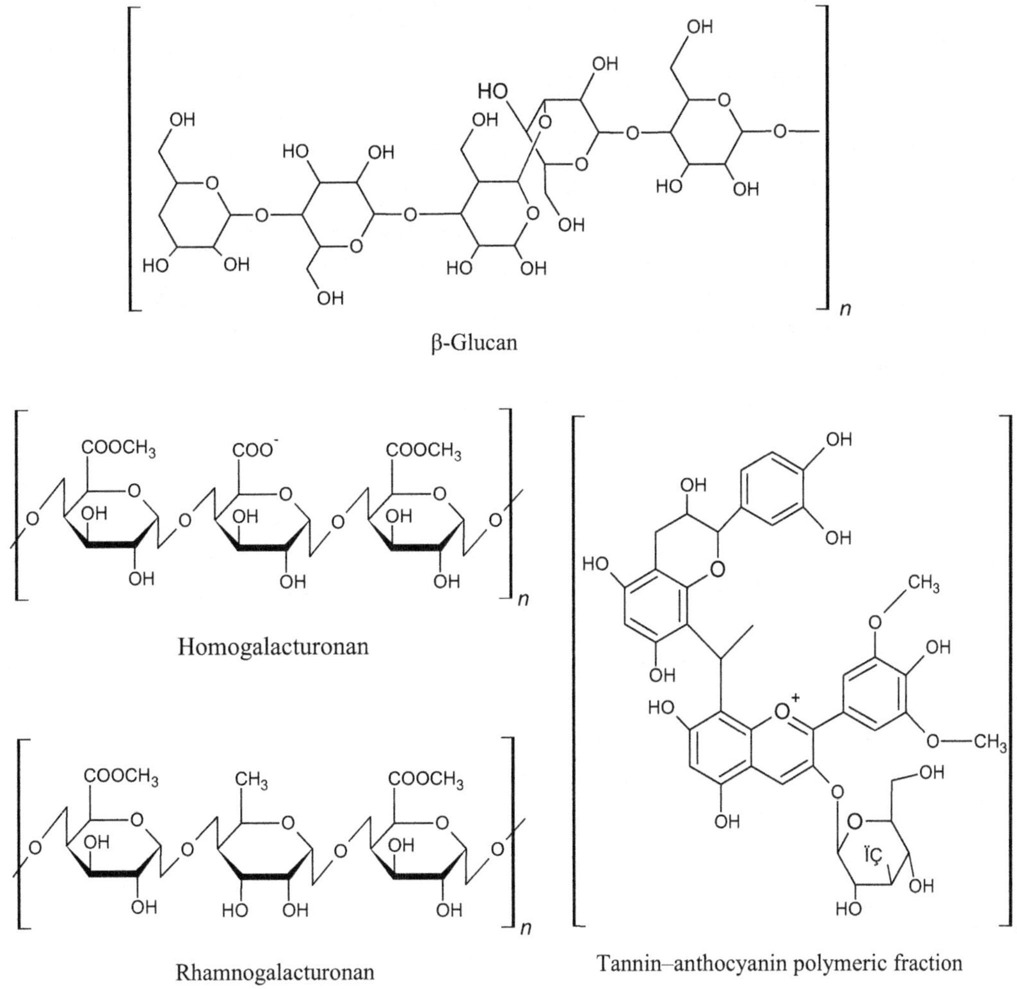

β-Glucan

Homogalacturonan

Rhamnogalacturonan

Tannin–anthocyanin polymeric fraction

FIGURE 3.2 Examples of Macromolecules Found in Food Wastes

Adapted from Galanakis, 2015

charged due to the demethylation of carboxylic groups (Fig. 3.2). Rhamnogalacturonans have fewer carboxylic groups inside their molecule compared with homogalacturonans, due to the repeated α-L-rhamnose-$(1\rightarrow4)$-α-D-galacturonic acids. Molecular weight and the methylation degree of pectin play a key role in processes like membrane filtration, as they can affect the overall recovery yield and equipment operation. Other examples of dietary fibers are β-glucan molecules: linear homopolysaccharides composed of continuant (1,4)-linked β-D-glucose fragments, which are separated by single (1,3) linkages (Fig. 3.2). The latter restrict alignment of glucose moieties and increase solubility in water (Lazaridou and Biliaderis, 2004).

Protein is another group of interest comprising polymeric chains rich in amino acids. Polypeptides have an amphoteric nature that is dependent on their pI. For example, cheese whey is acidic (pH = 4.8 ± 0.1) and contains several proteins. Thereby, immunoglobulin (150–1000 kDa and 5.5 < pI < 8.3) and ovine serum albumin (66 kDa and pI = 5.0 ± 0.1) are positively charged, whereas smaller α-lactalbumin (14 kDa and pI = 4.5 ± 0.3) and β-lactoglobulin (18 kDa and pI = 5.3 ± 0.1) are expected to be weakly negative and positive, respectively (Zydney, 1998; Lawrence et al., 2006; Bhattacharjee et al., 2006). Similarly, oat proteins, such as globulin (20–35 kDa and pI = 5.5), albumin (14–17 kDa, 4.0 < pI < 7.0), and prolamin (17–34 kDa, 5.0 < pI < 9.0) (Klose and Arendt, 2012) are positively charged in the acidic nature (pH = 4.5) of the respective mill processing waste. In the above examples, charge could be used for the recovery of proteins using isoelectric solubilization or precipitation.

Polymeric anthocyanins derived from grape by-products include malvidin 3-glucoside and respective pyruvic acid derivatives as well as pigments of anthocyanins linked to a catechin unit, a procyanidin dimer, or to a 4-vinylphenol group (Remy et al., 2000; Mateus et al., 2002). Anthocyanins are weakly positive and their polymerization can start from simple acetaldehyde malvidin 3-glucoside dimers (Atanasova et al., 2002). Polymerization degree and charge of anthocyanins can affect their alcohol precipitation yield as well as ultrafiltration performance, i.e. larger molecules cannot pass through small pores, whereas they can be adsorbed in hydrophilic membrane materials.

3.2.3 STRUCTURAL CHARACTERISTICS OF TARGET MICROMOLECULES

Table 3.2 provides some examples of common micromolecules found in food wastes and their characteristics in terms of chemical structure, MW, number of aromatic rings, and hydroxyl, carboxylic, and methylation groups. Aromatic rings and aliphatic chains provide a hydrophobic behavior, whereas an increased number of hydroxyl and carboxylic groups lead to intermolecular polarity in acidic mediums and volume increase due to water molecule attraction. This fact can affect particular processes like membrane filtration, as permeation through membrane pores could be restricted due to the "polarity resistance" phenomenon.

Monosaccharides (i.e. glucose) are very polar due to the presence of numerous hydroxyl groups, whereas their structure includes eventually an aromatic ring (their open chain exists in equilibrium with several cyclic isomers). Disaccharides like lactose have similar characteristics, but they contain more hydroxyl groups, i.e. eight instead of five compared with glucose. Polyphenols represent a wide variety of water soluble antioxidants occurring in fruits and vegetables and can be divided in two main groups: nonflavonoids and flavonoids. The major nonflavonoid phenolics include hydroxycinnamic acid derivatives of low MW (150–350 Da) and o-diphenols (e.g. hydroxytyrosol). Nonflavonoid phenolics are less polar than sugars (Galanakis et al., 2013b, 2015; Heng et al., 2015). Phenolic alcohols (e.g. γ-resorcylic acid) and aldehydes (e.g. isovanillic acid) found in grape derivatives and olive mill wastewater have similar MW as nonflavonoids. Flavonols (e.g. procyanidin B2) and flavones (e.g. apigenin) are larger molecules due to the existence of three (at least) aromatic rings surrounded by multiple hydroxyl groups. Anthocyanins' structure (e.g. malvidin) varies due to their partial polymerization. On the other hand, carotenoids (e.g. lycopene, astaxanthin, lutein) have a large aliphatic ring, minimum number of hydroxyl groups, and thus very high octanol/water partition coefficients (>8.5 compared with <3.5 for phenolics). Thereby, they are insoluble in water and soluble in nonpolar solvents.

Table 3.2 Structural Characteristics of Micromolecules Found in Food Wastes

Group of Micro-molecules	Compound	Molecular Weight (Da)	Chemical Group					Molecular Properties[a]			Molecular Type
			Aromatic Rings	–OH	–COOH	–CH₃		Octanol–Water Partition Coefficient (log P)	TPSA (Å)	Molecular Volume (Å³)	
Monosaccha-rides	Glucose	180	0–1	5	0–1	0		−3.22	138	164	
Disaccharides	Lactose	342	2	8	0	0		−4.45	190	284	
Hydroxy cinnamic acid derivatives	Cinnamic acid	148	1	0	1	0		1.91	37	138	
o-Diphenols	Hydro-xytyrosol	154	1	3	0	0		0.52	61	142	
Phenolic alcohols	γ-Resorcylic acid	154	1	2	1	0		1.39	78	127	
Phenolic aldehydes	Isovanillic acid	152	1	1	1	1		1.19	67	145	

(Continued)

Table 3.2 Structural Characteristics of Micromolecules Found in Food Wastes *(cont.)*

Group of Micromolecules	Compound	Molecular Weight (Da)	Aromatic Rings	Chemical Group			Molecular Properties[a]			Molecular Type
				–OH	–COOH	–CH$_3$	Octanol–Water Partition Coefficient (log P)	TPSA (Å)	Molecular Volume (Å^3)	
Flavonols	Procyanidin B2	579	6	10	0	0	3.91	201	484	
Flavones	Apigenin	270	3	3	0	0	2.46	91	224	
Anthocyanins	Malvidin	331	3	4	0	2	−0.42	111	278	
Carotenoid hydrocarbon	Lycopene	537	0	0	0	10	9.98	–	602	
Carotenoid ketones	Astaxanthin	597	2	2	0	10	8.60	75	612	
Carotenoid alcohols	Lutein	569	2	2	0	9	9.31	40	608	

[a]Molecular properties were calculated using the software miLogP2.2 (Molinspiration Cheminformatics, 2005).

3.3 **SUBSTRATE MACRO- AND MICROSTRUCTURE**

The prediction and control of the substrate properties require an understanding of the location of the various components, their interactions, and finally characterization of the matrix microstructure. For example, it is well known that the food microstructure affects the bioaccessibility and bioavailability of nutrients such as antioxidants (Palafox-Carlos et al., 2011). Similarly, disruption of the food waste matrix created during processing may influence the release, transformation, and ultimately functionality of target compounds in the final product.

Structures in food-related materials are generally organized hierarchically from molecules into (self- or forced) assemblies and organelles that are later compartmentalized into cells and tissues. Self-assembly expresses the spontaneous formation of small structural elements, such as aggregates, fibrils, or micelles through molecularly attractive properties. Forced assembly leads to the creation of larger and more controlled structures. Fibrous structures (e.g. muscles) are assembled from macromolecules into tissues and held together at different levels by specific interfacial interactions. Fleshy structures (fruits, vegetables, and tubers) are composites of hydrate cells that exhibit turgor pressure and are bonded together at the cell walls. Plant embryos (e.g. grains and pulses) contain a dispersion of starch, protein, and lipids assembled into discrete packets. Dispersions include either solids in a liquid medium (usually aqueous solution) or liquid droplets in another liquid (emulsions). The smaller dispersed particles (10 nm–10 μm) may possess a colloidal nature and their formation is strongly influenced by both self-assembly and superimposed process conditions, i.e. equilibrium between Brownian motion (thermal randomizing force) and interparticle forces (Zhou et al., 2001; Qin and Zaman 2003). The flowing behavior of larger, noncolloidal particles (>10 μm) is affected by hydrodynamic forces. The rheological properties (viscosity, yield stress, and modulus) of smaller and larger dispersions can be described by several theoretical equations that evaluate the effect of particle volume fraction and size, interparticle forces, or fractal dimension (Genovese et al., 2007).

Processed food waste materials are multicomponent matrices formed by the physical and chemical changes induced by individual components (proteins, dietary fibers, polysaccharides, antioxidants, sugars, and lipids) during processing (e.g. homogenization, physical or thermal treatment). For example, antioxidants could be entrapped in a complex macromolecular matrix of swollen starch granules and protein (e.g. isoflavones in baked products) or be bound to plant organelles (e.g. carotenoids in carrots). Proteins (e.g. β-lactoglobulin) form enthalpy-driven complexes with oppositely charged polysaccharides (e.g. acacia gum, pectin, alginate, or chitosan) due to electrostatic interactions (Girard et al., 2003; Schmitt et al., 2005; Harnsilawat et al., 2006). To reduce further the free energy of the system, these soluble complexes further aggregate until their size and surface properties lead to insolubilization (Schmitt and Turgeon, 2011). Starch granules are gelatinized when heat processing is involved in the presence of water. Due to this biophysical phenomenon, the granule structure changes from a crystalline to a disordered structure that is more easily accessible to the enzymes (Goñi et al., 2000; Osorio-Díaz et al., 2003). On the other hand, if gelatinized granules are stored for sufficient time (i.e. days), the linear regions of amylose and amylopectin chains tend to interact by hydrogen bonding, losing water from the fine structure and undergoing an incomplete recrystallization (retrogradation).

As shown above, the molecular properties of individual components induce particular responses in microscopic level that lead to changes in macroscopic level such as rheological properties of the matrix. All of these responses result in the formation of nonedible structures that typically include colloidal dispersions, emulsions, amorphous phases, crystalline phases, gel networks, and microstructural elements (<100 μm) such as cell walls, starch granules, water and oil droplets, fat crystals, and gas bubbles. The links between the molecular, microscopic, and macroscopic levels can be explored using a combination of rheological, compositional, and structural data. In turn, such knowledge should lead to the better designing of the recovery strategy for each target component separately.

Qualitative information about the physical state, surface and internal structures of food matrices can be obtained using microscopic methods, i.e. light microscopy (LM) or scanning electron microscopy (SEM), where electrons are reflected off the specimen. The use of transmission electron microscopy (TEM) allowed the molecular structure of foods to be probed for the first time (Morris and Groves, 2013). SEM and TEM require drying the sample prior measurement and yield two-dimensional nanoscale results. Confocal laser scanning microscopy (CLSM) is a newer technique that provides three-dimensional images of the fluorescence of hydrophobic components bound to fluorescent dyes. This technique is particularly useful for high-fat materials (El-Bakry and Sheeha, 2014). Electron microscopy techniques allow a much higher resolution imaging of the food matrix components of the food matrix components in comparison to classic LM and CSLM. On the other hand, environmental scanning electron microscopy allows the imaging of "wet" and "soft" matrices, although the achievable resolution still lags behind that obtained by conventional SEM (Stokes, 2003; Morris and Groves, 2013). Atomic force microscopy is another, well-studied technique that generates images by monitoring changes in van der Waals' forces between the probe and the surface of the sample. This technique requires minimal sample preparation and produces nanoscale results (Yang et al., 2007).

3.4 SELECTION OF THE APPROPRIATE SOLVENT
3.4.1 COMPOUND SOLUBILITY IN DIFFERENT SOLVENTS

The recovery of valuable compounds from food wastes usually includes the solubilization of solutes into one or more solvents that provide a physical carrier to transfer them between different phases (i.e. solid, liquid, and vapor). Water is the most known physical carrier, but its polar nature does not allow the solubilization of nonpolar compounds. Nevertheless, organic compounds have generally low water solubility and thus polar protic or aprotic, as well as nonpolar, mediums have been implemented for their recovery from natural sources. Only some small and very polar compounds (e.g. gallic acid) are preferably solubilized in water (Galanakis et al., 2013a). As a rule of thumb, solvents should solubilize target compounds in high amounts and at the same time should not solubilize nontarget compounds and impurities. Since the recovered compounds are in most cases destined for food applications, the selected solvent should additionally meet the following requirements:

1. cheap, abundant, and easily accessible in the food industry,
2. possess "GRAS" status (generally recognized as safe),
3. reusable and recyclable,
4. inhibit enzyme activity where it is appropriate,
5. inhibit oxidation and preserve the functional properties of target compounds.

Alcohols (polar protic mediums) are among the solvents that meet many of the above requirements. For example, the presence of a hydroxyl group and an aliphatic part within their molecule allow the solubilization of natural products with intermediate polarity. Moreover, they are known to reduce polyphenoloxidase activity (Abad-Garcia et al., 2007). Methanol contains a smaller and more flexible aliphatic fragment compared with other alcohols and thus it surrounds polyphenols with substituted carbons inside their aromatic ring more easily (e.g. ferulic, vanillic, syringic, and synaptic acids that contain three or four substituted carbons). However, the major disadvantage of methanol is that is toxic and thus it should be completely removed from the extract. Bigger phenolics (e.g. oleuropein and rosmarinic acid) as well as acids with two antidiametric substituted carbons (e.g. *p*-hydroxybenzoic and *p*-hydroxyphenyl acetic) or longer aliphatic fragments (e.g. tyrosol and hydroxytyrosol) are preferably solubilized in ethanol. This medium could better surround the above compounds since it can "cover" the gaps between the hydrogen bonds (Galanakis et al., 2013b). Absolute ethanol has been inferred to inhibit enzyme activity by precipitation and low pH conditions (pH 2–3) similar to other alcohols (Queimada et al., 2009). More importantly, it is potable, cheap (taxes are removed in case of industrial applications), recyclable, and corresponding extracts could be applied directly in the beverage industry.

Indeed, the fact that ethanol is mixed well with water allows the generation of powerful cosolvents. Water swells the plant material and allows the solvent to increase extractability by penetrating solid matrices more easily. Hydroethanolic mixtures have been widely used for the recovery of several polyphenols, whereas different phenolic fractions can be obtained based on polarity by varying the alcohol concentration in the mixture (Tsakona et al., 2012). The only disadvantage of hydroethanolic mixtures is that their efficacy is extended to other extractable compounds (e.g. sugars, organic acids) and thus additional purification steps are required (Obied et al., 2005; Oreopoulou and Tzia, 2007; Galanakis et al., 2010c, d).

Polar aprotic mediums (e.g. dichloromethane) are typically used for the recovery of less polar compounds, i.e. cinnamic acid that has a carboxyl instead of hydroxyl group. In this case, solubilization occurs via dipole/dipole interactions developed between the more electronegative fragment of the solvent (chlorine atoms) and the more electropositive fragment (hydrogen protons) of the acid (Galanakis et al., 2013a). The selection of the appropriate solvent is much more difficult for the case of lipophilic compounds, i.e. carotenoids. For instance, lycopene of tomato paste is more liposoluble and thus polar aprotic (e.g. acetone) and nonpolar (e.g. ethyl acetate) mediums are preferred (Strati and Oreopoulou, 2011). These kind of solvents are not food grade and should be totally removed from the extract prior to its reutilization in food formulations.

Hydrotropic solvents are an alternative to conventional solvents and have raised interest for their ability to dissolve sparingly soluble organic compounds in aqueous solutions. Hydrotropic solvents are amphiphilic organic salts that are highly soluble in water. They contain a hydrophobic chain and hydrophilic head that allow them to aggregate and form micelles. This way, they can precipitate the hydrophobic solutes out of the solution after its dilution with water, and thus enable the direct recovery of the dissolved solutes. Moreover, hydrotropic solution can be recycled using a simple evaporation process. In addition, hydrotropes are readily biodegradable under aerobic conditions and trace amounts contained in the final product can be easily washed out with water. Hydrotropic solvents have been inferred to extract several compounds such as piperine from pepper (Raman and Gaikar 2002), curcumin from turmeric (Dandekar and Gaikar, 2003), reserpine from *Rauwolfia vomitoria* (Sharma and Gaikar, 2012), and limonin from sour orange seeds (Dandekar et al., 2008).

Designing a recovery strategy also requires information regarding solute solubility as a function of temperature. Other parameters, such as pressure, also play an important role in pressurized processes, fermentations, or supercritical fluids extraction using gaseous carbon dioxide (Galanakis, 2012; Galanakis et al., 2012b). Calculation of compound solubility in a given solvent, under certain conditions, can be conducted with empirical equations based on the respective physical properties. The most typical examples include the quantitative structure–property relationship (Rytting et al., 2004) where the equations are described as a function of polarity or group contribution methods (Klopman and Zhu, 2001).

3.4.2 THERMODYNAMIC PREDICTION FOR THE PREFERENCE OF TARGET COMPOUNDS IN DIFFERENT SOLVENTS

Although the determination of compound solubility is very important for the designing of the recovery strategy, it is practically impossible to screen all temperature and pressure conditions in combination with the possible solvents and their mixtures. Moreover, calculation of compound solubility requires a huge amount of experimental data in order to estimate the coefficients of the involved equations. For these reasons, thermodynamic models that allow the prediction of molecular characteristics and solubility data in different conditions are of great interest.

The tendency of each compound to be solubilized, transferred, or diffused into a given solvent is governed by thermodynamics. One of the primary parameters describing this tendency is the activity coefficient, which can be calculated directly using several models, i.e. nonrandom two-liquid segment activity coefficient (Chen, 2006), COSMO-RS (Klamt et al., 2002), or the well-established UNIFAC (Fredenslund et al., 1975, 1977). UNIFAC is developed on the following principles:

1. fragmentation of each individual molecule to chemical subgroups,
2. estimation of the structural characteristics of each subgroup based on their geometrical shape and size,
3. estimation of the energy interaction between the subgroups and the tested molecule taking into account key parameters such as temperature,
4. calculation of the molecule's activity coefficient.

The basic properties of the subgroups that are estimated include the relative volume (R_k), relative surface (Q_k), and the interaction parameter α_{mk} (in Kelvin degrees) between them. The activity coefficient of an ingredient is calculated by a group of 12 equations using the values of R_k, Q_k, and α_{mk} found in literature for a number of subgroups (Fredenslund et al., 1977). Calculation can also be conducted more easily using a Microsoft Excel logistic tally (xlUNIFAC, version 1.0, GNU Public Licence, GRL, USA) designed for the calculation of activity coefficients and partial vapor pressures (Randhol and Engelien, 2000). This software provides the possibility of examining the effect of each parameter in different temperatures given in degrees Kelvin . Calculations are only applicable to condensable nonelectrolytes, while pressure and temperature should be less than 5 bar and 150°C, respectively. Moreover, assayed molecules should not contain more than 10 functional groups. If a subgroup does not appear within the tally, R_k and Q_k can be found in the literature and introduced to the system. Activity coefficients of more complex molecules can be simulated by a part of them with molecules of similar structure. For example, the activity coefficient of oleuropein (a well-known antioxidant found in olive mill waste) has been predicted by simulating its glycoside moiety with a sugar (e.g. maltose) and implementing the corresponding values of R_k and Q_k within the overall calculation (Cooke et al., 2001; Galanakis et al., 2013a).

The UNIFAC model has been implemented to estimate activity coefficients in nonelectrolyte liquid mixtures and it has been further implemented to include polymers (Liu and Cheng, 2005; Cheng and Li, 2013; Galanakis et al., 2013a) and common sugars (e.g. fructose and glucose) in aqueous and nonaqueous solutions (Ferreira et al., 2003). Moreover, it has also been used to simulate the aqueous solubility of phenolic antioxidants (e.g. tyrosol, ellagic, protocatechuic, syringic, and o-coumaric acids), together with an excess of the Gibbs energy equation and the cubic-plus-association equation of state (Queimada et al., 2009). More recently, Galanakis et al. (2013a) have shown that the prediction of activity coefficients using the UNIFAC can be applied for the direct prediction of numerous phenolic compound preferences in different solvents and temperatures. This theoretical consideration is based on the simplified assumption that the lower activity coefficient corresponds to higher solubility and solvent recovery preference. Following the denoted data of activity coefficients, solvents were classified for their ability to dissolve polar (e.g. gallic acid) and nonpolar phenols, whereas suggestions for the recovery of phenols in different classes were also demonstrated. This classification matched with bibliographic data obtained in practice for the extraction of the target phenols from agricultural wastes. For example, the recovery of phenolic compounds was proposed to be initially conducted with hydroalcoholic mixtures and then to progress sequential extraction steps with solvents of reducing polarity in order to separate the compounds of interest for each case.

The described approach provides two major advantages. First, it runs within the thermodynamics framework with the aim of predicting rather than performing tedious, time-consuming and costly experimental work. Second, it does not only lead to qualitative conclusions. Indeed, it provides quantification and interpretation of the data through the estimation of thermodynamic characteristics as shown in Section 3.2.

3.5 **SELECTION OF THE RECOVERY STAGES**

The recovery of valuable compounds from food by-products follows typically the principles of analytical chemistry. Indeed, since the target compounds usually exist in smaller amounts compared with the initial sources (i.e. a major part of the valuable compounds is contained in the produced food material), the recovery process follows the rules of advanced analytical chemistry such as substrate preparation, extraction, and purification of the target compounds.

In addition, modifications are introduced in the recovery strategy with a final purpose to:

1. maximize the yield of the target compounds,
2. adapt to the demands of industrial processing,
3. purify the high added-value ingredients from co-extracted compounds, impurities, and toxic substances,
4. avoid deterioration, autoxidation, and diminution of compounds' functional properties,
5. ensure the edibility (food grade characteristics) of the final product,
6. ensure sustainability of the process within the food industry.

Recovery of downstream processing could be accomplished in five distinct stages, although depending on the case, one or two steps can be removed and/or change order. The most important issue in the recovery process is to effectively separate the compounds of the food waste matrix. A simple way to accomplish this procedure is to progress separation from the macroscopic to the macromolecular and then to the micromolecular level. Thereafter, a clarification or isolation step is required and finally product formation or encapsulation of the target compounds is required (Fig. 3.3).

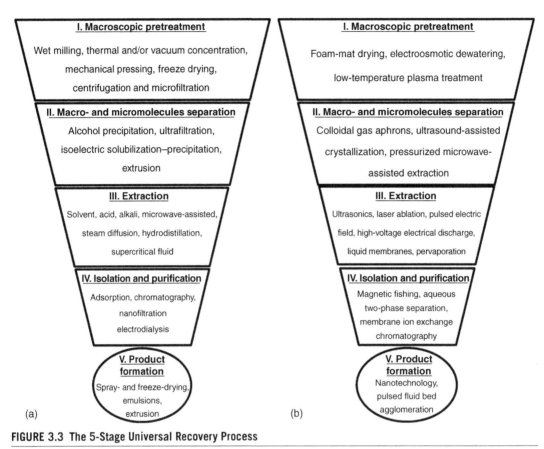

FIGURE 3.3 The 5-Stage Universal Recovery Process

(a) Conventional and (b) emerging technologies (Galanakis, 2012).

This recovery strategy has the advantage that it can be applied for simultaneous recovery of several ingredients in different streams, i.e. a micromolecule (e.g. polyphenols) using the ethanolic extract and macromolecules (e.g. pectin) in the ethanol insoluble residue. On the other hand, when the target compound is only a macromolecule (e.g. protein), the second stage could be omitted. Each step can be accomplished with different conventional (Fig. 3.3a) or emerging (Fig. 3.3b) technologies depending on processing cost, convenience, and specific restrictions that include:

1. low recovery yield,
2. overheating of the food matrix,
3. generation of solvent wastes,
4. high energy consumption, high capital, and operating costs,
5. loss of functionality,
6. poor stability of the final product,
7. accomplishment of increasingly stringent legal requirements on materials safety.

In general, emerging technologies promise to overcome problems such as overheating and loss of functionality and instability, but the positive effects on general cost and safety issues are still under debate (Deng et al., 2015).

3.6 SELECTION OF THE APPROPRIATE TECHNOLOGIES

3.6.1 PREPARATION OF THE MATERIAL

The first step of the recovery strategy is the macroscopic pretreatment of the food waste matrix. This step aims at the adjustment of the water, solids and fats content, activation or deactivation of enzymes, moderation of the microbial load, and finally increase in the permeability of the matrix. It typically involves only one process, which depends on the nature and structure of the substrate (e.g. solid, sludge, or wastewater). For instance, if the substrate is wastewater, partial water removal is necessary and thus concentration processes (e.g. thermal or vacuum) are applied in order to increase the content of valuable components in the matrix. Thermal concentration is the simplest process that has been applied in several cases (e.g. sugar beet pulp, olive mill wastewater, and apple pomace), while nonthermal dewatering (e.g. mechanical pressing or freeze drying) has been implemented in order to avoid loss of thermal labile compounds and loss of functionality. The thermal process also causes activation (e.g. at low temperature 40–60°C) or deactivation (e.g. at higher temperature >70–80°C) of enzymes (e.g. pectin methyl esterase and polyphenoloxidase) and thus affects the yield and functionality of respective compounds, i.e. pectin and polyphenols (Galanakis et al., 2010d, e). On the other hand, nonthermal dewatering has disadvantages that are related to increased cost and absence of microbial pasteurization, which lead to low shelf-life of the treated matrix. For this purpose, new techniques have recently been introduced. For example, foam mat drying has been used for the removal of water from heat-sensitive and viscous substrates, i.e. mango pulp or apple puree (Rajkumar et al., 2007). Its main advantage is that it requires lower temperatures and shorter drying time. In addition, electrically-assisted mechanical dewatering is an alternative technique applied to liquids that do not form readily or collapse during the drying process. This process combines the conventional pressure consolidation with electrostatic effects and has been applied to save energy during dewatering of a biscuit sewage sludge or tomato pomace (Jumah et al., 2005).

Other processes such as centrifugation or microfiltration, can also be applied in the pretreatment stage since they are able to remove solids, oils, and fats. Solids can restrict mechanical processing, such as substrate flow, mixing, and homogenization, while oil and fats can cause deterioration of the substrate via their autoxidation (Díaz et al., 2004; Galanakis et al., 2010a). Likewise, in the case of fruit or vegetable by-products, wet milling is often implemented in order to enhance diffusion of extractants inside via swelling and softening of the bioresource matrix. Also, the recovery of compounds with particular characteristics requires a different pretreatment approach, i.e. the recapture of phytosterols requires an additional hydrolysis step under high pressure and temperature (1.5–50 MPa and 200–260°C) or saponification with alkaline solution (Fernandes and Cabral, 2007).

3.6.2 REMOVAL OF MACROMOLECULES

Concerning the second recovery step, separation of small compounds (i.e. antioxidants, acids, or ions) from macromolecules (i.e. proteins or dietary fibers) is typically conducted using alcohol precipitation.

This method is cheap, easy to use, and works by collecting the larger molecules in a precipitate, namely alcohol insoluble residue. However, it is not selective and cannot separate the complexes between the smaller (e.g. phenols) and the larger (e.g. proteins) molecules. A more efficient separation of macro- from micromolecules can be performed with membranes, e.g. ultrafiltration (Galanakis, 2015; Galanakis & Schieber, 2014; Galanakis et al., 2014). Membrane procedures are nondestructive and easy going, but they are very sensitive in the variation of the feed content causing fouling problems. Isoelectric precipitation is a similar technique that allows the selective precipitation of proteins (e.g. from meat, fish, or marine by-products) by shifting pH value up to the proteins' isoelectric point (Gehring et al., 2011). Protein segregation has also been conducted via sonocrystallization that provides a process considered in terms of nucleation and crystal growth. Sonocrystallization provides faster and a more uniform crystal growth compared with conventional processes. For instance, it has been applied to accelerate whey protein removal during lactose recovery (De Castro and Priego-Capote, 2007; Patel and Murthy, 2010). Besides, the implementation of colloidal gas aphrons with cationic and nonionic surfactants has recently raised scientific interest for the selective separation of reverse charged macro- or micromolecules from waste liquids without mechanical aid. Nevertheless, the main drawback of this technique is the presence of surfactants in the product stream. Other combined methods (e.g. pressurized microwave-assisted extraction) have been suggested to accelerate the recovery of metabolites with different structures and polarities (e.g. terpenes, flavonoids, or pectin from orange albedo and peels). However, pressurized processes are difficult to control and also cause degradation of thermolabile ingredients.

3.6.3 DISSOCIATION OF MOLECULAR CLUSTERS AND COMPLEXES

Macro- and micromolecules exist either in free or bounded forms within bioresources. For example, polyphenols are known to bind both dietary fibers of plant materials (Bravo et al., 1994) and dietary proteins (noncovalently) in cheese products (Rawel et al., 2005). The extraction process (which is the third and most important stage of the recovery process) aims at solubilizing free molecules as well as the dissociation and solubilization of bounded compounds. Usually, solvents are employed in order to separate the target ingredients from the pretreated waste material and transfer them to a particular stream. Sequential solvent extractions increase the yield, but also the consumed time and cost of the process. Pressurized processes are sometimes used to accelerate the extraction of compounds like phenols or carotenoids, whereas distillation is used to recover flavorings. In addition, enzyme-assisted extraction has been used to soften the structural integrity of the assayed material (Sowbhagya and Chitra, 2010). Dietary fiber extraction (e.g. pectin, β-glucan, and hemicelluloses) is typically conducted using acid or alkali treatment after ethanol precipitation (Koubala et al., 2008). Extraction has also been performed prior to the precipitation step in particular applications such as pectin and phenol recovery from olive mill wastewater and mango peels. Inversion of the second and third processing steps is suggested in order to utilize a part of the added ethanol as surfactant. For instance, ethanol is able to penetrate the capillary porous structure of fruit tissues and dissociate molecular clusters or complexes with a subsequent precipitation of macromolecules in the alcohol insoluble residue and solubilization of micromolecules in the ethanolic extract (Berardini et al., 2005; Galanakis et al., 2010a, b).

Microwave-assisted extraction has recently raised interest due to its effectiveness, easy handling, and moderate solvent requirements. Moreover, it has been applied together with other technologies (e.g. steam diffusion or hydrodistillation) to enhance extractability of volatile compounds or essential oils from citrus by-products (Farhat et al., 2011). The disadvantages of steam diffusion include the thermal deterioration of the substrate and the difficult removal of solvent from the extract. On the other hand, hydrodistillation is a milder procedure as the substrate is completely immersed in boiling water, which acts as a barrier and prevents overheating of the essential oils. Pervaporation is a low energy technique that has also been implemented to avoid heat damage of sensitive aromas. Moreover, it has been proposed to deodorize food industry effluents (e.g. cauliflower blanching water) with a simultaneous recovery of flavorings (Souchon et al., 2002). The elution of aromas has also been suggested using liquid membranes. Liquid membranes show high selectivity and use energy efficiently compared with other separation techniques; however, their low stability restricts their industrial exploitation (San Román et al., 2009).

Supercritical fluid extraction is a modern and established technology, which involves the use of a gas (e.g. CO_2) above its critical temperature and pressure. In this case, the carrier shows physico-chemical properties that exhibit the behavior of a phase between a liquid and a gas. This technique is applied in difficult separation processes of compounds that are found in low contents within the waste material. It requires low solvent consumption (even if cosolvents are sometimes required) and has high selectivity (Sowbhagya and Chitra, 2010). On the other hand, operation handling is difficult due to the optimization of many parameters. Besides, the addition of cosolvents or modifiers as well as the co-extraction of other compounds cannot be avoided. Ultrasound extraction is another technology that has been employed to accelerate heat and mass transfer via its cavitational effect, which disrupts the plant cell walls and other complexes that exist in food wastes. Advantages include the accomplishment of the process in minutes, high reproducibility, and low solvent consumption.

Accelerated mass transfer can also be implied by pulsed electric fields. This emerging technique is nowadays used for an increasing number of cases such as the extraction of phenols from grape seeds, betalains (water soluble pigments) from red beetroot, and pectin from apple pomace (Vorobiev and Lebovka, 2010; Galanakis, 2012, 2013). Similar electrically induced extraction technologies (e.g. pulsed ohmic heating and high voltage electrical discharges) and "laser ablation" have also been proposed, although their application in the field is rather limited. High voltage electric discharge does not require the addition of organic solvent, but it demands high air generation capacity (Vorobiev and Lebovka, 2010; Roselló-Soto et al., 2015). On the other hand, the photodynamic effect induced to materials by laser irradiation promises minimal heating and tailor-made extraction of aromas, anthocyanins, polysaccharides, and proteins from food wastes. Laser ablation does not require solvents and is easily automated (Panchev et al. 2011), but perhaps it is too sophisticated for its implementation only in the extraction stage.

3.6.4 REMOVAL OF CO-EXTRACTED IMPURITIES

The fourth stage of recovery includes the clarification of the target compounds from co-extracted impurities. Adsorption is a simple process utilized for this purpose, as it is able to isolate selected low molecular weight compounds (e.g. antioxidants) from diluted solutions with high capacity as well as insensitivity to toxic substances (Soto et al., 2011). However, it is time consuming and requires further exploitation of the sorption behavior of each individual component separately. Similarly, ion exchange and affinity chromatography have been applied for the partitioning of polyvalent and charged whey proteins or phenols from olive mill wastewater (Fernández-Bolaños et al., 2002; El-Sayed and Chase, 2011; Rahmanian et al., 2014). These methodologies are laboratory-intensive, and solvent and time consuming, but they can ensure the isolation of highly purified compounds for pharmaceutical applications. Aqueous two-phase separation has been shown to be effective for the isolation of proteins and enzymes from crude cell extracts, while it has recently been applied for the fractionation of whey β-lactoglobulin and α-lactalbumin (Jara and Pilosof, 2011). This technique proceeds with mild conditions and thus is appropriate for the isolation of labile compounds, but it requires a long separation time and numerous processing steps. Isolation of proteins has also been performed with so-called "magnetic fishing". In this case, charged molecules bind to magnetic particles (e.g. ion-exchange groups or immobilized affinity ligands), which are eluted after washing out any co-isolated impurities (El-Sayed and Chase, 2011).

Membrane processes can perform direct separations of a different nature. Thereby, reverse osmosis allows only water molecules to pass through membrane pores, while nanofiltration is applied when monovalent salt permeation is desirable (Li et al., 2008; González et al., 2008; Galanakis et al., 2012a). For example, it has been employed to purify lactic acid with a simultaneous recovery of whey proteins in the concentrate. Moreover, electrodialysis has been implemented to demineralize oligosaccharides extracted from soybean sheet slurry and separate peptides from snow crab by-product hydrolysate (Wang et al., 2009; Doyen et al., 2011). Bhattacharjee et al. (2006) purified whey proteins up to 90% using two stages of ultrafiltration coupled with ion-exchange chromatography. This combined technique is very selective, but much slower compared with conventional membrane filtration. The drawbacks of membrane processes concern their restricted stability and short lifetime of the membranes.

3.6.5 OBTAINMENT OF THE FINAL PRODUCT

The fifth and final stage of the recovery strategy always involves an encapsulation or drying process. This is not actually a recovery procedure, but it should be taken into account since extracts, residues, and enriched elutions cannot be delivered in the market without ensuring that the valuable compounds will preserve their properties within a reasonable period. Also, other problems, such as transportation and cost may arise during commercialization. Encapsulation is known to entrap valuable compounds inside a coating material and this way it ensures their stability, masks undesirable organoleptic characteristics, and finally protects them against environmental stresses. In addition, it prevents nonfunctional interactions with food matrixes during their implementation as additives and improves their delivery into foods. Polysaccharides (starch, cellulose, cyclodextrin, inulin, pectin, gums, carrageenans, alginate, etc.) and proteins are typically used as coating materials in encapsulation processes. Thereby, when the compounds of interest are macromolecules, encapsulation is replaced with a direct drying process.

Spray drying is the most known encapsulation technique in the food industry because it is an easy to handle, continuous, and economic operation. Recently, it has been applied to the encapsulation of tomato carotenoids (lycopene, α-carotene, and β-carotene) from industrial residues and phenolics (e.g. myricetin, quercetin, quercetin-3-β-glucoside, caffeic acid, and p-coumaric acid) and from wine lees (Pérez-Serradilla and de Castro, 2011; Galanakis, 2012). The disadvantage of spray drying is the thermal destruction of labile antioxidants, e.g. volatile low molecular weight phenols. Freeze drying is a milder process that allows the preservation of labile antioxidants. Nevertheless, it consumes more time and energy compared with spray drying since the whole process is conducted under vacuum conditions. Melt extrusion is another technology used to increase palatability of polysaccharides (e.g. starch) and encapsulate flavors or nutrients. Extrusion requires shorter residence time and lower chemical and water consumption, but it has generally a low yield. Liposomes and emulsions are used to entrap lipophilic compounds (e.g. tomato carotenes) or hydrophilic antioxidants (e.g. phenols from potato peel) prior to their implementation in rapeseed oil mixtures (Habeebullah et al., 2010). Advanced encapsulation is today conducted using nanoemulsions, which are much more stable and provide moisture- and pH-triggered controlled release (Jaeger et al., 2010; McClements and Rao, 2011). For example, the preparation of nanoemulsions has been reported to improve the dispersion ability of liposoluble β-carotene in water and enhance its intestinal bioavailability (Silva et al., 2011). The obtained formulations could be added as natural colorings (dark orange to yellow) in water-based foods or as mask substances to hide taste and odor of tuna fish oil (rich in ω-3 fatty acids) during its implementation into bread (Neethirajan and Jayas, 2011).

REFERENCES

Abad-Garcia, B., Barrueta, L.A., Lopez-Marquez, D.M., Crespo-Ferrer, I., Gallo, B., Vicente, F., 2007. Optimization and validation of a methodology based on solvent extraction and liquid chromatography for the simultaneous determination of several polyphenolic families in fruit juices. J. Chromatogr. A 1154, 87–96.

Atanasova, V., Fulcrand, H., Le Guernevé, C., Cheynier, V., Moutounet, M., 2002. Structure of a new dimeric acetaldehyde malvidin 3-glucoside condensation product. Tetrahedron Lett. 43, 6151–6153.

Berardini, N., Knödler, M., Scieber, A., Carle, R., 2005. Utilization of mango peels as a source of pectin and polyphenolics. Innov. Food Sci. Emerg. Technol. 6, 442–452.

Bhattacharjee, S., Bhattacharjee, C., Datta, S., 2006. Studies on the fractionation of beta-lactoglobulin from casein whey using ultrafiltration and ion exchange membrane chromatography. J. Membr. Sci. 275, 141–150.

Bravo, L., Abia, R., Saura-Calixto, F., 1994. Polyphenols as dietary fiber associated compounds. Comparative study on *in vivo* and *in vitro* properties. J. Agr. Food Chem. 42, 1481–1487.

Chen, C.C., 2006. Correlation and prediction of drug molecule solubility in mixed solvent systems with the nonrandom two-liquid segment activity coefficient (NRTL-SAC) model. Ind. Eng. Chem. Res. 45, 4816–4824.

Cheng, Z.F., Li, C.Y., 2013. A modified UNIFAC model for polymer solutions. Asian J. Chem. 25, 731–734.

Cooke, S.A., Jónsdóttir, S.Ó., Westh, P., 2001. Phase equilibria of carbohydrates: the study of a series of glucose oligomers from glucose to maltopentaose in aqueous solution – experimental versus predicted data using various UNIQUAC/UNIFAC models. Fluid Phase Equilibr. 194, 947–956.

Dandekar, D.V., Gaikar, D.G., 2003. Hydrotropic extraction of curcuminoids from turmeric. Separ. Sci. Technol. 38, 1185–1215.

Dandekar, D.V., Jayaprakasha, G.K., Patil, B.S., 2008. Hydrotropic extraction of bioactive limonin from sour orange seeds. Food Chem. 109, 515–520.

De Castro, M.D.L., Priego-Capote, F., 2007. Ultrasound-assisted crystallization (sonocrystallization). Ultrason. Sonochem. 14, 717–724.

Deng, Q., Zinoviadou, K.G., Galanakis, C.M., Orlien, V., Grimi, N., Vorobiev, E., Lebovka, N., Barba, F.J., 2015. The effects of conventional and non-conventional processing on glucosinolates and its derived forms, isothiocyanates: extraction, degradation and applications. Food Eng. Rev., in press.

Díaz, O., Pereira, C.D., Cobos, A., 2004. Functional properties of ovine whey protein concentrates produced by membrane technology after clarification of cheese manufacture by-products. Food Hydrocolloid. 18, 601–610.

Doyen, A., Beaulieu, L., Saucier, L., Pouliot, Y., Bazinet, L., 2011. Impact of ultrafiltration membrane material on peptide separation from a snow crab byproduct hydrolysate by electrodialysis with ultrafiltration membranes. J. Agr. Food Chem. 59, 1784–1792.

El-Bakry, M., Sheeha, J., 2014. Analysing cheese microstructure: a review of recent developments. J. Food Eng. 125, 84–96.

El-Sayed, M.H.H., Chase, H.A., 2011. Trends in whey protein fractionation. Biotechnol. Lett. 33, 1501–1511.

Ertl, P., Rohde, B., Selzer, P., 2000. Fast calculation of molecular polar surface area as a sum of fragment based contributions and its application to the prediction of drug transport properties. J. Med. Chem. 43, 3714–3717.

Farhat, A., Fabiano-Tixier, A-.S., Maataoui, M.E., Maingonnat, J-.F., Romdhane, M., Chemat, F., 2011. Microwave steam diffusion for extraction of essential oil from orange peel: kinetic data, extract's global yield and mechanism. Food Chem. 125, 255–261.

Fernandes, P., Cabral, J.M.S., 2007. Phytosterols: applications and recovery methods. Bioresource Technol. 98, 2335–2350.

Fernández-Bolaños, J., Heredia, A., Rodríguez, G., Rodríguez, R., Guillén, R., Jiménez, A., 2002. Method for obtaining purified hydroxytyrosol from products and by-products derived from the olive tree. World Intellectual Property Organization. Patent Application. WO/2002/064537.

Ferreira, O., Brignole, E.A., Macedo, E.A., 2003. Phase equilibria in sugar solution using a UNIFAC model. Ind. Eng. Chem. Res. 42, 6212–6222.

Fredenslund, A., Gmehling, J., Michelsen, M.L., Rasmussen, P., Prausnitz, J.M., 1977. Computerized design of multicomponent distillation columns using the UNIFAC group contribution method for calculation of activity coefficients. Ind. Eng. Chem Process Design Devel. 16, 450–462.

Fredenslund, A., Jones, R.L., Prausnitz, J.M., 1975. Group-contribution estimation of activity coefficients in nonideal liquid mixtures. AIChe J. 21, 1086–1099.

Galanakis, C.M., 2011. Olive fruit and dietary fibers: components, recovery and applications. Trends Food Sci. Technol. 22, 175–184.

Galanakis, C.M., 2012. Recovery of high added-value components from food wastes: conventional, emerging technologies and commercialized applications. Trends Food Sci. Technol. 26, 68–87.

Galanakis, C.M., 2013. Emerging technologies for the production of nutraceuticals from agricultural by-products: a viewpoint of opportunities and challenges. Food Bioprod. Process. 91, 575–579.

Galanakis, C.M., 2015. Separation of functional macromolecules and micromolecules: from ultrafiltration to the border of nanofiltration. Trends Food Sci. Technol. 42, 44–63.

Galanakis, C.M., Schieber, A., 2014. Editorial. Special issue on recovery and utilization of valuable compounds from food processing by-products. Food Res. Int. 65, 299–484.

Galanakis, C.M., Tornberg, E., Gekas, V., 2010a. A study of the recovery of the dietary fibres from olive mill wastewater and the gelling ability of the soluble fibre fraction. LWT– Food Sci. Technol. 43, 1009–1017.

Galanakis, C.M., Tornberg, E., Gekas, V., 2010b. Dietary fiber suspensions from olive mill wastewater as potential fat replacements in meatballs. LWT– Food Sci. Technol. 43, 1018–1025.

Galanakis, C.M., Tornberg, E., Gekas, V., 2010c. Clarification of high-added value products from olive mill wastewater. J. Food Eng. 99, 190–197.

Galanakis, C.M., Tornberg, E., Gekas, V., 2010d. Recovery and preservation of phenols from olive waste in ethanolic extracts. J. Chem. Technol. Biotechnol. 85, 1148–1155.

Galanakis, C.M., Tornberg, E., Gekas, V., 2010e. The effect of heat processing on the functional properties of pectin contained in olive mill wastewater. LWT-Food Sci. Technol. 43, 1001–1008.

Galanakis, C.M., Kanellaki, M., Koutinas, A.A., Bekatorou, A., Lycourghiotis, A., Kordoulis, C.H., 2012b. Effect of pressure and temperature on alcoholic fermentation by Saccharomyces cerevisiae immobilized on γ-alumina pellets. Bioresour. Technol. 114, 492–498.

Galanakis, C.M., Fountoulis, G., Gekas, V., 2012a. Nanofiltration of brackish groundwater by using a polypiperazine membrane. Desalination 286, 277–284.

Galanakis, C.M., Goulas, V., Tsakona, S., Manganaris, G.A., Gekas, V., 2013a. A knowledge base for the recovery of natural phenols with different solvents. Int. J. Food Prop. 16, 382–396.

Galanakis, C.M., Markouli, E., Gekas, V., 2013b. Fractionation and recovery of different phenolic classes from winery sludge via membrane filtration. Separ Purif Technol. 107, 245–251.

Galanakis, C.M., Chasiotis, S., Botsaris, G., Gekas, V., 2014. Separation and recovery of proteins and sugars from Halloumi cheese whey. Food Res. Int. 65, 477–483.

Galanakis, C.M., Kotanidis, A., Dianellou, M., Gekas, V., 2015a. Phenolic content and antioxidant capacity of Cypriot wines. Czech J. Food Sci. 33, 126–136.

Gehring, C.K., Gigliotti, J.C., Moritza, J.S., Tou, J.C., Jaczynski, J., 2011. Functional and nutritional characteristics of proteins and lipids recovered by isoelectric processing of fish by-products and low-calue fish: a review. Food Chem. 124, 422–431.

Genovese, D.B., Lozano, J.E., Rao, M.A., 2007. The rheology of colloidal and noncolloidal food dispersions. J. Food Sci. 72, R11–R20.

Girard, M., Turgeon, S.L., Gauthier, S.F., 2003. Thermodynamic parameters of β-lactoglobulin-pectin complexes assessed by isothermal titration calorimetry. J. Agr. Food Chem. 51, 4450–4455.

Goñi, I., Valdivieso, L., Garcia-Alonso, A., 2000. Nori seaweed consumption modifies glycemic response in healthy volunteers. Nutr. Res. 20, 1367–1375.

González, M.I., Alvarez, S., Riera, F.A., Álvarez, R., 2008. Lactic acid recovery from whey ultrafiltrate fermentation broths and artificial solutions by nanofiltration. Desalination 228, 84–96.

Habeebullah, S.F.K., Nielsen, N.S., Jacobsen, C., 2010. Antioxidant activity of potato peel extracts in a fish-rapeseed oil mixture and in oil-in-water emulsions. J. Am. Oil Chem. Soc. 87, 1319–1332.

Harnsilawat, T., Pongsawatmanit, R., McClements, D.J., 2006. Characterization of β-lactoglobulin-chitosan interactions in aqueous solutions: a calorimetry, light scattering, electrophoretic mobility and solubility study. Food Hydrocolloid. 20, 124–131.

Heng, W.W., Xiong, L.W., Ramanan, R.N., Hong, T.L., Kong, K.W., Galanakis, C.M., Prasad, K.N., 2015. Two level factorial design for the optimization of phenolics and flavonoids recovery from palm kernel by-product. Ind. Crop. Prod. 63, 238–248.

Jaeger, H., Janositz, A., Knorr, D., 2010. The Maillard reaction and its control during processing. The potential of emerging technologies. Pathologie 58, 207–213.

Jara, F., Pilosof, A.M.R., 2011. Partitioning of alpha-lactalbumin and beta-lactoglobulin in whey protein concentrate/hydroxypropyl methyl cellulose aqueous two-phase systems. Food Hydrocolloid. 25, 374–380.

Jumah, R., Al-Asheh, S., Banat, F., Al-Zoubi, K., 2005. Electroosmotic dewatering of tomato paste suspension under AC electric field. Dry. Technol. 23, 1465–1475.

Klamt, A., Eckert, F., Hornig, M., Beck, M.E., Bürger, T., 2002. Prediction of aqueous solubility of drugs and pesticides with COSMO-RS. J. Comp. Chem. 23, 275–281.

Klopman, G., Zhu, H., 2001. Estimation of the aqueous solubility of organic molecules by the group contribution approach. J. Chem. Inform. Comp. Sci. 41, 439–445.

Klose, C., Arendt, E.K., 2012. Proteins in oats; their synthesis and changes during germination: a review. Crit. Rev. Food Sci. Nutr. 52, 629–639.

Koubala, B.B., Mbome, L.I., Kansci, G., Tchouanguep Mbiapo, F., Crepeau, M.-J., Thibault, J.-F., Ralet, M.C., 2008. Physicochemical properties of pectins from ambarella peels (*Spondias cytherea*) obtained using different extraction conditions. Food Chem. 106, 1202–1207.

Lawrence, N.D., Perera, J.M., Iyer, M., Hickey, M.W., Stevens, G.W., 2006. The use of streaming potential measurements to study the fouling and cleaning of ultrafiltration membranes. Sep. Sci. Technol. 48, 106–112.

Lazaridou, A., Biliaderis, C.G., 2004. Cryogelation of cereal β-glucans: structure and molecular size effects. Food Hydrocolloid. 18, 933–947.

Li, Y., Shahbazi, A., Williams, K., Wan, C., 2008. Separate and concentrate lactic acid using combination of nanofiltration and reverse osmosis membranes. Appl. Biochem. Biotechnol. 147, 1–9.

Liu, Q.L., Cheng, Z.F., 2005. A modified UNIFAC model for the prediction of phase equilibrium for polymer solutions. J. Polym. Sci.: Part B: Polym. Phys. 43, 2541–2547.

Mateus, N., de Pascual-Teresa, S., Rivas-Gonzalo, J.C., Santos-Buelga, C., de Freitas, V., 2002. Structural diversity of anthocyanin-derived pigments in port wines. Food Chem. 76, 335–342.

McClements, D.J., Rao, J., 2011. Food-grade nanoemulsions: formulation, fabrication, properties, performance, biological fate, and potential toxicity. Crit. Rev. Food Sci. Nutr. 51, 285–330.

McNeil, M., Darvill, A.G., Fry, S.C., Albersheim, P., 1984. Structure and function of the primary cell walls of plants. Annu. Rev. Biochem. 53, 625–663.

Molinspiration Cheminformatics, 2005. Available from: http://www.molinspiration.com.

Morris, V., Groves, K., 2013. Food microstructures: microscopy, measurement and modelling. Woodhead Publishing, Philadelphia, USA.

Neethirajan, S., Jayas, D.S., 2011. Nanotechnology for the food and bioprocessing industries. Food Bioprocess Technol. 4, 39–47.

Obied, H.K., Allen, M.S., Bedgood, D.R., Prenzler, P.D., Robards, K., 2005. Investigation of Australian olive mill waste for recovery of biophenols. J. Agr. Food Chem. 53, 9911–9920.

Oreopoulou, V., Tzia, C., 2007. Utilization of plant by-products for the recovery of proteins, dietary fibers, antioxidants, and colorants. In: Oreopoulou, V., Russ, W. (Eds.), Utilization of By-Products and Treatment of Waste in the Food Industry. Springer Science and Business Media, New York, pp. 209–232.

Osorio-Díaz, P., Bello-Pérez, L.A., Sáyago-Ayerdi, S.G., Benítez-Reyes, M., Tovar, J., Paredes-López, O., 2003. Effect of processing and storage time on in vitro digestibility and resistant starch content of two bean (Phaseolus vulgaris L) varieties. J. Sci. Food Agr. 83, 1283–1288.

Palafox-Carlos, H., Ayala-Zavala, J.F., González-Aguilar, G.A., 2011. The role of dietary fiber in the bioaccessibility and bioavailability of fruit and vegetable antioxidants. J. Food Sci. 76, R6–R15.

Panchev, I.N., Kirtchev, N.A., Dimitrov, D.D., 2011. Possibilities for application of laser ablation in food technologies. Innov. Food Sci. Emerg. Technol. 12, 369–374.

Patel, S.R., Murthy, Z.V.P., 2010. Optimization of process parameters by Taguchi method in the recovery of lactose from whey using sonocrystallization. Cryst. Res. Technol. 45, 747–752.

Patsioura, A., Galanakis, C.M., Gekas, V., 2011. Ultrafiltration optimization for the recovery of β-glucan from oat mill waste. J. Membr. Sci. 373, 53–63.

Pérez-Serradilla, J.A., de Castro, M.D.L., 2011. Microwave-assisted extraction of phenolic compounds from wine lees and spray-drying of the extract. Food Chem. 124, 1652–1659.

Qin, K., Zaman, A.A., 2003. Viscosity of concentrated colloidal suspensions: comparison of bidisperse models. J. Colloid Interf. Sci. 266, 461–467.

Queimada, A.J., Mota, F.L., Pinho, S.P., Macedo, E.A., 2009. Solubilities of biologically active phenolic compounds: measurements and modeling. J. Phys. Chem. B 113, 3469–3476.

Rahmanian, N., Jafari, S.M., Galanakis, C.M., 2014. Recovery and removal of phenolic compounds from olive mill wastewater. J. Am. Oil Chem. Soc. 91, 1–18.

Rajkumar, P., Kailappan, R., Viswanathan, R., Raghavan, G.S.V., Ratti, C., 2007. Foam mat drying of alphonso mango pulp. Dry. Technol. 25, 357–365.

Raman, G., Gaikar, V.G., 2002. Extraction of piperine from *Piper nigrum* (black pepper) by hydrotropic solubilization. Ind. Eng. Chem. Res. 41, 2966–2976.

Randhol, P., Engelien, H.K., 2000. xlUNIFAC version 1.0, a computer program for calculation of liquid activity coefficients using the UNIFAC Model. Available from: http://www.pvv.org/~randhol/xlunifac/.

Rawel, H.M., Meidtner, K., Kroll, J., 2005. Binding of selected phenolic compounds to proteins. J. Agr. Food Chem. 53, 4228–4235.

Remy, S., Fulcrand, H., Labarbe, B., Cheynier, V., Moutounet, M., 2000. First confirmation in red wine of products resulting from direct anthocyanin-tannin reactions. J. Sci. Food Agr. 80, 745–751.

Roselló-Soto, E., Barba, F.J., Parniakov, O., Galanakis, C.M., Grimi, N., Lebovka, N., Vorobiev, E., 2015. High voltage electrical discharges, pulsed electric field and ultrasounds assisted extraction of protein and phenolic compounds from olive kernel. Food Bioproc. Technol. 8, 885–894.

Rytting, E., Lentz, K.A., Chen, X.-Q., Qian, F., Venkatesh, S., 2004. A quantitative structure-property relationship for predicting drug solubility in PEG 400/water cosolvent systems. Pharmaceut. Res. 21, 237–244.

San Román, M.F., Bringas, E., Ibañez, R., Ortiz, I., 2009. Liquid membrane technology: fundamentals and review of its applications. J. Chem. Technol. Biotechnol. 85, 2–10.

Schmitt, C., Turgeon, S.L., 2011. Protein/polysaccharide complexes and coacervates in food systems. Adv. Colloid Interf. Sci. 167, 63–70.

Schmitt, C., Palma da Silva, T., Bovay, C., Rami-Shojaei, S., Frossard, P., Kolodziejczyk, E., Leser, M.E., 2005. Effect of time on the interfacial and foaming properties of β-lactoglobulin/acacia electrostatic complexes and coacervates at pH 4.2. Langmuir 21, 7786–7795.

Sharma, R.A., Gaikar, V.G., 2012. Hydrotropic extraction of reserpine from *Rauwolfia vomitoria* roots. Separ. Sci. Technol. 47, 827–833.

Silva, H.D., Cerqueira, M.A., Souza, B.W.S., Ribeiro, C., Avides, M.C., Quintas, M.A.C., Coimbra, J.S.R., Carneiro-da-Cunha, M.G., Vicente, A.A., 2011. Nanoemulsions of b-carotene using a high-energy emulsification–evaporation technique. J. Food Eng. 102, 130–135.

Soto, M.L., Moure, A., Domínguez, H., Parajó, J.C., 2011. Recovery, concentration and purification of phenolic compounds by adsorption: a review. J. Food Eng. 105, 1–27.

Souchon, I., Pierre, F.X., Athes-Dutour, V., Mari, M., 2002. Pervaporation as a deodorization process applied to food industry effluents: recovery and valorisation of aroma compounds from cauliflower blanching water. Desalination 148, 79–85.

Sowbhagya, H.B., Chitra, V.N., 2010. Enzyme-assisted extraction of flavorings and colorants from plant materials. Crit. Rev. Food Sci. Nutr. 50, 146–161.

Stokes, D.J., 2003. Recent advances in electron imaging, image interpretation and applications: environmental scanning electron microscopy. Philos. Trans. R.Soc. A Math. Phys. Eng. Sci. 361, 2771–2787.

Strati, I.F., Oreopoulou, V., 2011. Effect of extraction parameters on the carotenoid recovery from tomato waste. Int. J. Food Sci. Technol. 46, 23–29.

Tsakona, S., Galanakis, C.M., Gekas, V., 2012. Hydro-ethanolic mixtures for the recovery of phenols from Mediterranean plant material. Food Bioprocess Technol. 5, 1384–1393.

Vorobiev, E., Lebovka, N., 2010. Enhanced extraction from solid foods and biosuspensions by pulsed electrical energy. Food Eng. Rev. 2, 95–108.

Wang, Q., Ying, T., Jiang, T., Yang, D., Jahangir, M.M., 2009. Demineralization of soybean oligosaccharides extract from sweet slurry by conventional electrodialysis. J. Food Eng. 95, 410–415.

Yang, H., Wang, Y., Lai, S., An, H., Li, Y., Chen, F., 2007. Application of atomic force microscopy as a nanotechnology tool in food science. J. Food Sci. 72, R65–R75.

Zhou, Z., Scales, P.J., Boger, D.V., 2001. Chemical and physical control of the rheology of concentrated metal oxide suspensions. Chem. Eng. Sci. 56, 2901–2920.

Zydney, A.L., 1998. Protein separations using membrane filtration: new opportunities for whey fractionation. Int. Dairy J. 8, 243–250.

SECTION

CONVENTIONAL TECHNIQUES

CONVENTIONAL MACROSCOPIC PRETREATMENT

4

Lia Noemi Gerschenson*, Qian Deng**, Alfredo Cassano†

*Industry Department, Natural and Exact Sciences School (FCEN), Buenos Aires University (UBA), Buenos Aires, Argentina; National Scientific and Technical Research Council of Argentina (CONICET), Buenos Aires, Argentina; **Milne Fruit Products Inc., Prosser, Washington, USA; †Institute on Membrane Technology-Consiglio Nazionale delle Ricerche, University of Calabria, Rende Cosenza, Italy

4.1 INTRODUCTION

The first step for the recuperation of valuable compounds from food wastes is macroscopic pretreatment. This stage is composed of one straightforward step to adjust raw material format in order to make it more adequate for the following operations such as separation, extraction, purification, and product formation. Conventional macro-pretreatments are considered to be safe and low in capital investment but during their development, heat-sensitive components can be destroyed (Galanakis, 2012, 2013). Depending on the composition of raw materials (slurry, wastewater, or solid waste), conventional macro-pretreatment may involve the reduction of water content through thermal or vacuum concentration, the elimination of water through freeze drying, the reduction of particle size to enhance recuperation or to facilitate waste treatment, the separation of different phases through centrifugation, the generation of solids through pressing, or the reduction of the polluting load through microfiltration.

This chapter aims to assist in making decisions related to the appropriate macroscopic pretreatment of food waste to assure process efficiency, which involves high yield and low energy expenditure. For that purpose, previously mentioned processes are analyzed, engineering principles are discussed, and literature examples are summarized.

4.2 SIZE REDUCTION OF SOLIDS

4.2.1 SIZE REDUCTION PRINCIPLES

Raw materials often occur in sizes that are too large to be used and, therefore, they must be reduced in size. The operation for solids is called grinding or cutting. In the grinding process, materials are reduced in size by fracturing them. The mechanism of fracture is not fully understood, but in the process, the material is stressed by the action of mechanical moving parts in the grinding machine.

The force applied may be compressive, impact, or shear, and both the magnitude of the force and the time of application affect the extent of grinding achieved. For efficient grinding, the energy applied to the material should exceed, by as small a margin as possible, the minimum energy needed to rupture the material. Excess energy is lost as heat and this loss should be kept as low as practicable.

It is not easy to calculate the minimum energy required for a given reduction process, but some theories have been advanced that are useful. These theories depend upon the basic assumption that the energy required to produce a change dL in a particle of a typical size dimension L is a simple power function of L:

$$dE/dL = KL^n, \tag{4.1}$$

where dE is the differential energy required, dL is the change in a typical dimension, L is the magnitude of a typical length dimension, while K and n are constants.

The ratio of final size and initial size of the particles is called the size reduction ratio (Earle and Earle, 2004).

4.2.2 EQUIPMENT FOR SIZE REDUCTION OF SOLIDS

Grinding equipment can be divided into two classes – crushers and grinders. In the first class the major action is compressive, whereas grinders combine shear and impact with compressive forces. In general, mills are used as grinders. If the feed material is wet the process is called wet milling (Brennan et al., 1990a). Power consumption is generally high with wet grinding.

Friable and crystalline materials may fracture easily along cleavage planes. Fibers tend to increase toughness by relieving stress concentrations at the ends of the cracks. Disc mills, pin-disc mills, or cutting devices are used to break down fibrous materials. If the moisture content of the feed is too high, the efficiency of a mill may be adversely affected.

A considerable amount of heat may be generated in a mill, particularly if it operates at high speed. This heat can cause the temperature of the feed to rise significantly and a loss in quality could result. As a consequence, some mills are equipped with cooling jackets to reduce these effects.

Cryogenic milling involves mixing solid carbon dioxide or liquid nitrogen with the feed. This reduces undesirable heating effects and facilitates the milling of fibrous materials, such as plant tissues, into fine particles (Earle and Earle, 2004).

The hammer mill is an impact mill, which is used for hard, friable, fibrous, and sticky materials. A rotor mounted on a horizontal shaft turns at high speed inside a casing. The rotor carries hammers that pass within a small clearance of the casing (Fig. 4.1a). In the comminuting mill, knives replace the hammers or bars. Such mills are used for comminuting relatively soft materials, such as fruit and vegetable matter.

In the single-disc attrition mill, a disc rotates in close proximity to a stationary disc with matching grooves. The feed is introduced through the center of the stationary disc and makes its way outwards between the discs and is discharged from the mill via a screen (Fig. 4.1b). In the case of the double-disc attrition mill, two counter-rotating discs are located close to each other in a casing. In both cases, shear forces are the principal cause of the breakdown of the material. In general, these mills are used for milling fibrous materials such as corn and rice. The colloidal mill is an attrition mill, which is used for the preparation of pastes and purées (Brennan, 2006a; Earle and Earle, 2004).

4.2.3 WET MILLING APPLICATIONS

Size reduction of food wastes is necessary previous to solvent extraction and drying to increase surface area and contact and thus increase the yield of the following stages (Oreopoulou and Tzia, 2007;

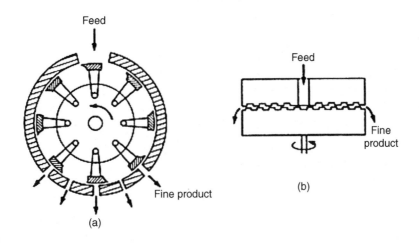

FIGURE 4.1

(a) Hammer mill and (b) attrition mill.

Reprinted with permission from Earle and Earle (2004)

Ayala-Zavala et al., 2011). Since the majority of food wastes contain an aqueous phase, this process is accomplished through a wet milling step. The latter is usually helpful for the preparation of fiber-rich fruit by-products, as it is able to minimize the losses of associated bioactive compounds (i.e. flavonoids, polyphenols, carotenes), which may exert higher health promoting effects than the dietary fiber itself (Larrauri, 1999). Too small particle size in the fresh raw material is not appropriate because a high amount of water can be held during the washing step, which in turn is detrimental for the drying process. Losses in the milled raw material producing lower yields during the separation of water may also occur. On the other hand, excessively high particle sizes do not facilitate the removal of the undesired components (such as sugars) during the washing step and, because of this, a longer drying time might be also needed. Hammer mills with a variety of screen sizes are preferred to colloidal mills in order to obtain a good control of particle size.

4.3 THERMAL AND VACUUM CONCENTRATION
4.3.1 GENERAL

The depression of water content in foods is typically used for food preservation. When removal of water is only partial and a concentrated solution, dispersion, or semisolid product with water contents in excess of 20% is obtained, the process is called a concentration process (Karel and Lund, 2003). In the frame of macroscopic pretreatment, the purpose of concentration is mainly to increase the content of target compounds within the initial substrate. For food waste concentration, the partial removal of the solvent can be performed by vaporization. Vacuum operations increase the cost of the process due to the need of low pressure maintenance, but they are necessary if heat-sensitive compounds are present in wastes.

4.3.2 EVAPORATION

Vaporization is typically conducted by boiling off the solvent, which means that the procedure occurs at the boiling point of the solution or dispersion. This process involves simultaneous heat and mass transfer to vaporize the solvent and remove vapor.

The combination of mass an energy balances allows calculation of the amount of steam required per pound of water removed for a given degree of concentration and a given feed temperature.

If it is supposed that the feed acquires instantaneously the temperature of the boiling point and that the product leaves the evaporator at this same temperature (T_2), the driving force for heat transfer is the temperature difference between the steam (T_1) and the product (T_2). The size of the exchanger capable of supporting the necessary overall rate of heat transfer can be calculated from the following equation:

$$q = UA(T_1 - T_2),\tag{4.2}$$

where U is the overall heat transfer coefficient, A is heat transfer area, and q is overall heat transfer rate.

The coefficient U is equal to $1/R$, with R being the total resistance to heat transfer, which means:

$$U = 1/R = 1/h_1 + r_w + 1/h_2,\tag{4.3}$$

where h_1 is the heat transfer coefficient in steam phase, r_w is the resistance of heat exchanger wall, and h_2 is the heat transfer coefficient in the liquid phase. For well-designed evaporators, U can be considered equal to h_2, because the overall resistance R is dominated by the resistance in the boiling liquid (Karel and Lund, 2003).

When a food-related material is concentrated, its viscosity increases and this can affect heat transfer and pumping requirements. The formation of foams due to the presence of surface active compounds and protein and polysaccharide deposition can reduce heat transfer efficiency.

Evaporators can be classified according to their operating pressure (atmospheric or vacuum), type of operation (batch or continuous), number of effects (single or multiple), type of convection (natural or forced), and type of design (plates or tubular).

The more common types of evaporators include: batch pans, plate evaporators, long tube evaporators, short tube evaporators, forced circulation evaporators, and evaporators for heat-sensitive liquids.

The long tube evaporators consist of tall slender vertical tubes, which may have a length to diameter ratio of the order of 100:1.These tubes pass vertically upward inside the steam chest (Fig. 4.2). The liquid may either pass down through the tubes (falling-film evaporator) or be carried up by the evaporating liquor (climbing-film evaporator). Evaporation occurs on the walls of the tubes. Because circulation rates are high and the surface films are thin, good conditions are obtained for the concentration of heat-sensitive liquids due to high heat transfer rates and short heating times. Generally, if sufficient evaporation does not occur in one pass, the liquid is fed to another pass. In the climbing-film evaporator, as the liquid boils on the inside of the tube, vapor carries up the remaining liquid, which continues to boil (Earle and Earle, 2004).

The plate-type evaporators were developed as an alternative to the traditional tubular system design. They offer accessibility to the heat transfer surfaces and capacity can be increased by adding more plate units (Fig. 4.3). The reduction in residence time with capacity increase results in a higher quality concentrate. This kind of evaporator has a more compact design and low installation cost. Different arrangements exist such as rising, falling, or rising/falling film. These evaporators are available as

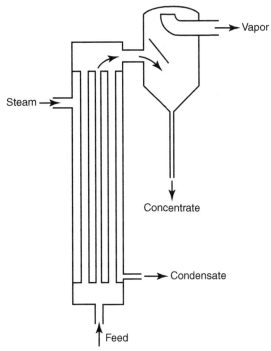

FIGURE 4.2 Long Tube Evaporator

Reprinted with permission from Earle and Earle (2004)

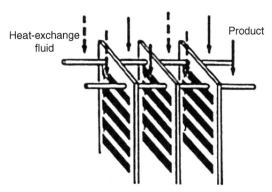

FIGURE 4.3 Plate Evaporator

Reprinted with permission from Earle and Earle (2004)

multieffect and/or multistage systems to allow relatively high concentration ratios. The falling-film plate has shorter residence time and larger evaporation capabilities, and is gaining wide acceptance for the concentration of heat-sensitive products (APV, 2008).

The falling-film evaporator, either plate or tubular, provides the highest heat transfer coefficients. It is usually the most economic device, but is not suitable for the evaporation of products with viscosities over 300 cP or if the products foul heavily.

The forced circulation evaporators can be operated up to viscosities of over 5000 cP and will significantly reduce fouling but both capital and operating costs are high (APV, 2008).

4.3.3 EVAPORATION OF HEAT-SENSITIVE LIQUIDS

When the vapor pressure of the liquid reaches the pressure of its surroundings, the liquid boils promoting concentration. But flavors can be lost and other compounds can be damaged due to the application of heat. Galanakis et al. (2010a, b, c, d, e) studied the recovery of phenols from fresh olive mill wastewater (OMW) through the extraction mediated by ethanol and evaluated phenol content and antioxidant capacity of the extracts. They observed that a preheating step of oil mill wastewater at 50–60°C as well as at 80°C resulted in reduction of the phenol concentrations and antioxidant activities of the extracts, probably due to the generation of enzymatic and nonenzymatic reactions, respectively. Also, heat processing of OMW at 60°C for 180 min can activate endogenous pectin methyl esterase that promotes demethylation resulting in loss of gelling functionality (Galanakis et al., 2010e).

For the evaporation of liquid streams that are adversely affected by high temperatures, it may be necessary to reduce the boiling temperature. The latter is accomplished by operating under vacuum conditions. The reduced pressures are obtained by mechanical or steam jet ejector vacuum pumps, combined generally with condensers for the vapors from the evaporator. Mechanical vacuum pumps are generally cheaper in running costs but more expensive in terms of capital compared with steam jet ejectors. The condensed liquid can either be pumped from the system or discharged through a tall barometric column in which a static column of liquid balances the atmospheric pressure. Vacuum pumps are used to discharge the noncondensable phase to the atmosphere. On the other hand, short contact times with the hot surfaces can also preserve the quality of heat-sensitive streams. For solutions of low viscosity, climbing- and falling-film evaporators, tubular, or plate types can be used. When the viscosity increases at higher concentrations, mechanically scraped surfaces or the flow of the solutions over heated spinning surfaces can help to reduce contact times (Earle and Earle, 2004).

4.3.4 STEAM ECONOMY

In a single-stage evaporator, the vapor generated can be used for preheating the entering feed, thus reducing the amount of steam required for evaporation.

In the case of multiple effect evaporators, the energy contained in the vapor discharged from one unit can be partially recovered by using the vapor as a heat source to another unit operating at a lower pressure (Karel and Lund, 2003).

The thermovapor recompression (TVR) can also be used for steam economy. In this case, a portion of the steam evaporated from the product is recompressed by a steam jet venturi and returned to the steam chest of the evaporator. The mechanical vapor recompression takes the vapor that has been evaporated from the product, compresses the vapor mechanically, and then uses the higher pressure vapor in the steam chest. Vapor compression is carried out by a radial-type fan or a compressor (APV, 2008).

4.3.5 APPLICATION OF EVAPORATION

GEA (2014) reports that the use of a multiple turbofan, mechanical vapor recompressor, heated falling-film pre-evaporator acting as a single effect system coupled to a single effect, forced circulation, thermal vapor recompressor, heated high concentrator is adequate for the recovery of compounds from a soy processing effluent (~2% w/w total solids) with a high discharge volume (>1660 gallons/min) and holding. The condensate produced from the evaporator can be recycled as clean water into the upstream processes.

Whey is a by-product derived from the dairy industry, and particularly cheese production. Whey products are today increasingly used in the food industry due to the nutritional and functional value of their components. Whey processing involves a concentration and drying process. The concentration stage traditionally takes place under vacuum in a falling-film evaporator with two or more stages. Evaporators with up to seven stages have been used since the mid-1970s to compensate whey streams, but the main disadvantage is the increasing energy cost demands. For this purpose, mechanical and thermal vapor compression has been introduced in most evaporators to reduce evaporation costs. Evaporation of water continues typically down to 45–65% total solids, and further water removal is typically performed with spray drying (Tetra Pak, 2003).

4.4 MECHANICAL SEPARATION (CENTRIFUGATION/MECHANICAL EXPRESSION)

4.4.1 CENTRIFUGATION

4.4.1.1 Centrifuge principle and equipment

The separation of two immiscible liquids (or of a liquid and a solid) by sedimentation can be accomplished through the application of centrifugal forces, allowing food wastes to be prepared for valuable compound isolation.

The centrifugation principle is based on Stoke's law in which the velocity of particles (v) under a centrifugal field is expressed as:

$$v = \frac{\left(\rho_p - \rho_m\right)d_p^2}{18\mu}\omega^2 r, \tag{4.4}$$

where v is the velocity of the target particle (also called sedimentation rate), d_p is the diameter of the particle, ρ_p is the target compound density, ρ_m is the density of surrounding medium, μ is the viscosity of the medium, ω is the angular velocity, and r is the distance from particle to the center of centrifugation. Particle velocity depends on the angular rotor velocity, relative density, particle size, and viscosity of the medium (Saravacos and Kostaropoulos, 2002). For the continuous centrifuge systems, feeding flow rate also impacts the separated particle size settling in the sediment (Fellows, 2000).

At the laboratory setting, bench-top cone centrifuges are popular among scientists. At the industrial scale, continuous centrifuges are used more frequently. Specifically, decanter centrifuge is suitable to be implemented at the pretreatment stage because of its capability of handling high solid content (3–60% w/w) and big particle size (5–50,000 μm), while other types of centrifuges such as disc bowl centrifuge (0.5–500 μm particle size and less than 10% solid content) are popular in the later stage of the process in order to further clarify liquid (Fellows, 2000).

Table 4.1 Centrifugation Used in Macroscopic Pretreatment of Agro-By-Product Recovery

Centrifuge Types and Models	Waste	Purpose	Centrifugation Condition	References
Bench-top; batch; cone	Olive mill wastewater	Remove fat	3,000g, 20°C, 30 min; fat was eliminated on the top	Galanakis et al. (2010c)
Bench-top; batch; cone	Beef lung	Obtain alkaline soluble crude protein	10,000g, 15 min; supernatant was stored and further subjected to acid precipitation	Darine et al. (2010)
Bench-top; batch; cone	Herring processed waste (heads and gonads)	Optimize the recovery of soluble proteins	2,560g, 15 min after heating; three layers were obtained including crude lipids (upper), soluble hydrophilic protein (middle), and solid residues (bones, skins, insoluble proteins)	Sathivel et al. (2004)
Bench-top; batch; cone	Shrimp processing waste	Recover protein, chitin, carotenoids, and glycosaminoglycans from shrimp processing waste	10,000g, 4°C, 10 min; the liquid was obtained for further extraction of glycosaminoglycans and carotenoids and the solid for chitin and chitosan	Cahú et al. (2012)
Bench-top, batch, cone	Citrus wastewater	Separation to three phases and production of high fructose citrus syrup	100 g sample and 6,000 rpm; 1–5% of total volume was upper layer containing limonene (discarded); the middle and the bottom layers were retained	Guerry-Kopecko et al. (1985)
Decanter	Fish by-product	Separation of fish meal, stickwater, and oil	Centrifuge capacity ranges from 12 to 300 tons per day	Dep (1986)

Most of the experiments in bench-top scale use batch cone centrifuges, whereas continuous systems are used at the industry scale (Table 4.1).

The primary purposes of centrifuges in the pretreatment stage include liquid/solid, liquid/liquid, and liquid/liquid/solid separations. However, only limited examples are available for the pretreatment of agro-by-products at the industrial production level. Nevertheless, some of the lab scale could be enlarged to industrial production by using decanter in the facility. For instance, decanter could be introduced to suspend oil from the OMW mentioned in Table 4.1. Alternatively, mixture solids can be removed mechanically by pressing.

4.4.2 MECHANICAL EXPRESSION

Mechanical expression or pressing is one of the more often used processes for the pretreatment of agro-by-products, but is also widely used for juice recovery and oil expelling. Mechanical expression acts by applying pressure on compressible solids in order to release liquid from disrupted cells of the matrix. Its efficiency is associated with viscosity of outlet liquid, yield stress of loaded material, applied pressure, temperature, loaded sample particle size, and loaded thickness of food solids (Brennan et al., 1990b; Bargale et al., 1999).

Table 4.2 Presses Used in Macroscopic Pretreatment Steps of Agro-By-Product Recovery

Press	Waste	Purpose	Pressing Condition Yield	Yield	References
Batch, bench-top, friction screw press	Onion waste	Separation of roots into juice, paste, and solid for characterization purpose	Moist onion waste was fed into the press; other conditions were not specified	N/A	Roldán et al. (2008)
Continuous, bench-top, single screw press/ expeller, KOMET CA 59 G, 5–8 kg input/h	Grape seed	Expression of oil in the cold press manner	Grape seeds were reduced to 10–15% moisture content and less than 0.5 mm particle size before expression	8.8–9.1% of oil yield	Fernández et al. (2010)
Continuous, pilot scale and small production, screw press/expeller, "Hander" S-52, 30–40 kg input/h	Wine grape seeds	Expelling seed oil	Three varieties of grape seeds were separated from the marc, dried to lower than 5% and 30 kg of each kind was pressed	4.2–9.1% of oil yield	Jordan (2002)
Continuous, industrial scale, twin-screw press	Precooked fish by-product	Separation of solid from liquid	Fish by-product is precooked before being transferred to the press	About 50% moisture content in press cake	Dep (1986)

Pressing can be categorized in batch (cage presses and box presses) and continuous presses (screw presses, roller presses, and belt presses) (Saravacos and Kostaropoulos, 2002). Screw presses are the most prevalent types in food processing facilities especially in recovering agro-by-products. The basic design of the screw press contains a tapered screw(s) and a curb, which generate pressure from sample inlet to press cake outlet. At the same time, liquid is expelled through the screen. Screw presses are also manufactured in different capacities to suit production needs, from bench top to industrial scale.

Compared with centrifugation, pressing is used more often during the by-product pretreatment stage of agro-by-products since it generates solids with less moisture content and fewer emulsified liquids (Dep, 1986). Table 4.2 illustrates the presses used for the pretreatment of agro-by-products. Among the different cases, three categories are of particular interest:

1. *Juice and winery pomace*: The pressing step in the fruit juice industry requires gentle handling in order to avoid the release of bitter compounds. Thereby, the resulting waste contains high content of phytochemicals, and low sugar content, making it ideal for the recovery of functional ingredients. After collecting the juice, water is added in the pomace and a wet milling process follows in order to reduce the particle size, which enables a higher contact surface for further extraction. The press is introduced after the wet milling stage (Oreopoulou and Tzia, 2007). The friction screw press was reported to be used to separate onion waste roots into juice, paste, and solid (Roldán et al., 2008). A similar manner could be applied to different wastes. Waste from the citrus canning industry contains 85–90% moisture content and the rest is composed of

peels, pulps, rags, seeds, etc. The solid waste is dried for further utilization as animal feed. A tremendous amount of wastewater is pressed out in the early stage and the press cakes undergo the typical drying process. Meanwhile, the citrus wastewater is used to produce essential oil or concentrate (Pulley, 1949).

2. *Animal processed waste*: The processed fish waste is precooked until fish protein coagulation occurs, which results in the expelling of oil and moisture from the solid. After cooking, the mixture is subjected to a continuous twin-screw press for further separation of solid (insoluble protein and bones) and liquid suspensions (soluble protein, oil, and water) (Dep, 1986).

3. *Oilseed*: In an example of the cold press oil scheme, described by Jariené et al. (2008), seeds from processed by-products, such as apricot stones, bilberry, pumpkin seed, avocado, cotton seed, raspberry, elderberry, and blackcurrant, were fed into the screw press after cleaning, whereas the output suspension of water and crude oil was further refined by centrifugation or filtration. Before going into the press for oil production, it is necessary to adjust the moisture content of any waste material. For example, Fernández et al. (2010) used 10% moisture content and size-reduced grape seeds to produce cold press oil using a bench-top scale unit. Apricot kernel was also studied to extract oil by table oil expeller (Gupta et al., 2012). The pressing step is typically followed by solvent extraction and enzymatic treatment to increase the yield of oil (Dominguez et al., 1994; Bhosle and Subramanian, 2005).

4.5 FREEZE DRYING

4.5.1 FREEZE DRYING PRINCIPLES AND EQUIPMENT

Freeze drying (or so-called lyophilization) can remove liquid from substances by a sublimation process during which liquid escapes from a substance by changing from solid phase to gas phase (Fig. 4.4). Three stages of drying are involved: (i) freezing samples below the triple point of water, (ii) primary drying to sublimate water from ice stage to gas stage, and (iii) secondary drying to dry the rest of the unfrozen water. At the freezing step, food (solid or aqueous) is first placed in the containers and frozen quickly at a low temperature. The low freezing temperature in combination with the precooled containers could promote the fast formation of smaller ice crystals, which minimizes the damage of cell structure in the sample. In order to produce high quality freeze-dried products, at least 95% of water transformation into ice is required. During the primary drying stage, dryer pressure is taken lower than the ice vapor pressure and subsequently ice sublimates to gas (Fig. 4.4). Pressure and temperature controls are extremely important at this stage, therefore, a vacuum pump is an essential component of the freeze dryer. Thereafter, the gas is trapped in a condenser in order to remove moisture from the chambers. This stage takes the longest and drying rate depends on many variables such as freeze dryer performance (vacuum pump and condenser), pressure, loaded sample thickness, and physicochemical properties of the sample. The secondary stage reduces further the noncrystallized water by involving thermal drying (Brennan, 2006b).

A manifold freeze dryer is typically used for laboratory applications. It has flasks attached to the dryer port and the sublimation heat is provided by the surrounding environment (LABCONCO, 2004). Batch, tunnel (semicontinuous), continuous, and vacuum spray dryers have been used for different industrial applications, such as meat, coffee powder, juice, microencapsulation, etc. (Brennan, 2006b; Heinzelmann and Franke, 1999).

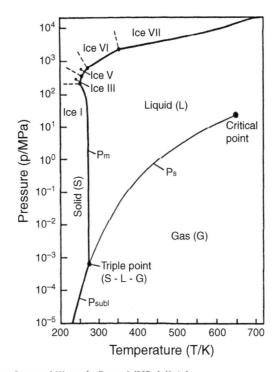

FIGURE 4.4 Phase-Boundary Curve of Water in Pascal (MPa)-Kelvin

Reprinted with permission from Wagner et al. (1994). Copyright © 1994, AIP Publishing LLC

4.5.2 FREEZE DRYING APPLICATIONS

Due to its high operational cost and the potential of microbial hazard, freeze drying technology is not appreciably implemented for the industrial pretreatment of by-products. Only a few references have reported the application of freeze drying technology for the early stage treatment of lemon by-products and goldenberry pomace at the laboratory scale (Masmoudi et al., 2008; Ramadan et al., 2008). However, there are a couple of reasons for freeze drying application.

First, freeze drying can reduce moisture content in the preparation of raw materials and at the same time reserve the heat-sensitive functional compounds. Second, freeze drying has been used as a referenced drying process. Even though it is economically unacceptable, it is well accepted as the best drying technology so far, due to its minimum impact on sensitive compounds, colors, and textures of food-related materials (Ratti, 2001; Dennis et al., 2009). Therefore, a lot of studies compare different technologies with freeze drying in order to find the appropriate drying approaches that retain the high amounts of functional compounds as those in freeze-dried products, while still being economically doable. For instance, it has been shown that freeze drying has no advantage over oven drying of brewers' spent grains. Even though retaining less arabinose than freeze drying, oven drying at 60°C preserved the same amount of xylose and hydroxycinnamic acids in the brewers' spent grain as the freeze drying

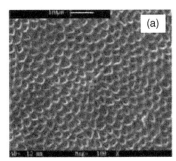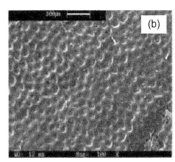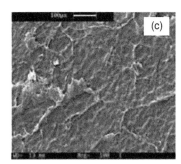

FIGURE 4.5 Tomato Peel Under Scanning Electron Microscope (100× magnification)

(a) Fresh peel; (b) freeze-dried peel; (c) cabinet-dried peel.

Reprinted with permission from Sarkar and Kaul (2014). Copyright © 2014, John Wiley and Sons Inc.

process, thus oven drying is more appropriate for drying brewers' spent grain (Bartolomé et al., 2002). Besides, low temperature air and vacuum drying of pomegranate peels (at 40°C and 60°C) had comparably high extractable polyphenols compared with freeze drying (Al-Rawahi et al., 2013). On the other hand, peel and seed meal from tomato processed waste were experimentally dried under a cabinet dryer (70°C for 8 h) and freeze dryer (-90°C and 40 mPa), separately. Freeze-dried tomato peel had a similar amount of lycopene to the fresh tomato peel, while the lycopene in cabinet-dried tomato peel was reduced by 26%. The latter result could be explained by the small morphological change in the freeze-dried peel and the extensive cell shrinkage observed in cabinet-dried peel (Fig. 4.5). In this case, tomato peel was better processed by the freeze dryer because a high value of lycopene content could justify the higher processing cost. On the other hand, tomato seed could be dried optimally under a cabinet dryer, as the contained proteins are not so sensitive to heat (Sarkar and Kaul, 2014).

Wine grape pomace has also been studied under different drying techniques using freeze drying as a reference. For instance, freeze-dried red wine grape skins had high retentions of volatile compounds (terpenes, sesquiterpenes, norisoprenoids, C_6 alcohols) without increasing the amount of browning derivatives. Meanwhile, both freeze drying and oven drying decreased significantly anthocyanins and flavonols compared with fresh product, but oven-dried pomace showed lower content of polyphenolic compounds (de Torres et al., 2010). In conclusion, depending on the application and the targeted compounds, freeze drying can be an alternative for thermal drying if the operational cost and food safety concerns are taken into account.

Freeze drying is more frequently used in other stages of the Universal Recovery Process, such as column purification (Chapter 6) and product formation (Chapter 8).

4.6 MICROFILTRATION
4.6.1 MEMBRANE SEPARATION PRINCIPLES

Pressure-driven membrane operations, such as microfiltration (MF), ultrafiltration (UF), nanofiltration (NF), and reverse osmosis, have gained today special recognition in food processing offering potential sustainable solutions for the separation and purification of bioactive compounds from food wastes in the first, second, and third stages of the Universal Recovery Process (Daufin et al., 2001;

Li and Chase, 2010; Moresi and Lo Presti, 2003; Patsioura et al., 2011; Muro et al., 2012; Galanakis et al., 2010d, 2012, 2013, 2014, 2015, Galanakis, 2015). All these processes are based on the use of a permselective membrane through which fluids and solutes are selectively transported under a specific hydrostatic pressure applied between the two sides of the membrane. As a result the feed solution is converted in a permeate containing all components that have permeated the membrane and a retentate containing all compounds retained by the membrane.

Membrane separation technologies are widely used on industrial-scale applications due to their key advantages over conventional separation techniques (precipitation, centrifugation, extraction, ion-exchange, adsorption). The latter include: (i) improved product quality, (ii) higher yields, (iii) better process economy, (iv) fewer chemical additives, (v) cleaner and simpler process solution, and (vi) lower energy consumption. In addition, the separation is performed at ambient temperature thereby reducing the thermal damage of thermosensitive compounds.

Among pressure-driven membrane operations MF most closely resembles conventional coarse filtration. Membranes used in MF are characterized by pore sizes in the range 0.05–1 μm and capable of retaining species with molecular weight greater than 200 kDa. The separation mechanism is based on a sieving effect and particles are separated according to their dimensions, although the separation is influenced by interactions between the membrane itself and particles being filtered (Petrus and Nijhuis, 1993). Materials for fabrication of commercial MF membranes include synthetic polymers (polypropylene, perfluoropolymers, polyamides, polysulfones, etc.), cellulose derivatives, ceramics, inorganics, and metals. Basic characteristics of the MF process are summarized in Table 4.3.

Table 4.3 Characteristics of the MF Process

Concept		
Membrane type	Structure	Symmetric or asymmetric
	Thickness	10–150 μm
	Pore sizes	0.05–10 μm
	Material	Polymeric, ceramic, inorganic, metal
Driving force	Pressure difference	1–5 bar
Separation principle	Sieving mechanisms	
Industrial applications	Water treatment	Wastewater treatment
	Food industry	Dairy and milk processing, fruit juice processing, wine and beer clarification, color and particle removal in sugar industry, preparation of emulsions
	Biotechnology and pharmaceutical	Downstream processing, pharmaceuticals from cell cultures
	Biomedical	Hemodialysis, biohybrid organs, analytical and diagnostic devices

4.6.2 MICROFILTRATION APPLICATIONS

MF plays an important role in the pretreatment of food processing wastewaters for the reduction of the polluting load and/or for the recovery of valuable compounds. For example, the presence of oils and fats in spent wastewaters derived during margarine production creates several problems in their biological treatment. Problems include high costs for aeration and sludge disposal, flotation and coating in the treatment plant, and saponification of the fats in the equalization tank. Thereby, the treatment of these effluents with 0.2 μm ceramic MF membranes reduces the initial chemical oxygen demand of 5000–10,000 mg/L below 250 mg/L. The produced permeate can be mixed with waters of low and medium contamination and then submitted to a biological treatment. A concentrated product can also be recovered from the MF retentate and reused for soap production after a skimming oil treatment (Chmiel et al., 2003).

In the dairy industry, pressure-driven membrane operations are largely used due to their ability to separate milk components, increase cheese yields, and minimize volumes of dairy wastes. In particular, MF membranes can be used to remove bacteria and fat and produce protein isolates and concentrates, without affecting the functional properties of proteins due to the low applied temperatures (Lipnizki, 2010).

Whey protein concentrates are widely used in the food industry in a large variety of formulated products such as dairy, meat, beverage, bakery, and infant formula due to their excellent functional properties (Morr and Ha, 1993).

MF with an Alfa-Laval MFS-7 fitted with Ceraver ceramic membranes of 1.4 and 0.8 μm porosities can remove 80% of residual lipids from Cheddar cheese whey. The obtained whey protein concentrates showed significant improvements in foaming (but not gelation) properties in proportion to the decrease in lipid content (Pearce et al., 1992).

The removal of residual lipids from whey by MF membranes involves the heat treatment and/or pH adjustment of whey in order to aggregate phospholipids by calcium binding and to allow their precipitation (thermocalcic precipitation). The resulted precipitate can be separated from whey by using MF membranes with a pore size of 0.14 μm (Fauquant et al., 1985; Gesan et al., 1995).

MF can also be used to isolate immunoglobulin (IgG) from colostrum whey. Piot et al. (2004) found that the treatment of different types of colostrum (bovine, equine, or goat) with 0.1 μm – MF membranes allows at least 80% of IgG in the permeate to be recovered. The microfiltered liquid, named serocolostrum, was free of fat globules and casein micelles.

Wastewaters generated in fish meal production contain a large amount of potentially valuable proteins. The combination of MF and UF membranes can be exploited to recycle proteins in the fish meal process with significant benefits in terms of pollution control and recovery of valuable raw materials and water. In the approach proposed by Afonso and Bórquez (2002), wastewaters produced in a fish meal factory were pretreated with MF membranes having a pore size of 5–10 μm. MF reduced drastically the oil and grease content and the suspended matter of the raw effluent. The following UF step performed with a ceramic membrane (Carbosep M2, tubular, 15 kDa molecular weight cut-off (MWCO)) reduced the organic load of the MF permeate allowing the recovery of valuable raw materials including proteins.

The utilization of membrane technologies for the separation, purification, and concentration of bioactive phenolic compounds from OMW is a field of growing interest, too (Galanakis, 2011; Rahmanian et al., 2014). In particular, pressure-driven membrane operations, mostly in a sequential form, have been recently investigated for the recovery of polyphenols from OMW with regard to their specific MWCO values (Cassano et al., 2013; Paraskeva et al., 2007; Takaç and Karakaya, 2009).

In these approaches MF systems represent an efficient tool to remove suspended solids and other impurities from the raw wastewaters with a production of a permeate stream enriched in the compounds of interest. Pizzichini and Russo (2005) proposed a fractionating process of OMWs allowing the recovery of polyphenolic compounds, the reuse of the concentrate residues for the production of fertilizers and biogas and the production of a purified aqueous solution of interest as a basic component for beverages. In this approach ceramic MF membranes with molecular size ranging between 0.1 μm and 1.4 μm operate on acidified (at pH 3–4.5 in order to prevent oxidation of polyphenols) and depectinized OMW producing an MF permeate, which is then treated by UF, NF, and RO membranes to produce concentrated phenolic fractions. The utilization of a diafiltration configuration is useful for further recovery of polyphenols in the retentate of the MF unit.

Table 4.4 MF Membrane Applications for the Treatment of Food Processing Wastewaters

Food Waste	Target Compounds	Membrane Type	Operating Conditions	References
Effluents of a fish meal plant	Proteins	Whatman filter No. 1, 5–10 μm	–	Afonso and Bórquez (2002)
Olive mill wastewaters	Polyphenols	Tubular ceramic membrane (Inopor GmbH), 0.2 μm	TMP, 0.72 bar; axial feed flow rate, 760 L/h; T, 22°C	Garcia-Castello et al. (2010)
Olive mill wastewaters	Polyphenols	Sunflower-shaped ceramic membrane (Tami), 0.1–1.4 μm	TMP, 1.7 bar; T, 25–30°C	Pizzichini and Russo (2005)
Olive mill wastewaters	Polyphenols	Ceramic membranes (Tami), 0.45 μm	TMP, 1.5 bar; T, 20–25°C; feed flow rate, 4000 L/h; VCR 3	Russo (2007)
Colostrum whey	Immunoglobulins	Membralox membrane (Pall Exekia), 0.1 μm	T <40°C; TMP, 0.4–0.5 bar; continuous diafiltration	Piot et al. (2004)
Wastewaters from margarine manufacture	Oils and fats	Ceramic membranes, 0.2 μm	TMP, 0.8 bar; VCR, 15	Chmiel et al. (2003)
Winery wastewaters	Oligomeric proanthocyanidins	FP200, tubular PVDF (PCI membrane systems)	TMP, 0.2 bar; diafiltration mode	Santamaria et al. (2002)
Whey	Phospholipids	M14 Carbosep (Techsep), 0.14 μm, composite membrane (ZrO_2/TiO_2 filtering layer on a carbon support)	TMP, 2.1 bar; stationary VCR, 5.1; T, 50°C	Gesan et al. (1995)
Whey	Whey protein concentrates	Ceramic membranes, 1.4 and 0.8 μm	–	Pearce et al. (1992)
Whey	Whey protein concentrates	Hollow fiber membranes (A/G™ Technology Corporation), 0.2 and 0.65 μm	TMP, 1.6–1.8 bar; feed flow rate, 2000 L/h; VCR, 5	Pereira et al. (2002)

TMP, transmembrane pressure; VCR, volume concentration ratio.

Winery wastewaters contain important amounts of biodegradable organic compounds together with small concentrations of valuable compounds such as phenolic compounds, sugars, organic acids, and nutrients. Santamaria et al. (2002) evaluated the performance of an integrated membrane process for the fractionation of polyphenolic extracts from winery wastes with the aim of obtaining proanthocyanidin fractions with different degrees of polymerization. A sequence of four different membrane operations based on the use of NF (rejection $CaCl_2$ 60%), UF (20 kDa), MF (200 kDa), and UF (8 kDa) membranes was implemented on a pilot-plant scale. In this process, the MF membrane was used in the fourth stage of the Universal Recovery Process to obtain a purified solution of oligomeric proanthocyanidins as permeate stream starting from a UF retentate. Table 4.4 summarizes some selected MF applications for the recovery of valuable compounds from food processing wastewaters.

REFERENCES

Afonso, M.D., Bórquez, R., 2002. Review of treatment of seafood processing wastewaters and recovery of proteins therein by membrane separation processes – prospects of the ultrafiltration of wastewaters from the fish meal industry. Desalination 142, 29–45.

Al-Rawahi, A.S., Rahman, M.S., Guizani, N., Essa, M.M., 2013. Chemical composition, water sorption isotherm, and phenolic contents in fresh and dried pomegranate peels. Dry. Technol. 31, 257–263.

APV, 2008. Evaporator Handbook, fourth ed. APV Americas, Engineered Systems Separation Technology, New York, USA. http://www. apv.com (accessed 24.05.2014.).

Ayala-Zavala, J.F., Vega-Vega, V., Rosas-Domínguez, C., Palafox-Carlos, H., Villa-Rodriguez, J.A., Siddiqui, M.W., Dávila-Aviña, J.E., González-Aguilar, G.A., 2011. Agro-industrial potential of exotic fruit by-products as a source of food additives. Food Res. Int. 44, 1866–1874.

Bargale, P.C., Ford, R.J., Sosulski, F.W., Wulfsohn, D., Irudayaraj, J., 1999. Mechanical oil expression from extruded soybean samples. J. Am. Oil Chem. Soc. 76, 223–229.

Bartolomé, B., Santos, M., Jiménez, J.J., Del Nozal, M.J., Gómez-Cordovés, C., 2002. Pentoses and hydroxycinnamic acids in brewer's spent grain. J. Cereal Sci. 36, 51–58.

Bhosle, B.M., Subramanian, R., 2005. New approaches in deacidification of edible oils – a review. J. Food Eng. 69, 481–494.

Brennan, J.G., 2006a. Mixing, emulsification and size reduction. In: Brennan, J.G. (Ed.), Food Processing Handbook. Wiley, Weinheim, pp. 513–558.

Brennan, J.G., 2006b. Evaporation and dehydration. In: Brennan, J.G. (Ed.), Food Processing Handbook. Wiley, Weinheim, pp. 71–124.

Brennan, J.G., Butters, J.R., Cowell, N.D., Lilly, A.E.V., 1990a. Size reduction and screening of solids. In: Brennan, J.G., Butters, J.R., Cowell, N.D., Lilly, A.E.V. (Eds.), Food Engineering Operations. second ed. Elsevier Applied Science, London, pp. 67–90.

Brennan, J.G., Butters, J.R., Cowell, N.D., Lilly, A.E.V., 1990b. Solid-liquid extraction and expression. In: Brennan, J.G., Butters, J.R., Cowell, N.D., Lilly, A.E.V. (Eds.), Food Engineering Operations. second ed. Elsevier Applied Science, London, pp. 199–236.

Cahú, T.B., Santos, S.D., Mendes, A., Córdula, C.R., Chavante, S.F., Carvalho, Jr., L.B., Nader, H.B., Bezerra, R.S., 2012. Recovery of protein, chitin, carotenoids and glycosaminoglycans from Pacific white shrimp (Litopenaeus vannamei) processing waste. Process Biochem. 47, 570–577.

Cassano, A., Conidi, C., Giorno, L., Drioli, E., 2013. Fractionation of olive mill wastewaters by membrane separation techniques. J. Hazard. Mater., 248–249, 185-193.

Chmiel, H., Kaschek, M., Blöcher, C., Noronha, M., Mavrov, V., 2003. Concepts for the treatment of spent process water in the food and beverage industries. Desalination 152, 307–314.

Darine, S., Christophe, V., Gholamreza, D., 2010. Production and functional properties of beef lung protein concentrates. Meat Sci. 84, 315–322.

Daufin, G., Escudier, J.P., Carrère, H., Bérot, S., Fillaudeau, L., Decloux, M., 2001. Recent and emerging applications of membrane processes in the food and dairy industry. Trans. IChemE 79, 89–102.

de Torres, C., Díaz-Maroto, M.C., Hermosín-Gutiérrez, I., Pérez-Coello, M.S., 2010. Effect of freeze-drying and oven-drying on volatiles and phenolics composition of grape skin. Anal. Chim. Acta 660, 177–182.

Dennis, C., Aguilera, J.M., Satin, M., 2009. Technologies shaping the future. In: Da Silva, C., Baker, D., Shepherd, A.W., Jenane, C., Miranda-da-Cruz, S. (Eds.), Agro-Industries for Development. CAB International, Wallingford, UK, pp. 92–135.

Dep, F.F., 1986. The Production of Fish Meal and Oil. Food and Agriculture Organization of the United Nations, Rome.

Dominguez, H., Nunez, M.J., Lema, J.M., 1994. Enzymatic pretreatment to enhance oil extraction from fruits and oilseeds: a review. Food Chem. 49, 271–286.

Earle, R.L., Earle M.D., 2004. Unit operations in food processing, web ed. In: The New Zealand Institute of Food Science & Technology Inc. Palmerston North, New Zealand. http://www.nzifst.org.nz/unitoperations. (accessed 24.05.2014).

Fauquant, J., Vieco, E., Brule, G., Maubois, J.L., 1985. Clarification of sweet cheese whey by thermocalcic aggregation of residual fat. Lait 65, 1–20.

Fellows, P., 2000. Separation and concentration of food components. In: Fellows, P. (Ed.), Food Processing Technology Principle and Practice, second ed. CRC Press, Cambridge, pp. 140–169.

Fernández, C.M., Ramos, M.J., Pérez, Á., Rodríguez, J.F., 2010. Production of biodiesel from winery waste: extraction, refining and transesterification of grape seed oil. Bioresource Technol. 101, 7030–7035.

Galanakis, C.M., 2011. Olive fruit and dietary fibers: components, recovery and applications. Trends Food Sci. Technol. 22, 175–184.

Galanakis, C., 2012. Recovery of high added-value components from food wastes: conventional, emerging technologies and commercialized applications. Trends Food Sci. Technol. 26, 68–87.

Galanakis, C.M., 2013. Emerging technologies for the production of nutraceuticals from agricultural by-products: a viewpoint of opportunities and challenges. Food Bioprod. Process. 91, 575–579.

Galanakis, C.M., 2015. Separation of functional macromolecules and micromolecules: from ultrafiltration to the border of nanofiltration. Trends Food Sci. Technol. 42 (1), 44–63.

Galanakis, C.M., Tornberg, E., Gekas, V., 2010a. A study of the recovery of the dietary fibres from olive mill wastewater and the gelling ability of the soluble fibre fraction. LWT-Food Sci. Technol. 43, 1009–1017.

Galanakis, C.M., Tornberg, E., Gekas, V., 2010b. Clarification of high-added value products from olive mill wastewater. J. Food Eng. 99, 190–197.

Galanakis, C.M., Tornberg, E., Gekas, V., 2010c. Dietary fiber suspensions from olive mill wastewater as potential fat replacements in meatballs. LWT-Food Sci. Technol. 43, 1018–1025.

Galanakis, C., Tornberg, E., Gekas, V., 2010d. Recovery and preservation of phenols from olive waste in ethanolic extracts. J. Chem. Technol. Biotechnol. 85 (8), 1148–1155.

Galanakis, C., Tornberg, E., Gekas, V., 2010e. The effect of heat processing on the functional properties of pectin contained in olive mill wastewater. LWT-Food Sci. Technol. 43 (7), 1001–1008.

Galanakis, C.M., Fountoulis, G., Gekas, V., 2012. Nanofiltration of brackish groundwater by using a polypiperazine membrane. Desalination 286, 277–284.

Galanakis, C.M., Markouli, E., Gekas, V., 2013. Fractionation and recovery of different phenolic classes from winery sludge via membrane filtration. Separ. Purif. Technol. 107, 245–251.

Galanakis, C.M., Chasiotis, S., Botsaris, G., Gekas, V., 2014. Separation and recovery of proteins and sugars from Halloumi cheese whey. Food Res. Int. 65, 477–483.

Galanakis, C.M., Kotanidis, A., Dianellou, M., Gekas, V., 2015. Phenolic content and antioxidant capacity of Cypriot wines. Czech J. Food Sci. 33, 126–136.

Garcia-Castello, E., Cassano, A., Criscuoli, A., Conidi, C., Drioli, E., 2010. Recovery and concentration of polyphenols from olive mill wastewaters by integrated membrane system. Water Res. 44, 3883–3892.

GEA Process Engineering Inc., Columbia, MD, USA. http://www.niroinc.com/evaporators_crystallizers/evaporation_systems.asp (accessed 10.05.2014.).

Gesan, G., Daufin, G., Merin, U., Labbe, J.P., Quemerala, J., 1995. Microfiltration performance – physicochemical aspects of whey pretreatment. J. Dairy Res. 62, 269–279.

Guerry-Kopecko, P., Koeble-Smith, C., Milch, R.A., Sybert, E.M., 1985., Washington, DC, US Patent and Trademark Office. US Patent No. 4,547,226.

Gupta, A., Sharma, P.C., Tilakratne, B.M.K.S., Verma, A.K., 2012. Studies on physico-chemical characteristics and fatty acid composition of wild apricot (*Prunus armeniaca* Linn.) kernel oil. Indian J. Nat. Prod. Res. 3, 366–370.

Heinzelmann, K., Franke, K., 1999. Using freezing and drying techniques of emulsions for the microencapsulation of fish oil to improve oxidation stability. Colloid. Surface. B 12, 223–229.

Jarienė, E., Danilčenko, H., Aleknvičienė, P., Kulaitienė, J., 2008. Expression–extraction of pumpkin oil. In: Jarienė, E., Danilčenko, H., Aleknvičienė, P., Kulaitienė, J. (Eds.), Experiments in Unit Operations and Processing of Foods. Springer, New York, pp. 53–61.

Jordan, R., 2002. Grape Marc Utilisation – Cold Pressed Grapeseed Oil and Meal. Technical report for the Cooperative Research Centre for International Food Manufacture and Packaging Science, Melbourne, Australia.

Karel, M., Lund, D.B., 2003. Concentration. In: Karel, M., Lund, D.B. (Eds.), Principles of Food Preservation. CRC Press, Marcel Dekker, New York, pp. 330–377.

LABCONCO, 2004. A Guide to Freeze Drying for the Laboratory. An Industry Service Publication. Kansas, Missouri.

Larrauri, J.A., 1999. New approaches in the preparation of high dietary fibre powders from fruit by-products. Trends Food Sci. Technol. 10, 3–8.

Li, J., Chase, H.A., 2010. Applications of membrane technique for purification of natural products. Biotechnol. Lett. 32, 601–608.

Lipnizki, F., 2010. Cross-flow membrane application in the food industry. In: Peinemann, K.V., Pereira Nunes, S., Giorno, L. (Eds.), Membranes for Food Applications, vol. 3. Wiley-VCH, Weinheim, pp. 1–24.

Masmoudi, M., Besbes, S., Chaabouni, M., Robert, C., Paquot, M., Blecker, C., Attia, H., 2008. Optimization of pectin extraction from lemon by-product with acidified date juice using response surface methodology. Carbohyd. Polym. 74, 185–192.

Moresi, M., Lo Presti, S., 2003. Present and potential applications of membrane processing in the food industry. Italian J. Food Sci. 15, 3–34.

Morr, C.V., Ha, E.Y.W., 1993. Whey protein concentrates and isolates: processing and functional properties. Crit. Rev. Food Sci. Nutr. 33, 431–476.

Muro, C., Riera, F., del Carmen Diaz, M., 2012. Membrane separation process in wastewater treatment of food industry. In: Valdez, B. (Ed.), Food Industrial Processes. Methods and Equipment. InTech, Rijeka, pp. 253–280.

Oreopoulou, V., Tzia, C., 2007. Utilization of plant by-products for the recovery of proteins, dietary fibers, antioxidants, and colorants. In: Oreopoulou, V., Russ, W. (Eds.), Utilization of By-Products and Treatment of Waste in the Food Industry, vol. 3, Springer Science, New York, pp. 209–232.

Paraskeva, C.A., Papadakis, V.G., Tsarouchi, E., Kanellopoulou, D.G., Koutsoukos, P.G., 2007. Membrane processing for olive mill wastewater fractionation. Desalination 213, 218–229.

Patsioura, A., Galanakis, C.M., Gekas, V., 2011. Ultrafiltration optimization for the recovery of β-glucan from oat mill waste. J. Membr. Sci. 373, 53–63.

Pearce, R.J., Marshall, S.C., Dunkerley, J.A., 1992. Reduction of lipids in whey protein concentrates by microfiltration – effect on functional properties. In: de Boer, R., Jelen, P., Puhan, Z. (Eds.), New Applications of Membrane Processes, Special Issue No. 201. International Dairy Federation, Brussels, pp. 118–129.

Pereira, C.D., Díaz, O., Cobos, A., 2002. Valorization of by-products from ovine cheese manufacture: clarification by thermocalcic precipitation/microfiltration before ultrafiltration. Int. Dairy J. 12, 773–783.

Petrus, C.F., Nijhuis, H.H., 1993. Application of membrane technology to food processing. Trends Food Sci. Technol. 4, 277–282.

Piot, M., Fauquant, J., Madec, M.N., Maubois, J.L., 2004. Preparation of "serocolostrum" by membrane microfiltration. Lait 84, 333–342.

Pizzichini, M., Russo, C., 2005. Process for recovering the components of olive mill wastewater with membrane technologies. Int. Patent. WO 2005/123603.

Pulley, G.N., 1949. Washington, DC, US Patent and Trademark Office. US Patent No. 2,471,893.

Rahmanian, N., Jafari, S.M., Galanakis, C.M., 2014. Recovery and removal of phenolic compounds from olive mill wastewater. J. Am. Oil Chem. Soc. 91, 1–18.

Ramadan, M.F., Sitohy, M.Z., Moersel, J.T., 2008. Solvent and enzyme-aided aqueous extraction of goldenberry (*Physalis peruviana* L.) pomace oil: impact of processing on composition and quality of oil and meal. European Food Res. Technol. 226, 1445–1458.

Ratti, C., 2001. Hot air and freeze-drying of high-value foods: a review. J. Food Eng. 49, 311–319.

Roldán, E., Sanchez-Moreno, C., de Ancos, B., Cano, M.P., 2008. Characterisation of onion (*Allium cepa* L.) by-products as food ingredients with antioxidant and antibrowning properties. Food Chem. 108, 907–916.

Russo, C., 2007. A new membrane process for the selective fractionation and total recovery of polyphenols, water and organic substances from vegetation waters (VW). J. Membr. Sci. 288, 239–246.

Santamaria, B., Salazar, G., Beltrán, S., Cabezas, J.L., 2002. Membrane sequences for fractionation of polyphenolic extracts from defatted milled grape seeds. Desalination 148, 103–109.

Saravacos, G.D., Kostaropoulos, A.E., 2002. Mechanical separation equipment. In: Saravacos, G.D., Kostaropoulos, A.E. (Eds.), Handbook of Food Processing Equipment. Springer, New York, pp. 207–259.

Sarkar, A., Kaul, P., 2014. Evaluation of tomato processing by-products: a comparative study in a pilot scale setup. J. Food Process Eng. 37, 1–9.

Sathivel, S., Bechtel, P.J., Babbitt, J., Prinyawiwatkul, W., Negulescu, I.I., Reppond, K.D., 2004. Properties of protein powders from arrowtooth flounder (Atheresthes stomias) and herring (Clupea harengus) byproducts. J. Agr. Food Chem. 52, 5040–5046.

Takaç, S., Karakaya, A., 2009. Recovery of phenolic antioxidants from olive mill wastewater. Recent Patents Chem. Eng. 2, 230–237.

Tetra Pak, 2003. Whey processing. In: Grafiska, L.P., Lund, A.B. (Eds.), Tetra Pak Processing Systems AB. Dairy Processing Handbook, pp. 331–352.

Wagner, W., Saul, A., Pruss, A., 1994. International equations for the pressure along the melting and along the sublimation curve of ordinary water substance. J. Phys. Chem. Ref. Data 23, 515–527.

CONVENTIONAL MACRO- AND MICROMOLECULES SEPARATION

Chiranjib Bhattacharjee*, Arijit Nath*, Alfredo Cassano**, Reza Tahergorabi[†], Sudip Chakraborty[§]

*Department of Chemical Engineering, Jadavpur University, Kolkata, West Bengal, India; **Institute on Membrane Technology-Consiglio Nazionale delle Ricerche, University of Calabria, Rende Cosenza, Italy; [†]Department of Family and Consumer Sciences, Food and Nutritional Sciences, North Carolina Agricultural & Technical State University, Greensboro, North Carolina, USA; [§]Institute on Membrane Technology, University of Calabria, ITM-CNR, Rende, Italy

5.1 INTRODUCTION

The second step for the recapture of valuable compounds from food wastes is macro- and micromolecules separation. Agricultural substrates and food wastes are mixtures of different physical forms (i.e. liquid, solids) and thereby respective bioseparations concern different situations such as (Ghosh, 2006):

1. removal of particles from a liquid,
2. separation of several particles within a liquid medium,
3. separation of particles from liquid medium containing target solutes,
4. removal of solutes from a solvent,
5. separation of different solutes in a liquid medium,
6. liquid/liquid separation, etc.

Separation of macro- from micromolecules fits typically in the category of solutes removal from a solvent or solutes separation within a liquid medium, as particles and solids have been totally or partially removed during macroscopic pretreatment. Among the different implemented technologies, some specific ones (e.g. alcohol precipitation, ultrafiltration (UF), and isoelectric solubilization/precipitation) target the separation of larger molecules (e.g. proteins, oligosaccharides, polysaccharides, dietary fibers, hydrocolloids) from the smaller molecules (e.g. sugars, acids, polyphenols) (Galanakis, 2012; Deng et al., 2015; Roselló-Soto et al., 2015). Likewise, extrusion has been proposed for food waste valorization purposes. These technologies are described in detail in this chapter.

5.2 ETHANOL PRECIPITATION

5.2.1 GENERAL

Antisolvent precipitation is the most known method for the separation of smaller biomolecules (e.g. antioxidants, acids or ions, lactose, etc.) from a crude extract (Patel and Murthy, 2012; Heng et al., 2015). With this method, solutes are precipitated in the presence of an antisolvent agent (e.g. ethanol), following their polarity and solubility characteristics (Galanakis et al., 2010d; Galanakis et al., 2013a; Tsakona et al., 2012). Precipitated solutes may be in a crystalline or other form, whereas the nucleation characteristics of the crystalline products depend on the conditions of the process. Hence, product yield and form could vary according to the application of several operational parameters, i.e. doses of antisolvent agent, solute concentration, temperature, pH, and coprecipitated impurities (Zadow, 1984; Holsinger, 1999). Ethanol precipitation is broadly used because it is cheap, nontoxic, and easy to use. Nevertheless, it is neither selective nor able to separate complexes between the smaller (e.g. phenols) and the larger (e.g. proteins, pectin) molecules. This means that it cannot be used as a stand-alone technique for the recovery of valuable compounds from food wastes, but it could be perfectly adapted in *"5-Stage Universal Recovery Process"* (described in Chapter 3). Particular case studies of alcohol precipitation in food wastes are presented in subsequent sections.

5.2.2 PRECIPITATION OF DIETARY FIBERS FROM FRUIT WASTES

Dietary fiber is a plant-derived material that is resistant to digestion by human alimentary enzymes (Rodríguez et al., 2006). Dietary fibers may be divided into two parts with different physiological effects when they are dispersed in water: a soluble (e.g. pectin, β-glucans) and an insoluble fraction (e.g. lignin, cellulose, hemicellulose) (Galanakis, 2011). Pectin is a heteropolysaccharide, mostly used as a gelling or thickening agent as well as a stabilizer in foods. The initial step of pectin recovery often involves the preparation of an alcohol insoluble material, with the purpose of removing low molecular weight compounds. Rest components of dietary fibers are also precipitated together with pectin. Therefore, cell wall isolation protocols from by-products include the addition of hot ethanol as a precipitant, followed by alkali, acid, or solvent treatment (Galanakis et al., 2010a, c, e). Alkali is used to solubilize hemicelluloses, whereas acids are used to solubilize pectin. Following this methodology, pectin has been extracted from several waste sources with different yields, i.e. 2.2% from raw papaya peel (Boonrod et al., 2006) or 77.6% from citrus waste (Bafrani, 2010). Among the different acids used to solubilize pectin (before or after ethanol precipitation), sulfuric and hydrochloric acid in low concentrations (e.g. 0.05 N) are the most popular choices (Abbaszadeh, 2008; Sudhakar and Mainp, 2000), whereas Galanakis et al. (2010a) suggested the addition of citric acid instead, due to its food grade nature.

5.2.3 PURIFICATION OF PHARMACEUTICALS FROM MARINE BIOMASS

Ethanol precipitation has also been employed to recover medicinal added-value compounds (i.e. chondroitin sulfate, hyaluronic acid) from marine biomass. In this case, the effectiveness of alcohol precipitation depends on the type of added alcohol, its concentration, and processing temperature (Tadashi, 2006). An ethanol, at concentrations of 40–60%, is typically used for this purpose (Tadashi, 2006; Murado et al., 2010; Takai and Kono, 2003). Besides, 40% isopropanol has been used to recapture chondroitin sulfate from scapular cartilage of Shortfin mako shark (Kim et al., 2012). Chondroitin sulfate has also been recovered from waste of *Scyliorhinus canicula*, using a three stages process. In particular, cartilage tissue was denatured at 65°C prior to the addition of ethanol and the precipitation of glycosaminoglycan polymers. Ultimately, ion-exchange gel chromatography was employed to purify and recover chondroitin sulfate with a yield of 1.5% (Gargiulo et al., 2009). More recently, a combination of alkaline proteolysis and selective precipitation using alkaline hydroalcoholic solutions has been used for the recovery of chondroitin sulfate from ray cartilage (Murado et al., 2010). Other processes combining alcohol precipitation and cation exchange separation have been patented for the purpose, too, using salmon nasal cartilage as the initial source (Takai and Kono, 2003; Nishigori et al., 2000). In addition, hyaluronic acid has been extracted from marine biomass with water, precipitated with ethanol and purified using DEAE-cellulose chromatography (Gao et al., 1996).

5.2.4 PURIFICATION OF BROMELAIN FROM PINEAPPLE WASTE

Bromelain is a group of proteolytic enzymes that can be found in the leaves and bark of pineapple (*Ananas comosus*). These enzymes have antiinflammatory, antithrombotic, and fibrinolytic properties (Soares et al., 2012). Bromelain has been recovered from ground pineapple stem using ethanol precipitation at low temperature either in a batch mode reactor or in two fed-batch pilot tanks, a glass and a stainless steel one (Silva et al., 2010). When concentration of ethanol varies from 30% to 70%, a purification factor of 2.28 could be achieved. This fact leads to the recovery of 98% of the total enzymatic activity. The effect of temperature on the extraction process (mainly when ethanol is used as antisolvent) has also been reported. Temperature plays a crucial role on enzyme activity similar to fermentation procedures (Galanakis et al., 2012b, 2015b). For instance, Martins et al. (2014) carried out ethanol precipitation of bromelain from pineapple stem, bark, and leaves under cold conditions (4°C). Bromelain has been precipitated successfully using the above ethanolic concentration range, with a purification factor of 2.34-fold and >98% recovery of enzyme activity. Besides, bromelain did not precipitate when ethanol concentration was below 30%. Increase of ethanol concentrations over 65% also resulted in increase in enzymatic activity (Martins et al., 2014).

5.3 ULTRAFILTRATION
5.3.1 GENERAL

Ultrafiltration (UF) is a pressure-driven membrane process, in which a hydrostatic pressure difference is applied between the two sides of a permselective barrier in order to separate specific components from a feed solution. As a result, the feed solution is separated in two streams: a permeate containing all the components passed through the membrane and a retentate containing all the compounds rejected by the membrane. Rejected species include dissolved macromolecules (e.g. proteins, sugars, polymers, biomolecules) and colloidal particles, whilst molecules (small particles, solvent, and salts) pass through the membrane. UF membranes are characterized by their molecular weight cut-off (MWCO), defined as the equivalent molecular weight of the smallest species that exhibit 90% rejection. The MWCO of UF membranes is between 1 and 100 kDa, whereas hydrostatic pressures range between 1 and 5 bar. Separation is based on the molecular size via a sieving mechanism, although secondary factors such as molecule shape and charge can play an important role, too (Baker, 2000; Galanakis, 2015). Polymeric UF membranes are asymmetric, with pore sizes in the skin layer of 2–10 nm, and are mainly prepared through the phase inversion technique. Materials for fabrication of commercial membranes include synthetic polymers, such as polypropylene, polyvinylidene fluoride (PVDF), polyamide, polysulfone (PS), polytetrafluoroethylene, polyethersulfone (PES), polyethylene, and cellulose derivatives. Ceramic membranes made of alumina, titania, silica, and zirconia as well as metallic membranes are also commercially available. UF membranes are often used in combination with other pressure-driven membrane operations such as microfiltration (MF), nanofiltration (NF), and reverse osmosis (RO) (Daufin et al., 2001; Li and Chase, 2010; Galanakis et al., 2012a). MF and NF are described in detail in Chapters 4 and 7, respectively. Table 5.1 summarizes some selected applications of UF for the recovery of valuable compounds from food processing wastewaters.

Table 5.1 Use of UF Membranes in the Treatment of Food Processing Wastewaters

Food Waste	Target Compounds	Membrane Type	Operating Conditions TMP (bar)	Crossflow Velocity (mL/s)	Feed Flow Rate (L/h)	Temperature (°C)	References
Fish meal effluents	Proteins	Carbosep M2 (Rhodia Orelis), tubular, ceramic, 15 kDa	4	4	–	Ambient	Afonso et al. (2004)
Fishery washing water	Proteins	Tubular, PS, 20 kDa (PCI)		0.47	–	–	Mameri et al. (1996)
Poultry processing wastewater	Proteins	Minitan-S (Millipore), flat-sheet, PS, 30 kDa	0.96	–	41	–	Lo et al. (2005)
Artichoke wastewaters	Polyphenols, sugars	DCQ-III 006C (China Blue Star Corporation), hollow-fiber, PS, 100 kDa	0.31	–	556	24	Conidi et al. (2014)
Olive mill wastewater	Polyphenols	PS (100 and 50 kDa), PES (10 and 2 kDa)	2–8	38	0.3–3.4	25	Galanakis et al. (2010b)
		Spiracel WS P005 (Microdyn Nadir) spiral-wound, cellulose acetate, 5 kDa	2.9–3.9	–	4000	15–25	Pizzichini and Russo (2005)
		Spiral-wound, PS, 80, and 20 kDa (GE Osmonics); Spiral-wound, PES, 6 kDa (GE Osmonics)	2.5–4.5	–	3000–5000	20–25	Russo (2007)
		Ceramic material (zirconium oxide), multichannel, 1 kDa (Tami)	1.5	–	4000	20–25	Russo (2007)
		HFS (Toray) hollow-fiber, PVDF, 0.02 µm	0.43	–	433	18	Cassano et al. (2013)
		Etna 01PP (Alfa Laval), flat-sheet, composite fluoro polymer, 1 kDa	9	–	600	30	Cassano et al. (2013)

Colostrum whey	Immunoglobulins	Ceramic material (zirconium oxide), multichannel, 100 nm	1–2.25	–	3400–4300	15–35	Paraskeva et al. (2007)
		Carbosep M1 (Rhodia Orelis), tubular, ceramic, 100 kDa	0.8–2.8	–	–	20–41	Piot et al. (2004)
Winery wastewaters	Oligomeric proanthocyanidins	PU608 and PU120 (PCI Membrane Systems), tubular, PS, 8 and 20 kDa	2 (PU 120) and 5 bar (PU 608)	–	–	–	Santamaria et al. (2002)
Winery sludge	Phenolic compounds	ETNA01PP (Alfa Laval), flat-sheet, composite fluoropolymer, 1 kDa	3–5	–	85	25	Galanakis et al. (2013b)
Whey	WPCs	FS10 (Zoltek Rt Mavibran), flat-sheet, PVDF, 6–8 kDa	1, 3, 5, 8	–	–	30, 40, 50	Atra et al. (2005)

TMP, transmembrane pressure.

5.3.2 APPLICATION IN THE DAIRY INDUSTRY

Whey protein concentrates (WPCs) are today produced using UF membranes with an MWCO between 10–20 kDa. These membranes allow the separation of lactose and minerals in the permeate from whey proteins in the retentate. The retentate can be further processed with evaporation and spray drying. The protein content of the final product is affected by the degree of concentration. For instance, WPCs of 35–60% protein content can be obtained at a volume concentration ratio ranging from 4.5 to 20, respectively (Zydney, 1998). Atra et al. (2005) obtained WPCs of 8–10% protein content, using PVDF membranes of 6–8 kDa. These concentrates can be reused for cheese manufacture. UF permeates, containing 0.1–0.5% of proteins and 5% of lactose, can be further loaded to NF units in order to obtain concentrated lactose for confectionary applications. In addition, purified immunoglobulins (up to 90%) can be obtained after a two-step membrane process including a macroscopic MF pretreatment to remove fat globules and casein micelles prior to UF treatment with a membrane of 100 kDa (Piot et al., 2004).

5.3.3 APPLICATION IN THE FISHERY INDUSTRY

Wastewaters from the fish processing industry contain a large amount of potentially valuable proteins (Massé et al., 2008). Tubular ceramic UF membranes with an MWCO of 15 kDa have been used to remove the organic load from fish meal wastewaters, allowing the recovery of valuable raw materials such as proteins (Afonso et al., 2004). In particular, operation of UF at 4 bar with permeate fluxes of 28 L/m^{2}h led to protein rejections of 62%. Tubular UF membranes with an MWCO of 20 kDa have also shown protein rejections of 70–80% during the treatment of fishery washing water. At the same time, the biochemical oxygen demand of the waste stream was reduced by 80% (Mameri et al., 1996).

5.3.4 APPLICATION IN THE POULTRY INDUSTRY

The recovery of proteins using UF has also been investigated in the case of poultry processing wastewaters. In the approach investigated by Lo et al. (2005), poultry wastewaters were pretreated by dissolved air flotation in order to remove fat substances. Thereafter, the defatted solution was ultrafiltered using polysulfone membranes with an MWCO of 30 kDa. These membranes retained almost all crude proteins and at the same time reduced the chemical oxygen demand of the effluent to <200 mg/L. In optimized operating conditions (feed flow rate of 683 mL/min and transmembrane pressure of 0.96 bar), the obtained permeate fluxes reached the 200 L/m^{2}h.

5.3.5 APPLICATION IN THE AGRICULTURAL INDUSTRY

Soybean processing wastewaters contain valuable compounds such as protein and sugars. Jiang and Wang (2013) proposed an integrated membrane process to extract proteins, oligosaccharides, and isoflavones from yellow bean product wastewaters. In this process, UF membranes were used to retain soybean proteins, whereas the concentrated solution was spray dried to obtain a pure protein powder. Soy isoflavones contained in the UF permeate were adsorbed on weak polar macroporous resins and then eluted with ethanol. In another case, an integrated membrane process (using both UF and NF) was investigated for the recapture of phenolic compounds and sugars from artichoke wastewaters (Conidi et al., 2014). The preliminary UF process removed mainly suspended solids and macromolecules from the artichoke wastewaters, whereas the permeate stream was assessed in a two-step NF process to fractionate sugars and phenolic compounds.

UF has been used for the filtration of olive mill wastewater (OMW), too (Rahmanian et al., 2014). For instance, Russo (2007) evaluated the removal efficiencies of three spiral-wound UF polymeric (80, 20, and 6 kDa) and one ceramic (1 kDa) membranes for the purification of polyphenolic components contained in an MF permeate, derived from OMW. According to the results of the study, the membranes of 1 and 6 kDa showed the same selectivity among the polyphenolic molecules, but they showed different rejection values. UF permeates were further treated with NF and RO in order to produce a concentrate with low molecular weight polyphenols. The latter product could be suitable for food and pharmaceutical industries in different formulations (e.g. liquid, frozen, dried, or lyophilized). Galanakis et al. (2010b) clarified two high added-value products (pectin containing solution and phenol containing beverage), recovered from OMW, using four types of UF membranes (100, 25, 10, and 2 kDa). In particular, the membranes of 25 and 100 kDa were able to separate pectin from concentrated cations and phenols. The membrane of 25 kDa was also able to partially remove the heavier fragments of hydroxycinnamic acid derivatives and flavonols, and simultaneously to sustain the antioxidant properties of the phenol containing beverage in the permeate stream. In a similar application, the UF membrane of 100 kDa was able to concentrate β-glucan in a feed solution derived from oat mill waste (Patsioura et al., 2011). More recently, Cassano et al. (2013) evaluated the performance of two sequential UF membranes (the first one with a nominal pore size of 0.02 μm and the second one with an MWCO of 1 kDa), followed by a final NF step for the recovery of phenolic compounds. Indeed, rejections of the two investigated membranes towards phenolic compounds reached 26% and 31%, respectively. However, in the second UF step, most of the organic substances were separated from phenolic compounds as showed by the high rejection (>70%) of the membrane towards total organic compounds.

5.3.6 APPLICATION IN WINERIES

Winery wastewaters contain huge amounts of biodegradable organic compounds together with small concentrations of valuable compounds, such as phenolic compounds, sugars, organic acids, and nutrients. For this reason, membrane operations such as MF, UF, and NF have been used to fractionate polyphenolic extracts from defatted milled grape seeds (Santamaria et al., 2002). In particular, polysulfone UF membranes (20 and 8 kDa) allowed the removal of monomeric and oligomeric proanthocyanidins from different streams generated in the process. In another application, UF membranes were used to fractionate phenolic compounds from hydroethanolic extracts of winery sludge (Galanakis et al., 2013b). Among the investigated membranes (100 and 20 kDa polysulfone, 1 kDa fluoropolymer), the nonpolar fluoropolymer membrane separated hydroxycinnamic acids successfully from anthocyanins and flavonols since the observed rejection towards acids was twofold higher than that observed for other phenolic compounds. Similar results have also been reported in diluted wine samples (Galanakis et al., 2015a).

5.4 ISOELECTRIC SOLUBILIZATION/PRECIPITATION
5.4.1 GENERAL

Isoelectric solubilization/precipitation (ISP) is a technique that allows the selective solubility of proteins from muscle food processing by-products (surimi wastewater, frames, heads, and viscera of meat, fish, or marine products) with concurrent removal of muscular lipids (fats and oils), bones, or skin (Tahergorabi and Jaczynski, 2014). ISP was first proposed by food scientists at Massachusetts University (Hultin and Kelleher, 1999, 2000, 2001, 2002). Following these pioneering developments along with earlier work by Meinke et al. (1972) and Meinke and Mattil (1973), several food science laboratories were actively researching the ISP field.

In fish muscle homogenates, myofibrillar proteins are present as aggregates that are held together by weak protein/protein hydrophobic interactions (Undeland et al., 2003). However, depending on the conditions that the fish muscle proteins are subjected to, the protein side chains can assume different electrostatic charges (Fig. 5.1). This means that the solubility of fish muscle proteins can be "turned" on or off, by providing conditions that either favor or disfavor protein solubility, respectively. When acid is added to a solution, it dissociates and releases hydronium ions (H_3O^+). Protonation of negatively charged side chains on glutamyl or aspartyl residues results in an increased net positive surface charge. Similarly, when base (OH^-) is added to a solution, deprotonation of side chains on tyrosyl, tryptophanyl, cysteinyl, lysyl, argininyl, or histidinyl residues contributes to an increased net negative surface charge (Fig. 5.1).

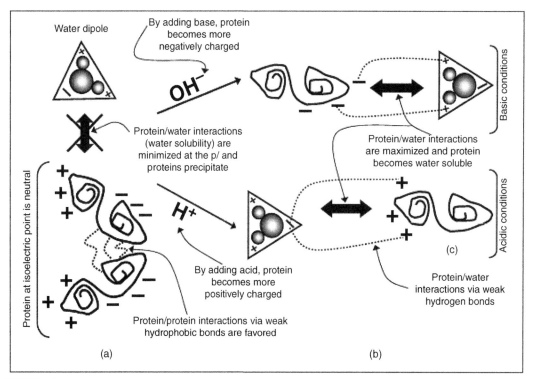

FIGURE 5.1

A Protein at its Isoelectric Point (pI) has a zero net electrostatic charge (a) At its pI, protein/water interactions are at their minimum, while protein/protein interactions via weak hydrophobic bonds are at their maximum, causing protein precipitation. (b) Protein/water interactions prevail under acidic or basic conditions far from the pI, resulting in protein solubility (Tahergorabi et al., 2014). Reprinted with permission.

Consequently, solubilization of fish muscle proteins is ascribed to protonation of aspartyl and glutamyl (pK_a = 3.8 and 4.2, respectively) residues at acidic pH, whereas deprotonation of lysyl, tyrosyl, and cysteinyl (pK_a = 9.5–10.5, 9.1–10.8, and 9.1–10.8, respectively) is ascribed to residues at basic pH. When the charge equilibrium is reached and protein solution attains homeostasis, the final status of protein surface electrostatic charge at a given pH is referred as the net charge. The accumulation of a net positive or negative charge induces protein/protein electrostatic repulsion and an increased hydrodynamic volume, due to expansion and swelling (Undeland et al., 2003; Kristinsson et al., 2005).

As proteins assume more positive or negative net charge, they gradually start electrostatic interactions with water (e.g. protein/water interactions). Due to increased protein/water interactions, the protein/protein hydrophobic interactions attenuate. Therefore, as the protein molecules turn to their charged form, more water associates on and around the protein surface and thus proteins become water soluble. However, it is possible to adjust the pH of a protein solution so that the number of negative charges on the protein's surface is equal to the number of positive charges. At this case, the protein molecule assumes a zero net electrostatic charge. The pH at which the net electrostatic charge of a protein is equal to zero is called the isoelectric point (pI) (Fig. 5.1). As the charges on a protein's surface diminish, protein/water interactions attenuate and hence protein/water solubility and water-holding capacity are reduced. In addition, proteins gel poorly at their pI. On the other hand, the hydrophobic protein/protein interactions are favored at the pI. Therefore, proteins at their pI achieve minimum solubility and typically precipitate. This pH-mediated behavior of protein allows the modification of its solubility/precipitation characteristics, as a function of pH adjustment (Thawornchinsombut and Park 2004; Chen and Jaczynski 2007a).

5.4.2 RECOVERY OF PROTEINS AT THE LABORATORY SCALE

The isoelectric behavior of fish muscle proteins can be used to recover proteins from aquatic animal processing by-products as well as low-value aquatic species. The proteins recovered by this approach can retain their functional properties such as gel-forming ability (Kristinsson and Hultin, 2003). While muscle proteins are in a soluble form (protein/water interactions are favored), the insoluble components (bone, skin, scale, etc.) can be removed from the solution by centrifugation, followed by protein precipitation and recovery at the pI (protein/protein interactions are favored). Functional muscle proteins from fish have so far been recovered at the laboratory (Kim et al., 2003; Kristinsson and Hultin, 2003; Tahergorabi et al., 2012) and pilot plant, batch mode scale (Mireles DeWitt et al., 2007). ISP processing in a continuous mode has also been applied to fish processing by-products (Chen and Jaczynski, 2007b; Chen et al., 2007), krill (Chen and Jaczynski, 2007a; Chen et al., 2009), whole fish (Taskaya et al., 2009a, b), and chicken by-products (Tahergorabi et al., 2011).

5.4.3 RECOVERY OF PROTEINS AT THE PILOT SCALE

The principle of the pI has been used in cheese making and manufacture of soy protein isolates. For instance, the action of rennet and/or lactic acid bacteria acidifies milk to pH = 4.6, and then casein precipitates at its pI. Thus, a casein curd is formed with a subsequent draining of water soluble protein, contained in whey (Galanakis et al., 2014). The same principle may be applied to muscle proteins. In this case, the precipitated fraction is composed of myofibril and stromal proteins, and the soluble fraction is composed of sarcoplasmic proteins (the casein and the whey fraction of cheese, respectively). The pI of fish muscle proteins is 5.5. Therefore, the fish muscle proteins precipitate at pH 5.5 and become water soluble gradually, as the pH becomes more acidic or alkaline.

In general, there are five steps in recovering proteins and lipids from fish processing by-products using ISP (Fig. 5.2). The first step is to homogenize a solution of by-products in water (1:6 per weight), in order to provide reaction medium and increase available surface area for the subsequent reaction of protein solubilization. In the second step, the fish muscle proteins are solubilized under acidic or alkaline conditions. As the pH moves further away from the pI, the fish muscle proteins assume a more uniform negative or positive surface charge, respectively (Fig. 5.2). Charge shift results in weaker protein/protein hydrophobic

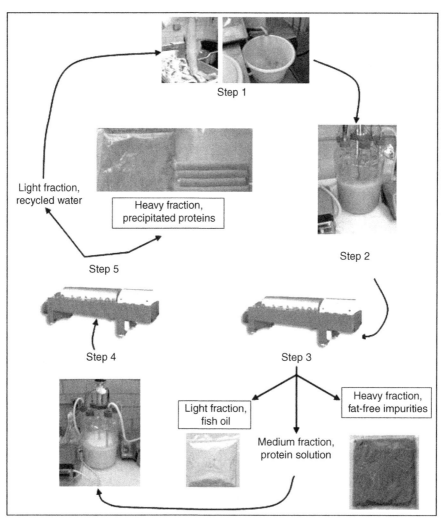

FIGURE 5.2 Diagram of the Isoelectric Precipitation and Solubilization Technology with Concurrent Oil Separation Proposed for Processing Fish By-Products

Materials in boxes are fractions to be further processed into food and other applications (Tahergorabi et al., 2014). Step 1: Homogenization of by-products with water (1:6, w/w). Step 2: Homogenization of by-products with water (1:6, w/w). Step 3: First separation. Step 4: Second pH adjustment, proteins precipitate. Step 5: Second separation. Reprinted with permission.

interactions, while protein/protein electrostatic repulsion becomes more predominant and leads to protein/water interaction (e.g. water solubility). When proteins begin to interact with water, a drastic increase of viscosity occurs. The viscosity decreases sharply as soon as the proteins become water soluble. Undeland et al. (2003) as well as Kristinsson and Ingadottir (2006) attributed this fact to the break-up of existing aggregates/myofibrillar assemblies or deprotonation of the ε-amino groups of lysyl residues. The viscosity increase is an important processing parameter that may result in mixing limitations, i.e. pH and protein solubility gradients, foam formation, etc. One way to compensate is to maintain the solution continuously at the desired pH, i.e. in a continuous protein and lipid recovery system (Torres et al., 2007).

Separation (the third step) is applied during interaction of muscle proteins with water. Typically, centrifugation separates the solution to light, medium, and heavy fractions containing fish oil, solubilized muscle proteins and impurities (bones, scale, skin, insoluble proteins, etc.), respectively. The hydrophobic triglycerides are fairly easy to separate from the solution, while the membrane phospholipids are relatively persistent because they are amphiphilic. Indeed, more than 50% of the membrane phospholipids are retained with the solubilized proteins after Step 3 (Undeland et al., 2002; Liang and Hultin, 2005). Although membrane phospholipids are present in smaller amounts in the fish muscle than triglycerides, the membrane phospholipids have been demonstrated to contribute more to rancidity (Vannuccini, 2004). Therefore, it is desirable to remove as much lipid as possible during the separation step. The latter results in the recapture of crude fish oil that is rich in ω-3 fatty acids and can be further processed for numerous food and nonfood applications (Chen et al., 2007). The heavy fraction is rich in minerals (e.g. Ca, Mg, and P) and therefore it can be implemented for the development of animal feeds and value-added pet foods (Chen et al., 2007).

The medium fraction (containing the water soluble fish muscle proteins) is recovered after a second pH adjustment in Step 4. Specifically, the pH is adjusted to the average p*I* of the fish muscle proteins (pH 5.5) and subsequently fish muscle proteins precipitate due to increased protein/protein hydrophobic interactions, decreased protein/water interactions and protein/protein electrostatic repulsion. Similar to the first pH adjustment, as the proteins gradually stop interacting with water dipoles, the viscosity increases significantly. This viscosity issue can be overcome by maintaining continuously the pH at 5.5. Typically, the precipitated fish muscle proteins are separated from the process water by centrifugation. The muscle proteins retain their gel-forming ability and thus can be used as a functional ingredient in surimi seafood (commonly referred to as crab-flavored seafood). In order to preserve protein functionality, the respective isolates obtained using ISP must include cryoprotectants (Thawornchinsombut and Park, 2004) for frozen storage. Following ISP, the process water can be reused in a continuous system in order to reduce processing cost. However, the purity of process water is greatly dependent on processing parameters.

Torres et al. (2007) proposed a continuous bioreactor system for fish processing by-products and whole fish, based on ISP principles. The bioreactors were equipped with built-in pumps for various processing additives such as antifoams, coagulants, flocculants, etc. The flocculants enabled more efficient separation of the precipitated fish muscle proteins from the process water and could consequently facilitate process scale-up from laboratory to pilot/industrial scale (Taskaya and Jaczynski, 2009). Following homogenization (Fig. 5.2, Step 1), the homogenate is pumped to the first bioreactor for a 10-min solubilization reaction (Step 2). The bioreactor continuously controls and maintains the pH of the medium. Since pH of the incoming homogenate is close to neutral (~6.6–7.0), a base is rapidly pumped into the vessel to adjust the pH to 11. Bioreactors are also equipped with mixing baffles to prevent pH gradients and excessive foaming. A refrigerant is used to maintain constant temperature, while small pumps are used to inject food-grade emulsion breakers and antifoam agents. The experimental recovery system

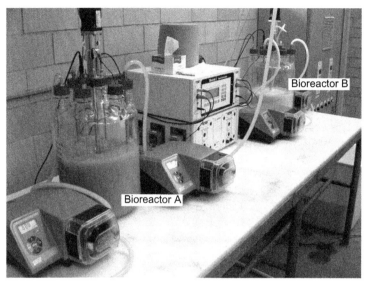

FIGURE 5.3 Bioreactors Equipped with Automatic pH and Temperature Controls, Continuous Pumping of Feed and Treated Stream, and Dosing of Food-Grade Additives such as Emulsion Breakers, Protein Flocculants, and Antifoaming Agents

Bioreactor A is used for protein solubilization (Step 2), while Bioreactor B is used for isoelectric precipitation (Step 4). A control box is placed between both bioreactors. This configuration is working in a continuous mode at flow rate of 300 L/h (Tahergorabi et al., 2014). Reprinted with permission.

works at 300 L/h and can process ~43 kg/h of fish by-products (Fig. 5.3). Although these small-scale bioreactors are manufactured from glass and stainless steel components, industrial strength high density polyethylene can be used in a fish processing plant. Based on the experimental system (Fig. 5.3), a modular 600-L bioreactor system has been designed to process 12 ton/day of fish processing by-products.

Following the pH adjustment in Step 2, the solution is pumped to a decanter centrifuge working typically below 4000g. Decanters are commonly used in surimi processing plants. However, surimi technology does not work under acidic or basic pH and therefore the assessment of an available decanter should be performed prior usage with ISP. On the other hand, there are no pH issues when separating proteins (Step 5). However, this process can be relatively slow unless the particle size of the precipitated muscle proteins is increased. The latter can be conducted by promoting protein/protein hydrophobic interactions with an extended precipitation time in Step 4 (~24 h).

The particle settling velocity under the centrifugal force (g) depends on the density differential between phases ($\Delta \rho$), viscosity (μ), and particle size expressed as equivalent diameter (D):

$$S = \frac{\Delta \rho g D^2}{\mu} \tag{5.1}$$

The D is the only variable that a processor can modify in ISP. Using only 10 min in Step 4, protein particle size can be increased by adding commercially available flocculants.

5.5 EXTRUSION

5.5.1 GENERAL

Extrusion is one of the most promising technologies in food processing and has been used since the mid-1930s for the production of ready-to-eat snack foods, breakfast cereals, etc. (Riaz et al., 2009). Extrusion of food waste deals with extrusion of ground material at barothermal conditions. With the help of shear energy (typically exerted by a rotating screw) and extra heating of the barrel, the feed materials are heated to their plasticizing or melting point. Thereafter, they are conveyed under high pressure through a series of dies. As a result of the latter process, the product (extrudate) expands to its final shape (Moscicki et al., 2007; Van Zuilichem, 1992), which provides different physical and chemical properties compared with the raw materials. For instance, extrudates improve digestibility (Singh et al., 2010) and nutrient bioavailability compared with products cooked with other processes. Moreover, extrusion is preferred over other cooking process due to its high productivity, energy efficiency, shorter cooking times and low operating cost. Moreover, it is able to develop a range of products with a distinct texture (Dehghan-Shoar et al., 2010).

5.5.2 EXTRUSION-COOKING TECHNIQUE

Extrusion is usually carried out in food extruders, in which the principal operating system is one screw or a pair of screws fitted in a barrel. At barothermal processing (pressure up to 20 MPa, temperature 200°C), the material is mixed, compressed, melted, and plasticized in the end part of the machine. The range of chemical and physical changes in the processed food materials depends mainly on the parameters of the extrusion process and the construction of the extruder. In Fig. 5.4, a schematic diagram of an extruder is presented. Food extruders may be classified by three major factors:

1. The method of generating mechanical friction energy, which is converted into heat. This method can be divided in three classes:
   a. isothermic (heated),
   b. polytropic (mixed),
   c. autogenic (source of heat is the friction of the material particles caused by the screw rotating at high speed).
2. The amount of mechanical energy generated. There are two types of extruders:
   a. low-pressure extruders producing relatively limited shear rate,
   b. high-pressure extruders generating huge amounts of mechanical energy and shear.
3. The construction of the plasticizing unit, where both the screw and the barrel may be designed as a uniform, integrated body, or fixed with separate modules (Moscicki and van Zuilichem, 2011).

Generally, the main ingredients of food waste are carbohydrate, protein, fatty acids, vitamins, etc. The effects of extrusion processing on bioactives (carbohydrate, protein, fatty acids, vitamins, etc.) are discussed in detail in subsequent sections.

5.5.3 EFFECT OF EXTRUSION PROCESSING FACTORS ON TARGET COMPOUNDS

In extrusion processing, there are many parameters affecting the composition of final products (Brennan et al., 2008). These include the raw material characteristics, mixing and conditioning of raw material,

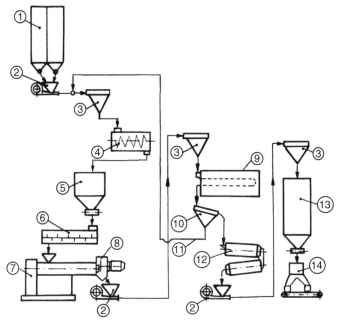

FIGURE 5.4 A Diagram of the Set-Up for the Production of Multiflavor Cereal Snacks

1, a silo with raw materials; 2, pneumatic conveyor; 3, collector; 4, mixer; 5, weigher; 6, conditioner; 7, extruder; 8, cutter; 9, dryer; 10, screen; 11, recycling of dust; 12, coating drums; 13, silos of finished product; 14, packing machine.

barrel temperature, pressure, screw speed, moisture content, flow rate, energy input, residence type, and screw configuration. Processing parameters have both positive and negative effects on the bioactive compounds of the extrudates, but in general, influence mainly macromolecules. Smaller molecules may also be affected either by extrusion itself or by induced changes in macromolecules.

For instance, sugars do not only provide sweetness, but they are also a great source of "quick" energy. In many cases, concentration of sugar is reduced after extrusion (Noguchi et al., 1982; Camire et al., 1990). Other compounds such as polysaccharides (e.g. amylose and amylopectin) contribute to increase viscosity and gel formation of the cooked paste, respectively. Most of the extruded carbohydrate is derived from rice, wheat, and corn. Extrusion of starch-based products is somewhat unique because gelatinization occurs at much lower moisture (12–22%) (Qu and Wang, 1994). Addition of sucrose, salt or fiber to corn meal may affect gelatinization (Jin et al., 1994). Low temperature (160 vs 185°C) and feed moisture (16% vs 20%) during extrusion reduce significantly the average molecular weight of starch in wheat flour (Politz et al., 1994). Moreover, low feed moisture (19%) and barrel temperature (110–140°C) promote high amylose/lipid complex formation, considering stearic acid and normal corn starch as feedstock (Bhatnagar and Hanna, 1994). Also, if acid and alkaline treatment are applied prior to extrusion, an increase in the soluble fiber fraction can be obtained (Ning et al., 1991). Indeed, an increase of processing temperature (150–200°C) can increase the solubilization of dietary fibers (Vasanthan et al., 2002) and the digestibility of proteins, due to their denaturation (Srihara and Alexander, 1984; Hakansson et al., 1987; Colonna et al., 1989; Areas, 1992).

Lower feed moisture content and longer residence time can lead to the acceleration of antinutritional factors such as trypsin inhibitors (Bjorck and Asp, 1983; Singh et al., 2000; Lorenz and Jansen, 1980; Asp and Bjorck, 1989). Other process variables such as length to diameter ratio and screw speed appear to be unimportant (Bhattacharya et al., 1988).

On the other hand, extrusion of amino acids such as lysine should be performed at low temperature ($<180°C$) and low moisture content ($<25\%$) in order to improve their retention (Noguchi et al., 1982; Bjorck and Asp, 1983; Cheftel, 1986; Asp and Bjorck, 1989). Extrusion of high-fat materials (over 5–6%) reduces extruder performance (Camire, 2000a; Nierle et al., 1980). Thereby, it is highly recommended to remove lipids during macroscopic pretreatment of food wastes. Nevertheless, lipid oxidation, which has negative impact on sensory and nutritional qualities of foods and feeds, does not take place during extrusion (Semwal et al., 1994; Arora and Camire, 1994).

Extrusion reduces measurable bioactive compounds (e.g. polyphenols, anthocyanins, tannins, carotene, and antioxidant activity) in food products up to 80% (Dlamini et al., 2007; Korus et al., 2007; Delgado-Licon et al., 2009; Shih et al., 2009; Repo-Carrasco-Valencia et al., 2009a, b). On the other hand, the partial destruction of polyphenols during heat treatment increases mineral absorption (Alonso et al., 2001). Total phenolic content is decreased significantly by increasing both barrel temperature and feed moisture content, too (Yagci and Gogus, 2008). Thereby, extrusion is not recommended where polyphenols are the target recovered compounds. Nevertheless, disruption of the food matrix induced by extrusion may increase the extraction of anthocyanin monomers and dimers (Khanal et al., 2009a, b). Moreover, in some cases (e.g. for wheat), it has been reported a significant increase of free and bound phenolic acids (e.g. syringic, coumaric) during extrusion, probably due to their solubilization (Zielinski et al., 2006). It has also been reported that the denaturation of grain proteins during extrusion leads to tannin/protein interaction and the formation of respective tannin/protein complexes that retain antioxidant activity (Riedl and Hagerman, 2001). Higher barrel temperatures, low feed moistures, and short-term time influence negatively the stability of fat soluble vitamins such as vitamins A and E (Harper, 1988; Killeit, 1994; Grela et al., 1999; Tiwari and Cummins, 2009).

5.5.4 RECOVERY OF MACROMOLECULES

The ability of extrusion to modify (physically or chemically) the physicochemical properties of macromolecules (polysaccharides, dietary fibers, or proteins) and micromolecules (sugars, polyphenols, vitamins, or minerals) has been used for particular applications of compound recovery from food wastes. For instance, extrusion has been used for onion waste derived from the white outer fleshy scale leaves. As a result, an increase in solubility of the cell wall pectic polymers and hemicelluloses was observed along with swelling of the cell wall material (Ng et al., 1999). Pectin and other polysaccharides (e.g. hemicelluloses or chitin) separation can also be performed by changing their microstructure via high mechanical energy input and extrusion technology (Zeitoun et al., 2010). Changes in the microstructure of dietary fibers cause an increase of substrate viscosity as well as enhancement of its gelling properties and thus the resultant products are typically used as confectionary gels. Also, starch containing materials are extruded not only in the food industry, but also in industry sectors such as animal feed, pulp and paper, and film and packaging material (Wolf, 2010). Another proposed application is sugar beet pulp processing in a twin-screw extruder under specific conditions. This process could change the confirmation of the noncellulosic cell wall materials without affecting thermal degradation. As a result of these conformational changes, sugar beet pulp was considered a cheap alternative source to thermoplastic starch for the processing of biodegradable materials (Rouilly et al., 2006a, b).

5.6 CONCLUSIONS

In this chapter different conventional technologies such as alcohol precipitation, UF, isoelectric solubilization/precipitation, and extrusion have been discussed in detail for their implementation in the second step of the *"5-Stage Universal Recovery Process"*. As antisolvents reduce the solubility of solute, macromolecules are separated from micromolecules in the presence of alcohols (e.g. ethanol or methanol). Although this technology is well accepted in the industry, it has several disadvantages with respect to product purity when different types of solutes are present in the recovery stream (Galanakis, 2012; Galanakis and Schieber, 2014). UF is a fine-tuned, pressure-driven, size exclusion-based membrane separation process. Solutes are rejected or permeate the stream depending on the MWCO of the membrane. The main disadvantage of UF is concentration polarization and fouling phenomena, which reduce the permeate flux. These problems raise the cost of the process. Further, UF has low selectivity upon the target compounds. Despite these disadvantages, UF is popular in industrial applications due to the physical nature of separation and continuous operation. To reduce the concentration polarization, development of new types of membranes and modules with high shear force is required. The ISP of food proteins has long been applied to recover milk and soy proteins. By thorough understanding of muscle food proteins and their isoelectric behavior, the food industry would be able to develop this technology for the efficient muscle food processing of by-products at a commercial scale. However, it will be necessary to develop final applications for the recovered materials. This process could be developed in the dairy and soybean processing industries, too. Extrusion could be a lucrative technology for food waste valorization. On the other hand, it provides low functional value end-products. Thereby, it can only be applied in particular applications, i.e. if macromolecules are the target compounds and also need modification. Furthermore, development of a new type of extruder will boost the process intensification strategies. Government agencies as well as academia should be actively engaged in the collaborative development of final food products.

REFERENCES

Abbaszadeh, A.H., 2008. Pectin and galacturonic acid from citrus wastes. This thesis comprises 30 ECTS credits and is a compulsory part in the Master of Science with a major in Industrial biotechnology No. 7/ 2008.

Afonso, M.D., Ferrer, J., Bórquez, R., 2004. An economic assessment of proteins recovery from fish meal effluents by ultrafiltration. Trends Food Sci. Technol. 15, 506–512.

Alonso, R., Rubio, L.A., Muzquiz, M., Marzo, F., 2001. The effect of extrusion cooking on mineral bioavailability in pea and kidney bean seed meals. Anim. Feed Sci. Technol. 94, 1–13.

Areas, J.A.G., 1992. Extrusion of proteins. Crit. Rev. Food Sci. Nutr. 32, 365–392.

Arora, A., Camire, M.E., 1994. Performance of potato peels in muffins and cookies. Food Res. Int. 27, 14–22.

Asp, N.G., Bjorck, I., 1989. Nutritional properties of extruded foods. In: Mercier, C., Linko, P., Harper, J.M. (Eds.), Extrusion Cooking. American Association of Cereal Chemists. Inc., St. Paul, Minnesota, pp. 399–434.

Atra, R., Vatai, G., Bekassy-Molnar, E., Balint, A., 2005. Investigation of ultra- and nanofiltration for utilization of whey protein and lactose. J. Food Eng. 67, 325–332.

Bafrani, M.P., 2010. Citrus waste biorefinery: process development, simulation and economic analysis. Department of Chemical and Biological Engineering, Chalmers University of Technology, Göteborg, Sweden.

Baker, W.R., 2000. Membrane Technologies and Applications. McGraw-Hill Companies, USA.

Bhatnagar, S., Hanna, M.A., 1994. Amylose–lipid complex formation during single-screw extrusion of various corn starches. Cereal Chem. 71, 582–587.

Bhattacharya, S., Das, H., Bose, A.N., 1988. Effect of extrusion process variables on in vitro protein digestibility of fish-wheat flour blends. Food Chem. 28, 225–231.

Bjorck, I., Asp, N.G., 1983. The effects of extrusion cooking on nutritional value. J. Food Eng. 2, 281–308.

Boonrod, D., Reanma, K., Niamsup, H., Mai, C., 2006. Extraction and physicochemical characteristics of acid-soluble pectin from raw papaya (*Carica papaya*) peel. J. Sci. 33, 129–135.

Brennan, M.A., Monro, J.A., Brennan, C.S., 2008. Effect of inclusion of soluble and insoluble fibres into extruded breakfast cereal products made with reverse screw configuration. Int. J. Food Sci. Technol. 43, 2278–2288.

Camire, M.E., Camire, A., Krumhar, K., 1990. Chemical and nutritional changes in foods during extrusion. Critical Rev. Food Sci. Nutr. 19, 35–57, 1040-8398.

Camire, M.E., 2000a. Chemical and nutritional changes in food during extrusion. In: Riaz, M.N. (Ed.), Extruders in Food Applications. CRC Press, Boca Raton, FL, pp. 127–147.

Cassano, A., Conidi, C., Giorno, L., Drioli, E., 2013. Fractionation of olive mill wastewaters by membrane separation techniques. J. Hazard. Mater., 248–249, 185–193.

Cheftel, J.C., 1986. Nutritional effects of extrusion cooking. Food Chem. 20, 263–283.

Chen, Y.C., Jaczynski, J., 2007a. Gelation of protein recovered from Antarctic krill (*Euphausia superba*) by isoelectric solubilization/precipitation as affected by functional additives. J. Agric. Food Chem. 55, 1814–1822.

Chen, Y.C., Jaczynski, J., 2007b. Protein recovery from rainbow trout (*Oncorhynchus mykiss*) processing by-products via isoelectric solubilization/ precipitation and its gelation properties as affected by functional additives. J. Agric. Food Chem. 55, 9079–9088.

Chen, Y.C., Tou, J.C., Jaczynski, J., 2007. Amino acid, fatty acid, and mineral profiles of materials recovered from rainbow trout (*Oncorhynchus mykiss*) processing by-products using isoelectric solubilization/precipitation. J. Food Sci. 72, C527–C535.

Chen, Y.C., Tou, J.C., Jaczynski, J., 2009. Amino acid and mineral composition of protein and other components and their recovery yields from whole Antarctic krill (*Euphausia superba*) using isoelectric solubilization/ precipitation. J. Food Sci. 74, H31–H39.

Colonna, P., Tayeb, J., Mercier, C., 1989. Extrusion cooking of starch and starchy products. In: Mercier, C., Linko, P., Harper, J.M. (Eds.), Extrusion Cooking. American Association of Cereal Chemists. Inc., St. Paul, Minnesota, pp. 247–319.

Conidi, C., Cassano, A., Garcia-Castello, E., 2014. Valorization of artichoke wastewaters by integrated membrane process. Water Res. 48, 363–374.

Daufin, G., Escudier, J.P., Carrère, H., Bérot, S., Fillaudeau, L., Decloux, M., 2001. Recent and emerging applications of membrane processes in the food and dairy industry. Trans. IChemE 79, 89–102.

Dehghan-Shoar, Z., Hardacre, A.K., Brennan, C.S., 2010. The physico-chemical characteristics of extruded snacks enriched with tomato lycopene. Food Chem. 123, 1117–1122.

Delgado-Licon, E., Ayala, A.L.M., Rocha-Guzman, N.E., Gallegos-Infante, J.A., Atienzo-Lazos, M., Drzewiecki, J., Martínez-Sánchez, C.E., Gorinstein, S., 2009. Influence of extrusion on the bioactive compounds and the antioxidant capacity of the bean/corn mixtures. Int. J. Food Sci. Nutr. 60, 522–532.

Deng, Q., Zinoviadou, K.G., Galanakis, C.M., Orlien, V., Grimi, N., Vorobiev, E., Lebovka, N., Barba, F.J., 2015. The effects of conventional and non-conventional processing on glucosinolates and its derived forms, isothiocyanates: extraction, degradation and applications. Food Eng. Rev. in press.

Dlamini, N.R., Taylor, J.R.N., Rooney, L.W., 2007. The effect of sorghum type and processing on the antioxidant properties of African sorghum-based foods. Food Chem. 105, 1412–1419.

Galanakis, C.M., 2011. Olive fruit and dietary fibers: components, recovery and applications. Trends Food Sci. Technol. 22, 175–184.

Galanakis, C.M., 2012. Recovery of high added-value components from food wastes: conventional, emerging technologies and commercialized applications. Trends Food Sci. Technol. 26, 68–87.

Galanakis, C.M., 2013. Emerging technologies for the production of nutraceuticals from agricultural by-products: a viewpoint of opportunities and challenges. Food Bioprod. Process. 91, 575–579.

Galanakis, C.M., 2015. Separation of functional macromolecules and micromolecules: from ultrafiltration to the border of nanofiltration. Trends Food Sci. Technol. 42, 44–63.

Galanakis, C.M., Schieber, A., 2014. Editorial. Special issue on recovery and utilization of valuable compounds from food processing by-products. Food Res. Int. 65, 299–484.

Galanakis, C.M., Tornberg, E., Gekas, V., 2010a. A study of the recovery of the dietary fibres from olive mill wastewater and the gelling ability of the soluble fibre fraction. LWT – Food Sci. Technol. 43, 1009–1017.

Galanakis, C.M., Tornberg, E., Gekas, V., 2010b. Clarification of high-added value products from olive mill wastewater. J. Food Eng. 99, 190–197.

Galanakis, C.M., Tornberg, E., Gekas, V., 2010c. Dietary fiber suspensions from olive mill wastewater as potential fat replacements in meatballs. LWT – Food Sci. Technol. 43, 1018–1025.

Galanakis, C.M., Tornberg, E., Gekas, V., 2010d. Recovery and preservation of phenols from olive waste in ethanolic extracts. J. Chem. Technol. Biotechnol. 85, 1148–1155.

Galanakis, C.M., Tornberg, E., Gekas, V., 2010e. The effect of heat processing on the functional properties of pectin contained in olive mill wastewater. LWT – Food Sci. Technol. 43, 1001–1008.

Galanakis, C.M., Fountoulis, G., Gekas, V., 2012a. Nanofiltration of brackish groundwater by using a polypiperazine membrane. Desalination 286, 277–284.

Galanakis, C.M., Kanellaki, M., Koutinas, A.A., Bekatorou, A., Lycourghiotis, A., Kordoulis, C.H., 2012b. Effect of pressure and temperature on alcoholic fermentation by *Saccharomyces cerevisiae* immobilized on γ-alumina pellets. Bioresour. Technol. 114, 492–498.

Galanakis, C.M., Goulas, V., Tsakona, S., Manganaris, G.A., Gekas, V., 2013a. A knowledge base for the recovery of natural phenols with different solvents. Int. J. Food Prop. 16, 382–396.

Galanakis, C.M., Markouli, E., Gekas, V., 2013b. Recovery and fractionation of different phenolic classes from winery sludge using ultrafiltration. Separ. Purif. Technol. 107, 245–251.

Galanakis, C.M., Chasiotis, S., Botsaris, G., Gekas, V., 2014. Separation and recovery of proteins and sugars from Halloumi cheese whey. Food Res. Int. 65, 477–483.

Galanakis, C.M., Kotanidis, A., Dianellou, M., Gekas, V., 2015a. Phenolic content and antioxidant capacity of Cypriot wines. Czech J. Food Sci. 33, 126–136.

Galanakis, C.M., Patsioura, A., Gekas, V., 2015b. Enzyme kinetics modeling as a tool to optimize food biotechnology applications: a pragmatic approach based on amylolytic enzymes. Crit. Rev. Food Sci. Technol., 55, 1758–1770.

Gao, Y.-Q., Liu, J.-H., Huo, X., Shan, Y.-L., Xu, Z.-X., 1996. The purification and identification of hyaluronic acid isolated from various tissues. J. Clin. Biochem. 12, 215–218.

Gargiulo, V., Lanzetta, R., Parrilli, M., de Castro, C., 2009. Structural analysis of chondroitin sulfate from *Scyliorhinus canicula*: a useful source of this polysaccharide. Glycobiology 19, 1485–1491.

Gehring, C.K., Gigliotti, J.C., Moritz, J.S., Tou, J.C., Jaczynski, J., 2011. Functional and nutritional characteristics of proteins and lipids recovered by isoelectric processing of fish by-products and low-value fish: a review. Food Chem. 124, 422–431.

Ghosh, R., 2006. Principles of Bioseparations Engineering. World Scientific Publishing Pte Ltd, Singapore.

Grela, E.R., Jensen, S.K., Jakobsen, K., 1999. Fatty acid composition and content of tocopherols and carotenoids in raw and extruded grass pea (*Lathyrus sativus* L). J. Sci. Food Agric. 79, 2075–2078.

Hakansson, B., Jagerstad, M., Oste, R., Akesson, B., Jonssson, L., 1987. The effects of various thermal processes on protein quality, vitamins and selenium content in whole-grain wheat and white flour. J. Cereal Sci. 6, 269–282.

Harper, J.M., 1988. Effects of extrusion processing on nutrients. In: Karmas, E., Harris, R.S. (Eds.), Nutritional Evaluation of Food Processing, third ed. Van Nostrand Reinhold Company, New York, NY, pp. 360–365.

Heng, W.W., Xiong, L.W., Ramanan, R.N., Hong, T.L., Kong, K.W., Galanakis, C.M., Prasad, K.N., 2015. Two level factorial design for the optimization of phenolics and flavonoids recovery from palm kernel by-product. Industr. Crop. Prod. 63, 238–248.

Holsinger, V.H., 1999. Lactose. Fundamentals of Dairy Chemistry, third ed. Van Nostrand Reinhold, New York, USA.

Hultin, H.O., Kelleher, S.D., 1999. Process for isolating a protein composition from a muscle source and protein composition. US Patent No. 6,005,073.

Hultin, H.O., Kelleher, S.D., 2000. High efficiency alkaline protein extraction. US Patent No. 6,136,959.

Hultin, H.O., Kelleher, S.D., 2001. Process for isolating a protein composition from a muscle source and protein composition. US Patent No. 6,288,216.

Hultin, H.O., Kelleher, S.D., 2002. Protein composition and process for isolating a protein composition from a muscle source. US Patent No. 6,451,975.

Jiang, B., Wang, J., 2013. Method for extracting soy protein, oligosaccharide and isoflavone from soybean wastewater by one-step process. Chinese Patent CN103265614-A.

Jin, Z., Hsieh, F., Huff, H.E., 1994. Extrusion cooking of cornmeal with soy fibre, salt, and sugar. Cereal Chem. 71, 227–234.

Khanal, R.C., Howard, L.R., Brownmiller, C.R., Prior, R.L., 2009a. Influence of extrusion processing on procyanidin composition and total anthocyanin contents of blueberry pomace. J. Food Sci. 74, H52–H58.

Khanal, R., Howard, L., Prior, R., 2009b. Procyanidin content of grape seed and pomace, and total anthocyanin content of grape pomace as affected by extrusion processing. J. Food Sci. 74, 174–182.

Killeit, U., 1994. Vitamin retention in extrusion cooking. Food Chem. 49, 149–155.

Kim, Y.S., Park, J.W., Choi, Y.J., 2003. New approaches for the effective recovery of fish proteins and their physicochemical characteristics. Fish. Sci. 69, 1231–1239.

Kim, S.-B., Ji, C.-I., Woo, J.-W., Do, J.-R., Cho, S.-M., Lee, Y.-B., Kang, S.-N., Park, J.-H., 2012. Simplified purification of chondroitin sulphate from scapular cartilage of shortfin mako shark (*Isurus oxyrinchus*). Int. J. Food Sci. Technol. 47, 91–99.

Korus, J., Gumul, D., Czechowska, K., 2007. Effect of extrusion on the phenolic composition and antioxidant activity of dry beans of *Phaseolus vulgaris* L. Food Technol. Biotechnol. 45, 139–146.

Kristinsson, H.G., Hultin, H.O., 2003. Changes in conformation and subunit assembly of cod myosin at low and high pH and after subsequent refolding. J. Agric. Food Chem. 51, 7187–7196.

Kristinsson, H.G., Ingadottir, B., 2006. Recovery and properties of muscle proteins extracted from tilapia (*Oreochromis niloticus*) light muscle by pH shift processing. J. Food Sci. 71, E132–E141.

Kristinsson, H.G., Theodore, A.E., Demir, N., Ingadottir, B., 2005. A comparative study between acid- and alkali-aided processing and surimi processing for the recovery of proteins from channel catfish muscle. J. Food Sci. 70, C298–C306.

Li, J., Chase, H.A., 2010. Applications of membrane technique for purification of natural products. Biotechnol. Lett. 32, 601–608.

Liang, Y., Hultin, H.O., 2005. Effect of pH on sedimentation of cod (*Gadus morhua*) muscle membranes. J. Food Sci. 70, C164–C172.

Liwei, Gu., House, S.E., Rooney, L.W., Prior, R.L., 2008. Sorghum extrusion increases bioavailability of catechins in weanling pigs. J. Agric. Food Chem. 56 (4), 1283–1288.

Lo, Y.M., Cao, D., Argin-Soyal, S., Wang, J., Hahm, T.S., 2005. Recovery of protein from poultry processing wastewater using membrane ultrafiltration. Bioresour. Technol. 96, 687–698.

Lorenz, K., Jansen, G.R., 1980. Nutrient stability of full-fat soy flour and corn-soy blends produced by low-cost extrusion. Cereal Foods World. 25, 161–172.

Mameri, N., Abdessemed, D., Belhocine, D., Lounici, H., 1996. Treatment of fishery washing water by ultrafiltration. J. Chem. Technol. Biotechnol. 67, 169–175.

Martins, B.C., Rescolino, R., Coelho, D.F., Zanchetta, B., Tambourg, E.B., Silveira, E., 2014. Characterization of bromelain from *Ananas comosus* agroindustrial residues purified by ethanol factional precipitation. Chem. Eng. Trans. 37, 781–786.

Massé, A., Vandanjon, L., Jaouen, P., Dumay, J., Kéchaou, E., Bourseau, P., 2008. Upgrading and pollution reduction of fishing industry process-waters by membrane technology. In: Bergé, J.P. (Ed.), Added Value to Fisheries Wastes. Transworld Research Network, Kerala, pp. 81–99.

Meinke, W.W., Mattil, K.F., 1973. Autolysis as a factor in the production of protein isolates from whole fish. Jo. Food Sci. 38, 864–866.

Meinke, W.W., Rahman, M.A., Mattil, K.F., 1972. Some factors influencing the production of protein isolates from whole fish. J. Food Sci. 37, 195–198.

Mireles DeWitt, C.A., Nabors, R.L., Kleinholz, C.W., 2007. Pilot plant scale production of protein from catfish treated by acid solubilization/isoelectric precipitation. J. Food Sci. 72, E351–E355.

Moscicki, L., van Zuilichem, D.J., 2011. Extrusion-cooking and related technique. In: Moscicki, L. (Ed.), Extrusion-Cooking Techniques: Applications, Theory and Sustainability. Wiley-VCH Publication, Singapore.

Moscicki, L., Mitrus, M., Wojtowicz, A., 2007. Technika ekstruzjiw przetworstwie rolno-spo zywczym. PWRiL, Warszawa (in Polish).

Murado, M.A., Fraguas, J., Montemayor, M.I., Vázquez, J.A., González, M.P., 2010. Preparation of highly purified chondroitin sulphate from skate (*Raja clavata*) cartilage by-products. Process optimization including a new procedure of alkaline hydroalcoholic hydrolysis. Biochem. Eng. J. 49, 126–132.

Ng, A., Lecain, S., Parker, M.L., Smith, A.C., Waldron, K.W., 1999. Modification of cell-wall polymers of onion waste III. Effect of extrusion-cooking on cell-wall material of outer fleshy tissues. Carbohyd. Polym. 39, 341–349.

Nierle, W., Elbaya, A.W., Seiler, K., Fretzdorff, B., Wolff, J., 1980. Veranderungen der getreideinhaltsstoffe wahrend der extrusion miteinem doppelschnecken extruder. Getreide Mehl. Brot. 34, 73–76.

Ning, L., Villota, R., Artz, W.E., 1991. Modification of corn fiber through chemical treatments in combination with twin-screw extrusion. Cereal Chem. 68, 632–636.

Nishigori, T., Takeda T., Ohori, T., 2000. Method for isolation and purification of chondroitin sulfate. Japanese Patent 2000273102-A.

Noguchi, A., Mosso, K., Aymanrd, C., Jevnink, J., Cheftel, J.C., 1982. Millard reactions during extrusion cooking of protein enriched biscuits. LWT 15, 105–110.

Paraskeva, C.A., Papadakis, V.G., Tsarouchi, E., Kanellopoulou, D.G., Koutsoukos, P.G., 2007. Membrane processing for olive mill wastewater fractionation. Desalination 213, 218–229.

Patel, S.R., Murthy, Z.V.P., 2012. Lactose recovery processes from whey: a comparative study based on sonocrystallization. Separ. Purif. Rev. 41, 251–266.

Patsioura, A., Galanakis, C.M., Gekas, V., 2011. Ultrafiltration optimization for the recovery of β-glucan from oat mill waste. J. Membr. Sci. 373, 53–63.

Piot, M., Fauquant, J., Madec, M.N., Maubois, J.L., 2004. Preparation of "serocolostrum" by membrane microfiltration. Lait 84, 333–342.

Pizzichini, M., Russo, C., Enea, C.R., Verdiana, S.R.I., 2005. Process for recovering the components of olive mill wastewater with membrane technologies. Int. Patent WO2. 0005, 123, 603.

Politz, M.L., Timpa, J.D., Wasserman, B.P., 1994. Quantitative measurement of extrusion-induced starch fragmentation products in maize flour using nonaqueous automated gel-permeation chromatography. Cereal Chem. 71, 532–536.

Qu, D., Wang, S.S., 1994. Kinetics of the formation of gelatinized and melted starch at extrusion cooking conditions. Starch/Starke 46, 225–229.

Rahmanian, N., Jafari, S.M., Galanakis, C.M., 2014. Recovery and removal of phenolic compounds from olive mill wastewater. J. Am. Oil Chem. Soc. 91, 1–18.

Repo-Carrasco-Valencia, R., de La Cruz, A.A., Alvarez, J.C.I., Kallio, H., 2009a. Chemical and functional characterization of kaiwa (*Chenopodium pallidicaule*) grain, extrudate and bran. Plant Foods Human Nutr. 64, 94–101.

Repo-Carrasco-Valencia, R., Pena, J., Kallio, H., Salminen, S., 2009b. Dietary fiber and other functional components in two varieties of crude and extruded kiwicha (*Amaranthus caudatus*). J. Cereal Sci. 49, 219–224.

Riaz, M., Asif, M., Ali, R., 2009. Stability of vitamins during extrusion. Crit. Rev. Food Sci. Nutr. 49, 361–368.

Riedl, K.M., Hagerman, A.E., 2001. Tannin-protein complexes as radical scavengers and radical sinks. J. Agric. Food Chem. 49, 4917–4923.

Rodrıguez, R., Jimenez, A., Fernandez-Bola~nos, J., Guillen, R., Heredia, A., 2006. Dietary fibre from vegetable products as source of functional ingredients. Trends Food Sci. Technol. 17, 3–15.

Roselló-Soto, E., Barba, F.J., Parniakov, O., Galanakis, C.M., Grimi, N., Lebovka, N., Vorobiev, E., 2015. High voltage electrical discharges, pulsed electric field and ultrasounds assisted extraction of protein and phenolic compounds from olive kernel. Food Bioprocess. Technol. 8, 885–894.

Rouilly, A., Jorda, J., Rigal, L., 2006a. Thermo-mechanical processing of sugar beet pulp. I. Twin-screw extrusion process. Carbohyd. Polym. 66, 81–87.

Rouilly, A., Jorda, J., Rigal, L., 2006b. Thermo-mechanical processing of sugar beet pulp. II. Thermal and rheological properties of thermoplastic SBP. Carbohyd. Polym. 66, 117–125.

Russo, C., 2007. A new membrane process for the selective fractionation and total recovery of polyphenols, water and organic substances from vegetation waters (VW). J. Membr. Sci. 288, 239–246.

Santamaria, B., Salazar, G., Beltrán, S., Cabezas, J.L., 2002. Membrane sequences for fractionation of polyphenolic extracts from defatted milled grape seeds. Desalination 148, 103–109.

Semwal, A.D., Sharma, A.K., Arya, S.S., 1994. Factors influencing lipid autoxidation in dehydrated precooked rice and Bengal Gram dhal. J. Food Sci. Technol. 31, 293–297.

Shih, M.C., Kuo, C.C., Chiang, W., 2009. Effects of drying and extrusion on colour, chemical composition, antioxidant activities and mitogenic response of spleen lymphocytes of sweet potatoes. Food Chem. 117, 114–121.

Silva, F.V. da., Santos, R.L. de. Andrade dos., Fujiki, T.L., Leite, M.S., Fileti, A.M.F., 2010. Design of automatic control system for the precipitation of bromelain from the extract of pineapple wastes. Ciênc. Tecnol. Aliment. 30, 1033–1040.

Singh, D., Chauhan, G.S., Suresh, I., Tyagi, S.M., 2000. Nutritional quality of extruded snakes developed from composite of rice broken and wheat bran. Int. J. Food Prop. 3, 421–431.

Singh, J., Dartois, A., Kaur, L., 2010. Starch digestibility in food matrix: a review. Trends Food Sci. Technol. 21, 168–180.

Soares, P.A.G., Vaz, A.F.M., Correia, M.T.S., Pessoa, Jr., A., Carneiro-da-Cunha, M.G., 2012. Purification of bromelain from pineapple wastes by ethanol precipitation. Separ. Purif. Technol. 98, 389–395.

Srihara, P., Alexander, J.C., 1984. Effect of heat treatment on nutritive quality of plant protein blends. Can. Inst. Food Sci. Technol. J. 17, 237–241.

Sudhakar, D.V., Mainp, S.B., 2000. Isolation and characterization of mango peel pectins. J. Food Process. Preser. 24, 209–227.

Tadashi, E., 2006. Sodium chondroitin sulfate, chondroitin-sulfate-containing material and processes for producing the same. US Patent No. 20,060,014,256.

Tahergorabi, R., Jaczynski, J., 2014. Protein and lipid recovery from seafood processing by-products with isoelectric solubilization/precipitation. In: Kim, S.-K. (Ed.), Seafood Processing By-Products: Trends and Applications. Springer, New York, US, pp. 101–123.

Tahergorabi, R., Beamer, S.K., Matak, K.E., Jaczynski, J., 2011. Effect of isoelectric solubilization/precipitation and titanium dioxide on whitening and texture of proteins recovered from dark chicken-meat processing by-products. LWT – Food Sci. Technol. 44, 896–903.

Tahergorabi, R., Beamer, S.K., Matak, K.E., Jaczynski, J., 2012. Functional food products made from fish protein isolate recovered with isoelectric solubilization/precipitation. LWT – Food Sci. Technol. 48, 89–95.

Takai, M., Kono, H., 2003. Salmon-Origin Chondroitin Sulphate. US Patent No. 20,030,162,744, 28 August.

Taskaya, L., Jaczynski, J., 2009. Flocculation-enhanced protein recovery from fish processing by-products by isoelectric solubilization/precipitation. LWT – Food Sci. Technol. 42, 570–575.

Taskaya, L., Chen, Y.C., Jaczynski, J., 2009a. Functional properties of proteins recovered from whole gutted silver carp (*Hypophthalmichthys molitrix*) by isoelectric solubilization/precipitation. LWT – Food Sci. Technol. 42, 1082–1089.

Taskaya, L., Chen, Y.C., Beamer, S., Jaczynski, J., 2009b. Texture and color properties of proteins recovered from whole gutted silver carp (*Hypophthalmichthys molitrix*) using isoelectric solubilisation/precipitation. J. Sci. Food Agric. 89, 349–358.

Thawornchinsombut, S., Park, J.W., 2004. Role of pH in solubility and conformational changes of pacific whiting muscle proteins. J. Food Biochem. 28, 135–154.

Tiwari, U., Cummins, E., 2009. Nutritional importance and effect of processing on tocols in cereals. Trends Food Sci. Technol. 20, 511–520.

Torres, J.A., Chen, Y.C., Rodrigo-Garcia, J., Jaczynski, J., 2007. Recovery of by-products from seafood processing streams. In: Shahidi, F. (Ed.), Maximising the Value of Marine By-Products. Woodhead Publishing Limited, Cambridge, UK, pp. 65–90.

Tsakona, S., Galanakis, C.M., Gekas, V., 2012. Hydro-ethanolic mixtures for the recovery of phenols from Mediterranean plant materials. Food Bioprocess. Technol. 5, 1384–1393.

Undeland, I., Kelleher, S.D., Hultin, H.O., 2002. Recovery of functional proteins from herring (*Clupea harengus*) light muscle by an acid or alkaline solubilization process. J. Agric. Food Chem. 50, 7371–7379.

Undeland, I., Kelleher, S.D., Hultin, H.O., McClements, J., Thongraung, C., 2003. Consistency and solubility changes in herring (*Clupea harengus*) light muscle homogenates as a function of pH. J. Agric. Food Chem. 51, 3992–3998.

Van Zuilichem, D.J., 1992. Extrusion cooking. Craft or science? PD thesis, Wageningen University, Netherlands.

Vannuccini, S., 2004. Overview of Fish Production, Utilization, Consumption and Trade. Food and Agriculture Organization Fisheries Information: Data and Statistics Unit, Rome.

Vasanthan, T., Gaosong, J., Yeung, J., Li, J., 2002. Dietary fibre profile of barley flour as affected by extrusion cooking. Food Chem. 77, 35–40.

Wolf, B., 2010. Polysaccharide functionality through extrusion processing. Curr. Opin. Colloid Interf. Sci. 15, 50–54.

Yagci, S., Gogus, F., 2008. Response surface methodology for evaluation of physical and functional properties of extruded snack foods developed from food-by-products. J. Food Eng. 86 (1), 122–132.

Zadow, J.G., 1984. Lactose: properties and uses. J. Dairy Sci. 67, 2654–2679.

Zeitoun, R., Pontalier, P.Y., Marechal, P., Rigal, L., 2010. Twin-screw extrusion for hemicelluloses recovery: influence on extract purity and purification performance. Bioresource Technol. 101, 9348–9354.

Zielinski, H., Michalska, A., Piskula, M.K., Kozlowska, H., 2006. Antioxidants in thermally treated buckwheat groats. Mol. Nutr. Food Res. 50, 824–832.

Zydney, A.L., 1998. Protein separations using membrane filtration: new opportunities for whey fractionation. Int. Dairy J. 8, 243–250.

CONVENTIONAL EXTRACTION

6

Juliana M. Prado*, Renata Vardanega**, Isabel C.N. Debien**, Maria Angela de Almeida Meireles**, Lia Noemi Gerschenson†, Halagur Bogegowda Sowbhagya‡, Smain Chemat§

*Centro de Ciências da Natureza (CCN), UFSCar (Federal University of São Carlos), Buri, Brazil; **LASEFI/DEA (Department of Food Engineering)/FEA (School of Food Engineering)/UNICAMP (University of Campinas), Campinas, São Paulo, Brazil; †Industry Department, Natural and Exact Sciences School (FCEN), Buenos Aires University (UBA), Buenos Aires, Argentina; National Scientific and Technical Research Council of Argentina (CONICET), Buenos Aires, Argentina; ‡Department of Plantation Products, Spices and Flavour Technology, CSIR – Central Food Technological Research Institute, Mysore, Karnataka, India; §Division Santé Centre de Recherches Scientifique et Technique en Analyses Physico-Chimiques (C.R.A.P.C), Bon-Ismail, Algeria

6.1 INTRODUCTION

Conventional extraction technologies to recover value-added products from plant materials include steam distillation, solvent extraction, and acid and alkali extraction. The application of supercritical fluids, microwaves, sonication, and enzymes as pretreatment or sole extraction methods has increased over the last decades, as these technologies have proved beneficial by improving the yield of target compounds from the raw materials, decreasing processing costs, and replacing toxic solvents with environmentally friendly ones (Meireles, 2009; Rostagno and Prado, 2013; Deng et al., 2015; Roselló-Soto et al., 2015; Galanakis, 2012; Galanakis et al., 2013a). All of the extraction techniques have advantages and drawbacks that should be carefully evaluated for each raw material, as the best extraction technique for one residue may not be the best solution for another.

Besides, the waste differs from one material to another. For instance, from the fruit industries, peels and kernels can be used as a source of pigments and flavonoids (Galanakis, 2011, 2012; Galanakis et al., 2013b, 2015b; Rahmanian et al., 2014; Heng et al., 2015), which can find applications as natural colorants and antioxidants, respectively. On the other hand, waste from the spice industry can be used as a source of protein and dietary fiber, whereas waste from the fish and whey industries can be processed to recover protein concentrates and pigments (Galanakis and Schieber, 2014; Galanakis et al., 2014). Within this context, each technology must be evaluated for each raw material separately in order to reach the highest extraction yield with feasible costs (Patsioura et al., 2011; Galanakis, 2013, 2015; Galanakis et al., 2014). Due to this feature, there are many data in the literature regarding the extraction of bioactive compounds from wastes of the food industry using several extraction techniques. In this chapter some examples are presented and discussed.

Food Waste Recovery. http://dx.doi.org/10.1016/B978-0-12-800351-0.00006-7

6.2 SOLVENT EXTRACTION

6.2.1 DESCRIPTION OF THE TECHNOLOGY

Solid/liquid extraction is used for the extraction of a constituent from a solid by means of a solvent. The process generally involves:

1. the change of phase of the solute as it dissolves in the solvent,
2. the solute diffusion from the solid pores to the outside of the particles, carried by the solvent, and
3. the transfer of the solute from the solution that is in contact with the particles to the main bulk of the solution.

Particle size, solvent type, temperature, and agitation are important factors in this process. In liquid/liquid extraction, it is essential that the liquid feed and solvent are at least partially immiscible, and the three stages involved are:

1. contact between the feed mixture and the solvent,
2. separation of the resulting two phases, and
3. removal and recovery of the solvent from each phase.

The stages of solid/liquid and liquid/liquid extraction may be carried out either as a batch or as a continuous process. Figure 6.1 shows a deposit in a batch extractor with a false bottom. The solid to be extracted is placed on the grid that limits the false bottom. The solvent is distributed in the surface of the solid and it percolates, producing a solution created by the solvent and the target compound. The

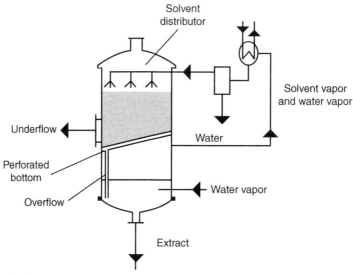

FIGURE 6.1 Batch Extractor

latter passes the false bottom and is recovered as the overflow. The underflow is the residue formed by the remaining solids and the solution retained within them. The overflow can be concentrated by boiling. For the sustainability of the process, one recycle step is generally introduced to recover part of the energy (Canovas and Ibarz, 2002). The extraction units are followed by distillation or a similar operation in order to separate the solvent the solute. As a consequence, not only the selective nature of the solvent but also the ease of its separation is an important factor to consider (Richardson et al., 2002).

6.2.2 APPLICATIONS

In general, food wastes from industrialization are important sources of valuable compounds, such as polyphenols, carotenoids, and carbohydrates in the case of plant wastes, and fatty acids from fish wastes. Solvent extraction is very convenient, as the solvent provides a physical carrier to transfer molecules between different phases (i.e. solid, liquid, and vapor). Phenols are easily solubilized in polar protic solvents like hydroalcoholic mixtures. Indeed, different fractions can be obtained by varying alcohol concentration based on polarity. Carotenoids (i.e. tomato lycopene) are in general hydrophobic and lipophylic, preferring polar aprotic or nonpolar mediums (i.e. acetone or ethyl acetate, respectively). In this case, the solvent should be removed completely from the extract prior to its reutilization in food products (Galanakis, 2012). Beyond the solvent used, other factors are important for extraction efficiency. For example, the size of the particle involved in the case of solid/liquid extraction, the solvent/solid ratio (S/F), and the use of multiple extraction stages. Results reported in the literature are diverse. Diankov et al. (2011) reported that the highest and faster extraction of natural antioxidants from lemon peels proceeded when using a particle size of 1 instead of 2 mm. Rajha et al. (2014) also concluded that the decrease of particle size increased the total phenolic content in the extracts obtained from grape wastes. Moreover, an increase in the ethanol:water/sample ratio resulted in a decrease of phenolic content in the extracts. On the other hand, particle size did not affect phenolic compound extraction from pistachio wastes (Mokhtarpour et al., 2014). When Stamatopoulos et al. (2014) used particle sizes ranging from 0.05 to 1.0 mm, solvent-to-feed (S/F) ratios ranging from 5:1 to 10:1, and a multistage extraction scheme, the particle size below 0.2 mm decreased the extraction yield of polyphenols from blanched olive leaves. This result was attributed to the fact that the small particles tend to agglomerate. Thereby, solvent penetration and mass transfer were decreased in the solid matrix. Indeed, the multiple stages of extraction could significantly increase the yield of polyphenols (Galanakis, 2012; Tsakona et al., 2012).

For the recovery of carotenoids from shrimp industrial waste, Sachindra et al. (2006) tested different mixtures of hexane and other solvents, different S/F ratios and different extraction cycles, and observed the highest yields (40.8 μg/g waste) using a mixture of 3:2 hexane:isopropyl alcohol with an S/F of 5:1 and with three cycles. According to this study, the carotenoid yield increased as a result of dried tissues and multiple extraction cycles. Sánchez-Camargo et al. (2011) used the same solvent for isolating carotenoids from Brazilian red-spotted shrimp (*Farfantepenaeus paulensis*) waste and obtained a yield of 53 mg/kg waste. Strati and Oreopoulou (2011) tested 15 mixtures of hexane and ethyl acetate varying from 10:90 to 80:20 (v/v), with S/F ratios varying from 3:1 to 10:1 (v/w), and particle size varying between 0.5 and 1.0 mm, for the isolation of carotenoids from tomato waste. The authors observed that the highest yield of nonpolar lycopene and β-carotene as well as the polar lutein (36.5 mg/kg) was obtained using 45% v/v hexane in ethyl acetate, S/F ratio of 9:1, and particle size of 0.56 mm.

The selection of the solvents used is directly related to the polarity and thus to the solubility of the target compounds. The selection of the S/F ratio and the size of the particles if the waste is a solid must be studied case by case.

6.2.3 SAFETY CONCERNS AND LOCAL REGULATIONS

Although the yield of the aforementioned technologies is important for industrial-scale processing, other factors such as product safety and general cost govern the final decision for the selected methodology. Ethanol has been declared as GRAS (generally recognized as safe) by the Food and Drug Administration (FDA, 2014) and is relatively cheap. The European Parliament approved a directive (EC, 2009) concerning the extraction solvents, which can be used during the processing of raw materials of foodstuffs, food components, or food ingredients. As can be observed, not only product safety but also local regulations must be considered. The use of additional stress factors like acid, alkali and enzymatic pretreatments can help to reduce the quantity of organic solvents used for extraction. Also the replacement of aggressive organic solvents with solvents of lower toxicity must be considered.

6.3 ACID, ALKALI, AND ENZYME EXTRACTION

The application of enzymes during extraction processes is becoming popular due to their ability to improve the yield of target compounds. Besides, the addition of edible acids (i.e. sulfur dioxide or citric acid) in a specific concentration has been used to enhance extraction of water soluble anthocyanins. Enzyme pretreatment of wastes, like grape skin or must after juice extraction, could be applied to enhance the extraction of pigments like anthocyanins. Cellulolytic enzymes (i.e. cellulase, hemicellulase, xylanase, pectinase, or a mixture of these enzymes) could be used for extraction purposes. These enzymes act on the cell wall of the plant waste, increasing the permeability of the cell wall and facilitating the leaching out of value-added cell components like pigments.

6.3.1 ANTHOCYANINS

In the grape juice industry and in other fruit juice industries (berry processing), skin or peel of the fruit waste is rich in anthocyanins. Therefore, after juice extraction the pressed residue containing peels (skin) and seeds can be further processed to obtain value-added products. Anthocyanins can be recovered from grape skin using acidic methanol or water as extraction solvent. Enzyme application before extraction can enhance the extraction efficiency of such pigments. The production of anthocyanin-rich extracts from the residue of the vinification of three *Vitis vinifera* grape varieties (Cabernet Sauvignon, Ribier, and Carmenere) has been reported. The method involved the addition of an active pectolytic enzyme preparation in an aqueous medium, which resulted in reduced time required for maceration, setting, and filtration. Enzymatic complexes studied were Pectinex BE3-L (pectinesterase, pectinlyase, hemicellulase, and cellulase from *Aspergillus niger*), Vinozyme C (pectinase and cellulase from *A. niger* and *Trichoderma longibrachiatum*), and Vinozym G (pectinlyase, polygalacturonase, hemicellulase, and cellulase from *A. niger*). The treatment of Ribier grape skin with Vinozyme C for 2 h gave optimum extraction (Munoz et al., 2004). A schematic diagram for the extraction of value-added products from the grape industry is shown in Fig. 6.2.

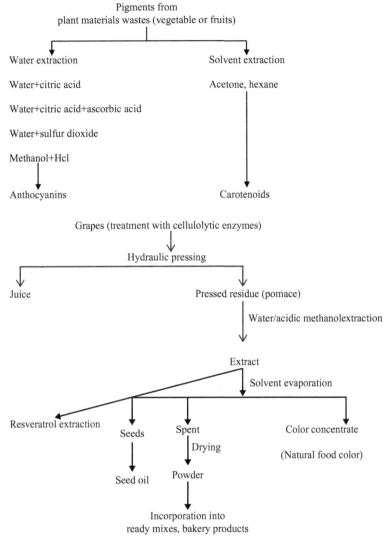

FIGURE 6.2 Recovery of Nutraceuticals from Grape Juice Industry Wastes: Natural Pigments, Seed-Oil, Pomace Powder Rich in Fiber, Flavanoids and Resveratrol

6.3.2 DIETARY FIBERS

The method for the extraction of soluble and insoluble fiber involves digestion of the sample successively by the enzymes amylase, pepsin and pancreatin followed by precipitation with alcohol (Asp et al., 1983). Highly valued grape seed oil as well as aromas can be fully recovered in their natural composition from seeds using enzyme treatment followed by distillation. Different extraction techniques can be applied to recover dietary fibers, protein, and starch from spice processing wastes. For instance, acid precipitation using trichloroacetic acid can be applied to remove protein. Thereafter, acid extraction and enzyme extraction with amylase, pepsin, and pancreatin have been employed to recover soluble and insoluble fiber fractions.

6.3.3 FERULIC ACID

Saponification of fruit peels with alkali (2 M NaOH) has been used for the extraction of esterified ferulic acid from agricultural waste. The process was standardized using response surface methodology and resulted in a 1.3-fold extraction increase compared with the conventional method. By optimized alkaline hydrolysis of maize bran, extraction of esterified ferulic acid was 1.9 g vs 1.5 g/100 g by conventional method. Peels of pomegranate, pineapple, and orange were found to contain 0.192, 0.018, and 0.021 g esterified ferulic acid/100 g, respectively.

6.3.4 PROTEINS

The recovery of proteins from fish meal using acid precipitation and sodium hexametaphosphate has been investigated (Batista, 1999). The method for the solubilization of protein was optimized with respect to the effect of pH, type of alkali used (NaOH, $Ca(OH)_2$), salt concentration, S/F, and extraction temperature (Batista, 1999). Hake and monkfish "sawdust" from a filleting factory was used for the study. The acid precipitation of proteins from the extract was carried out with hydrochloric acid. The solubilization of hake proteins was higher than that of monkfish and an increase in solubility was more towards the alkaline side due to the content of connective tissue in the second case. The amino acid compositions of hake waste proteins obtained by acid precipitation showed reductions in proline and glycine contents, while phenyl alanine content increased with alkaline precipitation of protein. There are 16–18 amino acids present in fish proteins that can be obtained either by enzymatic or chemical processes. In the enzymatic method, hydrolysis of protein substrates with enzymes such as alcalase, neutrase, carboxypeptidase, chymotrypsin, pepsin, and trypsin is carried out. In the chemical process, acid or alkali is used for the breakdown of protein to amino acids. The main disadvantage of the chemical method is the complete destruction of tryptophan and cysteine and partial destruction of tyrosine, serine, and threonine. The amino acids present in the fish can be utilized in animal feed in the form of fishmeal and sauce or can be used in the production of various pharmaceuticals.

6.4 MICROWAVE-ASSISTED EXTRACTION
6.4.1 DESCRIPTION OF THE TECHNOLOGY

Increasing interest to offset food waste burden has led to different technologies for the acceleration of valuable compound extraction from wastes by means of microwave-assisted extraction. This trend has initiated new research and developments in the field of natural product extraction in a desire to maximize yield and selectivity towards added-value compounds. Initially reported by Ganzler et al. (1986), microwave technology was further developed by Paré et al. into a family of technologies for the extraction of various chemical compounds, which were patented as microwave-assisted processes (Paré et al., 1990).

Microwave irradiation uses an electromagnetic field with a frequency that ranges from 300 MHz to 300 GHz. However, 0.915 and 2.45 GHz represent the most used frequencies worldwide among the few frequencies allowed for industrial, scientific, and medical uses. It is important to note that the energy of a microwave photon at 2.45 GHz (corresponding to a wavelength of 12.2 cm in air) is close to 0.00001 eV. This energy is too weak to break down hydrogen bonds. Therefore, it is important to insist that the efficiency of microwave irradiation relies on efficient conversion of electromagnetic energy to heat.

The basic components of a microwave applicator or an oven are:

1. a microwave source (magnetron), which is characterized by frequency and output power,
2. a waveguide that allows connection of the source to the cavity and permits wave propagation towards the applicator,
3. a microwave cavity, which is a metallic box of various shapes and sizes where the samples can be held for irradiation.

Depending on the nature of the target compounds, two different approaches can be used to achieve their recovery. In the first approach, a polar solvent (ethanol, methanol, water, etc.) absorbs all the microwave energy and dissipates it into heat. This way, the heating process becomes more volumetric and faster, penetration into the sample matrix is improved, and dissolution of target components takes place. The second approach is based on the intrinsic potential of polar compounds to transform electromagnetic energy into heat due to their high dielectric constant. In this case, solvent superheating is believed to increase high temperature and pressure points (called hot spots), where heat is directly transferred to the solid. This fact causes faster extraction of target compounds through cell walls of the material into the bulk medium. In addition, the application of microwaves is believed to disrupt cell walls as a consequence of a continuous dehydration process favored by fast exudation of content into the bulk medium (Chemat and Esveld, 2013).

6.4.2 APPLICATIONS AND FURTHER DEVELOPMENTS

Generally, traditional techniques require the use of a large amount of solvent and extended processing time to achieve optimum recovery. In this respect, microwave extraction stands as an interesting option for the extraction of valuable compounds from food wastes, as it promises efficiency, energy saving, higher yields, product quality, reduced extraction time, and, finally, low solvent consumption that leads to increased sustainability and consumer adoption.

Several classes of compounds such as essential oils, flavors, pigments, antioxidants, glycosides, and terpenes have been extracted successively by different microwave extraction techniques (Huma et al., 2011). This success triggered innovations and delivered derivatives that improved its performance. These include:

1. *Microwave-assisted distillation (MAD)*: Usually applicable for the isolation of essential oils from herbs and spices. It was further developed to vacuum microwave hydrodistillation (VMHD), elaborated and patented by Archimex in the 1990s (Mengal et al., 1993). The latter is based on selective heating by microwaves combined with application of sequential vacuum (between 100 and 200 mbar) to evaporate the azeotropic mixture of water-oil from the biological matrix.
2. *Microwave-assisted solvent extraction (MASE)*: In this case, the energy of microwaves is used to produce molecular movement and rotation of dipoles, causing a rapid heating of the solvent or the sample or both. Several studies using MASE for the recovery of food wastes were reported in the literature (Table 6.1). Prominent examples include total phenolics from vegetable

Table 6.1 Some Applications of Microwaves on the Recovery of Valuable Compounds from Food Wastes

Food Waste	Target Compound	Operational Conditions	Yield (Target Compound per Total Extract)	References
Yellow horn kernel (*Xanthoceras sorbifolia Bunge*)	Triterpene saponins	Microwave-assisted extraction (MAE) 51°C, 7 min, 900 W, 32 mL/g, 42% (v/v) ethanol and three cycles	11.62 ± 0.37% Radical-scavenging activity with an IC_{50} value of 0.782 mg/mL	Li et al. (2010)
Vitis vinifera seeds	Total polyphenols (TP) *o*-Diphenols (OD) Total flavonoids (TF)	Microwave-assisted extraction (MAE) Methanol at 110°C under nitrogen atmosphere and microwave irradiation (60 W) for 60 min	TP: 108.3 mg gallic acid equivalent/g OD: 47.0 mg gallic acid equivalent/g TF: 47.2 mg catechin equivalent/g DPPH-scavenging capacity of 78.6 μL of extract/μg DPPH	Alessandro et al. (2010)
Yellow onion (*Allium cepa L.*) wastes	Total quercitin	Vacuum microwave hydrodiffusion and gravity (VMHG) 500 W for 500 g of plant material (1 W/g)	662.27 mg/100 g of dry wastes DPPH-scavenging capacity with IC_{50} value of 4.39 mg/mL	Huma et al. (2011)
Spent filter coffee	Total polyphenols	Microwave-assisted extraction (MAE) 20% aqueous ethanol solution under 40 s of microwave radiation (80 W)	398.95 mg gallic acid equivalent/g DPPH-scavenging capacity with IC_{50} value of 3.75 μg/mL	Pavlović et al. (2013)
Grape red berries (Alphonse Lavallée, *Vitis vinifera*) from press cake waste	Total phenolics (TP) Total anthocyanins (TAC)	Solvent-free microwave hydrodiffusion and gravity (MHG) The press cake was heated for 20 min at atmospheric pressure and at a constant power density of 1 W/g without addition of solvent	TP: 21.41 ± 0.04 mg of gallic acid equivalent/g TAC: 4.49 ± 0.01 μg of malvidin-3-O-glucoside/g	Bittar et al. (2013)
Grape seed waste (*Vitis vinifera*)	Total phenolics (TP) Total flavonoids (TF) Total proanthocyanidins (TPA)	Microwave-assisted aqueous two-phase extraction (MAATPE) Solvent system: 32% (w/w) acetone/16% (w/w) ammonium citrate; microwave power of 650 W for 75 s	TP: 82.7 mg/g TF: 52.6 mg/g TPA: 30.7 mg/g	Dang et al. (2013)

Table 6.1 Some Applications of Microwaves on the Recovery of Valuable Compounds from Food Wastes *(cont.)*

Food Waste	Target Compound	Operational Conditions	Yield (Target Compound per Total Extract)	References
Citrullus lanatus fruit rinds	Pectin	Microwave-assisted extraction (MAE) Microwave power of 477 W, irradiation time of 128 s, pH of 1.52, S/F of 20.3 (v/w)	25.79 g of dry pectin per 100 g of *C. lanatus* fruit rinds peel waste	Prakash et al. (2014)
Wastes from: – asparagus – cauliflower – celery – chicory	Total phenolics	Microwave-assisted extraction (MAE) Microwave power of 750 W for 4 min Water as a solvent, S/F of 1:2 (w/w)	Content (mg gallic acid equivalent/kg of fresh material): – asparagus: 245 ± 16 – cauliflower: 486 ± 8 – celery: 155± 13 – chicory: 392 ± 32	Baiano et al. (2014)
Passion fruit peels (*Passiflora edulis* f. *flavicarpa*)	Pectin	Microwave-assisted extraction (MAE) Microwave power of 628 W for 9 min, using nitric, acetic, and tartaric acid	Expressed as g/100 g passion fruit peel flour Nitric acid: 13 Acetic acid: 12.9 Tartaric acid: 18.2	Seixas et al. (2014)

wastes (Alessandro et al., 2010; Baiano et al., 2014; Pavlovic et al., 2013) and pectin (Prakash et al., 2014; Seixas et al., 2014).

3. *Solvent-free microwave extraction*: Patented by Chemat et al. (2004), it stands as a distillation process combining microwave heating and distillation at atmospheric pressure without the addition of solvent to the matrix. Vapors released from the sample are condensed in a Clevenger-type system, in which essential oils are separated by phase density difference.

4. *Microwave hydrodiffusion and gravity (MHG)*: This is an original "upside-down" alembic system combining microwave heating and earth gravity at atmospheric pressure (Fig. 6.3). Based on the internal heating of in situ water present in biomass, hydrodiffusion occurs. This process leads to gland and oleiferous receptacle rupture as a consequence of microwave irradiation. In this case, a cooling system outside the microwave oven cools the extract continuously, which drops by gravity out of the microwave reactor and falls through the perforated Pyrex disc. Recent studies reported the successful application and higher yields exhibited by MHG for phenolics, flavonoids, and anthocyanins from waste sources like grape red berries (Bittar et al., 2013) (Table 6.1).

Furthermore, new techniques like microwave-integrated Soxhlet extraction (Virot et al., 2007), vacuum microwave hydrodiffusion and gravity (VMHG) (Huma et al., 2011), and combined ultrasound and microwave-assisted extraction (UMAE) have also emerged (Cravotto et al., 2008). These techniques achieved higher polyphenol yields compared with conventional techniques, without altering

1 L reactor at lab scale   150 L pilot scale (75 L of biomass)

FIGURE 6.3 Solvent-Free Microwave Hydrodiffusion and Gravity (NEOS GR) Developed by Milestones SRL

Courtesy of Prof. CHEMAT Farid, Université d'Avignon et des Pays de Vaucluse, France

their antioxidant potential. Pressurized microwave extraction with regard to its application to separate macro- and micromolecules is further discussed in Chapter 10.

6.5 STEAM DISTILLATION AND HYDRODISTILLATION

6.5.1 DESCRIPTION OF THE TECHNOLOGY

Steam distillation is a process used for the recovery of volatile compounds with high boiling point, from inert and complex matrices, solid or liquid, using saturated or superheated steam as separation and energy agent (Cerpa et al., 2008). This process is used for the extraction of essential oil from plants

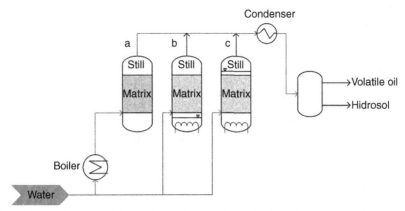

FIGURE 6.4 Generalized Flowsheet of the Different Types of Steam Distillation

Adapted from Cerpa et al., 2008

(Palma et al., 2013). In practice, the process uses water and/or steam as extracting agent to vaporize or liberate the volatile compounds from the raw material. The compounds are volatilized by absorbing heat from the steam, and are then transported to the steam where they are diffused. The resulting vapor phase is cooled and condensed prior to separating the water from the organic phase based on their immiscibility. In this process, two products are obtained: volatile oil and hydrosol. The volatile oil is in the upper phase and the hydrosol (water and some hydrolyzed compounds) is in the bottom phase of the decanter. According to the type of contact between the matrix and the water and/or steam, there are three variants of the steam distillation process (Fig. 6.4): dry steam distillation (a), direct steam distillation (b), and hydrodistillation (water distillation) (c).

In direct steam distillation (Fig. 6.4b), the matrix is supported on a perforated grid or screen inserted slightly above the bottom of the still. This scheme does not allow direct contact with water, whereas the boiler can be inside or outside the still. The low-pressure saturated steam flows up through the matrix, collecting the evaporated compounds. In hydrodistillation (Fig. 6.4c), the matrix is in direct contact with the boiling water either by floating or by being completely immersed depending on its density. The boiler is inside the still and agitation may be necessary to prevent agglutination. In dry steam distillation (Fig. 6.4a), the matrix is supported and steam flows through it. The differences are due to the steam being generated outside the still and superheated at moderate pressures (Cerpa et al., 2008).

Steam distillation is largely used because it presents several advantages compared with the other extraction processes:

1. the method generates organic solvent-free products,
2. there is no need for subsequent separation steps,
3. it has a large capacity for processing at the industrial scale,
4. the equipment is inexpensive, and
5. there is extensive know-how available for this technology.

On the other hand, the main drawbacks of this technology include:

1. sensitive compounds could be thermally degraded and/or hydrolyzed,

2. very long extraction times (1–5 h) are needed, and
3. high energy consumption (Palma et al., 2013).

Steam distillation can be combined with other extraction methods like microwave or ultrasound to increase efficiency. The combination of techniques may provide faster extraction kinetics at lower cost, besides reducing the environmental impact of the process and producing a similar product to those obtained by conventional hydrodistillation (Farhat et al., 2011; Ferhat et al., 2006; Palma et al., 2013).

6.5.2 APPLICATIONS

Steam distillation is the most used method to obtain essential oil from natural sources. These essential oils are used as food additives, natural flavorings, and/or preservatives, and in cosmetics and pharmaceutical industries, due to their notable antimicrobial, antioxidant, and anti-inflammatory properties (Imelouane et al., 2009; Kim et al., 2008; Yang et al., 2009).

On the other hand, there are few studies on the use of steam distillation to recover value-added compounds from food wastes. One of the reasons is that it is considered by most scientists to be a simple technique, not worthy of further study. It is generally used only as a reference method to be compared with novel and emerging extraction technologies. Therefore, this classical technology should be further developed, as it is inexpensive and simple to operate.

Nevertheless, some studies have explored the extraction of essential oil from citrus fruit waste by steam diffusion alone or in combination with another technique, such as microwaving (Table 6.2). Limonene (a monoterpene hydrocarbon) is the most abundant component present in the essential oil extracted from citrus fruit peel (Farhat et al., 2011; Ferhat et al., 2006; Yang et al., 2009) and it has antibacterial properties (Yang et al., 2009). Studies reported in Table 6.2 show that so far the yields of classical steam diffusion alone have been comparable to the yields of its combination with a more expensive and complex technique (i.e. microwave-assisted extraction) to recover essential oil from citrus fruit. The few data shown in the table, show that this technology has not been studied to its full extent.

Table 6.2 Some Applications of Steam Distillation on the Recovery of Valuable Compounds from Food Wastes

Food Waste	Target Compound	Operational Conditions	Yield (%)	References
Orange peel (*Citrus sinensis* L. *Osbeck*)	Essential oil	Microwave accelerated distillation (MAD) – 1000 W, 100°C, 30 min	MAD: 0.42 ± 0.02% HD: 0.39 ± 0.02%	Ferhat et al. (2006)
	Limonene	Hydrodistillation (HD) – 2 L of water, 180 min	MAD: 76.7% HD: 78.5%	
Orange peel (*Citrus sinensis* L. *Osbeck*)	Essential oil	Microwave steam distillation (MSD) – 1000 W, 100°C, 12 min	MSD: 1.54% SD: 1.51%	Farhat et al. (2011)
	Limonene	Dry steam distillation (SD) – 40 min	MSD: 94.88% SD: 95.03%	
Hallabong (*Citrus unshiu*)	Essential oil	Hydrodistillation – 360 min	1.0%	Yang et al. (2009)

6.6 SUPERCRITICAL FLUID EXTRACTION
6.6.1 DESCRIPTION OF THE TECHNOLOGY

Supercritical fluid extraction (SFE) is a process based on the use of solvents above or near their critical temperature and pressure to recover extracts from solid matrices. In the supercritical state, the solvent has intermediary properties between gases and liquids, which makes it useful as solvent for several compounds. This technology uses renewable solvents, such as CO_2, which is the most used supercritical fluid and has a critical point of 7.38 MPa and 304.2 K (Brunner, 1994; Rosa and Meireles, 2009).

SFE is characterized mainly by two steps: (i) extraction and (ii) separation of the extract from the solvent. Figure 6.5 presents a simplified scheme of the SFE process. The solvent (stored in R2 and pressurized by B1) continuously flows through the extractor filled with the raw material (E). If the addition of a cosolvent is necessary (stored in R3 and pumped by B2), it is mixed with the supercritical CO_2 in the mixer (MI) before the inlet of the extractor (E). The solutes are extracted due to convection and diffusion principles. After the extraction step (C2), a separation step follows where the pressure is reduced, leading to decrease in the solvent power of the fluid, and the precipitation of the solute in the separator (S). After the separation step, the solvent can be recirculated (C4) in the system. In industrial plants, the extraction process may be conducted at multiple stages, and the depressurization step may be fractionated in multiple separators, resulting in extracts with different compositions (Prado and Meireles, 2012).

6.6.2 ADVANTAGES AND APPLICATIONS

The use of SFE for the recovery of valuable compounds from different matrices has been extensively applied and has several advantages over traditional extraction techniques. The solvent can be easily

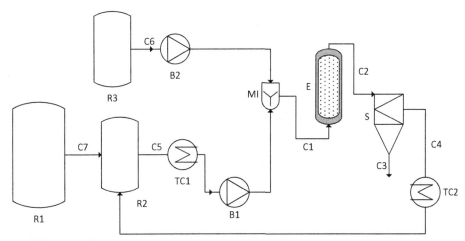

FIGURE 6.5 SFE Process Scheme

R1, CO_2 reservoir; R2, CO_2 buffer tank; R3, cosolvent reservoir; B1, CO_2 pump; B2, cosolvent pump; TC1 and TC2, heat exchangers; MI, mixer; E, extractor; S, separator. C1, solvent inlet stream in the extractor; C2, solvent + extract outlet stream from the extractor; C3, extract (+cosolvent) outlet stream; C4, CO_2 recycling stream; C5, CO_2 feeding stream; C6, cosolvent feeding stream; C7, CO_2 make-up stream.

removed from the mixture by pressure reduction and/or temperature adjustment. Moreover, the supercritical fluid extraction is relatively fast due to the low viscosity, high diffusivity, and tunable solvent power of the supercritical fluid. Also, due to the use of minimum organic solvent and recirculation of CO_2 into the system, supercritical fluid extraction is considered a safe and green technology (Cavalcanti et al., 2011).

Table 6.3 shows some examples of the application of supercritical fluid extraction for the recovery of valuable compounds. It can be noticed that there is a lot of interest in adding value to food wastes by using this technology. This happens mostly because of its environmental appeal, associated with the fact that this technology is economically feasible at the industrial level when the products have high value (Pereira et al., 2013).

Beyond the several advantages of using residues for extraction processes, the production costs with the purchase of raw material can also be diminished, when the supercritical fluid extraction plant and the residue generating industry are close to each other. It is an important advantage because with increasing processing capacity, the proportion of the cost of raw material increases in importance to the overall cost of the process (Farías-Campomanes et al., 2013).

6.7 SCALE-UP AND ECONOMIC ISSUES

The isolation of high value bioactive compounds from food wastes must be carefully analyzed because it is conditioned by the availability of large amounts of food wastes on a regular basis and by the high costs of isolating specific components often present in small amounts.

Steam diffusion and solvent extraction have always been the most commonly used processes to recover essential oils and raw extracts from condimentary, medicinal, and aromatic plants. At the industrial scale the most common methods are solvent extraction by agitation and dry steam distillation. The scale-up and economics of these processes have been well established for a long time. On the other hand, a major point to be considered at the industry level is energy consumption. At the end of the extraction process, the observed extract yield increase observed is very low compared with the beginning of extraction, leading to longer processing and higher energy consumption for a slow extraction rate (Cerpa et al., 2008).

The use of additional stress factors like microwaves, ultrasound, and enzymatic pretreatments can help to reduce the quantity of organic solvents used for extraction and thus improve the processes' economics while decreasing their environmental footprint. Therefore, when steam diffusion and solvent extraction are combined with microwave irradiations, one possible advantage is the reduced cost of the process in terms of energy, time, and environmental impact (Farhat et al., 2011; Ferhat et al., 2006).

Looking at its numerous advantages and the development of equipment for large-scale commercial operation, microwave-assisted extraction offers a very good perspective for the recovery of valuable compounds from food wastes. However, its industrial applications are for the moment limited, and concern mainly the extraction of essential oils or high value lipids used in cosmetics and pharmaceutical applications.

Pilot-scale or industrial plants revolve around two models, namely Radient Technologies (Radient Technologies, 2014) and SAIREM SAS (SAS, 2014). The first one is based on the technology developed by the Paré team through MAP patents and claims to offer dedicated large-scale commercial

Table 6.3 Some Applications of SFE on the Recovery of Valuable Compounds from Food Wastes

Type of Waste	Operational Conditions: Type of Solvent (s); Flow Rate (f); Time (t); Raw Material Moisture Content (h); Particle Size (d)	Temperature (T); Pressure (P)	Extract Yield	Target Compounds	References
Vegetable residues					
Broccoli leaves	$s = CO_2$ + methanol (20–35%); $f = 2$ mL/min; $t = 10$–50 min	$T = 50$–80°C $P = 100$–250 bar		Free amino acids (10 g/kg lyophilized sample)	Arnáiz et al. (2012)
Carob pulp kibbles	$s = CO_2$; CO_2 + ethanol:water, 80:20, v/v (0–12.35 % wt); $f = 0.28$–0.85 kg/h; $d = 0.27$–1.07 mm	$T = 40$–70°C $P = 150$–220 bar	0.2–0.6%	Phenolic compounds (10–160 mg gallic acid equivalent/kg carob pulp)	Bernardo-Gil et al. (2011)
Elderberry pomace	First step: $s = CO_2$; $f = 12.3 \pm 1.4 \times 10^{-5}$ kg/s; $t = 40$ min; $h = 5.2 \pm 0.1\%$ (w/w). Second step: $s = CO_2$ + ethanol: water (10–100%); $f = 12.3 \pm 1.4 \times 10^{-5}$ kg/s; $t = 45$ min; $h = 5.2 \pm 0.1\%$ (w/w)	$T = 40$°C $P = 200$ bar	First step: $17 \pm 0.3\%$ dry based (d.b.) Second step: 1.7–24.2% (d.b.)	Anthocyanins (0–24.2%, d.b.)	Seabra et al. (2010)
Grape bagasse	$s = CO_2$ + ethanol (10%, w/w); $f = 9.8 \pm 0.6 \times 10^{-5}$ kg/s–12.6 $8 \pm 0.6 \times 10^{-5}$ kg/s; $t = 300$–340 min; $h = 1.9 \pm 0.1\%$; $d = 0.83 \pm 0.02$ mm	$T = 40$°C $P = 200$ and 350 bar	$5.5 \pm 0.1\%$	Phenolic compounds (g/kg): gallic acid (3.0–4.5); protocatechuic acid (0.9–1.1); 3-ρ- hydroxybenzoic acid (1.4–1.8); vanillic acid (4.2–4.3); syringic acid (9.5–11); p-coumaric acid (0.52–0.7); quercetin (0.15–0.27).	Farías-Campomanes et al. (2013)
Jabuticaba residue	$s = CO_2$ + ethanol (20%, v/v); $f = 9.1 \times 10^{-5}$ kg/s	$T = 50$ and 60°C $P = 100$–300 bar	8.0–26.0% (d.b.)	Antioxidant compounds (4.2–15 %, d.b.)	Cavalcanti et al. (2011)
Olive husk	$s = CO_2$; CO_2 + ethanol (1–5%, v/v); $f = 1$ L/min; $t = 6$ h; $h = 6\%$; $d = 0.33$ mm	$T = 40$–60°C $P = 250$–350 bar	9.7–12.6% (g oil/100 g olive husk oil)	Tocopherol (113.8–157.9 ppm); carotenoids (8.0–16.4 ppm); chlorophylls (16.5 – 37.6 ppm)	Gracia et al. (2011)

(Continued)

Table 6.3 Some Applications of SFE on the Recovery of Valuable Compounds from Food Wastes *(cont.)*

Type of Waste	Operational Conditions: Type of Solvent (s); Flow Rate (f); Time (t); Raw Material Moisture Content (h); Particle Size (d)	Temperature (T); Pressure (P)	Extract Yield	Target Compounds	References
Orange pomace	$s = CO_2$ and CO_2 + ethanol (2, 5, and 8%); $f = 17 \pm 2$ g/min; $t = 300$ min; $h = 27 \pm 1\%$; $d = 0.597 \pm 0.006$ mm	$T = 40$ and $50°C$ $P = 100–300$ bar	$0.61 \pm 0.05–3.0 \pm 0.6\%$ (w/w)	Bioactive compounds	Benelli et al. (2010)
Rice by-products	$s = CO_2$; $f = 1.082$ g/min	$T = 40–80°C$ $P = 345–690$ bar	up to 24.65%	α-Tocopherol (1050.8–1279.3 mg/kg)	Perretti et al. (2003)
Spent coffee grounds and coffee husk	$s = CO_2$, CO_2 + ethanol (4 and 8%, w/w for coffee husk; 8 and 15%, w/w for spent coffee grounds); $t = 4.30$ h for coffee husk and 2.30 h for spent coffee; $h = 14 \pm 1\%$ for spent coffee and $13.04 \pm 0.02\%$ for coffee husk	$T = 40–60°C$ $P = 100–300$ bar	Coffee husk: 0.55–2.2% Spent coffee: 0.43–14%	Phenolic compounds in spent coffee (μg GAE/g): protocatechuic acid (0.6); chlorogenic acid (27.3–41.3); p- hydroxybenzoic acid (3.1); caffeic acid (0.1) Phenolic compounds in coffee husk (μg GAE/g): epicatechin (2.1); gallic acid (0.9); protocatechuic acid (0.8); chlorogenic acid (1745–9428)	Andrade et al. (2012)
Sugarcane residue	$s = CO_2$; $f = 0.120 \times 10^{-3}$ kg/s (lab scale) and 1.84×10^{-3} kg/s (pilot scale); $h = < 0.6$; $d = 0.769$ mm $t = 360$ min (lab scale) and 180 min (pilot scale)	$T = 60°C$ $P = 350$ bar	Lab scale: 2.64% Pilot scale: 3.0%	Policosanol	Prado et al. (2011)
Tomato waste (skin and seeds)	$s = CO_2$; $f = 0.26–1.18$ g/min; $h = 4.6$, 22.8 and 58.1%; $d = 0.15$, 0.36 and 0.72 mm	$T = 40–80°C$ $P = 200$ and 300 bar	9.7–23.0% (d.b.)	trans-Lycopene (recovery = 93%)	Nobre et al. (2009)

Animal residues

Fish skin	$s = CO_2$, CO_2 + ethanol (20%); $f = 2.0$ mL/min; $t = 6$ h; $h = 6.33\%$; $d = 0.2$–0.5 mm	$T = 45$–75°C $P = 200$–350 bar	11.6–53.2% (d.b.)	ω-3-Polyunsaturated fatty acids (27.07–29.3%)	Sahena et al. (2010)
Northern shrimp	$s = CO_2$; $f = 3$–5 L/min; $t = 90$ min	$T = 40$–50°C $P = 150$–350 bar	130 mg oil/g raw material	ω-3-Polyunsaturated fatty acids (157.5 mg/g oil)	Amiguet et al. (2012)
Pink shrimp	$s = CO_2$, CO_2 + hexane:isopropanol (50:50) (2–5%), CO_2 + sunflower oil (2–5%); $f = 8.3$ and 13.3 g/min; $h = 11.21$ and 46.3%; $d = 0.554$ cm	$T = 40$ and 60°C $P = 100$ and 300 bar	0.005 ± 0.002– 4.3 ± 0.1% (d.b.)	Carotenoids (9.6 ± 0.4– 1223 ± 19 μg/g)	Mezzomo et al. (2013)
Squid viscera	$s = CO_2$; $f = 22.0$ g/min; $t = 2.5$ h; $d = 0.7$ mm	$T = 35$–45°C $P = 150$–250 bar	0.25–0.34 g/g raw material	Lecithin (4.25%)	Uddin and Kishimura (2011)
Striped weakfish waste	$s = CO_2$; $f = 2.0 \times 10^{-4}$ kg/s; $t = 150$ min; $h = 5.2 \pm 0.2\%$; $d = 0.74$ mm	$T = 30$–60°C $P = 200$–300 bar	16.1 ± 0.3– 18.2 ± 0.1%	Polyunsaturated fatty acids (212.8 ± 0.5–303 ± 17 mg/g oil)	Aguiar et al. (2012)

microwave extraction (3 t/h) for industrial use. SAIREM (Lyon, France) has installed a platform for designing and constructing customized pilot plants that use LABOTRON modular systems (AMW 100 and RF 3000) for the extraction of high value compounds from biomass or biological streams. Based on its Internal Transmission Line technology delivering very high density of activation energy, it offers the possibility of using a pilot-scale test facility with microwave power levels up to 100 kW at 915 MHz with a throughput of 700 L/h (Luque de Castro et al., 2013).

The engineering parameters of supercritical fluid extraction processes are well known and described in the literature at laboratory scale, including economic aspects of the process. There are several studies dealing with scale-up issues as well. Despite the benefits and advantages of supercritical fluid extraction, just a few commercial plants are operating around the world, most of them dedicated to extraction and/or fractionation of biomolecules. Most companies still believe that supercritical technology is very expensive due to the high investment cost compared with classical low-pressure equipment. Yet, this is far from true when large volumes of materials are treated (Perrut, 2000). Studies have demonstrated that capital amortization sharply decreases when capacity increases (Farías-Campomanes et al., 2013; Pereira et al., 2013; Perrut, 2000; Prado et al., 2011). This is a strong incentive to use large capacity multi-product units in "time-sharing" rather than operating small capacity units dedicated to only one product (Perrut, 2000).

6.8 FUTURE PERSPECTIVES

Solvent, acid, and alkali extraction and steam distillation will always have a significant share of the extracts market due to their low equipment cost. Steam diffusion is an inexpensive technique that is still underexploited due to the lack of research. Its combination with microwave irradiation is a fast and green process of extraction that has also shown good results, especially when considering energy consumption.

Microwave-assisted methods provide a range of advantages such as rapidity, reductions in solvent, and energy consumption as well as the possibility of continuous processing backed up by online control, automation, and coupling with other techniques. In this respect, new variants of this process such as solvent-free MHG or VMHG portray sustainable food processes that offer opportunities to meet the growing demands for healthier food ingredients, and can be easily adopted at the industry scale for the extraction of polysaccharides, saponins, phenolics, flavonoids, and anthocyanins from fruit and vegetable wastes.

Extraction is the main application of supercritical fluids. However, its use has been extended to other sectors such as chromatography, impregnation, particle design, polymerization, and reaction, among others. The supercritical fluid processes are not always the best solution, but should be considered as potential alternatives among others, with their own advantages and limitations (Perrut, 2000). Due to its high investment cost, the product must have high added value to make the process feasible. In the field of processing food wastes, extraction of bioactive compounds with possible applications in the food, cosmetics, and pharmaceutical industries can always be chosen to be processed by SFE and thus its implementation tends to keep increasing over the years to come.

Independent from the selected technique, all of the referred technologies are now well established and tend to keep on developing. The most important thing to remember is that there is no generic solution in terms of recovery of high value compounds from food wastes; each one should be individually studied and optimized, and a different technology may be preferred for each product.

REFERENCES

Aguiar, A.C., Visentainer, J.V., Martínez, J., 2012. Extraction from striped weakfish (*Cynoscion striatus*) wastes with pressurized CO_2: global yield, composition, kinetics and cost estimation. J. Supercrit. Fluid. 71, 1–10.

Alessandro, A., Casazza, B.A., Mantegna, S., Cravotto, G., Perego, P., 2010. Extraction of phenolics from Vitis vinifera wastes using non-conventional technique. J. Food Eng. 100, 50–55.

Amiguet, V.T., Kramp, K.L., Mao, J.Q., McRae, C., Goulah, A., 2012. Supercritical carbon dioxide extraction of polyunsaturated fatty acids from Northern shrimp (*Pandalus borealis Kreyer*) processing by-products. Food Chem. 130, 853–858.

Andrade, K.S., Gonçalvez, R.T., Maraschin, M., Ribeiro-do-Valle, R.M., Martínez, J., Ferreira, S.R.S., 2012. Supercritical fluid extraction from spent coffee grounds and coffee husks: antioxidant activity and effect of operational variables on extract composition. Talanta 88, 544–552.

Arnáiz, E., Bernal, J., Martín, M.T., Nozal, M.J., Bernal, J.L., Toribio, L., 2012. Supercritical fluid extraction of free amino acids from broccoli leaves. J. Chromatog. A 1250, 49–53.

Asp, N., Johansson, C.G., Hallmer, H., Siljestrom, M., 1983. Rapid enzymatic assay of insoluble and soluble dietary fiber. J. Agr. Food Chem. 31, 346–482.

Baiano, A., Bevilacqua, L., Terracone, C., Contò, F., Del Nobil, M.A., 2014. Single and interactive effects of process variables on microwave-assisted and conventional extractions of antioxidants from vegetable solid wastes. J. Food Eng. 120, 135–145.

Batista, I., 1999. Recovery of proteins from fish waste products by alkaline extraction. European Food Res. Technol. 210, 84–89.

Benelli, P., Riehl, C.A.S., Smania, A., Smania, E.F.A., Ferreira, S.R.S., 2010. Bioactive extracts of orange (*Citrus sinensis* L. *Osbeck*) pomace obtained by SFE and low pressure techniques: mathematical modeling and extract composition. J. Supercrit. Fluid. 55, 132–141.

Bernardo-Gil, M.G., Roque, R., Roseiro, L.B., Duarte, L.C., Gírio, F., 2011. Supercritical extraction of carob kibbles (*Ceratonia siliqua* L.). J. Supercrit. Fluid. 59, 36–42.

Bittar, S.A., Périno-Issartier, S., Dangles, O., Chemat, F., 2013. An innovative grape juice enriched in polyphenols by microwave-assisted extraction. Food Chem. 141, 3268–3272.

Brunner, G., 1994. Gas Extraction: An Introduction to Fundamentals of Supercritical Fluids and the Application to Separation Processes Darmstadt. Steinkopff.

Canovas, G.V.B., Ibarz, A., 2002. Solid-liquid extraction. Unit Operations in Food Engineering. CRC Press (Taylor and Francis Group), Boca Raton, pp. 773–821, Chapter 21.

Cavalcanti, R.N., Veggi, P.C., Meireles, M.A.A., 2011. Supercritical fluid extraction with modifier of antioxidant compounds from jabuticaba (Myrciaria cauliflora) by-products: economic viability. Procedia Food Sci. 1, 1672–1678.

Cerpa, M.G., Rafael, B., Mato, M., Cocero, J., Ceriani, R., Meirelles, A.J.A., Prado, J.M., Patrícia, F.Leal., Takeuchi, T.M., Meireles, M.A.A., 2008. Steam distillation applied to the food industry. In: Meireles, M.A.A. (Ed.), Extracting Bioactive Compounds for Food Products: Theory and Applications. CRC Press, Boca Raton.

Chemat, S., Esveld, E.D.C., 2013. Contribution of microwaves or ultrasonics on carvone and limonene recovery from dill fruits (*Anethum graveolens* L.). Innov. Food Sci. Emerg. Technol. 17, 114–119.

Chemat, F., Lucchesi, M.E., Smadja, J., 2004. Solvent-free microwave extraction of volatile natural substances. Vol. US 04/0187340 A1.

Cravotto, G., Boffa, L., Mantegna, S., Perego, P., Avogadro, M., Cintas, P., 2008. Improved extraction of vegetable oils under high-intensity ultrasound and/or microwaves. Ultrason. Sonochem. 15, 898–902.

Dang, Y.Y., Zhang, H., Xiu, Z.K. 2013. Microwave-assisted aqueous two-phase extraction of phenolics from grape (*Vitis vinifera*) seed. J. Chem. Technol. Biotechnol. 89 (10), 1576–1581. doi: 10.1002/jctb.4241.

Deng, Q., Zinoviadou, K.G., Galanakis, C.M., Orlien, V., Grimi, N., Vorobiev, E., Lebovka, N., Barba, F.J., 2015. The effects of conventional and non-conventional processing on glucosinolates and its derived forms, isothiocyanates: extraction, degradation and applications. Food Eng. Rev., in press.

Diankov, S., Karsheva, M., Hinkov, I., 2011. Extraction of natural antioxidants from lemon peels. Kinetics and antioxidant capacity. J. Univ. Chem. Technol. Metal. 46, 315–319.

EC 2009. Directive 2009/32/EC of the European Parliament and of the Council on the approximation of the laws of the Member States on extraction solvents used in the production of foodstuffs and food ingredients. EU Publication Office, Brussels.

Farhat, A., Fabiano-Tixier, A.-S., Maataoui, M.E., Maingonnat, J.-F., Romdhane, M., Chemat, F., 2011. Microwave steam diffusion for extraction of essential oil from orange peel: kinetic data, extract's global yield and mechanism. Food Chem. 125, 255–261.

Farías-Campomanes, A.M., Rostagno, M.A., Meireles, M.A.A., 2013. Production of polyphenol extracts from grape bagasse using supercritical fluids: yield, extract composition and economic evaluation. J. Supercrit. Fluid. 77, 70–78.

FDA, 2014. Listing of Additive Status.

Ferhat, M.A., Meklati, B.Y., Smadja, J., Chemat, F., 2006. An improved microwave Clevenger apparatus for distillation of essential oils from orange peel. J. Chromatog. A 1112, 121–126.

Galanakis, C.M., 2011. Olive fruit and dietary fibers: components, recovery and applications. Trends Food Sci. Technol. 22, 175–184.

Galanakis, C.M., 2012. Recovery of high added-value components from food wastes: conventional, emerging technologies and commercialized applications. Trends Food Sci. Technol. 26, 68–87.

Galanakis, C.M., 2013. Emerging technologies for the production of nutraceuticals from agricultural by-products: a viewpoint of opportunities and challenges. Food Bioprod. Process. 91, 575–579.

Galanakis, C.M., 2015. Separation of functional macromolecules and micromolecules: from ultrafiltration to the border of nanofiltration. Trends Food Sci. Technol. 42 (1), 44–63.

Galanakis, C.M., Schieber, A., 2014. Editorial. Special Issue on Recovery and utilization of valuable compounds from food processing by-products. Food Res. Int. 65, 230–299.

Galanakis, C.M., Goulas, V., Tsakona, S., Manganaris, G.A., Gekas, V., 2013a. A knowledge base for the recovery of natural phenols with different solvents. Int. J. Food Prop. 16, 382–396.

Galanakis, C.M., Markouli, E., Gekas, V., 2013b. Fractionation and recovery of different phenolic classes from winery sludge via membrane filtration. Separ. Purif. Technol. 107, 245–251.

Galanakis, C.M., Chasiotis, S., Botsaris, G., Gekas, V., 2014. Separation and recovery of proteins and sugars from Halloumi cheese whey. Food Res. Int. 65, 477–483.

Galanakis, C.M., Patsioura, A., Gekas, V., 2015b. Enzyme kinetics modeling as a tool to optimize food biotechnology applications: a pragmatic approach based on amylolytic enzymes. Crit. Rev. Food Sci. Technol. 55, 1758–1770.

Ganzler, K., Salgo, A., Valko, K.J., 1986. Microwave extraction: a novel sample preparation method for chromatography. J. Chromatog. 371, 229–306.

Gracia, I., Rodríguez, J.F., Lucas, A., Fernandez-Ronco, M.P., 2011. Optimization of supercritical CO_2 process for the concentration of tocopherol, carotenoids and chlorophylls from residual olive husk. J. Supercrit. Fluid. 59, 72–77.

Heng, W.W., Xiong, L.W., Ramanan, R.N., Hong, T.L., Kong, K.W., Galanakis, C.M., Prasad, K.N., 2015. Two level factorial design for the optimization of phenolics and flavonoids recovery from palm kernel by-product. Ind. Crop. Prod. 63, 238–248.

Huma, Z., Abert-Vian, M., Elmaataoui, M., Chemat, F., 2011. A novel idea in food extraction field: study of vacuum microwave hydrodiffusion technique for by-products extraction. J. Food Eng. 105, 351–360.

Imelouane, B., Elbachiri, A., Ankit, M., Benzeid, H., Khedid, K., 2009. Physico-chemical compositions and antimicrobial activity of essential oil of Eastern Moroccan Lavandula dentata. Int. J. Agric. Biol. 11, 113–118.

Kim, J.-Y., Oh, T.-H., Kim, B.J., Kim, S.-S., Lee, N.H., Hyun, C.-G., 2008. Chemical composition and anti-inflammatory effects of essential oil from Farfugium japonicum flower. J. Oleo Sci. 57, 623–628.

Li, J., Zu, Y.G., Fu, Y.J., Yang, Y.C., Li, S.M., Li, Z.N., Wink, M., 2010. Optimization of microwave-assisted extraction of triterpene saponins from defatted residue of yellow horn (Xanthoceras sorbifolia Bunge.) kernel and evaluation of its antioxidant activity. Innov. Food Sci. Technol. 11, 637–643.

Luque de Castro, M.D., Fernández-Peralbo, M.A., Linares-Zea, B., Linares, J., 2013. The role of microwaves in the extraction of fats and oils. In: Chemat, F., Cravotto, G. (Eds.), Microwave-Assisted Extraction of Bioactive Compounds. Springer, New York.

Meireles, M.A.A., 2009. Extracting Bioactive Compounds for Food Products: Theory and Applications. CRC Press/Taylor & Francis Group, Boca Raton, p. 464.

Mengal, P., Behn, D., Bellido, G.M., Monpon, B., 1993. VMHD (vacuum microwave hydrodistillation). Parfums Cosmétiques Aromes 114, 66–67.

Mezzomo, N., Martínez, J., Maraschin, M., Ferreira, S.R.S., 2013. Pink shrimp (*P. brasiliensis* and *P. paulensis*) residue: supercritical fluid extraction of carotenoid fraction. J. Supercrit. Fluid. 74, 22–33.

Mokhtarpour, A., Naserian, A., Valizadeh, R., Danesh Mesgaran, M., Pourmollae, F., 2014. Extraction of phenolic compounds and tannins from pistachio by-products. Annu. Res. Rev. Biol. 4, 1330–1338.

Munoz, O., Sepulveda, M., Schwartz, M., 2004. Effects of enzymatic treatment on anthocyanic pigments from grape skin from Chilean wine. Food Chem. 87, 487–490.

Nobre, B.P., Palavra, A.F., Pessoa, F.L.P., Mendes, R.L., 2009. Supercritical CO_2 extraction of trans-lycopene from Portuguese tomato industrial waste. Food Chem. 116, 680–685.

Palma, M., Barbero, G.F., Pinheiro, Z., Liazid, A., Barroso, C.G., Rostagno, M.A., Prado, J.M., Meireles, M.A.A., 2013. Extraction of natural products: principles and fundamental aspects. In: Rostagno, M.A., Prado, J.M. (Eds.), Natural Product Extraction: Principles and Applications. Royal Society of Chemistry, Cambridge, pp. 58–88.

Paré, J.R.J., Sigouin, M., Lapointe, J., 1990. Microwave-assisted natural products extraction, CA 1336968 C.

Patsioura, A., Galanakis, C.M., Gekas, V., 2011. Ultrafiltration optimization for the recovery of β-glucan from oat mill waste. J. Membr. Sci. 373, 53–63.

Pavlović, M.D., Buntić, A.V., Šiler-Marinković, S.S., Dimitrijević-Branković, S.I., 2013. Ethanol influenced fast microwave-assisted extraction for natural antioxidants obtaining from spent filter coffee. Separ. Purif. Technolo. 118, 503–510.

Pereira, C.G., Prado, J.M., Meireles, M.A.A., 2013. Economic evaluation of natural product extraction processes. In: Rostagno, M.A., Prado, J.M. (Eds.), Natural Product Extraction: Principles and Applications. Royal Society of Chemistry, Cambridge, pp. 442–471.

Perretti, G., Miniati, E., Montanari, L., Fantozzi, P., 2003. Improving the value of rice by-products by SFE. J. Supercrit. Fluid. 26, 63–71.

Perrut, M., 2000. Supercritical fluid applications: industrial developments and economic issues. Ind. Eng. Chem. Res. 39, 4531–4535.

Prado, J.M., Meireles, M.A.A., 2012. Production of valuable compounds by supercritical technology using residues from sugarcane processing. In: Bergeron, C., Carrier, D.J., Ramaswamy, S. (Eds.), Biorefinery Co-Products: Phytochemicals, Primary Metabolites and Value-Added Biomass Processing. Wiley, West Sussex, pp. 133–151.

Prado, J.M., Prado, G.H.C., Meireles, M.A.A., 2011. Scale-up study of supercritical fluid extraction process for clove and sugarcane residue. J. Supercrit. Fluid. 56, 231–237.

Prakash, M.J., Sivakumar, V., Thirugnanasambandham, K., Sridhar, R., 2014. Microwave assisted extraction of pectin from waste Citrullus lanatus fruit rinds. Carbohyd. Polym. 101, 786–791.

Radient Technologies, 2014. [Online]. Available from: www.radientinc.com.

Rahmanian, N., Jafari, S.M., Galanakis, C.M., 2014. Recovery and removal of phenolic compounds from olive mill wastewater. J. Am. Oil Chem. Soc. 91, 1–18.

Rajha, H., El Darra, N., Hobaika, H., Boussetta, N., Vorobiev, E., Maroun, R., Louka, N., 2014. Extraction of total phenolic compounds, flavonoids, anthocyanins and tannins from grape byproducts by response surface methodology. Influence of solid-liquid ratio, particle size, time, temperature and solvent mixtures on the optimization process. Food Nutr. Sci. 5, 397–409.

Richardson, J., Harker, J., Backhurst, J., 2002. Particle technology and separation processes. In: Coulson, J.M., Richardson, J.F. (Eds.), Chemical Engineering Design, vol. 2, Elsevier, Oxford.

Rosa, P.T.V., Meireles, M.A.A., 2009. Fundamentals of supercritical extraction from solid matrices. In: Meireles, M.A.A. (Ed.), Extracting Bioactive Compounds for Food Products: Theory and Application. CRC Press/ Taylor & Francis Group, Boca Raton, pp. 272–288.

Roselló-Soto, E., Barba, F.J., Parniakov, O., Galanakis, C.M., Grimi, N., Lebovka, N., Vorobiev, E., 2015. High voltage electrical discharges, pulsed electric field and ultrasounds assisted extraction of protein and phenolic compounds from olive kernel. Food Bioprocess. Technol. 8, 885–894.

Rostagno, M.A., Prado, J.M. 2013. Natural Product Extraction: Principles and Applications. Royal Society of Chemistry, Cambridge, p. 500.

Sachindra, N.M., Bhaskar, N., Mahendrakar, N., 2006. Recovery of carotenoids from shrimp waste in organic solvents. Waste Manage. 26, 1092–1098.

Sahena, F., Zaidul, I.S.M., Jinap, S., Jahurul, M.H.A., Khatib, A., Norulaini, N.A.N., 2010. Extraction of fish oil from the skin of Indian mackerel using supercritical fluids. J. Food Eng. 99, 63–69.

Sánchez-Camargo, A., Meireles, M.A.A., Lopes, B.F., Cabral, F.A., 2011. Proximate composition and extraction of carotenoids and lipids from Brazilian redspotted shrimp waste (*Farfantepenaeus paulensis*). J. Food Eng. 102, 87–93.

SAS, S., 2014. [Online]. Available from: www.sairem.com.

Seabra, I.J., Braga, M.E.M., Batista, M.T., Sousa, H.C., 2010. Effect of solvent (CO_2/ethanol/H_2O) on the fractionated enhanced solvent extraction of anthocyanins from elder berry pomace. J. Supercrit. Fluid. 54, 145–152.

Seixas, F.L., Fukud, D.L., Turbiani, F.R.B., Garcia, P.S., Petkowiczb, C.L.O., Jagadevan, S., Gimenes, M.L., 2014. Extraction of pectin from passion fruit peel (*Passiflora edulis* f. *flavicarpa*) by microwave-induced heating. Food Hydrocolloid. 38, 186–192.

Stamatopoulos, K., Chatzilazarou, A., Katsoyannos, E., 2014. Optimization of multistage extraction of olive leaves for recovery of phenolic compounds at moderated temperatures and short extraction times. Foods 3, 66–81.

Strati, I.F., Oreopoulou, I.F., 2011. Process optimisation for recovery of carotenoids from tomato waste. Food Chem. 129, 747–752.

Tsakona, S., Galanakis, C.M., Gekas, V., 2012. Hydro-ethanolic mixtures for the recovery of phenols from Mediterranean plant materials. Food Bioprocess. Technol. 5, 1384–1393.

Uddin, M.S., Kishimura, H., 2011. Isolation and characterization of lecithin from squid (*Todarodes pacificus*) viscera deoiled by supercritical carbon dioxide extraction. J. Food Sci. 76, C350–C354.

Virot, M., Tomao, V., Colnagui, G., Visinoni, F., Chemat, F., 2007. New microwave-integrated Soxhlet extraction. An advantageous tool for the extraction of lipids from food products. J. Chromatog. A 1174, 138–144.

Yang, E.-J., Kim, S.-S., Oh, T.-H., Baik, J.S., Lee, N.H., Hyun, C.-G., 2009. Essential oil of citrus fruit waste attenuates LPS-induced nitric oxide production and inhibits the growth of skin pathogens. Int. J. Agric. Biol. 11, 791–794.

CONVENTIONAL PURIFICATION AND ISOLATION

Lorenzo Bertin*, Dario Frascari*, Herminia Domínguez, Elena Falqué[†],
Francisco Amador Riera Rodriguez[††], Silvia Alvarez Blanco[§]**

**Department of Civil, Chemical, Environmental and Materials Engineering (DICAM), University of Bologna,
Bologna, Italy; **Department of Chemical Engineering, University of Vigo, Pontevedra, Spain; †Department of
Analytical Chemistry, University of Vigo, Pontevedra, Spain; ††Chemical Engineering and Environmental Technology
Department, University of Oviedo, Oviedo, Spain; §Department of Chemical and Nuclear Engineering, Universitat
Politècnica de València, Valencia, Spain*

7.1 INTRODUCTION

Adsorption, chromatography-based techniques, nanofiltration (NF), and electrodialysis (ED) are consolidated operations that could be applied in the fourth stage of the Universal Recovery Process (Chapter 3). Since engineering aspects of such conventional techniques are well known, they could be applied with a large range of raw materials, depending on the chemical/physical compositions of such matrices. Recovery yields generally depend on the concentration of target molecules and on their relative amounts with respect to those compounds having similar features (size, charge, chemical structure, etc.). The latter compounds compete with the former ones within the recovery processes. This chapter is dedicated to briefly describing the main features of the above-mentioned technologies and proposes practical suggestions for the development of processes aimed at the selective recovery of chemicals from wastes. Furthermore, a number of case studies and recent works are presented.

7.2 ADSORPTION

Adsorption is a separation operation that leads to the transfer of one or more components contained in a fluid phase (gaseous or liquid) to a porous solid sorbent. In the framework of food waste recovery, the fluid phase is generally a liquid stream generated during the processing of food products. Numerous solids can act as sorbents, but the typical ones applied to liquid streams are activated carbon, zeolites, resins, clays, lignin, and polysaccharide-based compounds (Agalias et al., 2007; Ahmaruzzaman, 2008; Soto et al., 2011; Bertin et al., 2011; Rahmanian et al., 2014). Recently, innovative approaches based on the use of food wastes such as areca waste, rice husk, coconut shell, tobacco fiber, and deoiled soya as sorbents were proposed (Demirbas, 2008; Zheng et al., 2008; Gupta et al., 2009). In the case of adsorption from liquids, the porous sorbent is usually packed in a column in order to form a fixed bed through which the liquid flows. A possible alternative consists in maintaining the sorbing particles suspended in the liquid. This option, known as fluidized-bed adsorption, is not widespread because it requires a subsequent unit operation of sedimentation or filtration, aimed at separating the sorbing particles from the liquid. Therefore, this chapter focuses on fixed-bed adsorption.

7.2.1 ADSORPTION EQUILIBRIA

At the microscopic scale, adsorption can be represented as the combination of three processes that occur in series: (i) the diffusion of the target compound from the bulk of the fluid phase to the external surface of a sorbing particle, (ii) the diffusion of the same compound through the pores inside the particle, and (iii) the actual binding to the sorbing surface. The last step is generally instantaneous. Thus, equilibrium is assumed to be between the sorbed concentration (c_S) and the concentration in the liquid just in contact with the surface (c_L^*). Such equilibrium is strongly dependent on temperature, and the corresponding equation is known as the "adsorption isotherm." The adsorption isotherm most frequently utilized is that proposed by Freundlich:

$$c_S = \alpha(c_L^*)^{1/\beta}$$

(7.1)

where c_S indicates the sorbed concentration and c_L^* the concentration in the liquid just in contact with the surface, while α and $1/\beta$ are constants, which express a measure of adsorption capacity and intensity of adsorption, respectively (the higher the $1/\beta$ value, the more favorable is the adsorption). Another isotherm that may be utilized in the case of liquids is the Langmuir one:

$$c_S = c_{S,max} \frac{\gamma c_L^*}{1+\left(\gamma c_L^*\right)}$$

(7.2)

where $c_{S,max}$ indicates the maximum sorbed concentration (saturation condition), while γ represents the affinity constant for the adsorption process. The recovery of chemicals from food waste often involves the adsorption of more than one compound. In this case, the sorbed compounds interfere with each other, since each one occupies a fraction of the available sites on the sorbing surface. Langmuir's equation can be easily modified so as to describe multicomponent adsorption of the generic compound i:

$$c_{S,i} = c_{S,max,i} \frac{\gamma_i c_{L,i}^*}{1+\sum_1^n\left(\gamma_j c_{L,i}^*\right)}$$

(7.3)

where $c_{S,i}$, $c_{S,max,i}$, γ_i, and $c_{L,i}^*$ indicate respectively the sorbed concentration, the maximum sorbed concentration, the affinity constant and the concentration in the liquid just in contact with the surface relative to compound i, and n indicates the number of sorbed compounds. A more detailed discussion of the adsorption isotherms is reported by Backhurst et al. (2002).

7.2.2 FIXED-BED ADSORPTION: PROCESS DESCRIPTION AND DESIGN PROCEDURE

In this case, the profiles of target compound concentration in the liquid and in the sorbing phase versus column height evolve with time. Typical concentration profiles versus column height and versus time at the column exit are shown in Fig. 7.1. During the initial stages of the process, all the mass transfer from the liquid to the solid occurs in the first portion of the column, whereas in the remaining portions the concentration in both the liquid and the sorbing phase is nearly zero (time t_1 in Fig. 7.1). After a

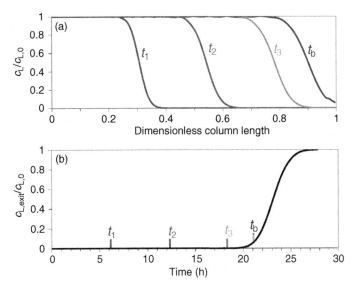

FIGURE 7.1 Fixed-Bed Adsorption

Typical liquid-phase normalized concentration profiles versus column length at different times (a), and versus time at the column exit (b).

certain time (t_2 and t_3 in Fig. 7.1), the liquid phase concentration is initially equal to the inlet value $c_{L,0}$ and the sorbed phase is in equilibrium with the liquid phase. The mass transfer zone is located in an intermediate position between the inlet and the outlet, whereas the last column portion and therefore the outlet stream has still a near-zero concentration. As the mass transfer zone continues to shift along the column, the outlet concentration of the target compound reaches the maximum value allowed by the process specification ($c_{L,b}$), at a specific time known as *breakthrough time* (t_b in Fig. 7.1). At this time, the adsorption process must be interrupted in order to allow desorption to take place. Desorption leads to the concentration of the sorbed chemicals in a solvent, from which they can be separated by distillation. Thus, in order to operate a continuous adsorption process, at least two columns must be available. A schematic flowsheet of a two-column continuous process, with simultaneous recovery by distillation of the desorbing solvent, is shown in Fig. 7.2.

The desorption solvent should be a food-grade compound, so as to avoid the presence of residual concentrations of toxic solvents in the recovered chemicals (Tsakona et al., 2012; Galanakis et al., 2010c, 2013a, b). The solvents most frequently employed are ethanol and isopropanol (Agalias et al., 2007; Scoma et al., 2012).

Once the sorbent and the desorbent have been selected, the second step in the design of the process includes the calculation of the optimal values of the column diameter D, length L, and breakthrough concentration of the key target compound $c_{L,b}$. In this simplified analysis it is assumed that the total number of columns has been set to two: at each instant, one is adsorbing and the other is desorbing. The calculation of L, D, and $c_{L,b}$ derives from a complex economical optimization, based on a balance between investment costs, operational costs, and gains deriving from the recovered chemicals. An

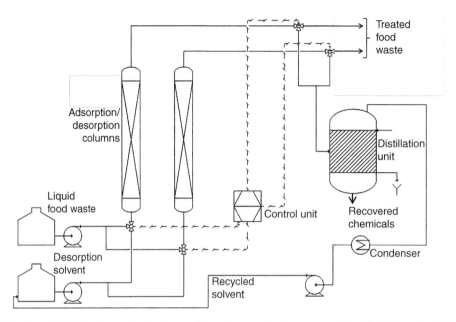

FIGURE 7.2 Simplified Flowsheet of a Continuous-Flow Plant for the Recovery of Chemicals from Food Waste via Adsorption/Desorption, with Recycle of the Desorption Solvent

important concept to be considered is the fraction of unutilized sorbent at t_b. Indeed, if $c_{L,b}$ is small in comparison with the concentration of the target chemical in the influent stream $c_{L,0}$ ($c_{L,b}/c_{L,0} = 0.05$–0.10 is a typical range), a fraction of the sorbing material remains unutilized when the column is switched from adsorption to desorption ($t = t_b$). This response determines a corresponding increase in investment cost. An increase of $c_{L,b}$ determines a decrease of the fraction of unutilized column and therefore of the investment cost. On the other hand, this approach results in a decrease of the amount of target chemical recovered and therefore of the gains.

As for the calculation of the column diameter D, given the flow rate of liquid food waste to be treated, a decrease of D determines a decrease of investment cost and an increase in superficial velocity u. The latter increase determines a proportional increase of pressure drop and therefore of operational cost. If the process is not controlled by the sorption kinetic (i.e. if the bulk of the liquid phase is in equilibrium with the sorbing phase everywhere in the column), the shape of the breakthrough curve and the fraction of unutilized column at the breakthrough time are not affected by the superficial velocity. On the other hand, if the process is controlled by the sorption kinetic, an increase of superficial velocity determines an increase of the mass transfer coefficient, which in turn leads to a steeper breakthrough curve and thus to a decrease of the fraction of unutilized column.

The increase of the column length L determines a decrease of the fraction of unutilized column and therefore of the investment cost. Besides, an increase of L leads to a corresponding increase of pressure drop. Once the optimal L has been determined, the breakthrough time t_b can be calculated accordingly. Relationships for the evaluation of t_b as a function of L and of the shape of the breakthrough curve as a function of u and L are reported by McCabe et al. (1993) and Backhurst et al. (2002).

7.2.3 SELECTIVE ADSORPTION BY MOLECULARLY IMPRINTED POLYMERS

Molecularly imprinted polymers (MIPs) are functional microparticles including cavities, whose size, shape, and surface chemistry replicate those of template molecules (Schirhagl, 2014). MIPs exert a high selectivity towards target substances and can be compared to that of enzymes. The synthesis of MIPs is generally obtained with a bulk polymerization of a functional monomer in the presence of the template molecule and a cross-linker. The latter induces a stable orientation of functional groups toward complementary ones occurring in the template. Such orientation persists after removal of the template molecule, thus favoring the subsequent incorporation of molecules of the template species. Schirhagl (2014) reviewed the main strategies to produce MIPs with specific functionalities. The definition of the polymer, imprinting protocol, and template are crucial factors, whereas particular attention has to be paid to interactions between monomer and template functional groups.

Conventional MIP applications do not include the extraction of added-value chemicals. Conversely, those polymers have been widely studied in the frame of chemical analysis, catalysis, separation and drug delivery control (Puoci et al., 2012; Schirhagl, 2014). However, first studies dedicated to the selective recovery of target substances occurring in actual site of agro-industrial liquid wastes by formulating have MIPs demonstrated the feasibility of such an approach. As an example, a selective and accurate extraction of gallic acid (recovery values of 91±6%) from olive mill wastewater has been achieved using MIPs (Puoci et al., 2012).

7.2.4 A CASE STUDY OF FOOD WASTE RECOVERY BY ADSORPTION

The separation of polyphenols from food waste represents a typical example of food waste recovery via adsorption (Monsanto et al., 2012). For this purpose, numerous adsorbents such as activated carbon, minerals, resins, fly ash, and biosorbents have been utilized (Soto et al., 2011). However, the recovery of polyphenols in purified form, with a minimum presence of other organic compounds, requires the use of adsorbents characterized by high selectivity, such as the Amberlite™ XAD16 and Amberlite™ FPX66 resins (Bertin et al., 2011; Dow, 2014). These resins, sold in the form of beads with diameter in the 0.5–0.7 mm range, are characterized by a high moisture holding capacity (60–70%) and a high surface area (700–800 m²/g). If the polyphenol-rich food waste contains suspended solids, continuous-flow column adsorption with these resins requires a pretreatment aimed at their removal, following the principles of the Universal Recovery Process (Chapter 3). As an example, centrifugation (8000 rpm) followed by filtration (10–20 μm) represents an effective solution (Chapter 4).

Polyphenol adsorption with the Amberlite™ XAD16 resin can be performed at superficial velocities in the 1–3 m/h range. For the subsequent desorption step, optimal results were obtained using ethanol (Scoma et al., 2012). Polyphenol-rich ethanolic extract is characterized by a significant increase in viscosity compared with pure ethanol, mainly due to the presence of small amounts of pectin (Galanakis et al., 2010a, b). Thereby, in order to avoid excessive increases in pressure drop, the superficial velocity maintained during desorption should be significantly lower than that of adsorption. An ethanol volume equal to 3–4 pore volumes leads to the nearly complete removal of the adsorbed phenols. The recovery of ethanol can be performed by vacuum distillation at 30–35°C to avoid the alteration of polyphenol properties.

7.3 CHROMATOGRAPHY

According to their widely differing characteristics on the basis of selectivity, mechanism of action or resolution, several chromatographic technologies have been applied to separate or purify valuable components from agro-industrial wastes.

Biomolecules are isolated, purified, and/or concentrated and characterized with chromatographic techniques according to the following different principles:

1. size (size-exclusion chromatography),
2. charge (anion- or cation-exchange chromatography),
3. biorecognition or ligand specificity (affinity chromatography),
4. hydrophobicity (hydrophobic interaction chromatography and reversed phase chromatography).

7.3.1 SIZE-EXCLUSION CHROMATOGRAPHY (SEC) OR GEL-FILTRATION CHROMATOGRAPHY (GFC)

SEC or GFC distinguishes the components based on their differing sizes. Retention and separation are determined by the diameter of the solute molecule as a function of pores in the packed column, i.e. pores are small enough to exclude large solute molecules (Harris, 2007).

The two most employed gels are Sephadex G (derived from dextran) and Bio-Gel P (porous poly-acrylamide beads). The smallest pore size in highly cross-linked gels exclude molecules bigger than 700 amu, whereas the largest pore sizes exclude molecules greater than 10^8 amu.

The hydrophilic high-performance liquid chromatography (HPLC) packaging used for molecular exclusion can be made of silica, polyvinyl alcohol), polyacrylamide, and sulfonated polystyrene. For HPLC applications with hydrophobic polymers, cross-linked polystyrene spheres are available.

7.3.2 ION-EXCHANGE CHROMATOGRAPHY (IEC)

IEC is applied for the separation of acidic or basic samples based on the dominant charges with varying pH. Specifically, the separation occurs through exchange or interchange of ions between the sample solution and the solid stationary phase or resin.

In anion-exchange chromatography (AEC), a positively charged support (anion exchanger) binds a compound with an overall negative charge:

$$R\text{-}A^+M^- + Y^- \leftrightarrow R\text{-}A^+Y^- + M^- \tag{7.4}$$

Conversely, in cation-exchange chromatography (CEC), a negatively charged support (cation exchanger) binds a compound with an overall positive charge:

$$R\text{-}A^-M^+ + X^+ \leftrightarrow R\text{-}A^-X^+ + M^+ \tag{7.5}$$

Ion exchangers can be classified as strongly or weakly acidic or basic (Harris, 2010). In strongly acidic resins (available commercially as Amberlite IR-120 or Dowex 50W), sulfonate groups remain ionized even in strongly acidic media. In weakly acidic resins (Amberlite IRC-50), carboxyl groups are protonated near pH ~4 and lose their cation-exchange capacity. Quaternary ammonium groups (Amberlite IRA-400 or Dowex 1) are strongly acidic and remain cationic at all values of pH. On the other hand, tertiary ammonium anion exchangers (Amberlite IRA-45 or Dowex 3) are weakly basic and are deprotonated under moderately basic conditions, losing the ability to bind anions.

Resins with a slight cross-linking degree permit rapid equilibration of solute between the inner and outer part of the gel. By increasing the cross-linking, the resin becomes more rigid and less porous. Resins are commonly employed for the separation of small molecules (molecular mass <500 amu), whereas gels are most adapted to bigger biomolecules (proteins or nucleic acids).

Cellulose and dextran ion exchangers possess larger pore sizes and lower charge densities than those of polystyrene resins. They are called "gels" because they are much softer than polystyrene resins. Dextran, cross-linked by glycerin, is sold under the name Sephadex. Other macroporous ion exchangers are based on the polysaccharide agarose and on polyacrylamide.

7.3.3 AFFINITY CHROMATOGRAPHY (AC)

AC can be used to separate an individual analyte from a complex mixture on the basis of a reversible interaction between the compound (or group of compounds) with a bio-specific ligand coupled to a chromatographic stationary phase. Separation is based on the specific interaction between the target molecule and the ligand, such as enzyme/substrate/coenzyme, antigen/antibody, and receptor/hormone. When the sample passes through the column, compounds that are complementary to the specific ligand will bind to the column. The rest of the solutes will tend to wash or elute from the column as a nonretained peak. The retained analytes are then eluted by applying a solvent that displaces them from the column or promotes dissociation of the solute/ligand complex specifically (using a competitive ligand) or nonspecifically (by changing the pH, ionic strength or polarity) (Harris, 2007).

A number of substances have been described and utilized as affinity matrices, including dextran, polyacrylamide, cellulose, controlled pore glass and Sepharose. The latter is the commercial name of the cross-linked beaded agarose and is the most widely used matrix. When the chromatographic support is a membrane, the technology is referred as affinity membrane chromatography.

7.3.4 HYDROPHOBIC INTERACTION CHROMATOGRAPHY (HIC), REVERSED-PHASE CHROMATOGRAPHY (RPC), AND REVERSED-PHASE HIGH-PERFORMANCE LIQUID CHROMATOGRAPHY (RP-HPLC)

HIC separates biomolecules based on their hydrophobicity. This mode of separation is suited for samples that have been precipitated by ammonium sulfate (a stage used for preliminary concentration and clean-up) or after IEC, because the sample contains elevated salt levels and can be applied directly to the HIC column (Gooding, 2005).

The most widely used supports are hydrophilic carbohydrates, such as cross-linked agarose and synthetic copolymer materials. The hydrophobic groups (phenyl, butyl, octyl, ether, or isopropyl) are attached to the stationary column.

HIC is sometimes referred to as a milder form of RPC. Both are based on interactions between hydrophobic moieties, but the bonded phase of the HIC supports consists of a hydrophilic matrix with inserted hydrophobic chains. The latter show generally low density in contrast to the higher-density organosilane chemicals used in RPC.

RPC and RP-HPLC involve the separation of molecules on the basis of hydrophobicity. The base matrix for the reversed phase media is generally composed of silica or a synthetic organic polymer such as polystyrene with linear hydrocarbon chains (C18, C4, C8, phenyl, and cyanopropyl ligands) (Aguilar, 2003). The porosity of the reversed phase beads is a key factor of the available capacity for solute binding by the medium. The more hydrophilic molecules (e.g. synthesized peptides and oligonucleotides) require strongly hydrophobic immobilized ligands (C18), whereas proteins and recombinant peptides are usually better separated on C8 ligands.

The solute mixture is initially applied to the sorbent in the presence of aqueous buffers. Thereafter, molecules bind to the hydrophobic matrix and then the solutes are eluted by the addition of organic solvent to the mobile phase. This process reduces the polarity of the mobile phase and the hydrophobic interaction between the solute and the solid support. The mobile phase is generally prepared with strong acids (e.g. trifluoroacetic or orthophosphoric) in order to maintain a low pH environment and suppress the ionization of the acidic groups in the solute molecules. In addition, the retention times of solutes can be modified by adding ion pairing agents (generally in the range 0.1–0.3%).

7.3.5 APPLICATIONS

All the chromatographic techniques described above have found both analytical and preparative applications since they offer high resolution and capacity for the analytes of interest. For instance, they have been applied for the separation of polysaccharides from vegetal, animal, and microbial wastes (Table 7.1). Fractionation can be achieved by ion exchange (anion or cation) and in some cases with a further size exclusion technique.

Table 7.1 Examples of Chromatographic Techniques for the Purification of Saccharides from Different Waste Fractions

Food Waste	Technique	Production/Extraction and Purification	Products	References
Chinese sturgeon discarded cartilage	CEC AEC SEC	Aqueous NaOH extraction DEAE-52 cellulose Sephadex G-100	Chondroitin sulfate	Zhao et al. (2013)
Mung bean hulls	IEC SEC	Microwave-assisted extraction DEAE-52 cellulose Sephadex G-100	Oligosaccharides Ara, Man, Gal Rha, Ara, Man, Gal	Zhong et al. (2012)
Soybean curd residue	AEC	Fermentation by *Ganoderma lucidum* DEAE Sephadex A-50	Oligosaccharides Ara, Rha, Xyl, Man, Glu Ara, Xyl, Glu Ara, Rha, Xyl, Gal, Man, Glu Ara, Rha, Fuc, Xyl, Man, Glu	Shi et al. (2013)
Tea plant waste	SEC	Boiling-water extraction Ethanol precipitation Sephadex G-100	Oligosaccharides Glu, Xyl, Rha, Gal Glu, Xyl, Rha, Ara	Quan et al. (2011)

AEC, anion exchange chromatography; CEC, cation exchange chromatography; SEC, size exclusion chromatography; IEC, ionic exchange chromatography; Ara, arabinose; Man, mannose; Gal, galactose; Rha, rhamnose; Xyl, xylose; Glu, glucose; Fuc, fucose.

Table 7.2 Recent Examples of Chromatographic Techniques for the Purification of Phenolic Compounds from Agricultural and Food Waste Fractions

Food Waste	Technique	Production/Extraction and Purification	Products	References
Barley husks	SEC	Autohydrolysis Ethyl acetate extraction Sephadex LH-20	Phenolic acids, aldehydes, and flavonoids	Conde et al. (2011)
Black bean canning wastewater	AC SEC	Macroporous resins (Diaion HP-20, Sepabeads SP-70, -207, -700, and -710)	Anthocyanins	Wang (2013)
Effluent of mustard protein isolation	IEC	Dowex 1X8 C1	Sinapic acid	Prapakornwiriya and Diosady (2014)
Olive mill waste water	SEC	Partial dehydration Ethyl acetate extraction Sephadex LH-20	Hydroxytyrosol, tyrosol, elenoic acid derivative linked with hydroxytyrosol	Angelino et al. (2011)
Olive mill waste water	SEC	Sephadex LH-20	Verbascoside, isoverbascoside, β-hydroxyverbascoside, β-hydroxyisoverbascoside, and various oxidized phenolics	Cardinali et al. (2012)
Olive oil solid waste	AC RP-HPLC	Hydrothermal treatment Amberlite XAD-16 Semipreparative RP-HPLC	3,4-Dihydroxyphenylglycol	Lama-Muñoz et al. (2013)
Rambutan rind	RP	Ethanol extraction Reverse-phase C18 column	Geraniin	Perera et al. (2012)
Sugar cane stillage	SEC	Sephadex LH-20	Esters of quinic acids	Caderby et al. (2013)
Sunflower protein extraction waste	AC	Amberlite XAD-16HP	Caffeoylquinic acids, caffeic acid, coumaroylquinic acid, feruloylquinic acid, ferulic acid	Weisz et al. (2013)

SEC, size exclusion chromatography; AC, adsorption chromatography; IEC, ion-exchange chromatography; RP-HPLC, reversed-phase high-performance liquid chromatography; RP, reversed-phase.

The data in Table 7.2 summarize some recent reported examples for the purification of phenolic compounds from food wastes. Gel filtration chromatographic separation using Sephadex LH 20 is quite selective and sensitive. Adsorption chromatography onto macroporous resins has been widely used, too, but only some of them are food grade to facilitate the production of nutraceuticals.

Chromatography is one of the last steps for the purification of proteins and enzymes (Table 7.3). For instance, IEC with a weak-type exchanger such as diethyl amino ethyl cellulose (DEAE cellulose), and elution with a NaCl increasingly linear gradient (0–0.5 M), is widely used. Also, gel filtration chromatography (using gels like Sephadex G-25, G-75, G-100, or Sephacryl S-200) has been applied.

Table 7.3 Examples of Utilization of Chromatographic Techniques for the Purification of Peptides and Enzymes from Different Waste Fractions

Food Waste	Technique	Production/Extraction and Purification	Products	References
Black pomfret viscera	IEC SEC	Hydrolysis by proteases DEAE-cellulose Sephadex G-25	Ala-Met-Thr-Gly-Leu-Glu-Ala	Ganesh et al. (2011)
Chicken intestine	IEC	Differential centrifugation DEAE and CM Sepharose	Aminopeptidases	Damle et al. (2010)
Chlorella vulgaris protein waste	SEC AEC RP-HPLC	Hydrolysis with pepsin Ammonium sulfate precipitation Sephacryl S-100 HR Q-Sepharose fast flow column ODS-3 semi-prep column	Hendecapeptide Val-Glu-Cys-Tyr-Gly-Pro-Asn-Arg-Pro-Glu-Phe	Sheih et al. (2009)
Crucian carp hepatopancreas	IEC SEC HIC	Extraction, ammonium sulfate fractionation DEAE-Sepharose Sephacryl 5-200 HR Phenyl-Sepharose and SP-Sepharose	Chymotrypsins A and B	Yang et al. (2009)
Fish processing waste	IEC SEC	Solubilization, centrifugation, ammonium sulfate precipitation DEAE-cellulose Sephadex G-75 and G-100 Sephacryl S-200 and S-200 HR	Fish pepsin	Zhao et al. (2011)
Pineapple peel	IEC	Ammonium sulfate precipitation Desalting and freeze-drying DEAE-Sepharose	Bromelain	Bresolin et al. (2013)
Sardinelle viscera	SEC RP-HPLC	Hydrolysis with proteases Sephadex G-25 Reversed-phase HPLC	Peptides Leu-His-Tyr; Leu-Ala-Arg-Leu; Gly-Gly-Glu, Gly-Ala-His; Gly-Ala-Trp-Ala; Pro-His-Tyr-Leu; Gly-Ala-Leu-Ala-Ala-His	Bougatef et al. (2010)
Sheep liver	AC	Sepharose-4B-L tyrosine-sulfanilamide	Carbonic anhydrase-II	Demirdag et al. (2012)
Wheat bran	IEC SEC	Cultivation of *Oidiodendron echinulatum* Ammonium sulfate precipitation DEAE-cellulose Sephadex G-100	Neutral pectin lyase	Yadav et al. (2012)
Whey	AC	Yellow HE-4R immobilized on Sepharose	Bovine lactoferrin	Baieli et al. (2014)
Whey	AEC or HIC	Capto Q or Octyl Sepharose 4 FF	Whey proteins	Nfor et al. (2012)

IEC, ionic exchange chromatography; SEC, size exclusion chromatography; AEC, anionic exchange chromatography; RP-HPLC, reversed-phase high-performance liquid chromatography; HIC, hydrophobic interaction chromatography; AC, affinity chromatography; Ala, alanine; Met, methionine; Thr, threonine; Gly, glycine; Leu, leucine; Glu, glutamic acid; Val, valine; Cys, cysteine; Tyr, tyrosine; Pro, proline; Asn, asparagine; Arg, arginine; Phe, phenylalanine; His, histidine; Trp, tryptophan.

Chromatographic methods have the advantages of eliminating additional steps, increasing yields, and improving process economics. However, they have particular limitations in regard to the biological origin of the ligands, as they tend to be fragile and associate with low binding capacities (Caramelo-Nunes et al., 2014).

7.4 NANOFILTRATION

Nanofiltration (NF) is an intermediate process between reverse osmosis (RO) and ultrafiltration (UF) since it is able to reject molecules with a size in the order of one nanometer. While rejection of RO is based mainly on solutes diffusion through the membrane, and microfiltration/ultrafiltration (MF/UF) is based on size exclusion mechanisms, transport in "tight UF" and NF can be explained by a combination of sieving, Donnan, and dielectric effects, and the importance of each mechanism depends on the nature of solutes (charged/uncharged), membrane charge, and the chemical environment (pH and ionic strength, mainly) (Hussain and Al-Rawajfeh, 2009; Patsioura et al., 2011; Galanakis et al., 2010b, 2013, 2014). The chemical, environmental, and solutes characteristics depend on the application of interest, while the surface charge depends on the membrane material. The different mechanisms involved in solutes transport make it difficult to predict membrane rejection or selectivity (when several solutes rejected or transmitted are present in the feed). In cases of uncharged solutes (e.g. carbohydrates and undissociated organic acids) transport mainly occurs by diffusion and convection, and solute retention can be estimated by Ferry's or similar equations. In the case of charged solutes, the Donnan exclusion mechanism can govern the transport and the equations to predict solute rejections are much more complex (Dey et al., 2012).

7.4.1 APPLICATIONS

For the rest of the membrane processes, NF is commonly used in the food industry, since mild conditions are possible, additives are not necessary and techniques are very flexible, allowing working in sequential form. The main applications of NF in the food industry are in the field of desalination, water reuse, and recovery of interesting compounds (Lipnizki, 2010; Salehi, 2014). In the case of food industry wastes, NF is being used to concentrate or purify valuable compounds (polyphenols, proteins, peptides, pectins, or carbohydrates) and fractionate mixtures of products of interest (organic acids and peptides) (Galanakis, 2012). The high transmission of monovalent salts through NF membranes is widely used to partially demineralize a number of final products (Suárez et al., 2006; Suárez et al., 2009; Galanakis et al., 2012). Separation of similar size molecules is difficult for all membrane processes, but in the case of NF, the influence of other solute characteristics such as charge and solutes/membrane interactions can be used to improve membrane selectivity. Thus, the most challenging applications in NF are those where chemical environment modifications affect the rejection of charged organic and inorganic compounds.

NF has similar disadvantages as the rest of membrane processes: concentration polarization and fouling. Concentration polarization can be reduced in cross-flow NF, if a good selection of process conditions is made (high feed velocity reduces the concentration polarization). However, when solutes concentration is high (last stages in industrial applications), the concentration polarization effect cannot be avoided. With respect to membrane fouling, the selection of the surface membrane material is crucial. The membrane should exhibit low affinity towards the solutes, thus hydrophilic membranes are recommended to treat aqueous or hydroalcoholic solutions. In the case of nonaqueous streams, hydrophobic membranes are recommended to obtain higher permeate flow rates. Fortunately, the materials developed in the last few years offer a wide range of opportunities (Van der Bruggen et al., 2008; Amoudi, 2010; Cheng et al., 2011).

Selected NF applications for the reuse of food industry wastes are presented in Sections 7.4.2, 7.4.3, and 7.4.4. Table 7.4 summarizes some recent studies and technical details about the recovery of some valuable products.

7.4.2 CONCENTRATION/PURIFICATION OF ANTIOXIDANTS

NF of olive mill wastewater has been extensively used for combining the reduction of the pollutant load with the recovery and purification of antioxidants such as polyphenols with a molecular weight (MW) between 200 and 400 Da (Takaç and Karakaya, 2009; Conidi et al., 2014). NF coupled with other membrane techniques such as UF and RO showed high polyphenols recovery (more than 99%) and a partial transmission of salts and saccharides (~55%) (Paraskeva et al., 2007; Coutinho et al., 2009; García-Castello et al., 2010). Extended removal of salts can be performed using diafiltration processes. According to several authors, polyphenols concentration in the feed can increase between six- and 10-fold using NF (Paraskeva et al., 2007; Russo, 2007; Díaz-Reinoso et al., 2009). Membrane fouling does not seem to be a great problem for these applications. In addition, initial permeability can be recovered up to 100% after cleaning (Salehi, 2014). However, it is necessary to follow previous stages of the Universal Recovery Process.

7.4.3 LACTIC ACID FROM CHEESE WHEY

Lactic acid recovery and purification from fermentation broths need complex and expensive sequential operations. These downstream processes account for about 50% of the total lactic acid production cost (Riera and Álvarez, 2012). NF is able to partially replace these steps, if membrane and process conditions are well selected, by rejecting sugars and permitting lactic acid transmission through the membrane (Sikder et al., 2012). NF can also be integrated with conventional fermenters (Pal and Dey, 2013), but after a MF step that is necessary to remove cells and reduce membrane fouling. The importance of pH for the recovery of lactic acid has been demonstrated using negatively charged membranes (González et al., 2008; Sikder et al., 2012): at pHs between 5.5 and 6, lactic acid is mainly found in its dissociated form (following the Henderson–Hasselbalch equation) and then lactate ion is highly rejected by the membrane by electrostatic repulsion (lactate rejection ~90% at pH 6 and ~30% at pH 2.7). However, the Donnan effect is attenuated at higher lactic acid concentration. In this case, sugars rejection is almost independent of the medium pH and they can be retained almost completely (rejection = 94%) (Sikder et al., 2012). One of the disadvantages of this purification scheme is that lactate ion must be converted into lactic acid after the purification step whereas alkali must be added before the final disposal. Nevertheless, economic studies have shown savings in the production of lactic acid using NF (González et al., 2007; Sikder et al., 2012).

7.4.4 BIOPEPTIDES FRACTIONATION

The most typical food wastes for the recovery of polypeptides are cheese whey, meat blood, fish, and shellfish wastes (Balti et al., 2010; Kim, 2013; Zanello et al., 2014). Biopeptides are

Table 7.4 Nanofiltration Applications in Recovery Valuable Products from Food Industries Residues

Source	NF Objective	Compounds of Interest	NF Characteristics	Main Results	Comments	References
Olive mill wastewaters	Concentration	Polyphenols (PP)	Polymeric spiral wound 200 Da, 2.5 m², 10–30 bar, 20°C	PP rejection >99.5% Initial conc.: 725 mg/l Final conc.: 9962 mg/l	NF feed with UF permeates	Paraskeva et al. (2007)
Olive mill wastewaters	Purification	Polyphenols (PP)	N30F spiral wound membrane (NADIR). Polyethersulphone 578 Da, 1.6 m², 8 bar, 20°C	5–3 kg/hm² at VCR (1–3) Sugar reductions~55.8% $R_{PP} \sim 5\%$	Treatment of MF permeates previous osmotic distillation (OD)	García-Castello et al. (2010)
Olive mill wastewaters	Purification/ concentration	Polyphenols (PP)	Ceramic tubular membranes 1000 Da, 0.35 m², 2.5–4.5 bar 20–25°C	Concentration until VCR 2.5	Purification/ preconcentration previous RO	Russo (2007)
Olive mill wastewaters	Concentration	Polyphenols (PP)	NF90 (Filmtec/Dow Chemical) spiral wound membranes R_{MgSO4} > 97%, 2.6 m² 5–9 bar, 22°C	14–2.7 l/hm² R_{PP} > 93%; R_{TOC} > 96% Initial PP conc.: 65.6 mg/l Final PP conc.: 86.2 mg/l	Previous MF and UF	Cassano et al. (2013)
Fermented ultrafiltered sweet whey	Purification	Lactic acid (LA)	– FE 2540SS (Filtration Tech.). Spiral wound, 300 Da, 1.8 m² – AFC80 (PCI) 0.68 nm pore size, 0.72 m² 10–40 bar, 40°C	R_{LA}: ~30% at pH 2.7 ~90% at pH 6	Strong influence of pH and LA concentration on the rejection	González et al. (2008)
Microfiltrate broth from sugar cane juice	Purification	Lactic acid (LA)	Several flat sheet membranes 150–300 Da, 0.01 m² 5–15 bar, 37°C	R_{LA}: 55–75%, R_{sugar}: 86–94% both depending on the pH	Strong influence of pH and membrane density charge on the LA and sugar rejection	Sikder et al. (2012)

(Continued)

Table 7.4 Nanofiltration Applications in Recovery Valuable Products from Food Industries Residues *(cont.)*

Source	NF Objective	Compounds of Interest	NF Characteristics	Main Results	Comments	References
Citrus peel	Production/purification	Flavonoids, anthocyanins	Several NF spiral wound membranes (NADIR, FILMTEC) 180–1000 Da, 1.6–2.6 m^2 6–20 bar, 20°C	Average rejection: $R_{flavonoids} = 95.4\%$ $R_{anthocyanin} = 96\%$ $R_{sugars} = 93.4\%$ R_{salts}: 55–95%	Membrane cut-off affects to the R_{salts} and less to the other components	Conidi et al. (2012)
Fermented grape pomace	Concentration	Antioxidants (AO)	Several organic/inorganic membranes 150–1000 Da, 2–8 bar, 20°C	R_{AO} between 30 and 90%. AO concentration between 3 and 6 times	AO concentration by UF/NF followed by extraction	Díaz-Reinoso et al. (2009)
Artichoke wastewaters	Concentration	Antioxidants (AO)	Microdyn Nadir (NP030) and GE Waters (Desal DL) Spiral wound 140–400 Da, 1.8–2–5 m^2 8 bar, 25°C	Low rejection (NP030) to glucose, sucrose, and fructose (lower than 10%). High rejection to AO 93.7% Desal 87.8% NP030	UF (clarification) + NF (concentration)	Conidi et al. (2014)

R, rejection.

generally produced after an enzymatic treatment of a previously purified substrate (Fernández and Riera, 2013a). Hydrolysates are a complex mixture of nonhydrolyzed proteins, peptides, amino acids, enzymes, carbohydrates, and salts, and NF is often preceded by clarification steps as MF and/or UF. Although biological activities of untreated hydrolysates have been measured by several authors (Otte et al., 2007; Lacroix and Li-Chan, 2013; Power et al., 2014), it is common to separate the pretreated hydrolysates into different MW fractions, as not all peptides show the same biological activity. NF and "narrow UF" are used to fractionate polypeptide mixtures from protein hydrolysates due to their low enough suitable MW cut-off and the electrochemically induced selectivity (Pouliot et al., 1999; Butylina et al., 2006; Kovacs and Samhaber, 2009). However, NF is usually used in combination with other technologies in order to increase selectivity of peptides fractionation (Yuanhui et al., 2007; Quian et al., 2011).

Membrane selectivity among peptides with similar MW can also be improved by modifying the chemical environment and taking advantage of the particular mechanisms of NF transmission (Fernández et al., 2013; Fernández and Riera, 2013a, b; Pouliot et al., 2000). For example, proteins are amphoteric electrolytes and their charge depends on the pH of the solution. When pH is higher than the peptides' isoelectric point, their charge is negative and vice versa (Fernández et al., 2013; Fernández and Riera, 2013a, b; Pouliot et al., 2000). The differences in retention between charged and noncharged solutes are due to the Donnan effect when solutes are diluted and ionic strength is low. According to the membrane charge, co-ions are rejected due to electrostatic effects (Tsuru et al., 1994; Garem et al., 1997). This effect is increased at higher membrane charge density and ion valence.

The ionic strength of the solution affects also the effective hydrodynamic volume of charged proteins and peptides, thus modifying their transmission through the membrane according to the following equation:

$$L_D = 0.304\,I^{-1/2} \tag{7.6}$$

where L_D is the Debye length (nanometers) and I is the ionic strength (mole per liter) (Zydney, 1998). The effective radius of some peptides can be increased more than 50% by reducing the ionic strength, which leads to a peptides rejection by the membrane. The effect of salts on the proteins transmission has been observed by several authors (Pujar and Zydney, 1994) and confirmed in the case of peptides (Fernández and Riera, 2013a). Indeed, due to the complexity of the hydrolyzed streams, it is common to classify the peptides according to their charge at a fixed pH as neutral (N), acidic (A), and basic (B) peptides instead of being studied individually. Selectivities between these peptides are strongly reduced at higher ionic strength and demineralization of hydrolyzed proteins will be recommendable to improve the fractionation process (Fernández and Riera, 2013a).

Most of the studies published involving amino acids and peptides have dealt with diluted solutions. When solutes concentration increases, membrane fractionation efficiency is reduced (Li et al., 2003). This happens probably due to the elimination of the Donnan effect, induced by the partial adsorption of peptides on the membrane surface (Fernández and Riera, 2012). As in other membrane applications described, fouling is a problem that remains unsolved. High solutes concentration is incompatible with high permeate flow rates in membrane technology (no matter the technique used) and thereby peptides fractionation should be done at low/medium concentration (5–15 g/L).

7.5 **ELECTRODIALYSIS**

Electrodialysis (ED) is a membrane process for the concentration or separation of ions, in which the driving force is an electric potential difference. The membranes employed in ED processes are ion-exchange materials in film form. Two types of ion-exchange membranes can be distinguished: (i) cation-exchange membranes (CEMs), which contain negatively charged groups, and (ii) anion-exchange membranes (AEMs), which contain positively charged groups. Due to the exclusion of the co-ions (Donnan exclusion), CEMs are mainly permeable to cations, while AEMs are preferentially permeable to anions.

The key properties required for ion-exchange membranes are high selectivity, high electrical conductivity, good mechanical stability, high chemical and thermal stability, and low production costs (Fidaleo and Moresi, 2006; Nagarale et al., 2006).

In commercial ED processes several hundreds of alternate CEMs and AEMs are placed between two electrodes. This set is called ED stack. Every couple of CEMs and AEMs are separated by a spacer, forming an individual cell. Figure 7.3 shows a schematic representation of an ED process. Under the effect of an electric field, the charged ions tend to migrate towards the oppositely charged electrode, whereas uncharged molecules are not affected.

In ED processes, the amount of ions transported through the membrane is directly proportional to the electrical current or current density. The latter depends on the applied voltage and the total resistance of the stack, according to Ohm's law. However, there is a maximum current density that can pass through a cell pair area without detrimental effects, the so-called limiting current density (LCD). If LCD is exceeded, the electric resistance of the diluted stream (diluate) will increase and water

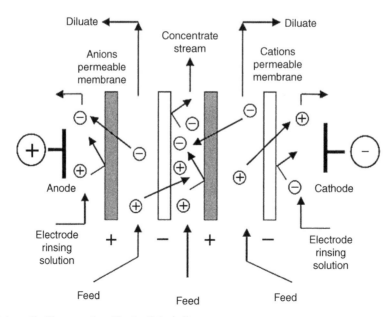

FIGURE 7.3 Schematic Diagram of an Electrodialysis Process

dissociation may occur at the membrane surface, causing loss of current utilization and changes in the pH of the solutions. In practical applications, LCD is usually empirically determined by the Cowan and Brown method (Cowan and Brown, 1959). The LCD corresponds to the minimum of the curve representing voltage/current density against the reversal of current density. LCD is closely related to concentration polarization and it is affected by cell geometry, diluate concentration, and operating conditions (Strathmann, 2010).

Another factor that affects the performance of the process is water transport from the diluate to the concentrate. Water transfer progresses due to the osmotic pressure difference between both solutions and to the coupling of water to the ions that are being transported. Moreover, Donnan exclusion becomes less effective at high ionic concentrations and greater energy consumption is required at high concentrations. All these factors limit the concentration that can be reached in the brine. Nevertheless, significantly higher brine concentrations are usually achieved by ED when compared with RO. On the other hand, ED is generally less competitive when extremely low salt concentrations are required due to the low LCD at low conductivities. As a general rule, the minimum conductivity that can be economically considered is 0.5 mS/cm (Velizarov et al., 2003).

The required membrane area for an ED plant can be calculated by:

$$A = \frac{QF(C_f - C_s)}{i\xi} \tag{7.7}$$

where A is the cell area, i is the current density for a cell pair, ξ is the current utilization, Q is the volume flow, F is the Faraday constant, and C_f and C_s are feed and diluate concentrations, respectively. Total costs of an ED plant reach a minimum at a certain current density, which has to be determined for each particular stack and process requirement (Strathmann, 2010). Processing costs depend on the required degree of demineralization. The degree of desalination that can be achieved is a function of the feed concentration, the applied current density, and the residence time of the solution in the stack. The plants can be operated in continuous (one pass flow), batch (recirculation of feed solution during a certain time), or feed and bleed (partial recirculation of feed solution) operating modes. The optimal temperature is a compromise between demineralization rates, which increase with temperature, and bacterial growth.

The major drawback of ED in food waste treatment is membrane fouling. Thus, feed clarification is usually required and the membranes must be periodically cleaned. ED reversal (EDR), where the polarity of the electric field is cyclically reversed, has been proposed for fouling removal, but it requires a much more sophisticated process control (Strathmann, 2010).

Apart from conventional ED, a number of ED-based processes have been developed: ED with bipolar membranes (EDBM), electrometathesis, electrodeionization, electro-electrodialysis, electro-ion injection-extraction, or electrodialysis with filtration membrane (EDFM), whereas some of them have been tested for food waste recovery (Strathmann, 2010; Xu, 2005). These processes can be used to recover, concentrate, or fractionate different ionic compounds. Some of these applications are summarized in Table 7.5.

7.5.1 INDUSTRIAL APPLICATIONS: WHEY DEMINERALIZATION

Whey can be used as an additive for a certain number of food products (ice creams, soups, candies, infant formulas, etc.). However, due to the high salt content it is not appropriate as a food ingredient.

Table 7.5 Some Applications of ED and ED-Related Processes in the Field of Food Waste Recovery

Substrate	Technology	Objective	References
Whey and whey protein isolate	Conventional ED	Demineralization	Gernigon et al. (2011)
	EDBM	Fractionation of proteins, production of phospholipids	Bazinet et al. (2004); Lin Teng Shee et al. (2007)
	EDFM	Separation of lactoferrin	Ndiaye et al. (2010)
Fermentation broths from food wastes	Conventional ED and EDBM	Recovery of organic acids	Riera and Álvarez (2012)
Protein hydrolysates	EDFM	Fractionation of peptides	Suwal et al. (2014); Poulin et al. (2006)

ED, electrodialysis; EDBM, electrodialysis with bipolar membrane; EDFM, electrodialysis with filtration membrane.

Some applications require demineralization degrees of 90–95% (infant formulas), while in other cases a 50–70% demineralization is enough. Whey demineralization is already conducted using ED at the industrial scale. The maximum demineralization degree that can be economically achieved by ED is about 90%. For this application, whey and acidified brine pass through alternated cells in the stack. Preconcentration to 20–30% dry matter is usually performed to increase the electrical conductivity of the solution and reduce the required membrane area and energy consumption. Whey should be clarified before the ED process to remove insoluble proteins and fat and reduce fouling. During whey ED multivalent ions are usually removed after the removal of monovalent ones. Temperatures between 30 and 40°C have been reported since proteins agglomerate at higher temperatures. At 90% demineralization degree, lactose losses up to 5–8% have been observed (Fidaleo and Moresi, 2006; Gernigon et al., 2011).

A critical factor is fouling due to the precipitation of calcium salts on the CEMs and proteins on the AEMs. Therefore cleaning must be periodically carried out. A typical cleaning sequence consists of water rinse, cleaning with an alkaline solution (up to pH 9.0), water rinse, cleaning with an acid solution (pH 1), and final water rinse. Casademont et al. (2009) reported that basic conditions prevented protein fouling on the AEMs while mineral fouling on the CEMs only occurred at basic pH values. Fidaleo and Moresi (2006) reported that EDR is the best option to perform whey demineralization due to fouling control. Fouling depends on the concentrate pH value. Batch ED is often used for demineralization levels greater than 70%, whereas continuous operation mode is more frequently used for lower demineralization degrees. As a general rule, increasing the demineralization level from 50–75% and from 75–90% doubles the operating costs (Gernigon et al., 2011).

7.5.2 INDUSTRIAL APPLICATIONS: LACTIC ACID RECOVERY

An example of recovery and concentration of ionic compounds by ED processes is the purification of lactic acid from food waste fermentation broths. This is an application under development at the pilot scale (Riera and Álvarez, 2012). Bipolar membrane electrodialysis has been tested for this purpose. It involves water splitting within the bipolar membrane, which is composed of a cation-exchange membrane laminated together with an anion-exchange membrane. The H^+ and OH^- ions generated by the bipolar membrane are combined with lactate anions and Na^+, respectively, obtaining lactic acid and NaOH at the same time. Therefore, no chemicals are needed to convert lactate into lactic acid and NaOH can be recycled to the fermentation tank.

All ED-based processes used to recover and purify lactic acid from fermentation broths need intensive pretreatments to reduce membrane fouling. MF or UF is often performed to remove cells, particles, colloids, or other suspended solids. Most of the works that use ED technologies to produce lactic acid follow a two-step scheme. In the first step, lactic acid is concentrated and the inorganic salts are removed by conventional ED, ion exchange, or NF. In the second step, EDBM is used to obtain lactic acid and alkali. Kim and Moon (2001) compared different alternatives. The best results were obtained by means of three compartment configurations that combine conventional ED membranes with bipolar membranes. The stack consisted of alternating bipolar, anionic-, and cationic-exchange membranes. The authors reported that by means of this process lactic acid can be recovered by 99% in one stage.

REFERENCES

Agalias, A., Magiatis, P., Skaltsounis, A.L., Mikros, E., Tsarbopoulos, A., Gikas, E., Spanos, I., Manios, T., 2007. A new process for the management of olive oil mill waste water and recovery of natural antioxidants. J. Agric. Food Chem. 55, 2671–2676.

Aguilar, M.I., 2003. HPLC of peptides and proteins: methods and protocols. In: Walker, J.M. (Ed.), Methods in Molecular Biology, vol. 251. Humana Press, Totowa, NJ, pp. 9–22.

Ahmaruzzaman, M., 2008. Adsorption of phenolic compounds on low-cost adsorbents: a review. Adv. Colloid Interf. Sci. 143, 48–67.

Amoudi, A.S., 2010. Factors affecting natural organic matter (NOM) and scaling fouling in NF membranes: a review. Desalination 259, 1–10.

Angelino, D., Gennari, L., Blasa, M., Selvaggini, R., Urbani, S., Esposto, S., Servili, M., Ninfali, P., 2011. Chemical and cellular antioxidant activity of phytochemicals purified from olive mill waste waters. J. Agric. Food Chem. 59 (5), 2011–2018.

Backhurst, J.R., Harker, J.H., Backhurst, J.R., 2002. Coulson and Richardson's Chemical Engineering – Volume 2: Particle Technology and Separation Processes, fifth ed. Butterworth-Heinemann, Oxford, pp. 970–1052.

Baieli, M.F., Urtasun, N., Miranda, M.V., Cascone, O., Wolman, F.J., 2014. Bovine lactoferrin purification from whey using Yellow HE-4R as the chromatographic affinity ligand. J. Separ. Sci. 37 (5), 484–487.

Balti, R., Bougatef, A., Ali, N.H.H., Zekri, D., Barkia, A., Nasri, M., 2010. Influence of degree of hydrolysis on functional properties and angiotensin I-converting enzyme–inhibitory activity of protein hydrolysates from cuttlefish (*Sepia officinalis*) by-products. J. Sci. Food Agric. 90, 2006–2014.

Bazinet, L., Ippersiel, D., Mahdavi, B., 2004. Fractionation of whey proteins by bipolar membrane electroacidification. Innov. Food Sci. Emerg. Technol. 5, 17–25.

Bertin, L., Ferri, F., Scoma, A., Marchetti, L., Fava, F., 2011. Recovery of high added value natural polyphenols from actual olive mill wastewater through solid phase extraction. Chem. Eng. J. 171, 1287–1293.

Bougatef, A., Nedjar-Arroume, N., Manni, L., Ravallec, R., Barkia, A., Guillochon, D., Nasri, M., 2010. Purification and identification of novel antioxidant peptides from enzymatic hydrolysates of sardinelle (*Sardinella aurita*) by-products proteins. Food Chem. 118 (3), 559–565.

Bresolin, I.R.A.P., Bresolin, I.T.L., Silveira, E., Tambourgi, E.B., Mazzola, P.G., 2013. Isolation and purification of bromelain from waste peel of pineapple for therapeutic application. Brazilian Arch. Biol. Technol. 56 (6), 971–979.

Butylina, S., Luque, S., Nyström, M., 2006. Fractionation of whey-derived peptides using a combination of ultrafiltration and nanofiltration. J. Membr. Sci. 280, 418–426.

Caderby, E., Baumberger, S., Hoareau, W., Fargues, C., Decloux, M., Maillard, M.N., 2013. Sugar cane stillage: a potential source of natural antioxidants. J. Agric. Food Chem. 61 (47), 11494–11501.

Caramelo-Nunes, C., Almeida, P., Marcos, J.C., Tomaz, C.T., 2014. Aromatic ligands for plasmid deoxyribonucleic acid chromatographic analysis and purification: an overview. J. Chromatogr. A 1327, 1–13.

Cardinali, A., Pati, S., Minervini, F., D'Antuono, I., Linsalata, V., Lattanzio, V., 2012. Verbascoside, isoverbascoside, and their derivatives recovered from olive mill wastewater as possible food antioxidants. J. Agric. Food Chem. 60 (7), 1822–1829.

Casademont, C., Sistat, P., Ruiz, B., Pourcelly, G., Bazinet, L., 2009. Electrodialysis of model salt solution containing whey proteins: enhancement by pulsed electric field and modified cell configuration. J. Membr. Sci. 328, 238–245.

Cassano, A., Conidi, C., Giorno, L., Drioli, E., 2013. Fractionation of olive mill wastewater by membrane separation techniques. J. Hazard. Mater., 248–249, 185–193.

Cheng, S., Oatley, D.L., Williams, P.M., Wright, C.J., 2011. Positively charged nanofiltration membranes: review of current fabrication methods and introduction of a novel approach. Adv. Colloid Interf. Sci. 164, 12–20.

Conde, E., Moure, A., Domínguez, H., Gordon, M.H., Parajó, J.C., 2011. Purified phenolics from hydrothermal treatments of biomass: ability to protect sunflower bulk oil and model food emulsions from oxidation. J. Agric. Food Chem. 59 (17), 9158–9165.

Conidi, C., Cassano, C., Drioli, E., 2012. Recovery of phenolic compounds from orange press liquor by nanofiltration. Food Bioprod. Process. 90, 867–874.

Conidi, C., Mazzei, R., Cassano, A., Giorno, L., 2014. Integrated membrane system for the production of phytotherapeutics from olive mill wastewaters. J. Membr. Sci. 454, 322–329.

Coutinho, C.M., Chiu, M.C., Basso, R.C., Ribeiro, A.P.B., 2009. State of art of the application of membrane technology to vegetable oils: a review. Food Res. Int. 42, 536–550.

Cowan, D.A., Brown, J.H., 1959. Effect of turbulence on limiting current in electrodialysis cells. Ind. Eng. Chem. 51, 1445–1448.

Damle, M., Harikumar, P., Jamdar, S., 2010. Chicken intestine: a source of aminopeptidases. Sci. Asia 36 (2), 137–141.

Demirbas, A., 2008. Heavy metal adsorption onto agro-based waste materials: a review. J. Hazard. Mater. 157, 220–229.

Demirdag, R., Yerlikaya, E., Kufrevioglu, O.I., 2012. Purification of carbonic anhydrase-II from sheep liver and inhibitory effects of some heavy metals on enzyme activity. J. Enzyme Inhib. Med. Chem. 27 (6), 795–799.

Dey, P., Linnanen, I., Pal, P., 2012. Separation of lactic acid from fermentation broth by cross flow nanofiltration: membrane characterization and transport modelling. Desalination 288, 47–57.

Díaz-Reinoso, B., Moure, A., Domínguez, H., Parajó, J.C., 2009. Ultra- and nanofiltration of aqueous extracts from distilled fermented grape pomace. J. Food Eng. 91, 587–593.

Dow, 2014. Dow water & process solutions: polyphenols. http://www.dowwaterandprocess.com/en/industries-and-applications/food_and_beverage/nutritionals/polyphenols. (accessed 24.06.14.).

Fernández, A., Riera, F.A., 2012. Membrane fractionation of α-lactoglobulin tryptic digest: effect of the hydrolysates concentration. Ind. Eng. Chem. Res. 51, 15738–15744.

Fernández, A., Riera, F.A., 2013a. Influence of ionic strength on peptide membrane fractionation. Separ. Purif. Technol. 119, 129–135.

Fernández, A., Riera, F., 2013b. β-Lactoglobulin tryptic digestion: a model approach for peptide release. Biochem. Eng. J. 70, 88–96.

Fernández, A., Suárez, A., Zhu, Y., Fitzgerald, R.J., Riera, F.A., 2013. Membrane fractionation of a β-lactoglobulin tryptic digest: effect of the pH. J. Food Eng. 114, 83–89.

Fidaleo, M., Moresi, M., 2006. Electrodialysis applications in the food industry. Adv. Food Nutr. Res. 51, 265–360.

Galanakis, C.M., 2012. Recovery of high added-value components from wastes: conventional, emerging technologies and commercialized applications. Trends Food Sci. Technol. 26, 68–87.

Galanakis, C.M., Tornberg, E., Gekas, V., 2010a. A study of the recovery of the dietary fibres from olive mill wastewater and the gelling ability of the soluble fibre fraction. LWT – Food Sci. Technol. 43, 1009–1017.

Galanakis, C.M., Tornberg, E., Gekas, V., 2010b. Clarification of high-added value products from olive mill wastewater. J. Food Eng. 99, 190–197.

Galanakis, C.M., Tornberg, E., Gekas, V., 2010c. Recovery and preservation of phenols from olive waste in ethanolic extracts. J. Chem. Technol. Biotechnol. 85, 1148–1155.

Galanakis, C.M., Fountoulis, G., Gekas, V., 2012. Nanofiltration of brackish groundwater by using a polypiperazine membrane. Desalination 286, 277–284.

Galanakis, C.M., Goulas, V., Tsakona, S., Manganaris, G.A., Gekas, V., 2013a. A knowledge base for the recovery of natural phenols with different solvents. Int. J. Food Prop. 16, 382–396.

Galanakis, C.M., Markouli, E., Gekas, V., 2013b. Fractionation and recovery of different phenolic classes from winery sludge via membrane filtration. Separ. Purif. Technol. 107, 245–251.

Galanakis, C.M., Chasiotis, S., Botsaris, G., Gekas, V., 2014. Separation and recovery of proteins and sugars from Halloumi cheese whey. Food Res. Int. 65, 477–483.

Ganesh, R.J., Nazeer, R.A., Kumar, N.S.S., 2011. Purification and identification of antioxidant peptide from black pomfret, *Parastromateus niger* (Bloch, 1975) viscera protein hydrolysate. Food Sci. Biotechnol. 20 (4), 1087–1094.

García-Castello, E., Cassano, A., Criscouli, A., Conidi, C., Drioli, E., 2010. Recovery and concentration of polyphenols from olive mill wastewaters by integrated membrane system. Water Res. 44, 3883–3892.

Garem, A., Daufin, G., Maubois, J.L., Léonil, J., 1997. Selective separation of amino acids with charged inorganic membrane: effect of physicochemical parameters on selectivity. Biotechnol. Bioeng. 20 (54), 291–302.

Gernigon, G., Schuck, P., Jeantet, R., 2011. Demineralization. In: Fuquay, J.W., Fox, P.F., McSweeney, P.L.H. (Eds.), Encyclopedia of Dairy Sciences. Academic Press, San Diego, pp. 738–743.

González, M.I., Álvarez, S., Riera, F.A., Álvarez, R., 2007. Economic evaluation of an integrated process for lactic production from ultrafiltered whey. J. Food Eng. 80 (2007), 553–561.

González, M.I., Álvarez, S., Riera, F.A., Álvarez, R., 2008. Lactic acid recovery from whey ultrafiltrate fermentation broths and artificial solutions by nanofiltration. Desalination 228, 84–96.

Gooding, K.M., 2005. Hydrophobic interaction chromatography. In: Cazes, J. (Ed.), Encyclopedia of Chromatography, vol. 1, second ed. Taylor & Francis, Boca Raton, FL, pp. 805–808.

Gupta, V.K., Mittal, A., Malviya, A., Mittal, J., 2009. Adsorption of carmoisine A from wastewater using waste materials – bottom ash and deoiled soya. J. Colloid Interf. Sci. 335, 24–33.

Harris, D.C., 2007. Análisis Químico Cuantitativo, third ed. Reverté Ed, Barcelona.

Harris, D.C., 2010. Quantitative Chemical Analysis, eighth ed. W. H. Freeman and Company, New York.

Hussain, A.A., Al-Rawajfeh, A.E., 2009. Recent patents of nanofiltration applications in oil processing, desalination, wastewater and food industries. Recent Patents Chem. Eng. 2, pp. 51–66.

Kim, S.K., 2013. Marine Proteins and Peptides: Biological Activities and Applications. John Wiley & Sons, New York.

Kim, Y.H., Moon, S.-H., 2001. Lactic acid recovery from fermentation broth using one-stage electrodialysis. J. Chem. Technol. Biotechnol. 76, 169–178.

Kovacs, Z., Samhaber, W., 2009. Nanofiltration of concentrated amino acid solutions. Desalination 240, 78–88.

Lacroix, I.M.E., Li-Chan, E.C.Y., 2013. Inhibition of dipeptidyl peptidase (DPP)-IV and α-glucosidase activities by pepsin treated whey protein. J. Agric. Food Chem. 61, 7500–7506.

Lama-Muñoz, A., Rodríguez-Gutiérrez, G., Rubio-Senent, F., Palacios-Díaz, R., Fernández-Bolaños, J., 2013. A study of the precursors of the natural antioxidant phenol 3,4-dihydroxyphenylglycol in olive oil waste. Food Chem. 140 (1–2), 154–160.

Li, S., Li, Ch., Liu, Y., Wang, X., Cao, Z., 2003. Separation of L-glutamine from fermentation broth by nanofiltration. J. Membr. Sci. 222, 191–201.

Lin Teng Shee, F., Angers, P., Bazinet, L., 2007. Delipidation of a whey protein concentrate by electroacidification with bipolar membranes (BMEA). J. Agric. Food Chem. 55, 3985–3989.

Lipnizki, F., 2010. Membrane Technology: Membranes for Food ApplicationsVol. 3Wiley-Verlag GmbH & Co kGaA, Weinheim.

McCabe, W.L., Smith, J.C., Harriott, P., 1993. Unit Operations of Chemical Engineering, fifth ed. McGraw-Hill, Singapore, pp. 810–837.

Monsanto, M., Zondervan, E., Trifunovic, O., Bongers, P.M.M., 2012. Integrated optimization of the adsorption of theaflavins from black tea on macroporous resins. In: Karimi, I.A., Rajagopalan Srinivasan, R. (Eds.), 11th International Symposium on Process Systems Engineering – PSE2012, 15–19 July 2012. Elsevier Science, Singapore, Amsterdam, pp. 725–729.

Nagarale, R.K., Gohil, G.S., Shahi, V.K., 2006. Recent developments on ion-exchange membranes and electro-membrane processes. Adv. Colloid Interf. Sci. 119, 97–130.

Ndiaye, N., Pouliot, Y., Saucier, L., Beaulieu, L., Bazinet, L., 2010. Electroseparation of bovine lactoferrin from model and whey solutions. Separ. Purif. Technol. 74, 93–99.

Nfor, B.K., Ripic, J., van der Padt, A., Jacobs, M., Ottens, M., 2012. Model-based high-throughput process development for chromatographic whey proteins separation. Biotechnol. J. 7 (10), 1221–1232.

Otte, J., Shalaby, S.M., Zakora, M., Pripp, A.H., El-Shabrawy, S.A., 2007. Angiotensin-converting enzyme inhibitory activity of milk protein hydrolysates: effect of substrate, enzyme and time of hydrolysis. Int. Dairy J. 17, 488–503.

Pal, P., Dey, P., 2013. Process intensification in lactic acid production by three stage membrane integrated hybrid reactor system. Chem. Eng. Process. 64, 1–9.

Paraskeva, C.A., Papadakis, V.G., Tsarouchi, E., Kanellopoulou, D.G., Koitsoukos, P.G., 2007. Membrane processing for olive mill wastewater fractionation. Desalination 213, 218–229.

Patsioura, A., Galanakis, C.M., Gekas, V., 2011. Ultrafiltration optimization for the recovery of β-glucan from oat mill waste. J. Membr. Sci. 373, 53–63.

Perera, A., Appleton, D., Ying, L.H., Elendran, S., Palanisamy, U.D., 2012. Large scale purification of geraniin from *Nephelium lappaceum* rind waste using reverse-phase chromatography. Separ. Purif. Technol. 98, 145–149.

Poulin, J.-F., Amiot, J., Bazinet, L., 2006. Simultaneous separation of acid and basic bioactive peptides by electrodialysis with ultrafiltration membrane. J. Biotechnol. 123, 314–328.

Pouliot, Y., Wijers, M.C., Gauthier, S.F., Nadeau, L., 1999. Fractionation of whey protein hydrolysates using charged UF/NF membranes. J. Membr. Sci. 158, 105–114.

Pouliot, Y., Gauthier, S.F., L'Heureux, J., 2000. Effect of peptide distribution on the fractionation of whey protein hydrolysates by nanofiltration membranes. Lait 80, 113–122.

Power, O., Nongonierma, A.B., Jakeman, P., Fitzgerald, R.J., 2014. Food protein hydrolysates as a source of dipeptidyl peptidase IV inhibitory peptides for the management of type 2 diabetes. Proc. Nutr. Soc. 73, 34–46.

Prapakornwiriya, N., Diosady, L.L.J., 2014. Recovery of sinapic acid from the waste effluent of mustard protein isolation by ion exchange chromatography. Am. Oil Soc. 91, 357–362.

Pujar, N.S., Zydney, A.L., 1994. Electrostatic and electrokinetic interactions during protein transport through narrow pore membrane. J. Ind. Eng. Chem. Res. 33, 2473–2482.

Puoci, F., Scoma, A., Cirillo, G., Bertin, L., Fava, F., Picci, N., 2012. Selective extraction and purification of gallic acid from actual site olive mill wastewaters by means of molecularly imprinted microparticles. Chem. Eng. J., 198–199, 529–535.

Quan, H., Qiong-Yao, Y., Jiang, S., Chang-Yun, X., Ze-Jie, L., Pu-Ming, H., 2011. Structural characterization and antioxidant activities of 2 water-soluble polysaccharide fractions purified from tea (*Camellia sinensis*) flower. J. Food Sci. 76 (3), C462–471.

Quian, B., Xing, M., Cui, L., Deng, Y., Xu, Y., Huang, M., Zhang, S., 2011. Antioxidant, antihypertensive and immunomodulatory activities of peptide fractions from fermented skim milk with *Lactobacillus delbrueckii* ssp. Bulgaricus LB340. J. Dairy Res. 78, 72–79.

Rahmanian, N., Mahdi Jafari, S., Galanakis, C.M., 2014. Recovery and removal of phenolic compounds from olive mill wastewater. J. Am. Oil Chem. Soc. 91, 1–18.

Riera, F.A., Álvarez, S., 2012. Purification of lactic acid obtained by fermentative processes by means of membrane techniques and ion exchange. In: Jiménez Migallón, A., Ruseckaite, R.A. (Eds.), Lactic Acid: Production, Properties and Health Effects. Nova Science Publishers, New York, pp. 1–45.

Russo, C., 2007. A new membrane process for the selective fractionation and total recovery of polyphenols, water and organic substances from vegetation waters (VW). J. Membr. Scie. 288, 239–246.

Salehi, F., 2014. Current and future applications for nanofiltration technology in the food processing. Food Bioprod. Process. 92, 161–177.

Schirhagl, R., 2014. Bioapplications for molecularly imprinted polymers. Anal. Chem. 86, 250–261.

Scoma, A., Pintucci, C., Bertin, L., Carlozzi, P., Fava, F., 2012. Increasing the large scale feasibility of a solid phase extraction procedure for the recovery of natural antioxidants from olive mill wastewaters. Chem. Eng. J., 198–199, 103–109.

Sheih, I.C., Fang, T.J., Wu, T.K., 2009. Isolation and characterisation of a novel angiotensin I-converting enzyme (ACE) inhibitory peptide from the algae protein waste. Food Chem. 115, 279–284.

Shi, M., Zhang, Z.Y., Yang, Y.N., 2013. Antioxidant and immunoregulatory activity of *Ganoderma lucidum* polysaccharide (GLP). Carbohyd. Polym. 95 (1), 200–206.

Sikder, J., Chakraborty, S., Pal, P., Drioli, E., Bhattacharjee, C., 2012. Purification of lactic acid from microfiltrate fermentation broth by cross-flow nanofiltration. Biochem. Eng. J. 69, 130–137.

Soto, M.L., Moure, A., Domínguez, H., Parajó, J.C., 2011. Recovery, concentration and purification of phenolic compounds by adsorption: a review. J. Food Eng. 105, 1–27.

Strathmann, H., 2010. Electrodialysis, a mature technology with a multitude of new applications. Desalination 264, 268–288.

Suárez, F., Lobo, A., Álvarez, S., Riera, F.A., Álvarez, R., 2006. Partial demineralization of whey and milk ultrafiltration permeate by nanofiltration at pilot plant scale. Desalination 198, 274–281.

Suárez, F., Lobo, A., Álvarez, S., Riera, F.A., Álvarez, R., 2009. Demineralization of whey and milk ultrafiltrate permeate by means of nanofiltration. Desalination 241, 272–280.

Suwal, S., Roblet, C., Doyen, A., Amiot, J., Beaulieu, L., Legault, J., Bazinet, L., 2014. Electrodialytic separation of peptides from snow crab by-product hydrolysate: effect of cell configuration on peptide selectivity and local electric field. Separ. Purif. Technol. 127, 29–38.

Takaç, S., Karakaya, A., 2009. Recovery of phenolic antioxidants from olive mill wastewater. Recent Patents Chem. Eng. 2, pp. 230–237.

Tsakona, S., Galanakis, C.M., Gekas, V., 2012. Hydro-ethanolic mixtures for the recovery of phenols from Mediterranean plant materials. Food Bioprocess Technol. 5, 1384–1393.

Tsuru, T., Shutou, T., Nakao, S.I., Kimura, S., 1994. Peptide and amino acid separation with nanofiltration membranes. Separ. Sci. Technol. 29, 971–984.

Van der Bruggen, B., Mänttäri, M., Nyström, M., 2008. Drawbacks of applying nanofiltration and how to avoid them: a review. Separ. Purif. Technol. 63, 251–263.

Velizarov, S., Reis, M.A., Crespo, J.G., 2003. Removal of trace mono-valent inorganic pollutants in an ion exchange membrane bioreactor: analysis of transport rate in a denitrification process. J. Membr. Sci. 217, 269–284.

Wang, X. 2013. Isolation and purification of anthocyanins from black bean canning wastewater using macroporous resins. Thesis Utah State University. http://digitalcommons.usu.edu/etd/1523.

Weisz, G.M., Carle, R., Kammerer, D.R., 2013. Sustainable sunflower processing – II. Recovery of phenolic compounds as a by-product of sunflower protein extraction. Innov. Food Sci. Emerg. Technol. 17, 169–179.

Xu, T., 2005. Ion exchange membranes: state of their development and perspective. J. Membr. Sci. 263, 1–29.

Yadav, S., Dubey, A.K., Anand, G., Yadav, D., 2012. Characterization of a neutral pectin lyase produced by *Oidiodendron echinulatum* MTCC 1356 in solid state fermentation. J. Basic Microbiol. 52 (6), 713–720.

Yang, F., Su, W.J., Lu, B.J., Wu, T., Sun, L.C., Hara, K., Cao, M.J., 2009. Purification and characterization of chymotrypsins from the hepatopancreas of crucian carp (*Carassius auratus*). Food Chem. 116 (4), 860–866.

Yuanhui, Z., Bafang, L., Zunying, L., Shiyuan, D., Xue, Z., Mingyong, Z., 2007. Antihypertensive effect and purification of an ACE inhibitory peptide from sea cucumber gelatine hydrolysate. Process Biochem. 42, 1586–1591.

Zanello, P.P., Sforza, S., Dossena, A., Lambertini, F., Bottesini, C., Nikolaev, I.V., Koroleva, O., Ciociola, T., Magliani, W., Conti, S., Polonelli, L., 2014. Antimicrobial activity of poultry bone and meat trimmings hydrolyzates in low-sodium turkey food. Food Function 5, 220–228.

Zhao, L., Budge, S.M., Ghaly, A.E., Brooks, M.S., Dave, D., 2011. Extraction, Purification and Characterization of Fish Pepsin: A Critical Review. J Food Process Technol 2:126. doi: 10.4172/2157-7110.1000126.

Zhao, T., Zhou, Y., Mao, G.H., Zou, Y., Zhao, J.L., Bai, S.Q., Yang, L.Q., Wu, X.Y., 2013. Extraction, purification and characterisation of chondroitin sulfate in Chinese sturgeon cartilage. J. Sci. Food Agric. 93 (7), 1633–1640.

Zheng, W., Li, X., Wang, F., Yang, Q., Deng, P., Zeng, G., 2008. Adsorption removal of cadmium and copper from aqueous solution by areca – a food waste. J. Hazard. Mater. 157, 490–495.

Zhong, K., Lin, W.J., Wang, Q., Zhou, S.M., 2012. Extraction and radicals scavenging activity of polysaccharides with microwave extraction from mung bean hulls. Int. J. Biol. Macromol. 51 (4), 612–617.

Zydney, A.L., 1998. Protein separations using membrane filtration: new opportunities for whey fractionation. Int. Dairy J. 8, 243–250.

CONVENTIONAL PRODUCT FORMATION

Paola Pittia*, Adem Gharsallaoui**

*Faculty of Bioscience and Technology for Food, Agriculture and Environment, University of Teramo, Teramo, Italy;
**BioDyMIA (Bioingénierie et Dynamique Microbienne aux Interfaces Alimentaires), Bourg en Bresse, France

8.1 INTRODUCTION

A wide range of conventional, membrane, and innovative technologies are nowadays applied to recover macronutrients (e.g. oils, proteins, polysaccharides, carbohydrates) as well as micronutrients (e.g. vitamins, polyphenols, glucosinolates, flavonoids, mineral salts, pigments, volatile compounds, lipid compounds) (Galanakis et al., 2010b, d, 2012a; Patsioura et al., 2011; Galanakis, 2012, 2013, 2015; Rahmanian et al., 2014; Roselló-Soto et al., 2015; Deng et al., 2015; Heng et al., 2015). There is also an increasing interest towards minor compounds whose presence and concentration may contribute to the health and nutritional quality of foods like some polyphenolic compounds and the ω-3 and ω-6 fatty acids.

To increase value, recovered compounds need to be processed to a food-grade state, following the removal of any toxic compound and contaminant present in the waste raw material, and transforming to a specific physical state such as liquid (solution or emulsion), semisolid, or solid. This state should be appropriate for its addition in food products and at the same time induce specific technological abilities as well as affect positively the quality of the final product and its sensory acceptability. Moreover, it should ensure the stability of the recovered compounds over storage, distribution and usage, as in many cases compounds are prone to degradation and loss of their functional properties.

The sensory properties of foods depend on various factors such as the structures formed by constituents due to their nature or processing. Interactions occurring among the molecules at different levels (e.g. from nano- to macroscale) create assemblies of molecules. The latter determine the structure of the food and thus its mechanical properties and texture upon consumption. In formulated products, the ingredients are assembled during manufacturing and the structure developed is governed by the application of physical, chemical, or biological actions (e.g. fermentation, enzymes) and their consequent effects (Galanakis, 2012; Galanakis et al., 2012b, 2015b). By the controlled choice of process parameters, the processors aim to generate products of predictable properties from materials with defined quality and technological properties at the lowest cost.

Usage and stability of food waste compounds could be overcome by the application of conventional processes (e.g. drying) as well as more modern technologies of encapsulation (micro-, nano-) where the sensitive bioactive compound is included within a secondary material acting as physical barrier towards degrading factors during storage. For example, microencapsulation is used to optimize technological performances of the specific component by modulating its delivery and decreasing any undesired interaction with other components in the food matrix. On the other hand, some food waste components exert structuring properties that allow them to be used as encapsulating agents (e.g. hydrocolloids, polysaccharides, and proteins).

Much research has been carried out to characterize the compounds recovered from by-products and explore their applications in real food products, by substituting existing ingredients and modifying as well as increasing health and nutritional quality. Table 8.1 briefly summarizes the main components that could be recovered from food wastes and by-products along with their general technological properties.

The use of food waste components as new ingredients in food product development requires knowledge of their specific technological properties as related to the given food matrix in which they are intended to be added. These include any potential interaction occurring or favored during processing that may either improve or impair the overall quality and stability of the product. The effects of raw materials, recovery, and isolation process conditions on the above technological properties have to be investigated along with those related to the process conditions applied in product manufacturing in order to enhance their performances in real food products. Finally, it is important to identify quality attributes that may obstacles to the usage for processing and food product design, that is, their sensory impact, safety (e.g. dosage limits), and stability upon processing and storage.

In this chapter emulsification and microencapsulation processes will be discussed as conventional technologies largely applied for ingredients production and development of conventional and innovative food products. The potential and actual use of compounds and materials recovered, extracted, or isolated from food by-products in food emulsions and microencapsulated matrices will be reported.

8.2 TECHNOLOGICAL FUNCTIONALITY AND QUALITY PROPERTIES OF FOOD WASTE COMPONENTS

Technological properties are those abilities exerted by a given component present in optimum concentration and subjected to processing at optimum parameters, which contribute to the expected desirable quality and sensory characteristics of a food product, usually by interacting with other food constituents (Sikorski and Piotrowska, 2007). In general these properties are also referred to as "functional" properties. They depend on intrinsic (molecular, chemical, physicochemical, physical) and extrinsic (e.g. temperature, pressure) factors as well as any interaction via hydrophobic, hydrogen, and covalent bonds with other components that may occur in the matrix where the compound is added. The identification and characterization of the technological properties of a food component are a prerequisite for its usage, especially when new components are recovered from conventional raw materials or known components are obtained from unconventional sources (e.g. food waste, by-products).

Proteins, hydrocolloids, and polysaccharides may present viscosity effects as well as interesting structuring and gelling properties, adhesion and film formation upon processing. Thereby, they could be used as encapsulating agents for microencapsulation of bioactive compounds. Amphiphilic molecules (i.e. proteins, peptides, phospholipids, etc.), because of their surface activity, play a major role in the dispersion of immiscible components in complex food matrices like emulsified and foamed

Table 8.1 General Technological Functionalities of the Main Compounds Present in Food Waste and By-Products

Compound	Origin/Source of Waste and By-Products	Solubility	Technological/Quality Functionality
Pectin	Fruit and vegetable extracts	Water	Gelling and structuring Surface activity
Proteins	Meat (animal) Milk Eggs Vegetables (legumes) Seeds	Water Amphiphilic behavior	Emulsifying and foaming activity Gelling and structuring Binding (aroma, lipids) Antioxidant properties
Peptides and amino acids	Meat (animal) Milk Vegetables Seeds	Water Amphiphilic behavior	Solubility Emulsifying and foaming activity Bioactivity Health properties
Oligosaccharides	Fruit and vegetables	Water	Solubility Healthy properties
Polysaccharides	Fruit and vegetables	Water	Water-holding and binding properties Gelling and structuring
Hydrocolloids and gums	Vegetables, seeds	Water	Gelling and structuring Water-holding capacity
Oils and fats	Animal, fish, seeds	Oil	Structure forming Binding (aroma, proteins) Sensory properties
Phenolic compounds	Plant and fruit extracts	Water-to-oil depending on chemical structure and molecular weight	Antioxidant Health properties Surface activity Sensory properties (color and taste)
Phytochemicals	Plant extracts	Water-to-oil depending on chemical structure and molecular weight Some have amphiphilic behavior	Solubility Surface activity Emulsifying properties Healthy properties
Pigments	Plant and fruit extracts Algae and seaweeds extracts Meat (myoglobin)	Water-to-oil depending on the compound	Color and sensory properties
Aroma compounds and essential oils	Plant and fruit extracts	Water-to-oil depending on the compound	Aroma and sensory properties

products. The use of emulsifying and foaming agents is actually critical for the dispersion of ingredients and compounds with no reciprocal phase affinity (i.e. water and lipids, water and air) into fine and homogeneous systems. Optimal formulation and process conditions (including pH, concentration, ionic force, presence of other solutes in both dispersing and dispersed phases) must be defined to allow surface-active compounds to exert their technological functionality in food emulsified and foamed products or, even better, to enhance them.

Several food waste and by-products have already been shown to be important sources of surface-active compounds. Whey derivatives and blood plasma obtained from milk and meat processing, respectively, due to the presence of specific proteins (β-lactoglobulin, serum albumin, globulins, and fibrinogen) are widely used in food products as emulsifying and foaming agents (Foegeding et al., 2002; Raeker and Johnson, 1995). Recent studies have shown that some phenolic compounds, like those present in olive oil and water olive wastes, exert surface-active properties. Their presence and concentration may affect either positively or negatively the dispersion degree and stability of model and real emulsified food products (e.g. mayonnaise) while exerting their bioactivity (Di Mattia et al., 2010, 2014, 2015; Souilem et al., 2014). Currently, the knowledge on bioactive ingredients effects on food structure is rather limited and to extend their use as food components, for both their bioactivity and technological performances, further studies are needed.

Technological properties may also contribute to the overall quality of the product including sensorial acceptability. When plant extracts are added in formulated foods, they may interfere with the sensory properties of the final product such as color and taste. The addition of polyphenolic compounds may, in fact, impart an astringent or bitter taste or give a brown color to the product (Wang and Bohn, 2012; Peng et al., 2010). For example, fish oils are of particular interest for their nutritional and health aspects but their application in formulated products is hindered due to their unpleasant flavor. Thus, formulation strategies need to be applied in order to avoid undesired effects and to this purpose encapsulation is currently used to reduce the sensory impact of some food ingredients.

Technological properties may also include those aimed at improving the stability and shelf-life of foods and thereby the overall quality. Antioxidant and antimicrobial abilities of food compounds are important to preserve foods and improve their safety during storage. Many plant extracts containing phenolic compounds and phytochemicals are able to inhibit lipid oxidation and other oxidative reactions due to their antioxidant activity exploited via diverse mechanisms (Shi and Noguchi, 2001; Tsakona et al., 2012). Essential oils and some phenolic compounds may also exert bacteriostatic and bactericidal effects against degrading and pathogenic microorganisms (Burt, 2004). The desire to incorporate essential oils in foods as preservatives is due to their recognition as safe natural compounds, and a potential natural alternative to produce foods free of synthetic additives. Beyond the antioxidant properties of bioactive grape-derived polyphenolic compounds (Galanakis et al., 2013b, 2015a), their in vitro and in situ (in real foods) inhibitory effects against pathogenic bacteria, viruses, and fungi have been evidenced and recently reviewed by Friedman (2014).

When referring to a health-promoting bioactive component, it is important to understand the relationship between its bioavailability, bioaccessibility, and food microstructure (Parada and Aguilera, 2007). In the case of biopolymers and structure-forming agents, the main relevance is the knowledge of its ability to favor or to hinder the creation at the different levels (micro-, meso-, and macro-) of the final product in relation to the desired sensory quality attributes. Process extraction from waste and by-products and food product manufacturing conditions along with the presence of other compounds may affect the functionality of the compounds in the systems and thereby increase or impair their performances. For instance, the presence of acids, salts, and other cosolutes in residues as well as the physicochemical and thermal processes of proteins affect their effective technological performances. Also, the addition of bioactive compounds is challenging as they can lose their antioxidative properties and bioactive functionality due to their sensitivity to oxygen, temperature, pH, and light. Furthermore, specific process actions and formulations could be needed in order to improve their solubilization and delivery in the specific matrix (Galanakis et al., 2013a).

More recent basic and applied research is increasingly focused in understanding the relationship between molecular characteristics, technological functionality, and quality properties of food components in food systems with the aim of enhancing their performances in complex food products and further increasing their use. The quantitative structure–function relationship is currently applied to the rational design and production of functional food ingredients, in the optimization of their delivery (Lesmes and McClements, 2009) and new product development.

8.3 **PRODUCT DESIGN BY EMULSIFICATION**
8.3.1 **GENERAL ASPECTS OF EMULSIONS**

An emulsion is formed of two or more immiscible liquids (usually oil and water), where one liquid (dispersed phase) is dispersed as small spherical droplets in the other (continuous phase). The average diameter of the droplets usually ranges between 100 nm and 100 µm, but in some systems it may be smaller or larger. For a two-phase liquid/liquid system, the more stable state is that corresponding to a minimum interfacial area between the dispersed phase and the dispersion medium. Consequently, emulsions are thermodynamically unstable systems, and over time an emulsion evolves towards the separation of the two phases (organic and aqueous) that constitute it. Thus, stability is governed by the interactions between the droplets derived from the surface forces. The latter are created by the droplets' membrane and/or the surrounding medium. Many small dispersed droplets can coalesce to form large aggregates until the separation of the two phases. This separation may take place due to the following phenomena: flocculation, coalescence, and creaming (Fig. 8.1). However, the creation of an emulsion stable within the time of observation is related to the slow down of the following destabilizing phenomena.

Flocculation: After emulsification, the formed droplets do not remain independent of each other, but tend to form "clusters". This step is considered a precursor of the creaming of the formed clusters. This phenomenon can be explained by the association of the droplets due to the competition between thermal agitation and attraction forces. Flocculation may also be induced by the presence of polymers or micelles in the continuous phase (depletion flocculation) or by adsorption of one polymer chain on two (or more) droplets simultaneously (bridging flocculation) (Guzey and McClements, 2006).

Creaming: In order to return to the system equilibrium, the molecules of the continuous phase are subjected to Brownian motion that agitates the droplets of the dispersed phase. At the same time, the gravity force applied to a droplet tends to impose an upward motion if its content is less dense than the continuous phase. This competition between the two forces leads to emulsion heterogeneity (stored without agitation) and creaming clarification of the aqueous phase (called serum phase) of oil-in-water emulsions.

Coalescence and phase separation: The two phenomena described previously (flocculation and creaming) are reversible and a simple stirring allows redispersion of associated droplets. On the other hand, coalescence is the final phase of emulsion degradation that results in the formation of large droplets by approximation and fusion of the droplets due to the instability of the interfacial membrane surrounding the droplets (Fig. 8.1). To prevent coalescence, it is important to form a sufficiently thick and elastic membrane around the droplets. Rheological properties of the

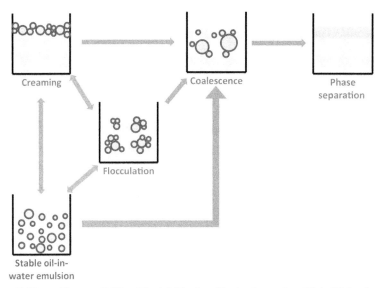

FIGURE 8.1 Reversibility and Irreversibility of Destabilization Mechanisms of an Oil-in-Water Emulsion

interfacial membrane depend on the structural and physicochemical properties (e.g. molecular dimensions, electric charge, compactness, etc.) of the emulsifier and its amount. It also depends on the nature and the intensity of the interactions between the droplets and the continuous phase.

8.3.2 USE OF WASTE COMPONENTS AS EMULSIFIERS

Emulsifiers are surface-active molecules that are used in emulsions to facilitate the formation of small droplets within the homogenizer and to prevent droplet aggregation and growth after formation (McClements, 2005). During homogenization, emulsifiers absorb to the oil/water interface, which leads to a reduction in the interfacial tension. Consequently, the division of the large droplets to form small ones is facilitated. Once they have adsorbed to the droplet surface, these surface-active molecules should prevent the droplets from flocculation and coalescence. This means they must form a protective interfacial membrane around the oil droplets. An effective emulsifier system should meet a number of criteria to prevent droplet coalescence and to ensure a good stability of the emulsion:

1. It must be present at a sufficiently high concentration to cover all of the oil/water surface formed.
2. It must form a coating around the droplet surfaces faster than the droplets can associate with each other.
3. The adsorbed emulsifier molecules must form a coating that prevents the droplets from coming into close proximity and coalescing (McClements, 2005).

There are a number of emulsifiers extracted from food wastes that fulfill these requirements, but they differ considerably in their effectiveness at forming stable emulsions.

Soybean lecithin: Lecithin is predominantly produced from soybean oilseeds due to their abundant availability and low cost. Crude soybean oil contains 1–3% phospholipids. These phospholipids are extracted as a by-product during oil refinement. Soy lecithin is a highly valued emulsifier because of its natural origin, abundant supply, and good emulsion stabilizing properties. The polarity of lecithin that belongs to phospholipids family makes it a useful emulsifier because it forms hydrogen bonds with water molecules and creates nonpolar interactions with the hydrocarbon chains of triglycerides. The major constituents of soybean lecithin are phosphatidylcholine, phosphatidylethanolamine, and lysophosphatidylcholine. Because commercial lecithins are mixtures of phospholipids and other substances, their surface activities are a combined result of all surface-active components. Phosphatidylcholine is the main constituent of lecithins; however, soybean lecithin is also rich in phosphatidylethanolamine (Dickinson, 1993). Lecithin use is often combined with synthetic surfactants to reduce the amount of the synthetic surfactants (Rust and Wildes, 2008).

Pea proteins: When the main objective is pea starch extraction, pea proteins can be considered as a by-product. However, in recent years manufacturers have developed efficient processes for extracting and purifying these proteins for usage as emulsifiers. Pea proteins can be categorized into two major classes: globulins (the salt soluble proteins) and albumins (the water soluble proteins). Albumin fraction accounts for approximately 30% of the total proteins (Schroeder, 1982). Emulsifying properties of albumins have been shown to be highest at acid pH (Lu et al., 2000). Pea globulins are subdivided into two major groups on the basis of their sedimentation coefficients: the 11S fraction (legumin) and the 7S fraction (vicilin, convicilin). The dissociation and unfolding pea globulins at pH 2.4 increase their hydrophobicity and thus improve their interfacial properties (Gharsallaoui et al., 2009). At low pH values, the relatively slow implementation of the interfacial membrane, followed by a reorganization of the protein subunits on the surface of oil droplets, improves emulsion properties. These include ageing stability, decrease of the oil droplets' size, increase of the particle net charge, increase of the interfacial tension lowering kinetics, and increase of the interfacial film viscoelasticity. Thus, when these proteins are used as an emulsifier, the suitable choice of pH during emulsion preparation allows the oil/water interfacial film to densify and therefore the emulsification capacity, the homogeneity of the droplet size distribution and the emulsion stability (Gharsallaoui et al., 2009).

Soybean proteins: Soy proteins are obtained after the extraction of soybean oil. Soy proteins are available in three major forms based on protein content: soy flours (~55% protein), soy protein concentrates (~70% protein), and soy protein isolates (~90% protein). The major storage proteins, namely, β-conglycinin and glycinin, possess a variety of functional properties for food applications such as emulsion and foam stabilizing properties. It has been reported that soy proteins are able to form emulsions by decreasing the interfacial tension between water and oil. However, soy proteins, because of their globular structure, do not unfold and adsorb at the interface, but rather form a thick interfacial layer that acts as a physical barrier to coalescence (Dickinson and Stainsby, 1988). Consequently, the use of soy proteins as emulsifier is limited to concentrated emulsions (Molina et al., 2001).

Cereal proteins: Wheat proteins are in abundance as a by-product of wheat starch production. These proteins can be used as an additive in the food industry because of both nutritional and functional properties as well as low cost. The two main protein types present in wheat gluten are gliadins and glutenins. Gluten utilization as a functional ingredient in the food industry is limited by its insolubility in water at neutral pH, which leads to poor emulsifying and water-holding properties (Mimouni et al., 1994). However, a number of studies have reported that modification of wheat gluten by the deamidation process markedly improves its solubility (Wong et al., 2012).

Gelatin: Collagen can be obtained from animal skins, bone extracts or skeletal muscles. Native collagen has few interesting properties for the food industry. However, collagen denaturation followed by partial hydrolysis can convert it to a protein (gelatin) with very good functional properties. Gelatin often produces relatively large droplet sizes during homogenization (Dickinson and Lopez, 2001), and consequently it is generally used in combination with other surfactants to improve its emulsifying effectiveness.

Whey proteins: Whey is the fluid by-product resulting from the acid and/or enzymatic precipitation of proteins in milk (Galanakis et al., 2014). Unlike caseins, whey proteins have globular structures that are stabilized by intramolecular disulfide bonds between cysteine residues (Dickinson, 2001). They can be used over a wide pH range although their solubility goes through a minimum at their isoelectric point (Kinsella and Whitehead, 1989). The major whey protein is β-lactoglobulin that is a compact globular protein containing 162 amino acid residues with two disulfide bonds. The isoelectric point of β-lactoglobulin is about 5.2. Thus, oil droplets stabilized by this protein display a global positive charge below this value and a negative charge above it. Once adsorbed at the interface, there is a structural reorganization of the protein in order to partially unfold and expose nonpolar groups toward the oily phase (Kulmyrzaev et al., 2000). Whey proteins are widely used in foods as emulsifiers because of their ability to facilitate the formation and stability of oil-in-water emulsions (Kinsella and Whitehead, 1989).

Pectin: Pectin is extracted from various agricultural by-products that contain high levels of pectic polysaccharides, including apple pomace, orange peel, lemon, olive mill waste, and sugar beet pulp. It is a water soluble polysaccharide widely used in food manufacturing for its thickening and gelling properties (Galanakis et al., 2010a, c, e; Galanakis, 2011). Native pectins are high methoxyl pectins in which the majority of carboxylic acid groups are esterified by methanol and consequently are nonionizable. However, low methoxyl pectins having a higher proportion of free ionized carboxylic acid groups favor electrostatic interactions between pectin chains or oil droplets stabilized by pectin as an emulsifier. The surface activity of pectin has its molecular origin in the presence of a protein component linked covalently or physically to the polysaccharide backbone (Dickinson, 2009). In fact, during emulsification, only a small part of the pectin, which is associated with most of the protein, was adsorbed onto the oil. Citrus and beet pectin are able to reduce the interfacial tension between the oil and water phase, and can be efficient for the preparation of emulsions (Leroux et al., 2003). In addition, pectin has been used as a second coating to improve the stability of oil-in-water emulsions stabilized by pea proteins (Gharsallaoui et al., 2010).

8.3.3 USE OF LIPIDS RECOVERED FROM FOOD WASTE AND BY-PRODUCTS

Food waste and by-products from meat, fish, vegetables, seeds from fruits and vegetables, cereals, algae, and seaweeds processing may be important sources of lipids with nutritional and/or technological functionalities (Galanakis, 2012; Galanakis and Schieber, 2014). Indeed, modern nutritional recommendations are currently suggesting the consumption of lipids with a high content of ω-3 fatty acids, mainly composed of eicosapentaenoic acid, C20:5 ω-3 (EPA), and docosahexaenoic acid, C22:6 ω-3 (DHA). The ω-3 fatty acids have showed beneficial bioactivities including prevention of atherosclerosis, protection against manic-depressive illness and various other health promoting properties (Arab-Tehrany et al., 2012).

Lipids play major roles in affecting the quality and stability of food products. They contribute to the development of structures (either as continuous or dispersed phases) and their rheology depending on the specific physical properties. The latter are related to the fatty acid composition and have the ability to interact with other components in the food matrix (e.g. proteins in dough). The contribution of oils and fats on the development of complex dispersed systems depends on the specific surface or interfacial tension, a physical property that in oils is related to the presence of amphiphilic compounds other than triacylglycerols. These surface-active compounds may originate from the raw material (e.g. phospholipids in palm oil, phenols, and particles in virgin olive oils) or be due to several chemical reactions (e.g. hydrolysis of triacylglycerols, oxidation) taking place during oil production and storage or its usage by the consumer (Dopierala et al., 2011).

The main limiting factor to the use of oils and fats in food processing and product development is their susceptibility toward oxidative deterioration favored by a relatively high content of polyunsaturated fatty acids. Oxidation has a negative impact on nutritional properties including the destruction of essential fatty acids and the lipid soluble vitamins A, D, E, and K and the decrease in caloric content. At the same time, it impairs the sensory quality (rancidity) due to off-flavors, color changes (darkening of fats and oils, lightening of pigments, etc.) and flavor loss. While the mechanisms triggering and affecting the oxidation rate and their effects have been widely studied in bulk systems, the information available on lipid oxidation in multicomponent and multiphasic systems has been until now limited. In dispersed lipidic systems like emulsions (both oil/water (o/w) and water/oil (w/o)) oxidation is considered as an interfacial phenomenon affected by the presence of antioxidant and pro-oxidant compounds in the aqueous and oil phases as well as by the interactions between the various ingredients of the system (Waraho et al., 2011). The presence of the aqueous phase can often decrease the activity of antioxidants since their hydrogen donation properties are weakened by the formation of hydrogen-bonded complexes with water (Frankel et al., 1994). In o/w emulsions, lipid oxidation chemistry is very dependent on the physical properties of the water/lipid interface, which thus results in playing a key role in the general stability of dispersed systems.

In Table 8.2 the general composition of oils and fats recovered from some food by-products is presented. Fish processing by-products are a main source of oils. The majority of fish oils are currently used to produce hydrogenated oil, while interest towards fish oils with a high content of ω-3 fatty acids (EPA and DHA) is increasing. However, the high content of polyunsaturated compounds make these oils prone to oxidative deterioration and production of volatile secondary oxidation products detrimental to consumer acceptance. The recovery of oils from seed by-products of fruits and vegetables and/or underutilized oilseeds is also of particular interest.

Grape seed oil recovered from wine processing is similar to sunflower oil with a relatively high quantity of polyunsaturated fatty acids. This provides the main nutritional value to this oil, but the high content of linoleic acid makes it more susceptible to oxidation and associated potential negative effects on human health (Matthaus, 2008). Nevertheless, the natural presence of a relatively high quantity of phenolic compounds and tannins makes it more naturally resistant to peroxidation (Cao and Ito, 2003). Tomato seed oil also has a relatively high content of ω-6 essential fatty acids, relevant for nutritional or industrial use (Botineş tean et al., 2015). Besides, mango seed kernel fats comprise a promising lipid ingredient to develop innovative o/w emulsified gels and a cocoa butter alternative according to EU chocolate directive 2000/36/EC (Sagiri et al., 2014). Finally, interesting amounts of EPA and DHA can be produced by the use of seaweeds and algae. The seaweed lipid content varies between 20% and 50%

Table 8.2 Oil and Fatty Acid Profiles of Some Oils and Fats Recovered from Food Wastes and By-Products

	Oil/Fat Type							
	Tuna Oil[a,b]	Menhaden Oil[c]	Rice Bran Oil[d]	Coffee Oil[e,f]	Grape Seeds[g,h]	Tomato Seeds[i,j]	Seaweed Oil Brown[l,j]	Mango Seed Kernel Fat[m]
Content	13–36%$_{wt}$	5–20%$_{wt}$	15–23%$_{wt}$	9–16%$_{db}$	14–17%$_{wt}$ (7–20%$_{db}$)	18–22%$_{wt}$	5–20 %$_{db}$	5–15%$_{db}$
C14:0	2–3	7.2–12.1	0.1–0.3	Trace–2.3	0.0–0.1	0.0–0.3	0.3–2.2	—
C15:0	0.4–1	0.4–2.3	—	Trace–1.7	—	—	0.2–0.4	—
C15:1	—	—	—	Trace–0.9	—	—	—	—
C16:0	13–22	15.3–25.6	12.8–21.6	16.8–38.6	7.0–9.4	13.0–24.0	13.0–40.0	3.0–10.0
C16:1	2.5–3.0	9.3–15.8	0.0–0.3	0.2–4.0	Trace	0.1–0.8	0.1–0.4	—
C17:0	0.5–1	0.2–3.0	—	Trace–0.6	Trace	0.1–0.3	0.0–0.2	—
C18:0	3.0–6.0	2.5–4.1	0.7–4.7	4.5–13.1	2.5–5.0	0.4–6.0	0.9–1.5	24.0–57.0
C18:1	11.0–21.0	8.3–13.8	32.4–43.4	7.6–18.9	13.9–29.4	8.0–22.0	6.0–20.5	34.0–56.0
C18:2	0.3–1.0	0.7–2.8	28.0–53.4	30.5–50.4	59.5–75.3	47.0–73.0	3.0–7.4	1.0–13.0
C18:3	0.32–1.0	0.8–2.3	0.2–1.6	0.3–6.0	0.3–1.1	2.2–3.2	0.3–11.2	0.0–2.3
C18:4	0.0–1.0	1.7–4.0	—	—	—	—	1.2–25.8	—
C20:0	0.0–0.2	0.1–0.6	0.5–1.4	0.7–6.7	0.0–0.2	0.3–0.6	0.0–0.4	1.0–4.0
C20:1	1.0–2.0	—	—	Trace–0.4	0.1–0.6	0.0–0.2	0.0–4.1	—
C20:4	—	1.5–2.7	—	0.7	Trace–0.1	—	5.3–13.3	—
C22:1	0.2–3.0	0.1–1.4	—	0–0.4	—	0.0–0.1	0.0–1.5	—
C20:5	6.0–7.8	11.1–16.3	—	—	—	—	13.2–42.4	—
C22:0	0.0–0.1	0.0–0.1	—	0–3.0	—	0.0–0.1	Trace	0.0–1.0
C22:1	0.0–0.2	0.1–1.4	—	0–0.3	—	—	0.0–1.7	—
C22:2	Trace	—	—	0–0.4	—	—	—	—
C22:5	0.0–2.0	1.3–3.8	—	—	—	—	Trace	—
C22:6	22.0–24.6	4.6–13.8	—	—	—	—	—	—

Other free fatty acids	C16:2 (0.2–2.8%); C16:3 (0.9–3.5%); C16:4 (0.5–2.8%) High content of ω-fatty acids	—	Traces of C23:0, C24:0, C28:0	—	Traces of C24:0	High levels of ω-3 and ω-6 polyunsaturated fatty acids	C24:0: 0.5%
Other lipophylic compounds		Tocotrienols, tocopherols, γ-oryzanols, phytosterols	Aroma compounds[d], diterpenes, sterols, tocopherols, phytosterols (squalene)	Tocotrienols, tocopherols, phenolic components, phytosterols (β-sitosterols)	Phytosterols (cholesterol, campesterol, stigmasterol, and Δ^5-avenasterol), tocopherols, β-carotene, lycopene, fatty acids (C16:1, C18:1 ω-9, and ω-7)	Phospholipid, sterols (fucosterol and fucoxanthin), pigments	

db, dry basis; wt, weight basis.
[a] *Suseno et al. (2014)*
[b] *Bimbo (2000)*
[c] *O'Brien (2004)*
[d] *Gopala Krishna et al. (2006)*
[e] *Speer and Kölling–Speer (2001)*
[f] *Calligaris et al. (2009)*
[g] *Lutterodt et al. (2011)*
[h] *Beveridge et al. (2005)*
[i] *Lazos et al. (1998)*
[j] *Botines tean et al. (2015)*
[k] *Ambrozova et al. (2014)*
[l] *Dawczynski et al. (2007)*
[m] *Jahurul et al. (2013)*

(Gosch et al., 2012) and is characterized by high levels of ω-3 and ω-6 polyunsaturated fatty acids (Miyashita et al., 2013). In addition, brown seaweed lipids are stable to oxidation due to the presence of these polyunsaturated fatty acids in their glycoglycerolipid forms.

8.3.4 DESIGN AND DEVELOPMENT OF EMULSIFIED FOOD INGREDIENTS USING WASTE COMPONENTS

Interest in food by-products for the generation of aqueous or lipidic fractions and the manufacturing of emulsified foods is growing. Nevertheless, these products have a limited use due to the presence of compounds that may raise safety concerns (e.g. mycotoxins, pesticides) and reduce the technological functionality (e.g. salts and lactose in whey, brown phenolic compounds in plant extracts) in food products. For instance, proteins could be partially or totally denatured. This fact could impair their amphiphilicity and limit their emulsifying ability. Similarly, aqueous fractions and unrefined oils are characterized by a complex composition, which includes undesired impurities, contaminants, and other minor compounds that could affect positively or negatively the technological functionality of the substance itself. However, unrefined oils contain small lipophilic compounds (e.g. volatile aroma compounds, phytochemicals, and apolar polyphenols) with important health, antioxidant, and technological properties. Some fatty compounds (e.g. free fatty acids, monoglycerides) may also exert an amphiphilic behavior and thus play a role in the formation of dispersed and colloidal systems. In emulsified food systems, other molecules (inherently existing or added) with surface-active could be present. These compounds complicate the surface functionality of the emulsifying agents by competing at the interface. Finally, the interaction of all hydrophilic and lipophilic components (sugars, solutes, fats or functional compounds) of the system should be considered, as they could affect the interfacial properties and thus influence chemical and physical stability (Di Mattia et al., 2011).

Despite the aforementioned obstacles, some compounds recovered from food by-products are already used in food emulsified products either as surface-active compounds or as dispersed phase (lipids). Among them, whey protein isolates and concentrates have been used in a variety of foods (e.g. in dairy, meat, and bakery formulated products) due to their superior emulsification and gelling abilities. On the other hand, recovered fats and oils could be directly mixed in formulated food products and dispersed in an emulsified state within a given complex food matrix (mayonnaise, sauce, meat products, etc.). They can also be used to prepare dispersed systems, which could be applied as food additives in the form of liquid micro- and nanoemulsions or in dried forms. Ongoing research aims to substitute conventional highly saturated oils and fats with recovered ones with improved nutritional value.

Emulsions could also be used as efficient edible delivery systems to disperse, encapsulate, protect, and release bioactive lipids (i.e. carotenoids, phytosterols, ω-3 fatty acids) within the food, medical, and pharmaceutical industries. In many cases, it is advantageous to deliver bioactive lipids in an aqueous medium since it increases their palatability, desirability, and bioactivity. Potential applications include the incorporation of bioactive lipids into a beverage or food to improve their health and nutritional quality. Using an interfacial engineering approach, oil-in-water emulsion droplets can be used to inhibit lipid oxidation by decreasing the interactions between the highly oxidizable oils and pro-oxidative species in the aqueous phase. Multilayer emulsions, where a thicker droplet interface is composed by one or multiple layers of biopolymers, can provide more physical stability and inhibit interactions between continuous phase components and the lipids in the emulsion droplet core (McClements et al., 2007). A higher chemical and physical stability of spray-dried powder of menhaden oil, highly rich in ω-3 fatty acids, into an aqueous system was achieved by the preliminary preparation of emulsions with a multi-layered interface made of lecithin and chitosan (Shaw et al., 2007).

8.4 PRODUCT DESIGN BY MICROENCAPSULATION
8.4.1 GENERAL ASPECTS OF MICROENCAPSULATION

Microencapsulation is defined as the process by which tiny particles or droplets are surrounded by a coating, or embedded in a homogeneous or heterogeneous matrix, to give small capsules with many useful properties. Microencapsulated ingredients are totally enveloped in a coating material, thereby conferring useful or eliminating useless properties to (or from) the original ingredient. Microencapsulation is a technology that allows sensitive ingredients to be physically enveloped in a protective matrix or "wall" material in order to protect them from degradative reactions, volatile loss, or nutritional deterioration. In addition to the primary roles of stabilization and protection, microcapsules should release their contents at controlled rates over prolonged periods of time. For these reasons, one of microencapsulation challenges is to preserve the stability of the encapsulated ingredients during processing and storage, as well as to release these ingredients at a given physicochemical condition.

Especially in the food field, microencapsulation is a technique by which liquid droplets, solid particles or gas compounds are entrapped into thin films of a food-grade microencapsulating agent. The core may be composed of just one or several ingredients and the wall may be single or multilayered. The retention of these cores is governed by their chemical functionality, solubility, polarity, and volatility. Shahidi and Han (1993) proposed six reasons to apply microencapsulation in the food industry:

1. Reducing the core reactivity with environmental factors,
2. Decreasing the transfer rate of the core material to the outside environment,
3. Promoting easier handling,
4. Controlling the release of the core material,
5. Masking the core taste,
6. Diluting the core material when it should be used in very small amounts.

Most microcapsules are small spheres with diameters of a few micrometers or millimeters. Depending on the physicochemical properties of the core, the wall composition, and the used microencapsulation technique, following different types of particles can be obtained (Gibbs et al., 1999):

1. Simple sphere surrounded by a coating of uniform thickness,
2. Particle containing an irregular-shaped core,
3. Several core particles embedded in a continuous matrix of wall material,
4. Several distinct cores within the same capsule, and
5. Multiwalled microcapsules.

These different types of microcapsules are produced by a large number of processes such as spray drying, spray cooling, spray chilling, air suspension coating, extrusion, centrifugal extrusion, freeze drying, coacervation, rotational suspension separation, cocrystallization, liposome entrapment, interfacial polymerization, and molecular inclusion. Spraying methods are the most common microencapsulation techniques used in the food industry and are employed to encapsulate a wide range of hydrophilic and lipophilic food ingredients. Spray chilling and fluidized-bed coating are generally used for microencapsulating water soluble molecules (e.g. vitamin C), whereas spray drying of emulsions is generally recommended for the microencapsulation of hydrophobic compounds such as polyunsaturated fatty acids, vitamins A, D, and E, lycopene, and β-carotene (McClements et al., 2007). Conventional oil-in-water emulsions are considered as the more important delivery system of lipophilic molecules, because of their relative ease of preparation and low cost (McClements et al., 2007). Emulsions can be spray dried to form powders, which may facilitate their storage, transport, and utilization in some applications.

8.4.2 MICROENCAPSULATION BY SPRAY DRYING

Spray drying is a unit operation by which a liquid product is atomized in a hot gas current to instantaneously obtain a powder. The generally used gas is air or more rarely inert nitrogen. The initial liquid feeding the sprayer can be a solution, an emulsion or a suspension. Spray drying produces very fine powders (10–50 μm) or large size particles (2–3 mm), depending on the starting feed material and operating conditions. By decreasing water content and water activity, spray drying is generally used in food industry to ensure the microbiological stability of products, avoid the risk of chemical or biological degradations, reduce the storage and transport costs, and finally obtain a product with specific properties like instantaneous solubility.

Although many techniques have been developed to microencapsulate food ingredients, spray drying has been the most common technology used in the food industry for decades due to low cost and available equipment. Compared with freeze drying, the cost of spray drying is 30–50 times cheaper (Desobry et al., 1997). The application of spray drying involves two basic steps: preparation of the dispersion or emulsion to be processed by homogenization and atomization of the mass into the drying chamber. The first stage is the formation of a fine and stable emulsion of the core material in the wall solution. The mixture to be atomized is prepared by dispersing the core material, which is usually of hydrophobic nature, into a solution of the coating agent. The dispersion must be homogenized, with or without the addition of an emulsifier depending on the emulsifying properties of the coating materials, as some of them have interfacial activities. Before the spray-drying step, the formed emulsion must be stable over a certain period of time, oil droplets should be rather small (1–100 μm), and viscosity should be low enough to prevent air inclusion in the dried particle and facilitate the liquid passage in the atomization nozzle. The obtained oil-in-water emulsion can be then mixed with the drying matrix (e.g. maltodextrins) and atomized into a heated air stream supplied to the drying chamber, leading to the evaporation of the solvent and the formation of microcapsules (Fig. 8.2). As the sprayed particles fall

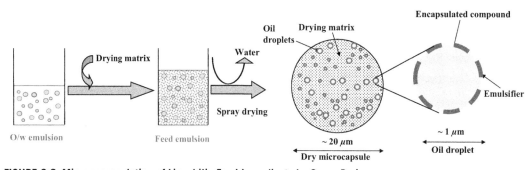

FIGURE 8.2 Microencapsulation of Lipophilic Food Ingredients by Spray Drying

through the gaseous medium, they assume a spherical shape with the oil encased in the aqueous phase. The short time exposition and the rapid evaporation of water keep the core temperature below 40°C, in spite of the high temperatures generally used in the process (~180°C).

8.4.3 USE OF WASTE COMPONENTS AS COATINGS

The number of available wall materials used as a microencapsulating agent by spray drying is limited. The wall material must be highly soluble in water and should possess good emulsification and film-forming properties. Typical wall materials are low molecular weight carbohydrates, milk, or soy proteins, gelatin, hydrocolloids (e.g. acacia gum and pectin), and carbohydrates (e.g. maltodextrins). They are considered to be good encapsulating agents because they exhibit low viscosities at high solids contents and good solubility. However, they lack interfacial properties required for high microencapsulation efficiency. In contrast, proteins have an amphiphilic character that offers the physicochemical and functional properties required to encapsulate hydrophobic core materials. Moreover, protein compounds such as sodium caseinate, soy protein isolate, and whey protein concentrates and isolates could also be expected to have good microencapsulating properties.

> *Starches*: Starches and corn syrup solids are good encapsulating agents because they exhibit low viscosities at high solids contents and good solubility, but most of them lack of the required interfacial properties. Soybean soluble polysaccharide was found to be a superior emulsifier over gum arabic to retain microencapsulated ethyl butyrate during spray drying (Yoshii et al., 2001). In addition, it is known that polysaccharides with gelling properties could stabilize emulsions towards flocculation and coalescence (Dalgleish, 2006).
>
> *Pectin*: Sugar beet pectin could be considered as a suitable wall material for microencapsulation of lipophilic food ingredients by spray drying. Fish oil has been successfully encapsulated in a wall system containing sugar beet pectin as coating agent and glucose syrup as a bulk agent (Drusch, 2007). Moreover, it was shown that spray drying had no effect on the majority of functional properties of pectin (Monsoor, 2005).
>
> *Whey proteins*: These have been successfully used as a wall system to encapsulate anhydrous milk fat by spray drying, obtaining an encapsulation yield greater than 90% (Young et al., 1993). According to the same authors, microencapsulation efficiency can be partially improved (50%) by replacing whey proteins with lactose. In fact, the incorporation of lactose in the whey protein-based wall system can limit the diffusion of nonpolar substances through this wall. Lactose in its amorphous state acts as a hydrophilic sealant that significantly limits diffusion of the hydrophobic core through the wall and thus leads to high microencapsulation efficiency values.
>
> *Gelatin*: It is a water soluble material with wall-forming ability. The characteristics and morphology of gelatin microparticles could be improved by the addition of mannitol (Bruschi et al., 2003). Based on the drying characteristic curves, Imagi et al. (1992) showed that gelatin had advanced properties of an effective entrapping agent compared with maltodextrin, pullulan, glucose, maltose, and mannitol. These include high emulsifying and stabilizing activity, as well as a tendency to form a fine dense network upon drying.
>
> *Cereal proteins*: Iwami et al. (1988) obtained spherical microcapsules of gliadin containing polyunsaturated lipid. These spray-dried microparticles were resistant to oxidative deterioration during long-term storage at various water activity values. Indeed, spray drying raised the antioxidant effect of gliadin without impairing its digestibility. In addition, although the

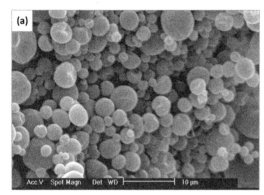 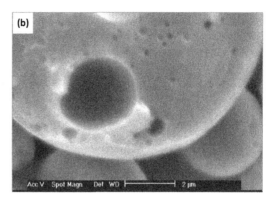

FIGURE 8.3

External (a) and internal (b) structures of dry emulsion (oil, 5 wt%) stabilized by pea protein isolate (0.25 Wt%) in the presence of glucose syrup DE 28 (11 wt%) (core/wall, 0.44).

spray-drying process is known to induce protein denaturation, in the case of gluten hydrolysates, it only modified interfacial protein properties (Linarès et al., 2001).

Pea proteins: Pea protein isolate has been used in combination with maltodextrins to encapsulate ascorbic acid by spray drying (Pereira et al. 2009), whereas spherical microcapsules were obtained with an ascorbic acid retention of 69%. In previous studies, microcapsules containing α-tocopherol (>77% retention) have been successfully obtained by Pierucci et al. (2007). In the latter two works, contrary to the encapsulation of hydrophobic molecules (Fig. 8.2), protein isolate, maltodextrins, and core material were directly mixed together to obtain spray-drying feed solutions. Recently, oil-in-water emulsions, stabilized by a pea protein amount just enough to cover oil droplets, were prepared at pH 7 (Gharsallaoui et al., 2012). Oil microcapsules were then successfully produced by spray drying the obtained emulsion using starch hydrolysates (maltodextrins) as drying matrices (Fig. 8.3).

Finally, it should be noted that there are sometimes some issues when using proteins as encapsulating agents. For example, labeling, allergy, and precipitation when protein-based microcapsules are added to food products having a pH near their isoelectric point. In addition, some religious and social (halal, kosher, vegetarian food choices) issues must be considered.

8.4.4 APPLICATIONS

Encapsulation is increasingly used in the development and manufacturing of food ingredients due to its beneficial effects on quality, health, and technological functionality of valuable compounds. Examples of existing and commercially available ingredients as microcapsules include flavoring agents, lipids (e.g. fish oil, milk fat, vegetable oils) and lipidic compounds, aroma and essential oils, pigments (e.g. carotenoids), enzymes and microorganisms (e.g. probiotic bacteria), amino acids and peptides, vitamins and minerals, antioxidants, polyphenols, phytonutrients, and soluble fibers. Concerning food waste and by-products, encapsulation has already been applied to manufacturing of commercially available ingredients made of fish oil and enriched ω-3 fatty acids oils, bioactive compounds (e.g. resveratrol recovered from grape skin, phytochemicals from vegetables and oil seeds) as well as essential oils (e.g. recovered from citrus peel). Recent applied studies have led to the production of microencapsulated Gac oil powder, containing a high content of β-carotene and lycopene produced using spray drying with whey protein concentrate and gum arabic (Kha et al., 2014). Saikia et al. (2015) have extracted phenolic compounds from carambola (or star fruit, *Averrhoa carambola*) pomace and successfully microencapsulated them into a maltodextrins-based system. Finally, food waste biopolymers (e.g. pectins, chitosan, proteins) have also shown potential as encapsulating agents, giving add value to underutilized and/or waste raw materials.

REFERENCES

Ambrozova, J.V., Misurcova, L., Vicha, R., Machu, L., Samek, D., Baron, M., Mlcek, J., Sochor, J., Jurikova, T., 2014. Influence of extractive solvents on lipid and fatty acids content of edible freshwater algal and seaweed products, the green microalga *Chlorella kessleri* and the cyanobacterium *Spirulina platensis*. Molecules 19, 2344–2360.

Arab-Tehrany, E., Jacquot, M., Gaiani, C., Imran, M., Desobry, S., Linder, M., 2012. Beneficial effects and oxidative stability of omega-3 long-chain polyunsaturated fatty acids. Trends Food Sci. Technol. 25, 24–33.

Beveridge, T.H.J., Girard, B., Kopp, T., Drover, J.C.G., 2005. Yield and composition of grape seed oils extracted by supercritical carbon dioxide and petroleum ether: varietal effects. J. Agric. Food Chem. 53, 1799–1804.

Bimbo, A.P., 2000. Fish meal and oil. In: Martin, R.E., Paine Carter, E., Flick, Jr., G.J., Davis, L.M. (Eds.), Marine and Freshwater Products Handbooks. Technomics Publishing Co, Lancaster (USA), pp. 541–581.

Botineştean, C., Gruia, A.T., Jianu, I., 2015. Utilization of seeds from tomato processing wastes as raw material for oil production. J. Mater. Cycles Waste Manage. 17, 118–124.

Bruschi, M.L., Cardoso, M.L.C., Lucchesi, M.B., Gremiao, M.P.D., 2003. Gelatin microparticles containing propolis obtained by spray-drying technique: preparation and characterization. Int. J. Pharm 264, 45–55.

Burt, S., 2004. Essential oils: their antibacterial properties and potential applications in foods: a review. Int. J. Food Microbiol. 94 (3), 223–253.

Calligaris, S., Munari, M., Arrighetti, G., Barba, L., 2009. Insights into the physicochemical properties of coffee oil. European J. Lipid Sci. Technol. 111, 1270–1277.

Cao, X., Ito, Y., 2003. Supercritical fluid extraction of grape seed oil and subsequent separation of free fatty acids by high-speed counter-current chromatography. J. Chromatogr. A 1021, 117–124.

Dalgleish, D.G., 2006. Food emulsions – their structures and structure forming properties. Food Hydrocolloid 20, 415–422.

Dawczynski, C., Schubert, R., Jahreis, G., 2007. Amino acids, fatty acids, and dietary fibre in edible seaweed products. Food Chem 103, 891–899.

Deng, Q., Zinoviadou, K.G., Galanakis, C.M., Orlien, V., Grimi, N., Vorobiev, E., Lebovka, N., Barba, F.J., 2015. The effects of conventional and non-conventional processing on glucosinolates and its derived forms, isothiocyanates: extraction, degradation and applications. Food Eng. Rev., in press. DOI 10.1007/s12393-014-9104-9

Desobry, S.A., Netto, F.M., Labuza, T.B., 1997. Comparison of spray-drying, drum drying and freeze-drying for (1→3, 1→4)-β-carotene encapsulation and preservation. J. Food Sci. 62, 1158–1162.

Di Mattia, C.D., Sacchetti, G., Mastrocola, D., Sarker, D.K., Pittia, P., 2010. Surface properties of phenolic compounds and their influence on the dispersion degree and oxidative stability of olive oil o/w emulsions. Food Hydrocolloid 24 (6–7), 652–658.

Di Mattia, C.D., Sacchetti, G., Pittia, P., 2011. Interfacial behavior and antioxidant efficiency of olive phenolic compounds in o/w olive oil emulsions as affected by surface active agent type. Food Biophys. 6, 295–302.

Di Mattia, C., Paradiso, V., Andrich, L., Giarnetti, M., Caponio, F., Pittia, P., 2014. Effect of olive oil phenolic compounds and maltodextrins on the physical properties and oxidative stability of olive oil o/w emulsions. Food Biophys. 9 (4), 396–405.

Di Mattia, C., Balestra, F., Sacchetti, G., Neri, L., Mastrocola, D., Pittia, P., 2015. Physical and structural properties of extra-virgin olive oil based mayonnaise. LWT-Food Sci. Technol., doi: 10.1016/j.lwt.2014.09.065.

Dickinson, E., 1993. Towards more natural emulsifiers. Trends Food Sci. Technol 4 (10), 330–334.

Dickinson E, 2001. Milk protein interfacial layers and the relationship to emulsion stability and rheology. Colloids Surf. B: Biointerfaces 20, 197–210.

Dickinson, E., 2009. Hydrocolloids as emulsifiers and emulsion stabilizers. Food Hydrocolloid 23 (6), 1473–1482.

Dickinson, E., Lopez, G., 2001. Comparison of the emulsifying properties of fish gelatin and commercial milk proteins. J. Food Sci. 66, 118–123.

Dickinson, E., Stainsby, G., 1988. Emulsion stability. In: Dickinson, E., Stainsby, G. (Eds.), Advances in Food Emulsions and Foams. Elsevier Applied Science, London and New York, pp. 1–44.

Dopierala, K., Javadi, A., Krägel, J., Schano, K.-H., Kalogianni, E.P., Leser, M.E., Miller, R., 2011. Dynamic interfacial tensions of dietary oils. Colloid. Surface. A 382, 261–265.

Drusch, S., 2007. Sugar beet pectin: a novel emulsifying wall component for microencapsulation of lipophilic food ingredients by F spray-drying. Food Hydrocolloid. 21 (7), 1223–1228.

Foegeding, E.A., Davis, J.P., Doucet, J., McGuffey, M.K., 2002. Advances in modifying and understanding whey protein functionality. Trends Food Sci. Technol. 13, 151–159.

Frankel, E.N., Huang, S.W., Kanner, J., German, J.B., 1994. Interfacial phenomena in the evaluation of antioxidants: bulk oils vs emulsions. J. Agric. Food Chem. 42, 1054–1059.

Friedman, M., 2014. Antibacterial, antiviral, and antifungal properties of wines and winery byproducts in relation to their flavonoid content. J. Agric. Food Chem. 62, 6025–6042.

Galanakis, C.M., 2011. Olive fruit and dietary fibers: components, recovery and applications. Trends Food Sci. Technol 22, 175–184.

Galanakis, C.M., 2012. Recovery of high added-value components from food wastes: conventional, emerging technologies and commercialized applications. Trends Food Sci. Technol. 26, 68–87.

Galanakis, C.M., 2013. Emerging technologies for the production of nutraceuticals from agricultural by-products: a viewpoint of opportunities and challenges. Food Bioprod. Process. 91, 575–579.

Galanakis, C.M., 2015. Separation of functional macromolecules and micromolecules: from ultrafiltration to the border of nanofiltration. Trends Food Sci. Technol. 42 (1), 44–63.

Galanakis, C.M., Schieber, A., 2014. Editorial of special issue on "Recovery and utilization of valuable compounds from food processing by-products". Food Res. Int. 65, 230–299.

Galanakis, C.M., Tornberg, E., Gekas, V., 2010a. A study of the recovery of the dietary fibres from olive mill wastewater and the gelling ability of the soluble fibre fraction. LWT-Food Sci. Technol 43, 1009–1017.

Galanakis, C.M., Tornberg, E., Gekas, V., 2010b. Clarification of high-added value products from olive mill wastewater. J. Food Eng. 99, 190–197.

Galanakis, C.M., Tornberg, E., Gekas, V., 2010c. Dietary fiber suspensions from olive mill wastewater as potential fat replacements in meatballs. LWT-Food Sci. Technol. 43, 1018–1025.

Galanakis, C.M., Tornberg, E., Gekas, V., 2010d. Recovery and preservation of phenols from olive waste in ethanolic extracts. J. Chem. Technol. Biotechnol. 85, 1148–1155.

Galanakis, C.M., Tornberg, E., Gekas, V., 2010e. The effect of heat processing on the functional properties of pectin contained in olive mill wastewater. LWT-Food Sci. Technol. 43, 1001–1008.

Galanakis, C.M., Fountoulis, G., Gekas, V., 2012a. Nanofiltration of brackish groundwater by using a polypiperazine membrane. Desalination 286, 277–284.

Galanakis, C.M., Kanellaki, M., Koutinas, A.A., Bekatorou, A., Lycourghiotis, A., Kordoulis, C.H., 2012b. Effect of pressure and temperature on alcoholic fermentation by *Saccharomyces cerevisiae* immobilized on γ-alumina pellets. Bioresource Technol. 114, 492–498.

Galanakis, C.M., Goulas, V., Tsakona, S., Manganaris, G.A., Gekas, V., 2013a. A knowledge base for the recovery of natural phenols with different solvents. Int. J. Food Prop. 16, 382–396.

Galanakis, C.M., Markouli, E., Gekas, V., 2013b. Fractionation and recovery of different phenolic classes from winery sludge via membrane filtration. Separ. Purif. Technol. 107, 245–251.

Galanakis, C.M., Chasiotis, S., Botsaris, G., Gekas, V., 2014. Separation and recovery of proteins and sugars from Halloumi cheese whey. Food Res. Int 65, 477–483.

Galanakis, C.M., Kotanidis, A., Dianellou, M., Gekas, V., 2015a. Phenolic content and antioxidant capacity of Cypriot wines. Czech J. Food Sci. 33, 126–136.

Galanakis, C.M., Patsioura, A., Gekas, V., 2015b. Enzyme kinetics modeling as a tool to optimize food biotechnology applications: a pragmatic approach based on amylolytic enzymes. Crit. Rev. Food Sci. Technol. 55, 1758–1770.

Gharsallaoui, A., Cases, E., Chambin, O., Saurel, R., 2009. Interfacial and emulsifying characteristics of acid-treated pea protein. Food Biophys. 4 (4), 273–280.

Gharsallaoui, A., Yamauchi, K., Chambin, O., Cases, E., Saurel, R., 2010. Effect of high methoxyl pectin on pea protein in aqueous solution and at oil/water interface. Carbohyd. Polym. 80, 817–827.

Gharsallaoui, A., Saurel, R., Chambin, O., Voilley, A., 2012. Pea (*Pisum sativum,* L.) protein isolate stabilized emulsions: a novel system for microencapsulation of lipophilic ingredients by spray drying. Food Bioprocess. Technol. 5 (6), 2211–2221.

Gibbs, B.F., Kermasha, S., Alli, I., Mulligan, C.N., 1999. Encapsulation in the food industry: a review. Int. J. Food Sci. Nutr. 50, 213–224.

Gopala Krishna, A.G., Hemakumar, K.H., Khatoon, S., 2006. Study on the composition of rice bran oil and its higher free fatty acids value. J. Am. Oil Chem. Soc. 83 (2), 117–121.

Gosch, B.J., Magnusson, M., Paul, N.A., de Nys, R., 2012. Total lipid and fatty acid composition of seaweeds for the selection of species for oil-based biofuel and bioproducts. GCB Bioenergy 4 (6), 919–930.

Guzey, D., McClements, D.J., 2006. Formation, stability and properties of multilayer emulsions for application in the food industry. Adv. Colloid Interf. Sci., 128–130, 227-248.

Heng, W.W., Xiong, L.W., Ramanan, R.N., Hong, T.L., Kong, K.W., Galanakis, C.M., Prasad, K.N., 2015. Two level factorial design for the optimization of phenolics and flavonoids recovery from palm kernel by-product. Ind. Crop. Prod. 63, 238–248.

Imagi, J., Yamanouchi, T., Okada, K., Tanimoto, M., Matsuno, R., 1992. Properties of agents that effectively entrap liquid lipids. Biosci. Biotechnol. Biochem. 56, 477–480.

Iwami, K., Hattori, M., Yasumi, T., Ibuki, F., 1988. Stability of gliadin-encapsulated unsaturated fatty acids against autoxidation. J. Agric. Food Chem. 36, 160–164.

Jahurul M.H.A., Zaidul I.S.M., Norulaini N.A.N., Sahena F., Jinap S., Azmir J., Sharif K.M., Mohd Omar A.K., 2013. Cocoa butter fats and possibilities of substitution in food products concerning cocoa varieties, alternative sources, extraction methods, composition, and characteristics. J. Food Eng. 117, 467–476.

Kha, T.C., Nguyen, M.H., Roach, P.D., Stathopoulos, C.E., 2014. Microencapsulation of Gac oil by spray drying: optimization of wall material concentration and oil load using response surface methodology. Drying Technol. 32 (4), 385–397.

Kinsella, J.E., Whitehead, D.M., 1989. Proteins in whey: chemical and physical and functional properties. Adv. Food Nutr. Res. 33, 343–438.

Kulmyrzaev, A., Bryant, C., McClements, D.J., 2000. Influence of sucrose on the thermal denaturation, gelation and emulsion stabilization of whey proteins. J. Agric. Food Chem. 48, 1593–1597.

Lazos, E.S., Tsaknis, J., Lalas, S., 1998. Characteristics and composition of tomato seed oil. Grasas y Aceites 49, 440–445.

Leroux, J., Langendorff, V., Schick, G., Vaishnav, V., Mazoyer, J., 2003. Emulsion stabilizing properties of pectin. Food Hydrocolloid. 17 (4), 455–462.

Lesmes, U., McClements, D.J., 2009. Structure-function relationships to guide rational design and fabrication of particulate food delivery systems. Trends Food Sci. Technol. 20, 448–457.

Linarès, E., Larré, C., Popineau, Y., 2001. Freeze- or spray-dried gluten hydrolysates. 1. Biochemical and emulsifying properties as a function of drying process. J.Food Eng. 48, 127–135.

Lu, B.-Y., Quillien, L., Popineau, Y., 2000. Foaming and emulsifying properties of pea albumin fractions and partial characterisation of surface-active components. J. Sci. Food Agric. 80 (13), 1964–1972.

Lutterodt, H., Slavin, M., Whent, M., Turner, E., Yu, L., 2011. Fatty acid composition, oxidative stability, antioxidant and antiproliferative properties of selected cold-pressed grape seed oils and flours. Food Chem. 128, 391–399.

Matthaus, B., 2008. Virgin grape seed oil: is it really a nutritional highlight? European J. Lipid Sci. Technol. 110, 645–650.

McClements, D.J., 2005. Food Emulsions: Principles, Practices, and Techniques, second ed. CRC Press, Boca Raton, Florida.

McClements, D.J., Decker, E.A., Weiss, J., 2007. Emulsion-based delivery systems for lipophilic bioactive component. J. Food Sci. 72 (8), R109–R124.

Mimouni, B., Raymond, J., Merle-Desnoyers, A.M., Azanza, J.L., Ducastaing, A., 1994. Combined acid deamidation and enzymic hydrolysis for improvement of the functional properties of wheat gluten. J. Cereal Sci. 20 (2), 153–165.

Miyashita, K., Mikami, N., Hosokawa, M., 2013. Chemical and nutritional characteristics of brown seaweed lipids: a review. J. Funct. Foods 5, 1507–1517.

Molina, E., Papadopoulou, A., Ledwar, D.A., 2001. Emulsifying properties of high pressure treated soy protein isolate and 7s and 11s globulins. Food Hydrocolloid. 15, 263–269.

Monsoor, M.A., 2005. Effect of drying methods on the functional properties of soy hull pectin. Carbohyd. Polym. 61, 362–367.

O'Brien, R.D., 2004. Raw materials. In: O'Brien, R.D. (Ed.), Fats and Oils: Formulating and Processing for Applications. CRC Press, Boca Raton, pp. 1–76.

Parada, J., Aguilera, J.M., 2007. Food microstructure affects the bioavailability of several nutrients. J. Food Sci. 72 (2), R21–R32.

Patsioura, A., Galanakis, C.M., Gekas, V., 2011. Ultrafiltration optimization for the recovery of β-glucan from oat mill waste. J. Membr. Sci. 373, 53–63.

Peng, X., Ma, J., Cheng, K.-W., Jiang, Y., Chen, F., Wang, M., 2010. The effects of grape seed extract fortification on the antioxidant activity and quality attributes of bread. Food Chem. 119, 49–53.

Pereira, H.V.R., Saraiva, K.P., Carvalho, L.M.J., Andrade, L.R., Pedrosa, C., Pierucci, A.P.T.R., 2009. Legumes seeds protein isolates in the production of ascorbic acid microparticles. Food Res. Int. 42 (1), 115–121.

Pierucci, A.P.T.R., Andrade, L.R., Farina, M., Pedrosa, C., Rocha-Leão, M.H.M., 2007. Comparison of α-tocopherol microparticles produced with different wall materials: pea protein a new interesting alternative. J. Microencapsul. 24 (3), 201–213.

Raeker, M.D., Johnson, L.A., 1995. Thermal and functional properties of bovine blood plasma and egg white proteins. J. Food Sci. 60 (4), 685–690.

Rahmanian, N., Jafari, S.M., Galanakis, C.M., 2014. Recovery and removal of phenolic compounds from olive mill wastewater. J. Am. Oil Chem. Soc. 91, 1–18.

Roselló-Soto, E., Barba, F.J., Parniakov, O., Galanakis, C.M., Grimi, N., Lebovka, N., Vorobiev, E., 2015. High voltage electrical discharges, pulsed electric field and ultrasounds assisted extraction of protein and phenolic compounds from olive kernel. Food Bioprocess. Technol. 8, 885–894.

Rust, D., Wildes, S., 2008. Surfactants: A Market Opportunity Study Update. United Soybean Board, Omni Tech International, Ltd, Midland.

Sagiri, S.S., Sharma, V., Basak, P., Pal, K., 2014. Mango butter emulsion gels as cocoa butter equivalents: physical, thermal, and mechanical analyses. J. Agric. Food Chem. 62, 11357–11368.

Saikia, S., Mahnot, N.K., Mahanta, C.L., 2015. Optimisation of phenolic extraction from Averrhoa carambola pomace by response surface methodology and its microencapsulation by spray and freeze drying. Food Chem. 171, 144–152.

Schroeder, H.E., 1982. Quantitative studies on the cotyledonary proteins in the genus Pisum. J. Sci. Food Agric. 33 (7), 623–633.

Shahidi, F., Han, X.Q., 1993. Encapsulation of food ingredients. Crit. Rev. Food Sci. Nutr. 33, 501–547.

Shaw, L.A., McClements, D.J., Decker, E.A., 2007. Spray-dried multilayered emulsions as a delivery method for ω-3 fatty acids into food systems. J. Agric. Food Chem. 55, 3112–3119.

Shi, H., Noguchi, N., 2001. Introducing natural antioxidants. In: Pokorny, J., Yanishlieva, N., Gordon, M. (Eds.), Antioxidants in Food. Practical Applications. CRC Press, Boca Raton, pp. 147–155.

Sikorski, Z.E., Piotrowska, B., 2007. Food components and quality. In: Zdzislaw, E. (Ed.), Chemical and Functional Properties of Food Components. CRC Press, Boca Raton, pp. 1–14.

Souilem, S., Kobayashi, I., Neves, M.A., Jlaiel, L., Isoda, H., Sayadi, S., Nakajima, M., 2014. Interfacial characteristics and microchannel emulsification of oleuropein-containing triglyceride oil–water systems. Food Res. Int. 62, 467–475.

Speer, K., Kölling-Speer, I., 2001. Lipids. In: Clarke, R.J., Vitzum, O.G. (Eds.), Coffee Recent Developments. Blackwell, Berlin, Germany, pp. 33–46.

Suseno, S.H., Haiati, S.S., Izaki, A.F., 2014. Fatty acid composition of some potential fish oil from production centers in Indonesia. Oriental J. Chem. 30 (3), 975–980.

Tsakona, S., Galanakis, C.M., Gekas, V., 2012. Hydro-ethanolic mixtures for the recovery of phenols from Mediterranean plant materials. Food Bioprocess. Technol. 5, 1384–1393.

Wang, L., Bohn, T., 2012. Health-promoting food ingredients and functional food processing. In: Bouayed, J., Bohn, T. (Eds.), Nutrition, Well-Being and Health, ISBN 978-953-51-0125-3 (Chapter 9). Available from: NTECH Open Science (http://www.intechopen.com/

Waraho, T., McClements, D.J., Decker, E.A., 2011. Mechanisms of lipid oxidation in food dispersions. Trends Food Sci. Technol. 22, 3–13.

Wong, B.T., Zhai, J., Hoffmann, S.V., Aguilar, M.-I., Augustin, M., Wooster, T.J., Day, L., 2012. Conformational changes to deaminated wheat gliadins and β-casein upon adsorption to oil–water emulsion interfaces. Food Hydrocolloid. 27 (1), 91–101.

Yoshii, H., Soottitantawat, A., Liu, X.-D., Atarashi, T., Furuta, T., Aishima, S., et al., 2001. Flavor release from spray-dried maltodextrin/gum arabic or soy matrices as a function of storage relative humidity. Innov. Food Sci. Emerg. Technol. 2, 55–61.

Young, S.L., Sarda, X., Rosenberg, M., 1993. Microencapsulation properties of whey proteins. 1. Microencapsulation of anhydrous milk fat. J. Dairy Sci. 76, 2868–2877.

EMERGING TECHNOLOGIES

III

EMERGING MACROSCOPIC PRETREATMENT

N.N. Misra*, Patrick J. Cullen*, Francisco J. Barba, Ching Lik Hii[†], Henry Jaeger[‡],
Julia Schmidt[‡], Attila Kovács[§], Hiroshi Yoshida[¶]**

[*]*School of Food Science and Environmental Health, Dublin Institute of Technology, Dublin, Ireland;* [**]*Department of Nutrition and Food Chemistry, Faculty of Pharmacy, Universitat de València, Valencia, Spain;* [†]*Department of Chemical and Environmental Engineering, University of Nottingham, Malaysia Campus, Malaysia;* [‡]*Department of Food Science and Technology, University of Natural Resources and Life Sciences (BOKU), Vienna, Austria;* [§]*Faculty of Agricultural and Food Sciences, Institute of Biosystems Engineering, University of West Hungary, Hungary;* [¶]*Department of Materials Chemistry and Bioengineering, Oyama National College of Technology, Oyama, Tochigi, Japan*

9.1 INTRODUCTION

The basic idea of macroscopic pretreatment is to prepare the food matrix for downstream processing. The pretreatment steps aim at one of the following:

1. adjustment of the different phases of the matrix,
2. moderation of enzyme activity,
3. protection from microbial growth,
4. concentration of the waste for economic extraction,
5. facilitation of storage of the waste substrate until further processing,
6. avoid loss of functionality down the processing chain.

The type of pretreatment to be applied is governed by the nature of the substrate, its state (solid or liquid), and structure. Often dewatering is performed for concentration of the substrate matrix, which traditionally employed heat. However, such thermal processes often cause inactivation of key enzymes and denaturation or destruction of many desirable components. While deactivation of certain enzymes and microorganisms is desirable, the destruction of desired components is certainly not. Thus, utilization of conventional pretreatment and recovery methods is often restricted by several problems that are difficult to overcome. These include overheating of the food matrix, high-energy consumption and general cost, loss of functionality and poor stability of the final product, and accomplishment of increasingly stringent legal requirements on materials safety (Galanakis, 2012; Galanakis and Schieber, 2014). Emerging technologies based on nonthermal concepts promise to surpass most of the previous challenges and optimize processing efficiency (Deng et al., 2015; Roselló-Soto et al., 2015). Typical examples of such novel nonthermal processes, which have the potential to overcome these disadvantages, include radio-frequency drying, foam-mat drying, and electro-osmotic drying (EOD). A recent alternative approach for inactivation of undesirable microorganisms and enzymes is the application of nonthermal plasma, which is being actively researched. The pretreatment methods are also employed at times to facilitate the diffusion process and increase mass-transfer efficiency. High-pressure processing is an emerging technology, which facilitates mass transfer from substrate into extraction medium through a multitude of mechanisms, including modulation of cell permeability (especially in plant food wastes).

This chapter provides an in-depth discussion of the novel and emerging macroscopic pretreatments that facilitate downstream processing and ease of handling of food wastes. The topics discussed include foam-mat drying, radio-frequency drying, and EOD, followed by nonthermal plasma technology and high-pressure processing. In many cases, emerging technologies have so far not been applied in the field, but they have been applied in other food processes. Thereby, this chapter provides the prospect of utilizing these methods in particular applications.

9.2 FOAM-MAT DRYING

The history of foam-mat drying can be traced back to 1917 from a patent filed for foam drying of evaporated milk by the Campbell food company (Campbell et al., 1917) as cited by Ratti and Kudra (2006). In general, foam-mat drying is a process whereby foams are generated with the aid of a forming agent through a mixing process in liquids or semiliquid food products. The foamed materials have density from almost 0 to 800 kg/m^3 depending on the degree of mixing between the gas bubbles and liquid (Ratti and Kudra, 2006). Figure 9.1 shows a typical foam volume profile during and after foam formation.

The foam can then be dried using various drying techniques such as hot air, freeze, microwave, belt conveyor, drum, and also spray (dispersed foams) drying. Foam-mat-dried products also have better reconstitution properties because of their honeycomb structure (Sharada, 2013). Foam-mat drying offers several benefits to the drying process as described in the literature (Table 9.1). However, there are some inherent drawbacks of this technique such as foam instability, low drying throughput, and some losses in aromatic components (Marques et al., 2006).

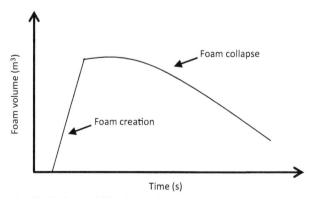

FIGURE 9.1 Foam Volume Profile During and After Formation

Redrawn from Muthukumaran et al. (2008)

Table 9.1 Benefits of Foamed-Mat Drying

Benefits	References
Fast drying rate, ability to dry hard-to-dry materials and retain volatiles	Ratti and Kudra (2006)
Reduces capital cost in belt conveyor and drum dryers and has higher drying efficiency	Kudra and Ratti (2006)
Produces porous structure and crispy product, and reduces volatile losses	Thuwapanichayanan et al. (2012)
Highly suitable for fruit and vegetable pulps and dried powders; has high reconstitutability in water	Sharada (2013)
Suitable for drying high-viscosity liquid or semiliquid food materials	Muthukumaran et al. (2008)

9.2.1 FOAMING AGENTS

Several types of foaming agents/additives can be used to create foams. The formation of foam should be both mechanically and thermally stable, so as not to collapse during handling (e.g., feeding, depositing, and conveying) and drying. As a guideline, foams that do not collapse for at least 1 h are considered at least mechanically stable (Bates, 1964; Ratti and Kudra, 2006). Table 9.2 shows the type of foaming agents used in various reported studies. Egg white is a popular choice of foaming agent since it is a natural product and has excellent ability in trapping air. In general, foams are usually thermodynamically unstable and therefore stabilizers are normally incorporated to improve the strength and stability of the foams. Muthukumaran et al. (2008) reported that the addition of stabilizer (propylene glycol alginate and xanthan gum) in egg white foam improved foam stability. Likewise, the absence of stabilizer showed

Table 9.2 Foaming Agents/Additives Used in Foam-Mat Drying

Foaming Agents/Additives	Products	References
Egg white with stabilizers (xanthan gum, propylene glycol alginate, and methyl cellulose)	Freeze-dried egg white foam	Muthukumaran et al. (2008)
Egg white with dextrin	Dried papaya foam	Widyastuti and Srianta (2011)
Dry egg albumin and methylcellulose	Dried apple juice foam	Kudra and Ratti (2006)
Egg white, soy protein isolate, and whey protein concentrate	Dried banana foam	Thuwapanichayanan et al. (2012)
Methyl cellulose	Dried papaya powder	Kandasamy et al. (2012)
Egg albumin	Dried tomato powder	Kadam and Balasubramanian (2011)
Milk	Dried mango powder	Kadam et al. (2010)
Egg albumen	Dried banana foam mat	Prakotmak et al. (2011)
Methocel	Dried star fruit puree	Abd Karim and Chee Wai (1999)
Xanthan gum	Dried shrimp foam	Azizpour et al. (2013)
Egg white and methyl cellulose	Dried carrageenaan	Djaeni et al. (2013)

a 50% reduction of the initial volume. The selection of stabilizer is crucial, too, as methyl cellulose produces sticky foam.

9.2.2 FOAM STABILITY

The density of the foam can be measured to determine its stability over time. This can be simply determined by measuring the mass of a fixed volume of the foam (Thuwapanichayanan et al., 2012). Lower foam density means more trapped air during the foaming process and vice versa. A typical range of densities used in foam-mat drying is within 300 and 600 kg/m^3 (Ratti and Kudra, 2006). Foam stability can also be determined by measuring the drainage volume (Eq. 9.1) (Kandasamy et al., 2012). Maximum foam stability is usually correlated with minimum drainage volume and density.

$$\text{Foam stability} = V_0 \frac{\Delta t}{\Delta V} \tag{9.1}$$

where V_0 = initial volume (m^3) and ΔV = change in foam volume during the time interval Δt (s).

9.2.3 DRYING KINETICS: RATES

In general, most food products exhibit two major drying periods, namely the constant and falling rate periods. The constant rate period is mostly associated with surface evaporation, while the falling rate period is diffusion controlled. A gradual transition of the single constant rate period to the falling rate period is usually observable during drying depending on operating conditions. However, some foamed materials exhibit more than one constant rate period during drying as cited by Ratti and Kudra (2006) in studies carried out for tomato paste (Lewicki, 1975), as shown in Fig. 9.2. This could be attributed to mechanical instability where cracks had occurred within the foam, which resulted in increased interfacial area. Hence, an additional constant rate period had occurred midway through the falling rate period. However, this mechanism is more obvious when the foam density is low and under mild drying conditions. Drying of foamed materials is usually faster than the nonfoamed materials due to the increased interfacial area and lower foam density. Therefore, energy consumption is lower due to higher

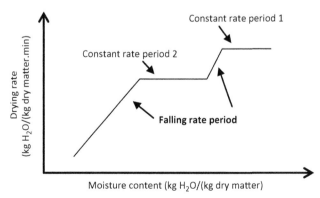

FIGURE 9.2 Sketch of Drying Rate Curve with Two Constant Rate Periods for Tomato Paste Foam

Redrawn from Ratti and Kudra (2006)

drying efficiency and shorter drying time. As a result, throughput could be higher, which in turn is translated to capital cost savings (Kudra and Ratti, 2006).

9.2.4 DRYING KINETICS: EFFECTIVE DIFFUSIVITY

The effective diffusivity (D_e) is an overall mass transport property of water in the drying material, which includes liquid diffusion, vapor diffusion, hydrodynamic flow, and other possible mass-transfer mechanisms (Karathanos et al., 1990). However, no single mechanism prevails throughout the drying process. Fick's second law model with the assumption of negligible shrinkage is usually used for the effective determination of diffusivity (Crank, 1975). D_e values can then be represented by an Arrhenius model as reported in foam-mat drying studies (Azizpour et al., 2013; Kadam and Balasubramanian, 2011; Djaeni et al., 2013):

$$D_e = D_0 \exp^{-\frac{E}{RT}} \tag{9.2}$$

where D_0 = Arrhenius constant (m^2/s), E = activation energy (J/mol), R = universal gas constant [8.314 J/(mol.K)], and T = temperature (K).

In general, shrinkage is unavoidable in many food products such as foamed materials, due to the inherent high moisture content. The shrinking thickness will provide a shorter and varying path for moisture diffusion during drying and hence results in variable diffusivity values. Thuwapanichayanan et al. (2012) reported the use of a variable diffusivity model to describe the change in effective diffusivity as a function of moisture content and temperature in foam-mat drying:

$$D_e = a\exp^{\left(\sum_{i=1}^{4} \alpha_i X^i\right)} \tag{9.3}$$

where a = constant (m^2/s), X = moisture content (kg H$_2$O/kg dry matter), and α = polynomial coefficients.

9.2.5 PRODUCT QUALITY

Thermal drying has a great impact on the quality of the end product due to the heated and highly aerated environment. A high degree of shrinkage, changes in color and texture, flavor, and nutritional losses are some of the typical quality deteriorations observed in the dried product. The effects of foam-mat drying on various quality attributes had been reported in several recently published works (Widyastuti and Srianta, 2011; Thuwapanichayanan et al., 2012; Kandasamy et al., 2012; Kadam and Balasubramanian (2011); Raharitsifa and Ratti, 2010; Kadam et al., 2010). Generally, foam-mat-dried product would show good reconstitution properties due to the porous structure. However, Akintoye and Oguntunde (1991) reported that reconstitution properties were inferior to those obtained from spray drying.

Undoubtedly, foaming agent plays also a role in affecting product quality such as color and texture. It was observed that banana purée foamed using egg white and whey protein concentrate showed a spongy texture as compared to those using soy protein isolate, which has lower foam stability (Thuwapanichayanan et al., 2012). Hardness was also found to decrease by increasing foaming agent concentration (egg albumin) in the dried banana foams (Thuwapanichayanan et al., 2008).

Flavor and nutritional losses are unavoidable in food drying. This is mostly due to the heat-sensitive nature of the nutrients and flavor volatiles. Similarly, foam-mat-dried products do suffer to a certain extent some losses, depending on the drying condition used. Kadam et al. (2010) had observed significant losses in ascorbic acids and total carotenes in foam-mat-dried mango powder, especially those dried at 90°C. Abd Karim and Chee Wai (1999) reported that the rehydrated foam-mat star fruit powder was slightly lacking in flavor compared to the freeze-dried powder. It must be noted that substantial losses of volatile substances could have occurred even during the foaming process prior to drying (Thuwapanichayanan et al., 2012.).

9.3 RADIO-FREQUENCY DRYING

Conventional drying of solid biomaterials strongly depends on heat and mass transfer (Fig. 9.3). Mass transfer is limited due to structural characteristics of the material. For instance, the presence of cellular structural barriers hinders the transport of water and water vapor inside the product. Heat transfer is the basis for supplying the latent heat of evaporation in order to induce the evaporation of water for its subsequent removal. Structural aspects such as the network of intercellular air space and related thermophysical material properties as well as the size of the product to be dried are the main impact factors. In traditional heating and drying processes, heat transfer is predominantly limited due to the supply of the product surface to its center. The locations of the slowest heating or drying region are therefore found in the thermal center of the treated product. This fact leads to overprocessing of surface regions.

On the other hand, the application of dielectric drying (radio-frequency or microwave) using electromagnetic energy and a carrier gas (to remove the evaporated liquid) improves the drying process and reduces processing times. Volumetric heating avoids resistance to heat transport and thereby enables quality improvement of dried products. However, radio-frequency heating is dependent on the product itself, its geometry, and thermophysical properties. As radio-frequency electromagnetic waves heat volumetrically, they can theoretically improve heating uniformity. Unfortunately, there are a number of well-known factors that cause nonuniform heating patterns.

The distribution of electromagnetic energy in radio-frequency and microwave heating systems is governed by Maxwell's equations with appropriate boundary conditions defined by the configuration of the systems and the interfaces between the treated materials and remaining space. The dielectric properties of the materials are the main parameters of Maxwell's equations, which have a significant impact on the transformation of electromagnetic into thermal energy within those materials.

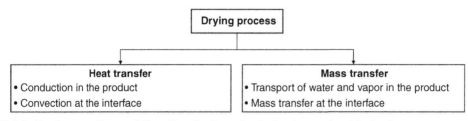

FIGURE 9.3 Simultaneous Heat and Mass Transfer Occurring During Conventional Drying Processes

9.3.1 PRINCIPLES OF DRYING OF BIOMATERIALS AND RADIO-FREQUENCY APPLICATION

Drying is one of the most common and diverse food (and chemical) engineering unit operations aimed at removing water from the material. By definition, drying (or dehydration) is "the application of heat under controlled conditions to remove the majority of the water normally present in a food by evaporation." Removal of water from food materials is usually understood to be accomplished by thermal evaporation. On the other hand, part of the water from "wet" materials (e.g., solid food wastes) can be removed by inexpensive nonthermal processes such as filtration, centrifugation, osmotic dehydration, or reverse osmosis. These processes are also called "dewatering," where liquid water is "drained" or "squeezed" out of the material (Chen and Mujumdar, 2008).

The conventional drying methods require heat energy to be generated externally and then transferred to the food product. The heat can be supplied in different ways: convection, radiation, conduction, microwave, radio-frequency, or even Joule (ohmic) heating. Nevertheless, in most cases the terms "drying" and "dehydration" are used differently. "Drying" is traditionally used for thermal removal of water to approximately the equilibrium moisture content (about 15–20% dry basis) of drying products at ambient air conditions. Meanwhile, the term "dehydration" is used for removal of water down to about 2–5% (some terminology defines it as no more than 2.5%) (Fellows, 2000).

Concerning waste recovery, the term "biodrying" cannot be avoided. Biodrying or biological drying is an option for the bioconversion reactor in mechanical/biological treatment plants to dry and partially stabilize residual wastes. The physical phenomena, however, do not differ from the conventional methods: in biodrying the main drying mechanism is convective evaporation, using heat from the aerobic biodegradation of waste components and facilitated by the mechanically supported airflow. In this way the moisture content of the waste matrix is reduced through two main steps: (i) water molecules evaporate (e.g., change phase from liquid to gaseous) from the surface of waste fragments into the surrounding air and (ii) the evaporated water is transported through the matrix by the airflow and removed with the exhaust gases (Velis et al., 2009).

Regardless of the processes, water removal reduces water activity of the materials. This results – among others things – in the inhibition of microbial growth and enzyme activities (Fellows, 2000).

Drying kinetics is the most important information when dealing with this operation. Typical representations of drying kinetics are the following functions (or curves): (i) drying curve (moisture content versus time); (ii) drying rate curve (drying rate versus time); (iii) time independent or Krischer curve (drying rate versus moisture content); and (iv) temperature–time plot (Kemp et al., 2001). The drying rate of a material depends on its properties such as bulk density and initial moisture content as well as its relation to the equilibrium moisture content under the drying conditions.

During drying, mass and thermal transport properties are strongly affected by the physical structure (porosity) of the material (Maroulis and Saravacos, 2003) and other physical factors such as temperature, humidity, surface area, air velocity, and atmospheric pressure (or vacuum) (Potter and Hotchkiss, 1998).

Besides the underlying fundamentals there are other factors that control the rate at which biological structures (e.g., foods) dehydrate or dry. These can be grouped into three categories: those related to (i) processing conditions; (ii) nature of the material; and (iii) dryer design (Maroulis and Saravacos, 2003). These factors also determine the quality of the final dried material and the efficiency of the process. Obtaining dried products that meet several specifications and quality standards is a challenging task (Valentas et al., 1997), too.

Radio-frequency radiation consists of electromagnetic waves and represents a nonionizing type of radiation that causes molecular motion through migration of ions and rotation of dipoles without altering the molecular structure. The radio-frequency region of the electromagnetic spectrum lies between 1 and 300 MHz (Orfeuil, 1987). Radio-frequencies of 13.56, 27.12, and 40.68 MHz are allocated in different countries for industrial, scientific, and medical applications corresponding to a wavelength of around 10 m.

A material can be heated by applying energy to it in the form of electromagnetic waves. The heating effect originates from the ability of an electric field to exert a force on charged particles. If the particles present in a material can move freely, a current will be induced. However, if the charge carriers in the material are bound to certain regions, they will only move until balanced by a counterforce resulting in dielectric polarization. Both conduction and dielectric polarization are sources of radio-frequency heating. The heating effect depends on the frequency and the applied power of the electromagnetic waves as well as on the material properties.

The dielectric properties of materials are defined by two parameters, the dielectric constant (ε') and the dielectric loss (ε''). The dielectric constant describes the ability of a molecule to be polarized by the electric field. At low frequencies, the dielectric constant peaks at the maximum amount of energy that can be stored in the material. Dielectric loss measures the efficiency with which the energy of the electromagnetic radiation is converted into heat. Dielectric loss moves towards a maximum as the dielectric constant decreases (Haswell and Kingston, 1997). The ratio of the dielectric loss factor to its dielectric constant is expressed as the dissipation factor $\tan \delta = \varepsilon''/\varepsilon'$.

While electromagnetic waves penetrate the sample, energy is absorbed depending on the dissipation factor of the material. The greater the dissipation factor of a sample, the less electromagnetic energy will penetrate into it since it is rapidly absorbed and dissipated. In order to characterize the penetration, the half-power depth is used. This magnitude represents the distance from the surface at which the initial power density is reduced by half. The half-power depth for water is about 12 mm at 2450 MHz and 75 mm at 100 MHz. Penetration depths of electromagnetic waves in the radio-frequency and microwave range are exemplarily given for plant materials in Table 9.3.

Radio-frequency electromagnetic energy is transformed into heat via two mechanisms that occur simultaneously: ionic conduction and dipole rotation. Ionic conduction refers to the conductive migra-

Table 9.3 Penetration Depths for Plant Materials Calculated from the Measured Dielectric Properties at 20°C for Radio-Frequency (27 MHz) and Microwaves (915 MHz) according to Wang et al. (2003)

Fruit	27 MHz	915 MHz
Apple	15.2	5.3
Almond	538.2	1.9
Cherry	8.5	2.8
Avocado	5.1	1.5
Persimmon	10.5	2.1
Walnut	653.6	3.1

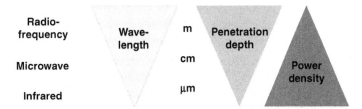

FIGURE 9.4 Characteristics of Infrared, Microwave, and Radio-Frequency Electromagnetic Waves

tion of dissolved ions in the electromagnetic field. The resulting current flow leads to heat dissipation due to the resistance of the product. The losses due to ionic migration depend on the size, charge, and conductivity of the dissolved molecules (Decareau, 1985). The dissipation factor changes with temperature, which affects ion mobility and concentration. The second mechanism (dipole rotation) refers to the alignment of molecules that have a permanent or induced dipole moment. An increase of the electric field leads to the polarization of the molecules. A decrease leads to the restoration of the thermally induced disorder.

The relative contribution of each energy conversion mechanism depends mainly on the temperature. For small molecules such as water, the contribution of dipole rotation decreases as sample temperature increases whereas the contribution of ionic conduction increases as the product temperature is increased. Depending on the mobility and concentration of ions and the relaxation time, the relative contribution of the two mechanisms may vary. If mobility and concentration of ions is low, dipole rotation will be the dominant mechanism. Increasing the ionic concentration enhances ionic conduction and makes the heating effect independent of the relaxation time of the sample. The dissipation factor will increase and the heating time will decrease.

Heating by microwaves is much faster compared to conventional conductive heating and, depending on dielectric properties of the sample or sample fractions, it is more uniform. However, because heating is much faster, substantial localized superheating can occur (Kingston and Jassie, 1998).

Energy delivery to the biomaterial by radio-frequency radiation depends on a number of variables such as power input, exposure time, and material properties and size of the sample. As mentioned earlier, initial temperature of the sample affects also the heating behavior of a sample by electromagnetic waves. Concerning sample size, the input frequency affects the penetration depth (see Fig. 9.4). However, the greater the dissipation factor of a sample, the less it will be penetrated by the electromagnetic wave. Hence, in large samples with high dissipation factors, heating of the central part may still depend on thermal conductance.

9.3.2 RADIO-FREQUENCY APPLICATIONS IN DRYING OF BIOMATERIALS

Radio-frequency heating is currently used primarily in nonfood industries, such as wood, paper, textiles, glass fibers, water-based glues, and pharmaceutical products (Zhao et al., 2000). The first reported applications of radio-frequency heating to foods included packed bread, blanching vegetables, thawing frozen foods, baking and postdrying snack foods, and pasteurizing and sterilizing processed meat products (Sun, 2005). The most advanced applications of radio-frequency energy to drying are wood and textile and postbaking applications (Sun, 2005). The most successful results were by applying a combination of two or more drying techniques (infrared, forced air, microwave, etc.): e.g., radio-frequency drying is used in various postbaking systems, followed by the commercial fuel-heated oven, which increases the production rate of cookies, biscuits, etc. by 30–50% (Saravacos and Maroulis, 2011).

Early applications of radio-frequency heating of foods (which dates back to the 1940s) included cooking meat products, heating bread, dehydrating and blanching vegetables, thawing frozen foods, and post-bake drying of cookies and snack foods (Zhao et al., 2000). Since the 1990s researchers have realized the possibility of food sterilizing or pasteurizing and disinfestations of fruits and seeds using radio-frequency technology (Marra et al., 2009).

In the agro-food industry the following applications are commercialized and offered as radio-frequency processing equipment: (i) defrosting (vegetables, meat, and fish); (ii) industrial baking (especially post-baking); (iii) pasteurization and sterilization; and (iv) disinfestation (pest control of dry agricultural products such as grains and nuts).

One advantage of radio-frequency technology in sterilizing and pasteurizing comes from the much lower time–temperature values than those required using conventional heating techniques. Another reported benefit of using radio-frequency energy rather than conventional heating comes from a potential selective killing effect on microorganisms (Sun, 2005).

Drying applications are somewhat constrained by the small number of equipment manufacturers and the limited market demand.

Combined pasteurization and predrying of semolina pasta is already industrially used for keeping intense taste and aroma as well as for technological reasons (Alberti et al., 2011). A comparative study subjecting fenugreek dehydration to conventional hot air, low humidity air drying (28–30% RH) and radio-frequency drying resulted in a 27% reduced drying time for low humidity drying and radio-frequency drying equally compared to hot air. However, with regard to green color retention, uniformity of structure, and even radical scavenging activity, low humidity air drying achieved much better performance than radio-frequency drying (Naidu et al., 2012).

Another advantage of radio-frequency drying is the feasibility of simultaneous disinfection for enhancing shelf-life and safety aspects. In relevant experiments, radio-frequency drying was shown to be suitable for keeping quality criteria of red and black peppers of different initial moisture levels (Jeong and Kang, 2014).

9.3.3 ENERGY CONSIDERATIONS

The efficiency of a convection dryer drops significantly as lower moisture levels are reached and the dried product surface becomes a greater thermal insulator. At this point, the radio-frequency dryer provides an energy-efficient means of achieving the desired moisture objectives. Typically, 1 kW of radio-frequency energy will evaporate 1 kg of water per hour. Additionally, since radio-frequency is a "direct" form of applying heat, no heat is wasted in the drying process (Devki Energy Consultancy et al., 2006). Drying with radio-frequency energy can be achieved in a shorter time, and provides higher energy efficiency and better product quality as compared to conventional hot air heating. In order to develop effective dielectric drying with radio-frequency (or megawatt) energy, knowledge of dielectric properties (most importantly the dielectric constant ε' and dielectric loss factor ε'') of the product is important. The latter influences the absorption of electromagnetic energy and its conversion to heat. As compared to megawatt, the penetration of radio-frequency (especially 100 MHz) is seldom a limiting factor, as it provides much better heating uniformity, too (Ramaswamy and Marcotte, 2006). The heat generation rate Q (W) per unit volume of a material is given by Eq. 9.4:

$$Q = 0.56 \times 10^{-10} E^2 \varepsilon' \omega \tan \delta \qquad (9.4)$$

where E (V) is the electric field strength, ω (1/s) is the frequency, ε' is the dielectric constant, and tan δ is the loss tangent of the material, defined as tan $\delta = \varepsilon''/\varepsilon'$, where ε'' is the dielectric loss (Saravacos and Maroulis, 2011; Maroulis and Saravacos, 2003).

9.4 ELECTRO-OSMOTIC DRYING

Dewatering of sludges or suspensions is very important, not only for recovery purposes, but also for reduction of transport and disposal expenditures as well as energy savings in post thermal processing treatments (drying and incineration). Mechanical dewatering methods using fluid pressure (filtration), compressive force (expression), and centrifugal force (centrifugation) have been widely used for sludge dewatering, and have so far been applied practically in many industries. However, colloidal particle and gelatinous sludge are very difficult to dewater efficiently by these mechanical dewatering methods (Yoshida, 1993).

Electro-osmotic drying (EOD), also called electro-dewatering, has been investigated by many researchers for over a century. It has been applied to various kinds of semisolid sludges, mainly inorganic materials such as peat, clay, soil, metal hydroxides, and mineral washery sludges, resulting in effectiveness of EOD for the hardly dewaterable sludge mentioned earlier (Rampacek, 1966; Lockhart, 1986; Sunderland, 1987; Danish, 1989). Practical applications of EOD have been performed in laboratory- or pilot-scale demonstrations and in several industrial fields for some time, as shown mainly in Table 9.4. However, in the present circumstances, it seems that the applications of EOD for sludge dewatering have not so much been realized practically and extensively in industrial fields.

Referring to biosludge, EOD has been examined for dewatering of waterworks and sewage processing sludges, biological wastes, and food processing products and wastes over the last several decades (Barton et al., 1999; Viji, 2004; Saveyn et al., 2005; Lee et al., 2007; Mujumadar and Yoshida, 2008; Tuan and Mika, 2010; Tuan et al., 2008, 2010, 2012; Mahmoud et al., 2010, 2011; Yang et al., 2011; Citeau et al., 2011, 2012). In particular, investigations into EOD of food materials have been done since around 1990s, and the operation of EOD has been attempted on food products and wastes such as sardine, seaweed, brewers' grain, apple pomace, vegetable waste, soybean residue "Okara," soy food "Tofu," tomato paste, and so on (Chen et al., 1996; Li et al., 1999; Xia et al., 2003; Al-Asheh et al., 2004; Jumah et al., 2005, 2007; Mujumadar and Yoshida, 2008; Ng et al., 2011). However, the application of EOD to food processing materials is rather limited, although it has been investigated from laboratory- to pilot-scale demonstrations.

9.4.1 PRINCIPLE OF EOD OF SEMI-SOLID MATERIAL

Electro-osmosis is an interfacial electrokinetic phenomenon that is based on an electrical double layer appearing at the interface between the surfaces of solid particles and liquid in a semisolid material like sludge. When an external electric field under a continuous direct current (DC) condition is applied to a liquid-filled medium placed between two electrodes, the negative ions in the liquid migrate electrically

Table 9.4 A Historical View of EOD in Fundamental and Industrial Investigations

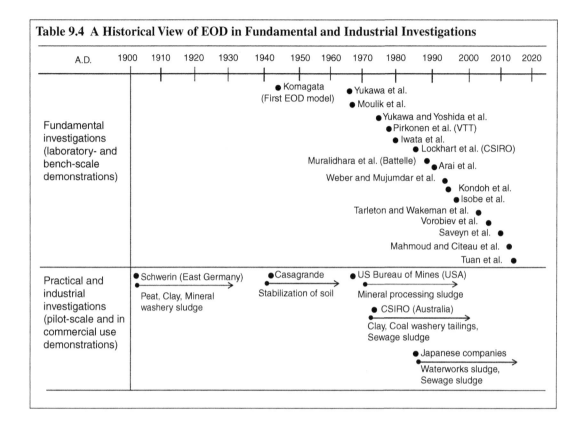

to the anode and the positive ions migrate to the cathode. If the positive ions are fixed chemically or electrochemically at the interface and the density of negative ions is higher than that of positive ions, negative ions migrate to the anode accompanying the surrounding liquid molecules, resulting in occurrence of an electro-osmotically net liquid flow in one direction. In this liquid flow, electrical zeta (ζ)-potential at a certain distance from the interface has a very important role for electro-osmotic phenomena. Thus, electro-osmosis caused by the action of an electric field is a liquid movement relative to the stationary solid phase in the solid/liquid system. Accordingly, when the liquid is not supplied to the medium from outside, it is possible to remove the liquid within the medium (deliquoring or dewatering).

Figure 9.5 exemplifies schematically a comparison between the processes of EOD and mechanical dewatering (Yoshida, 1993). In the process of mechanical dewatering of the solid particle bed filled with liquid, a layer of the particle bed in close vicinity to the filter medium is ordinarily compacted and porosity is reduced by fluid pressure. Then, filter fouling of the medium occurs, and consequently the rate of dewatering gradually decreases with the lapse of time. In such a situation, fine particle sludge like colloids, gelatinous sludge, and sludge digested biologically is difficult to dewater mechanically. On the contrary, in the process of EOD, porosity near the filter medium is negligibly changed electrically. Porosity decreases at the upper part of the bed by opposites, so that EOD can be effective for sludge samples, which are not successful in conventional mechanical dewatering. The mechanism of

EOD is essentially different from the widely used mechanical dewatering methods such as vacuum or pressure dewatering, centrifugal dewatering, and expression dewatering, hence, EOD has been well known to be particularly effective.

9.4.2 PRACTICAL AND INDUSTRIAL ASPECTS OF OPERATING EOD AND ITS PROBLEMS

EOD has been regarded as a complicated dewatering process (Mahmoud et al., 2010; Tuan et al., 2012). When an electric field under the DC condition is applied to a bed of semisolid material placed between two electrodes, several phenomena such as an electrophoretic motion of the solid particles, electrochemical reactions at the electrodes, and ohmic heating may also occur in the dewatering process. Due to the electrode reactions, corrosion of the electrode and the generation of gases (O_2 and H_2) and ions (H^+ and OH^-) caused by electrolysis may be produced with dewatering. In this case, electrode corrosion may cause contamination of the material. These ions affect the profile of pH gradient in the material and consequently ζ-potential may be locally changed by the variation of pH distribution. If the electrical conductivity of the material is large and ohmic heating occurs, the viscosity of the liquid in the material may also be varied by Joule's heat, resulting in a usual reduction of it. Thus, several physical and electrochemical variations occur simultaneously in the process of EOD.

In addition to the aforementioned problems, a major problem may be the inability to remove liquid throughout a bed of the material, as shown in Fig. 9.5. In the process of a batch EOD under the continuous DC condition, dewatering proceeds downwards in a vertical direction. Hence, water content is reduced remarkably near the upper side electrode opposite the drainage surface, resulting in an increase of electric resistance in lower water content layer and electrical contact resistance between the upper electrode and the bed. This situation hinders greatly the continuation of EOD (Mujumadar and Yoshida, 2008).

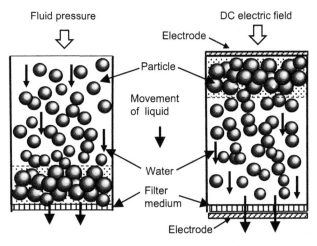

FIGURE 9.5 Difference of Dewatering Processes by EOD and Mechanical Dewatering

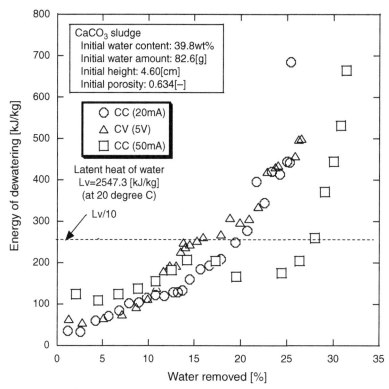

FIGURE 9.6 An Example of Variations of Energy Efficiency Plotted Against Water Removal in Operating EOD for CaCO₃ Sludge

The correlation between the observed energy consumption for water removal and the water removed in the process of EOD is exemplified in Fig. 9.6, which illustrates the variations of energy of dewatering plotted against the percent water removed. In this figure, constant current (CC) and constant voltage (CV) operations under DC electric field applied to $CaCO_3$ sludge were used. It is evident that the energy of dewatering is increased rapidly near the end of dewatering because of the foregoing electrical contact resistance. Accordingly the process of EOD has to be stopped forcibly as the applied voltage is too high at the CC operation. The process operated at the CV operation stops spontaneously because the electric current passing through the sludge bed gradually decreases (Yoshida et al., 2004). Compared with the energy of latent water heat (L_v) for the vaporization in thermal processing, the energy of dewatering using EOD can be sufficiently low (approximately one-fifth of L_v) until the end of dewatering (Fig. 9.6).

9.4.3 ESTIMATION OF ELECTRIC POWER APPLICATIONS AND MATERIALS SUITABLE FOR EOD

Electrical power application for operating the process of EOD is principally divided into two general categories involving the use of fixed electrodes and using machinery that holds moving feedstock between electrodes (Danish, 1989). The first category has been used for the dewatering of a large amount of mine tailing sediments and ponds as well as the stabilization and decontamination of soil. The second category has been used for the dewatering of sludge such as clay, coal, and biologically treated materials. This category may be more appropriate for the processing of food products and wastes.

Using the classical Helmholtz–Smoluchowski equation, the electro-osmotic velocity of liquid flow, u, through a porous medium is represented as follows:

$$u = \left(\frac{\zeta D}{4\Pi\mu} \right) E \tag{9.5}$$

where ζ is the ζ-potential of the solid phase, D is the dielectric constant of the liquid, μ is the viscosity of the liquid, and E is the strength of electric field applied to the medium. Equation 9.5 is based on an electrical double layer in a capillary model and is derived after assuming that the thickness of the electrical double layer is quite small compared to the diameter of a capillary or the pore size of the medium.

When applying Eq. 9.5 to a sludge bed consisting of solid particles and water, the rate of electro-osmotic flow, q, through the sludge bed is approximately expressed by:

$$q = \varepsilon u = \varepsilon \left(\frac{\zeta D}{4\Pi\mu} \right) E \tag{9.6}$$

where ε is the porosity (volumetric rate of water content) of the bed. The capillary diameter or the pore size is not included in Eq. 9.6. This fact suggests that the electro-osmotic flow is independent of capillary structure or pore size of the medium and consequently EOD is regarded effective for the sludge containing very fine particles such as colloids.

In Eq. 9.6, the electric field strength, E, applied to the sludge bed can be given by the following equation:

$$E = \frac{I}{\lambda} = \rho I = \frac{V}{L} \tag{9.7}$$

where λ and $\rho = (1/\lambda)$ are the superficial specific electrical conductivity and specific electrical resistance of the bed, respectively, L is the thickness of the bed, I is the electric current density passing through the cross-section of the bed, and V is the voltage applied to the bed placed between the electrodes.

Equations 9.6 and 9.7 indicate that q is proportional to the electrical properties such as ζ, D, and ρ, and also ε of the physical property. In addition, it is inversely proportional to λ of the electrical property, μ and L of the physical property. Accordingly, the relations between ζ and pH as well as ρ and ε (observed experimentally) are very important for estimating the application of EOD. Incidentally ε is related to $\rho(\lambda)$ and L. When EOD of a sludge bed proceeds, ε throughout the bed is reduced, λ and L decrease, and ρ is usually increased. These circumstances are advantageous for the implementation of EOD, but the influence of the electrical contact resistance on the process has to be taken into much consideration.

9.5 LOW-TEMPERATURE PLASMA

Sterilization processes play an important role in the management of wastes from food industries at one or the other stage, irrespective of their end usage. Even if food waste has to be recycled (e.g., for animal feeding) it should be free of microbial contamination. Conventional heat-based sterilization techniques are often associated with challenges such as complete or partial degradation of the target molecules in the substrate matrix. If the entity of interest is polymeric in nature (e.g., polyhydroxyalkanoates and poly-β-hydroxybutyric acid), chain breakage scission and cross-linking could be major undesirable effects. It is no exaggeration to state that there is currently no ideal method to achieve acceptable decontamination at ambient temperature. Amongst the various novel decontamination and sterilization approaches, cold plasma (CP) technology offers a great opportunity for sterilization at ambient temperature and pressure.

9.5.1 PLASMA SCIENCE AND TECHNOLOGY

The term "plasma" refers to a quasi-neutral ionized gas, primarily composed of photons, ions, and free electrons as well as atoms in their fundamental or excited states, with a net neutral charge. Plasma generation methods may differ substantially depending on the type of discharge applied – DC or AC power source, dielectric barrier discharge (DBD) (Misra et al., 2013), corona discharge, arc and gliding arc, spark microdischarges, and pulsed discharges of many kinds (e.g., pulsed-DBD and pulsed-corona discharge). The type of discharge determines the energy consumed by the device and the discharge stability. A detailed discussion on various CP sources has been provided by Bárdos and Baránková (2008, 2010). The concentrations in which the plasma agents occur in plasma depend greatly on the device set-up (reactor geometry), operating conditions (gas pressure, type, flow, frequency, and power of plasma excitation), and gas composition.

Atmospheric pressure CP in air (or within various gas mixtures generated by electrical discharges) presents considerable interest for a wide range of environmental, biomedical, and industrial applications. These include air pollution control, wastewater cleaning, biodecontamination and sterilization, material and surface treatment, electromagnetic wave shielding, carbon beneficiation and nanotube growth, and element analysis (Machala et al., 2007). However, from a food industry perspective, the sterilization and decontamination applications of plasma are important.

9.5.2 COLD PLASMA FOR DECONTAMINATION AND STERILIZATION

Plasma-based treatments, including the generation of ultraviolet (UV) radiation, can perform a reduction of microorganisms at low temperature and even at ambient pressure up to 1 atm (Brandenburg et al., 2007). The range of processes in which the CP decontamination and/or sterilization method can partially or wholly replace conventional approaches is very diverse and ranges from sterilization of culture media, plant and animal foods (Misra et al., 2014a, b), dairy products, seeds and grains, and processing equipment, to polymer surfaces. Important studies concerning inactivation or sterilization of microbial growth culture media and processing surfaces using CP technology are summarized in Table 9.5. It is clear that in all cases CP permits the procurement of acceptable levels of microbial reductions in model or real culture media and complex matrices (Misra et al., 2011).

The antimicrobial efficacy of CP stems from the myriad of reactive species constituting the plasma. The antimicrobial agents that could be generated in a gas discharge include reactive oxygen species (ROS), UV radiation, energetic ions, and charged particles. Among the ROS, ozone, atomic oxygen, superoxide, peroxide, and hydroxyl radicals are some of the important radicals when the plasma is generated in air. The ROS generated from various plasma sources can oxidize the cell membrane and cause leakage of cytoplasm (Xingmin et al., 2011). It has been confirmed that NTP causes acidification of lipid films (Helmke et al., 2009), though lipid peroxidation in cells remains a point of controversy

Table 9.5 Important Research Findings in the Area of CP-Based Decontamination of Growth Media and Processing Surfaces

Substrate Material	Microorganism	Plasma Source	Process Parameters	Salient Results	References
Rotating metallic cutting knife	*Listeria innocua*	DBD	Power: 1.8 W, 3.6 W, 0.36 kW; operation time: 0–350 s; gas: atmospheric air	5 log reduction of *L. innocua* is obtained after 340 s; knife temperature after treatment <30°C	Leipold et al. (2010)
Liquid media (water or diluted culture media)	*S. aureus*	Concentric electrode, microjet DC plasma source	Voltage: 400–600 V (DC); current: 20–35 mA; operation time: 0–20 min; gas: atmospheric air	100% inactivation after 16 min of plasma treatment	Liu et al. (2010)
Liquid culture media inside a sealed package	*E. coli*	DBD	Voltage: 40 kV (AC); frequency: 50 Hz; mode: indirect; operation time: 0–60 s; gas: atmospheric air	20 s of direct and 45 s of indirect plasma treatment resulted in complete bacterial inactivation (7 log/mL)	Ziuzina et al. (2013)
Cells dried in laminar flow hood and cells suspended in distilled water	*Deinococcus radiodurans*	Dielectric barrier discharge (DBD)	Voltage: 30 kV; power density: 1 W/cm^2; frequency: 10 kHz; operation time: 0–120 s	4 log reduction of CFU count in 15 s of the extremophile organism suspended in distilled water	Cooper et al. (2010)
Poultry wash water	*Escherichia coli* NCTC 9001, *Campylobacter jejuni* ATCC 33560, *Campylobacter coli* ATCC 33559, *Listeria monocytogenes* NCTC 9863, *Salmonella enterica* serovar Enteritidis ATCC 4931, *S. enterica* serovar Typhimurium ATCC 14028, and *Bacillus cereus* NCTC 11145 endospores.	Pulsed plasma gas discharge (PPGD)	Voltage: 40 kV (DC) pulsed; gas: nitrogen, carbon dioxide, oxygen, and air; temperature: 4°C	Rapid reductions in microbial numbers (by ≤8 log CFU/mL). Use of oxygen alone produced the greatest reductions. In general gram negative test bacteria were more susceptible	Rowan et al. (2007)
Agar media	*E. coli*, *Staphylococcus aureus*	Corona discharge	Voltage: 20 kV DC; temperature: 35–40°C; gas: air; operation time: 0–20 min	Changes of pH levels from alkaline to acid, upon plasma application to bacteria in water, does not play a predominant role in cell death	Korachi et al. (2010)

(Continued)

Table 9.5 Important Research Findings in the Area of CP-Based Decontamination of Growth Media and Processing Surfaces *(cont.)*

Substrate Material	Microorganism	Plasma Source	Process Parameters	Salient Results	References
Glass plate and petri dish	*E. coli*, *Bacillus subtilis*, *Candida albicans*, *S. aureus*	High-frequency capacitive discharge (CD) and barrier discharge (BD)	Power: 0.1 W/cm^3; pressure: 0.4 torr (CD), 0.4–0.5 torr (BD); frequency: 5.28 MHz; gas: air; operation time: 0–10 min (CD), 0–30 min (BD)	The most probable sterilization agents of the plasma generated were established to be "hot" and "cold" OH radicals, the excited electrically neutral N$_2$ and O$_2$ molecules, and the UV plasma radiation	Azharonok et al. (2009)
Peptone solution	*E. coli* (NCTC 9001), *S. aureus* (NCTC 4135), *S. enteritidis* (NCTC PT4), *B. cereus* (NCTC KD4)	Pulsed power plasma in liquid	Voltage: 28 kV (peak); pulse rate: 320 pps; gas: air, nitrogen, carbon dioxide; operation time: 0–50 s	Complete inactivation was observed	Marsili et al. (2002)
Sterile water with 10% glycerol	*D. radiodurans*, *Enterobacter aerogenes*, *Enterococcus faecium*, *E. coli*, *Geobacillus stearothermophilus*, *Neisseria sicca*, *Staphylococcus epidermidis*, *Stenotrophomonas maltophilia*, *Streptococcus sanguinis*, *C. albicans*	Corona discharge	Voltage: up to 10 kV; current: 0.5 mA; gas: air; operation time: 0–32 min; point to plane distance: 2–10 mm	Complete sterilization of *E. coli* was achieved within 120 s and *S. epidermidis* within 4–5 min	Scholtz et al. (2010)
Liquid media, stainless steel, polyethylene surface	*S. epidermidis*, *Leuconostoc mesenteroides*, *Hafnia alvei*	Quenched gliding arc plasma activated water	Mode: indirect treatment; pressure: atmospheric pressure; operation time: 0–30 min; gas: humid air	Inactivation was more effective for bacteria than for the yeast	Kamgang Youbi et al. (2009)
Aqueous solution	*E. coli* (IFO 3301), *L. citreum*	Low-frequency plasma jet	Voltage: −3.5 to +5.0 kV (pulsed); frequency: 13.9 kHz; operation time: 0–300 s; gas: air, helium	A critical pH value of about 4.7 was observed for the bactericidal effects	Ikawa et al. (2010)

(Leduc et al., 2010). Oxidation of lipid in microbial cell membranes increases membrane permeability, decreases fluidity, and the products emitted react with DNA (Marnett, 2002).

9.5.3 ENZYME INACTIVATION

The ability of CP to inactivate enzymes was only recently realized and since then it has been studied by several researchers. In one of the foremost studies, Bernard et al. (2006) reported the successful inactivation of hen egg-white lysozyme following CP treatments, using the flowing afterglow of nitrogen/oxygen plasma generated using microwave power source. A later study also confirmed the inactivation of lysozyme following CP treatments, using a low-frequency plasma jet operating in helium gas (Takai et al., 2012). In spite of the favorable results obtained in the previous studies, it could be noted that the inactivation of enzymes in some cases may be undesirable, especially if a valuable enzyme is to be isolated from a substance. Contrary to most reported works, Li et al. (2011) observed an increase in the activity of lipase (from *Candida rugosa*) after treatment with a radio-frequency plasma jet operating in air. Authors attributed the modulation of enzyme activity to changes in secondary and tertiary structure of proteins (observed through circular dichroism spectroscopy) and tryptophan residues of amino acid side chains (observed through fluorescence measurements).

9.5.4 ODOR CONTROL

Objectionable odors in the food industry are generally a result of the physical processing of foods in which biological or chemical reactions form volatile organic compounds (VOCs). These reactions are usually associated with thermal processing steps. In addition, malodors could also be produced from stored decaying materials (e.g., raw materials, food, and waste products) (Rappert and Muller, 2005). In addition to its application for decontamination and protein inactivation, CP shows potential for odor control, too. This is especially important if the raw material emanates off-odor either during storage or during the recovery process within a plant. The main advantages of low-temperature plasma technologies compared to conventional air treatment methods are their relatively low energy consumption, generally moderate cost, and, more importantly, their ability to treat air containing low concentrations of VOCs at relatively low operating temperatures (Hirota et al., 2004; Preis et al., 2013).

9.6 HIGH HYDROSTATIC PRESSURE

High pressure (HP) assisted extraction can be a useful tool to valorize wastes and byproducts, avoiding microbial contamination and facilitating the recovery of bioactive compounds (Galanakis, 2012, 2013; Barba et al., 2014).

Regarding the macroscopic level, the application of HP processing produces physical disruption of the plant tissue, thus inducing changes in the permeability of cell membranes (Gonzalez and Barrett, 2010). Knorr (1992) attributed these changes to modifications in the molecular organization of the lipid/peptide complex and disruptions in the phosphatidic acid bilayer membrane structure. Respective changes may play an important role in the raw material fragmentation, supporting intracellular valuable compounds extraction. HP treatment is characterized by three processing parameters: temperature T, pressure p, and exposure time t. The third processing parameter allows great variability in the design of the process (Heinz and Buckow, 2010). Therefore, at this stage of development, it is necessary to optimize HP process conditions to recover valuable compounds from plant food materials.

In a study conducted by Kato et al. (2002) HP impact on cell membrane damage was evaluated. By applying pressures below 100 MPa, a significant increase in membrane fluidity of the lipid bilayer was observed, thus facilitating reversible changes in transmembrane protein conformation. However, using pressures from 100 to 220 MPa, reversible phase transitions in parts of the lipid bilayer (from the liquid crystalline to the gel phase) as well as conformational changes in the protein subunits were observed. The application of pressures above 220 MPa destroyed and fragmented the membrane structure due to protein unfolding and interface separation. These pressure changes explain observed damage to cell organelles at approximately 200 to 300 MPa in plant cells and microorganisms.

HP can induce changes in membrane permeabilization, favoring the rupture of intracellular structures and consequent release of valuable compounds. Therefore, its use enhances mass transfer rates and increases cell permeability and secondary metabolite diffusion according to changes in phase transitions (Richard, 1992). This phenomenon can be used to favor the selective recovery of valuable compounds from intracellular compartments.

The degree of permeabilization clearly influences the recovery of intracellular compounds such as bioactive compounds, as they are located in different cell substructures, and their extraction is highly dependent on cell damage. In Fig. 9.7, an example of a vegetable cell and the location of some valuable compounds are shown. Another key factor is the location of different enzymes in the cell and tissue, which may contribute to the degradation of bioactive components. For instance, it is important to inactivate peroxidase and polyphenoloxidase as these are linked to degradation of bioactive compounds, especially polyphenols, and the subsequent enzymatic browning of fruits and vegetables (Bahçeci et al., 2005; Gökmen et al., 2005; Galanakis et al., 2010b, c, 2015a, b). There are several types

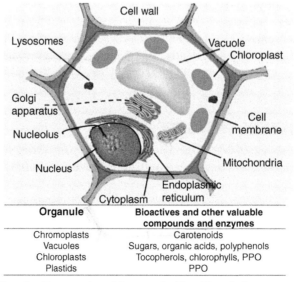

Organule	Bioactives and other valuable compounds and enzymes
Chromoplasts	Carotenoids
Vacuoles	Sugars, organic acids, polyphenols
Chloroplasts	Tocopherols, chlorophylls, PPO
Plastids	PPO

FIGURE 9.7 Location of Bioactive Compounds and Enzymes Inside a Plant Cell

of peroxidases in fruit and vegetable tissues with distinct features in terms of their molecular structure, cellular location, and function (Asada, 1992; Chen and Asada, 1992). Asada (1992) reported two major classes of peroxidases based on their function in plants: (i) ascorbate peroxidases, which are located in chloroplasts and cytosol, and (ii) guaiacol peroxidases, which can be found in cytosol, vacuole, cell wall, and extracellular space (Hu and Van Huystee, 1989; Asada, 1992) and mitochondria of maize (Prasad et al., 1994). Another important enzyme related to bioactive compounds degradation is myrosinase. This enzyme catalyzes the hydrolysis of glucosinolates and is found in different myrosin cells that are embedded in the tissue. Several authors have found that micronutrients and bioactive compounds in certain fruits and vegetables may be more extractable by HP treatments (Fig. 9.8).

For example, plant antioxidants tend to be water soluble, because they frequently appear combined as glycosides and they are located in the cell vacuole (Galanakis, 2013, 2015; Deng et al., 2015). Along these lines, different authors have studied the impact of HP on polyphenol recovery from tea leaves (Jun et al., 2009), *Rubus coreanus* fruits (Seo et al., 2011), longan fruit pericarp (Prasad et al., 2009a, 2010), and litchi fruit pericarp (Prasad et al., 2009b), among other plant food matrices. Overall, these studies concluded that HP processing is preferred, giving comparable or better yield recovery. In addition, HP is very time efficient and economic when it comes to solvents and the use of other laboratory facilities compared to the traditional methods.

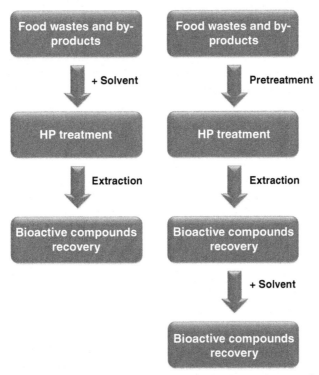

FIGURE 9.8 Flow Chart for Plant Food High-Pressure Extraction with Conventional or High Pressure (HP) Pretreatment Prior to Liquid/Solid Separation

Several research groups have also focused their interest on evaluating the impact of HP on carotenoid recovery from plant food materials. Some authors obtained an improvement in lycopene yield in HP processed tomato purée (Krebbers et al., 2003; Qiu et al., 2006) and high-pressure processing (HPP) tomato paste (Xi, 2006) compared to conventional methods. Sánchez-Moreno et al. (2004, 2005) attributed the increase in carotenoid yield in HPP orange products and persimmon purée (compared to conventional techniques) to the ability of HP for inducing cell permeabilization and facilitating denaturation of the carotenoid-binding protein induced by pressure. Moreover, McInerney et al. (2007) found that HP (600 MPa) had a significant impact on the availability of lutein in comparison to untreated samples or those processed at the lower pressure level (400 MPa). In another study, Barba et al. (2012) evaluated HP impact on tocopherols from orange derived products. These authors observed an increase in α-tocopherol after HP and suggested that HPP can facilitate the disruption of chloroplasts in orange samples, thus facilitating the extraction of α-tocopherol. Van Eylen et al. (2009) evaluated the impact of HP on glucosinolates and their degradation products (sulforaphane, glucoraphanin, iberin). These authors observed a significant negative effect of treatment time on glucosinolates degradation mainly attributed to leaching. In addition, they found an increase in glucosinolates hydrolysis when pressure was higher. This phenomenon was explained due to cell permeabilization during a pressure treatment.

9.7 CONCLUSIONS

This chapter looked at a range of emerging macroscopic pretreatments that have proven or potential applications in the recovery of valuable bioactives from food wastes. The operations specifically discussed drying methods such as foam-mat drying, EOD, and radio-frequency drying. All these methods appear to be commendable opportunities for food waste processing. The recently evolving CP technology provides ample scope for sterilization and inactivation of enzymes. High-pressure processing assisted gentle extraction, fractionation, and recovery of healthy, high added-value ingredients. The acceptability of these approaches will be largely governed by the cost ratio of the final product to that of the preparation process. Specific hallmark features, opportunities, and challenges for each of these technologies can be summarized as follows:

- Foam-mat drying is a promising method for the preservation of food products, allowing acceptable reconstitution properties. Cost analyses have indicated possible saving in capital cost and increase in production throughput. However, no conclusive success can be shown without further studies, especially regarding the retention of flavor, nutrients, and also the associated losses due to the foaming process.
- EOD has potential benefits for sludge dewatering and also has several advantages compared to conventional mechanical dewatering methods. EOD has been known to be effective for colloidal, gelatinous as well as biological materials, and has so far been applied to the food processing industry. The use of EOD to food processing products and wastes have been applied to various materials, mostly in the operation of EOD combined with mechanical dewatering. In addition, it is advisable to lower the electrical contact resistance between the electrode and the sludge bed and maintain the electric field strength applied to the bed as uniform and as large as possible. For this purpose, the above-mentioned combined operation of EOD with mechanical expression can be suggested. However, in such a combined operation, capital and running costs for equipment are respectively higher than those of EOD alone.

- Radio-frequency drying is an energy efficient technology due to rapid and direct uniform internal heating without surface damage, resulting in short process time and good product quality. The main drawback with the application of radio-frequency heating from a product application point of view is the lack of knowledge regarding dielectric properties of various foods as a function of composition, temperature, and frequency. The high initial investment capital cost of the specially designed equipment could also be the reason for the slow widespread use of this process. Understanding the fundamental principles and factors that influence the heating rate and uniformity of radio-frequency application will help improve the effectiveness and efficiency of the technology. The latest advances in using powerful computers and software for the modeling of thermo- and fluid-dynamic phenomena need to be further developed in order to have appropriate tools for the calculation of realistic electromagnetic fields and temperature distribution.
- CP is an upcoming sterilization technology for food and biomaterials. However, CP is not a universal solution to sterilization or decontamination of heat-sensitive materials, especially when the substrate is likely to undergo oxidation. The detrimental effects of plasma species can be regulated to a certain extent by changing the composition of gas in which the plasma is induced. In addition, if the recovery of any enzyme from the waste or raw material is of interest, then special care must be taken to assess the effects of CP treatments on enzyme activity. Alternatively, the downstream process should focus upon the recovery of the enzyme as a first step, followed by sterilization.
- The potential of HPP arises from the enormous amounts of waste materials discharged by the food industry. At this stage of development there is a need to optimize HP processing conditions. Moreover, there is a need to select the compounds that can be extracted by HP assisted extraction. A microscale analysis of cell substructures as well as functional cell characteristics should also be performed to improve HP assisted extraction of bioactive compounds from food byproducts.

REFERENCES

Abd Karim, A., Chee Wai, C., 1999. Foam-mat drying of starfruit (*Averrhoa carambola* L.) puree. Stability and air drying characteristics. Food Chem. 64, 337–343.

Akintoye, O.A., Oguntunde, A.O., 1991. Preliminary investigation on the effect of foam stabilizers on the physical characteristics and reconstitution properties of foam-mat dried soymilk. Drying Technol. 9 (1), 245–262.

Al-Asheh, S., Jumah, R., Banat, F., Al-Zou'Bi, K., 2004. Direct current electro-osmotic drying of tomato paste suspension. Food Bioprod. Process. 82 (C3), 193–200.

Alberti, F., Quaglia, N.C., Spremulli, L., Dambrosio, A., Todaro, E., Tamborrino, C., Lorusso, V., Celano, G.V., 2011. Radio-frequency technology for fresh stuffed pasta pasteurization/pre-drying processing. Preliminary results. Italian J. Food Sci. 23, 146–148.

Asada, K., 1992. Ascorbate peroxidase – a hydrogen peroxide-scavenging enzyme in plants. Plant Physiol. 85, 235–241.

Azharonok, V., Krat'ko, L., Nekrashevich, Y.I., Filatova, I., Mel'nikova, L., Dudchik, N., Yanetskaya, S., Bologa, M., 2009. Bactericidal action of the plasma of high-frequency capacitive and barrier discharges on microorganisms. J. Eng. Phys. Thermophys. 82 (3), 419–426.

Azizpour, M., Mohebbi, M., Khodaparast, M.H.H., Varidi, M., 2013. Foam-mat drying of shrimp: characterization and drying kinetics of foam. Agric. Eng. Int.: CIGR J. 15 (3), 159–165.

Bahçeci, K.S., Serpen, A., Gökmen, V., Acar, J., 2005. Study of lipoxygenase and peroxidase as indicator enzymes in green beans: change of enzyme activity, ascorbic acid and chlorophylls during frozen storage. J. Food Eng. 66, 187–192.

Barba, F.J., Esteve, M.J., Frigola, A., 2012. Impact of high-pressure processing on vitamin E (α-, γ-, and ϵ-tocopherol), vitamin D (cholecalciferol and ergocalciferol), and fatty acid profiles in liquid foods. J. Agric. Food Chem. 60, 3763–3768.

Barba, F.J., Esteve, M.J., Frígola, A., 2014. Bioactive components from leaf vegetable products. Studies Nat. Prod. Chem. 41, 321–346.

Bárdos, L., Baránková, H., 2008. Plasma processes at atmospheric and low pressures. Vacuum 83 (3), 522–527.

Bárdos, L., Baránková, H., 2010. Cold atmospheric plasma: sources, processes, and applications. Thin Solid Films 518 (23), 6705–6713.

Barton, W.A., Miller, S.A., Veal, C.J., 1999. The electrodewatering of sewage sludges. Drying Technol. 17 (3), 497–522.

Bates, R.P., 1964. Factors affecting foam production and stabilization of tropical fruit products. Food Technol. 18 (1), 93–96.

Bernard, C., Leduc, A., Barbeau, J., Saoudi, B., Yahia, L.H., De Crescenzo, G., 2006. Validation of cold plasma treatment for protein inactivation: a surface plasmon resonance-based biosensor study. J. Phys. D: Appl. Phys. 39 (16), 3470.

Brandenburg, R., Ehlbeck, J., Stieber, M., von Woedtke, T., Zeymer, J., Schlüter, O., Weltmann, K.D., 2007. Antimicrobial treatment of heat sensitive materials by means of atmospheric pressure RF-driven plasma jet. Contrib. Plasma Phys. 47 (1–2), 72–79.

Campbell, C.H. 1917. Drying milk. US Patent 1,250,427.

Chen, G.-X., Asada, K., 1992. Inactivation of ascorbate peroxidase by thiols requires hydrogen peroxide. Plant Cell Physiol. 33, 117–123.

Chen, H., Mujumdar, A.S., Raghavan, G.S.V., 1996. Laboratory experiments on electro-osmotic drying of vegetable sludge and mine tailings. Drying Technol. 14 (10), 2435–2445.

Chen, X.D., Mujumdar, A.S. (Eds.), 2008. Drying Technologies in Food Processing. Blackwell Publishing, Oxford, UK.

Citeau, M., Larue, O., Vorobiev, E., 2011. Influence of salt, pH and polyelectrolyte on the pressure electro-dewatering of sewage sludge. Water Res. 45 (6), 2167–2180.

Citeau, M., Olivier, J., Mahmoud, A., Vaxelaire, J., Larue, O., Vorobiev, E., 2012. Pressurised electro-osmotic drying of activated and anaerobically digested sludges: electrical variables analysis. Water Res. 46 (14), 4405–4416.

Cooper, M., Fridman, G., Fridman, A., Joshi, S.G., 2010. Biological responses of *Bacillus stratosphericus* to floating electrode-dielectric barrier discharge plasma treatment. J. Appl. Microbiol. 109 (6), 2039–2048.

Crank, J., 1975. The Mathematics of Diffusion, second ed. Clarendon Press, Oxford.

Danish, L.A., 1989. Investigation of electro-osmosis and its applications. Report for the Canadian Electrical Association 716 U 629. New Brunswick Research and Productivity Council, Fredericton.

Decareau, R.V., 1985. Microwaves in the Food Processing Industry. Academic Press Inc, Orlando, USA.

Deng, Q., Zinoviadou, K.G., Galanakis, C.M., Orlien, V., Grimi, N., Vorobiev, E., Lebovka, N., Barba, F.J., 2015. The effects of conventional and non-conventional processing on glucosinolates and its derived forms, isothiocyanates: extraction, degradation and applications. Food Eng. Rev. DOI: 10.1007/s12393-014-9104-9.

Devki Energy Consultancy Pvt. Ltd., 2006. Best Practice Manual: Dryers. Devki Energy Consultancy Pvt. Ltd., Vadodara, India.

Djaeni, M., Prasetyaningrum, A., Sasongko, S.B., Widayat, W., Hii, C.L., 2013. Application of foam-mat drying with egg white for carrageenan: drying rate and product quality aspects. J. Food Sci. Technol. 52, 1170–1175.

Fellows, P.J., 2000. Food Processing Technology – Principles and Practice, second ed. CRC Press, Boca Raton, FL.

Galanakis, C.M., 2012. Recovery of high added-value components from food wastes: conventional, emerging technologies and commercialized applications. Trends Food Sci. Technol. 26, 68–87.

Galanakis, C.M., 2015. Separation of functional macromolecules and micromolecules: from ultrafiltration to the border of nanofiltration. Trends Food Sci. Technol. 42, 44–63.

Galanakis, C.M., Goulas, V., Tsakona, S., Manganaris, G.A., Gekas, V., 2013. A knowledge base for the recovery of natural phenols with different solvents. Int. J. Food Prop. 16, 382–396.

Galanakis, C.M., Kotanidis, A., Dianellou, M., Gekas, V., 2015a. Phenolic content and antioxidant capacity of Cypriot Wines. Czech J. Food Sci. 33, 126–136.

Galanakis, C.M., Patsioura, A., Gekas, V., 2015b. Enzyme kinetics modeling as a tool to optimize food biotechnology applications: a pragmatic approach based on amylolytic enzymes. Crit. Rev. Food Sci. Technol. 55, 1758–1770.

Galanakis, C.M., Schieber, A., 2014. Editorial. Special issue on recovery and utilization of valuable compounds from food processing by-products. Food Res. Int. 65, 230–299.

Galanakis, C.M., Tornberg, E., Gekas, V., 2010a. A study of the recovery of the dietary fibres from olive mill wastewater and the gelling ability of the soluble fibre fraction. LWT – Food Sci. Technol. 43, 1009–1017.

Galanakis, C.M., Tornberg, E., Gekas, V., 2010b. Recovery and preservation of phenols from olive waste in ethanolic extracts. J. Chem. Technol. Biotechnol. 85, 1148–1155.

Galanakis, C.M., Tornberg, E., Gekas, V., 2010c. The effect of heat processing on the functional properties of pectin contained in olive mill wastewater. LWT – Food Sci. Technol. 43, 1001–1008.

Gökmen, V., Bahçeci, K.S., Serpen, A., Acar, J., 2005. Study of lipoxygenase and peroxidase as indicator enzymes in peas: change of enzyme activity, ascorbic acid and chlorophylls during frozen storage. LWT 38, 903–908.

Gonzalez, M.E., Barrett, D.M., 2010. Thermal, high pressure, and electric field processing effects on plant cell membrane integrity and relevance to fruit and vegetable quality. J. Food Sci. 75 (7), R121–R130.

Haswell, S.J., Kingston, H.M. (Eds.), 1997. Microwave-Enhanced Chemistry: Fundamentals. American Chemical Society, Washington, USA.

Heinz, V., Buckow, R., 2010. Food preservation by high pressure. J. Verbraucherschutz Lebensmittelsicherheit 5, 73–81.

Helmke, A., Hoffmeister, D., Mertens, N., Emmert, S., Schuette, J., Vioel, W., 2009. The acidification of lipid film surfaces by non-thermal DBD at atmospheric pressure in air. New J. Phys. 11 (11), 115025.

Hirota, K., Sakai, H., Washio, M., Kojima, T., 2004. Application of electron beams for the treatment of VOC streams. Ind. Eng. Chem. Res. 43 (5), 1185–1191.

Hu, C., Van Huystee, R.B., 1989. Role of carbohydrate moieties in peanut peroxidases. Biochem. J. 263, 129–135.

Ikawa, S., Kitano, K., Hamaguchi, S., 2010. Effects of pH on bacterial inactivation in aqueous solutions due to low-temperature atmospheric pressure plasma application. Plasma Process. Polym. 7 (1), 33–42.

Jeong, S.G., Kang, D.H., 2014. Influence of moisture content on inactivation of *Escherichia coli* O157:H7 and *Salmonella enterica* serovar *Typhimurium* in powdered red and black pepper spices by radio-frequency heating. Int. J. Food Microbiol. 175, 15–22.

Jumah, R., Al-Asheh, S., Banat, F., Al-Zoubi, K., 2005. Electroosmotic drying of tomato paste suspension under AC electric field. Drying Technol. 23 (7), 1465–1475.

Jumah, R., Al-Asheh, S., Banat, F., Al-Zoubi, K., 2007. Influence of salt, starch and pH on the electroosmosis dewatering of tomato paste suspension. J. Food Agric. Environ. 5 (1), 34–38.

Jun, X., Deji, S., Zhao, S., Lu, B., Li, Y., Zhang, R., 2009. Characterization of polyphenols from green tea leaves using a high hydrostatic pressure extraction. Int. J. Pharm. 382 (1–2), 139–143.

Kadam, D.M., Balasubramanian, S., 2011. Foam mat drying of tomato juice. J. Food Process. Pres. 35, 488–495.

Kadam, D.M., Wilson, R.A., Kaur, S., 2010. Determination of biochemical properties of foam-mat dried mango powder. Int. J. Food Sci. Technol. 45, 1626–1632.

Kamgang Youbi, G., Herry, J.M., Meylheuc, T., Brisset, J.L., Bellon Fontaine, M.N., Doubla, A., Naïtali, M., 2009. Microbial inactivation using plasma activated water obtained by gliding electric discharges. Lett. Appl. Microbiol. 48 (1), 13–18.

Kandasamy, P., Varadharaju, N., Kalemullah, S., Moitra, R., 2012. Asian J. Food Agro-Ind. 5 (05), 374–387.

Karathanos, V.T., Villalobos, G., Saravacos, G.D., 1990. Comparison of two methods of estimation of the effective moisture diffusivity from drying data. J. Food Sci. 55 (1), 218–231.

Kato, M., Hayashi, R., Tsuda, T., Taniguch, K., 2002. High pressure induced changes of biological membrane. Study on the membrane-bound Na+/K+-ATPase as a model system. European J. Biochem. 269, 110–118.

Kemp, I.C., Fyhr, B.C., Laurent, S., Roques, M.A., Groenewold, C.E., Tsotsas, E., Sereno, A.A., Bonazzi, C.B., Bimbenet, J.J., Kind, M., 2001. Methods for processing experimental drying kinetics data. Drying Technol. 19, 15–34.

Kingston, H., Jassie, L., 1998. Microwave-Enhanced Chemistry, Fundamentals, Sample Preparation and Applications. American Chemical Society, Washington, USA.

Knorr, D., 1992. High-pressure effects on plant derived foods. Baslny, C., Hayashi, R., Herman, K., Masson, P. (Eds.), High Pressure and Biotechnology, 24, John Libbey Euro-text, Montrouge, pp. 211–217.

Korachi, M., Gurol, C., Aslan, N., 2010. Atmospheric plasma discharge sterilization effects on whole cell fatty acid profiles of *Escherichia coli* and *Staphylococcus aureus*. J. Electrostat. 68 (6), 508–512.

Krebbers, B., Matser, A.M., Hoogerwerf, S.W., Moezelaar, R., Tomassen, M.M.M., Van den Berg, R.W., 2003. Combined high pressure and thermal treatments for processing of tomato puree: evaluation of microbial inactivation and quality parameters. Innov. Food Sci. Emerg. Technol. 4, 377–385.

Kudra, T., Ratti, C., 2006. Foam-mat drying: energy and cost analyses. Can. Biosys. Eng. 48, 3.27−3.32.

Leduc, M., Guay, D., Coulombe, S., Leask, R.L., 2010. Effects of non-thermal plasmas on DNA and mammalian cells. Plasma Process. Polym. 7 (11), 899–909.

Lee, J.E., Lee, J.K., Choi, H.K., 2007. Filter press for electrodewatering of waterworks sludge. Drying Technol. 25 (12), 1985–1993.

Leipold, F., Kusano, Y., Hansen, F., Jacobsen, T., 2010. Decontamination of a rotating cutting tool during operation by means of atmospheric pressure plasmas. Food Control 21 (8), 1194–1198.

Lewicki, P.P., 1975. Mechanisms concerned in foam-mat drying of tomato paste. Trans. Agric. Acad. Warsaw 55, 1–67, (in Polish).

Li, H.-P., Wang, L.-Y., Li, G., Jin, L.-H., Le, P.-S., Zhao, H.-X., Xing, X.-H., Bao, C.-Y., 2011. Manipulation of lipase activity by the helium radio-frequency. Atmospheric-pressure glow discharge plasma jet. Plasma Process. Polym. 8 (3), 224–229.

Li, L., Li, X., Uemura, K., Tatsumi, E., 1999. Electro-osmotic drying of okara in different electric fields. In: Proceedings of 99 International Conference on Agricultural Engineering. Beijing, China., pp. IV58–IV63.

Liu, F.X., Sun, P., Bai, N., Tian, Y., Zhou, H.X., Wei, S.C., Zhou, Y.H., Zhang, J., Zhu, W.D., Becker, K., Fang, J., 2010. Inactivation of bacteria in an aqueous environment by a direct-current, cold-atmospheric-pressure air plasma microjet. Plasma Process. Polym. 7 (3–4), 231–236.

Lockhart, N.C., 1986. Electro-dewatering of fine suspensions. In: Muralidhara, H.S. (Ed.), Advances in Solid−Liquid Separation. Battelle Press, Columbus, pp. 241–274.

Machala, Z., Janda, M., Hensel, K., Jedlovský, I., Leštinská, L., Foltin, V., Martišovitš, V., Morvová, M., 2007. Emission spectroscopy of atmospheric pressure plasmas for bio-medical and environmental applications. J. Mol. Spectrosc. 243 (2), 194–201.

Mahmoud, A., Olivier, J., Vaxelaire, J., Hoadley, A.F.A., 2010. Electrical field: a historical review of its application and contributions in wastewater sludge dewatering. Water Res. 44 (8), 2381–2407.

Mahmoud, A., Olivier, J., Vaxelaire, J., Hoadley, A.F.A., 2011. Electro-dewatering of wastewater sludge: influence of the operating conditions and their interactions effects. Water Res. 45 (9), 2795–2810.

Marnett, L.J., 2002. Oxy radicals, lipid peroxidation and DNA damage. Toxicology 181, 219–222.

Maroulis, Z.B., Saravacos, G.D., 2003. Food Process Design. Marcel Dekker, New York.

Marques, L.G., Silveira, A.M., Freire, J.T., 2006. Freeze-drying characteristics of tropical fruits. Drying Technol. 24 (4), 457–463.

Marra, F., Zhaug, L., Lyng, J.G., 2009. Radio frequency treatment of foods: review of recent advances. J. Food Eng. 91, 497–508.

Marsili, L., Espie, S., Anderson, J.G., MacGregor, S.J., 2002. Plasma inactivation of food-related microorganisms in liquids. Radiat. Phys. Chem. 65 (4–5), 507–513.

McInerney, J.K., Seccafien, C.A., Stewart, C.M., Bird, A.R., 2007. Effects of high pressure processing on antioxidant activity, and total carotenoid content and availability, in vegetables. Innov. Food Sci. Emerg. Technol. 8 (4), 543–548.

Misra, N.N., Keener, K.M., Bourke, P., Mosnier, J.P., Cullen, P.J., 2014a. In-package atmospheric pressure cold plasma treatment of cherry tomatoes. J. Biosci. Bioeng. 118 (2), 177–182.

Misra, N.N., Patil, S., Moiseev, T., Bourke, P., Mosnier, J.P., Keener, K.M., Cullen, P.J., 2014b. In-package atmospheric pressure cold plasma treatment of strawberries. J. Food Eng. 125, 131–138.

Misra, N.N., Tiwari, B.K., Raghavarao, K.S.M.S., Cullen, P.J., 2011. Nonthermal plasma inactivation of food-borne pathogens. Food Eng. Rev. 3 (3–4), 159–170.

Misra, N.N., Ziuzina, D., Cullen, P.J., Keener, K.M., 2013. Characterization of a novel atmospheric air cold plasma system for treatment of packaged biomaterials. Trans. ASABE 56 (3), 1011–1016.

Mujumadar, A.S., Yoshida, H., 2008. Electro-osmotic drying (EOD) of bio-materials. In: Vorobiev, E., Lebovka, N. (Eds.), Electrotechnologies for Extraction from Food Plants and Biomaterials. Springer, New York, pp. 121–154.

Muthukumaran, A., Ratti, C., Raghavan, V.G.S., 2008. Foam-mat freeze drying of egg white and mathematical modeling. Part I Optimization of egg white foam stability. Drying Technol. 26 (4), 508–512.

Naidu, M.M., Khanum, H., Sulochanamma, G., Sowbhagya, B., Hebbar, U.H., Prakash, M., Srinivas, P., 2012. Effect of drying methods on the quality characteristics of fenugreek (*Trigonella foenum-graecum*) greens. Drying Technol. 30, 808–816.

Ng, S.K., Plunkett, A., Stojceska, V., Ainsworth, P., Lamont-Black, J., Hall, J., White, C., Glendenning, S., Russell, D., 2011. Electro-kinetic technology as a low-cost method for dewatering food by-product. Drying Technol. 29 (14), 1721–1728.

Orfeuil, M., 1987. Electric Process Heating: Technologies/Equipment/Applications. Battelle Press, Columbus, USA.

Potter, N.N., Hotchkiss, J.H., 1998. Food science. Chapman & Hall Food Science Book. fifth ed. Aspen Publishers, Gaithersburg, MA.

Prakotmak, P., Soponronnarit, S., Prachayawarakorn, S., 2011. Effect of adsorption conditions on effective diffusivity and textural property of dry banana foam mat. Drying Technol. 29 (9), 1090–1100.

Prasad, K.N., Hao, J., Shi, J., Liu, T., Li, J., Wei, X., Qiu, S., Xue, S., Jiang, Y., 2009b. Antioxidant and anticancer activities of high pressure-assisted extract of longan (*Dimocarpus longan* Lour.) fruit pericarp. Innov. Food Sci. Emerg. Technol. 10, 413–419.

Prasad, K.N., Shi, J., Yua, C., Zhao, M., Xue, S., Jiang, Y., 2010. Enhanced antioxidant and antityrosinase activities of longan fruit pericarp by ultra-high-pressure-assisted extraction. J. Pharm. Biomed. Anal. 51, 471–477.

Prasad, N.K., Yang, B., Zhao, M.B., Wang, S., Chen, F., Jiang, Y., 2009a. Effects of high-pressure treatment on the extraction yield, phenolic content and antioxidant activity of litchi (*Litchi chinensis* Sonn.) fruit pericarp. Int. J. Food Sci. Technol. 44, 960–966.

Prasad, T.K., Anderson, M.D., Stewart, C.R., 1994. Acclimation, hydrogen peroxide, and abscissic acid protect mitochondria against irreversible chilling injury in maize seedlings. Plant Physiol. 105, 619–627.

Preis, S., Klauson, D., Gregor, A., 2013. Potential of electric discharge plasma methods in abatement of volatile organic compounds originating from the food industry. J. Environ. Manage. 114, 125–138.

Qiu, W., Jiang, H., Wang, H., Gao, Y., 2006. Effect of high hydrostatic pressure on lycopene stability. Food Chem. 97, 516–523.

Raharitsifa, N., Ratti, C., 2010. Foam-mat freeze-drying of apple juice. Part 2: Stability of dry products during storage. J. Food Process. Eng. 33, 341–364.

Ramaswamy, H., Marcotte, M., 2006. Food Processing – Principles and Applications. Taylor and Francis Group/CRC Press, Boca Raton, FL.

Rampacek, C., 1966. Electro-osmotic and electro-phoretic dewatering as applied to solid-liquid separation. In: Poole, J.B., Doyle, D. (Eds.), Solid–Liquid Separation – A Review and a Bibliography. Her Majesty's Stationery Office, London, pp. 100–108.

Rappert, S., Muller, R., 2005. Odor compounds in waste gas emissions from agricultural operations and food industries. Waste Manage. 25 (9), 887–907.

Ratti, C., Kudra, T., 2006. Drying of foamed biological materials: opportunities and challenges. Drying Technol. 24 (9), 1101–1108.

Richard, J.S., 1992. High Pressure Phase Behaviour of Multicomponent Fluid Mixtures. Elsevier, Amsterdam.

Roselló-Soto, E., Barba, F.J., Parniakov, O., Galanakis, C.M., Grimi, N., Lebovka, N., Vorobiev, E., 2015. High voltage electrical discharges, pulsed electric field and ultrasounds assisted extraction of protein and phenolic compounds from olive kernel. Food Bioprocess Technol. 8, 885–894.

Rowan, N., Espie, S., Harrower, J., Anderson, J., Marsili, L., MacGregor, S., 2007. Pulsed-plasma gas-discharge inactivation of microbial pathogens in chilled poultry wash water. J. Food Protect. 70 (12), 2805–2810.

Sánchez-Moreno, C., Plaza, L., De Ancos, B., Cano, M.P., 2004. Effect of combined treatments of high-pressure and natural additives on carotenoid extractability and antioxidant activity of tomato puree (*Lycopersicum esculentum* Mill). Eur. Food Res. Technol. 219 (2), 151–160.

Sánchez-Moreno, C., Plaza, L., Elez-Martinez, P., De Ancos, B., Martin-Belloso, O., Cano, M.P., 2005. Impact of high pressure and pulsed electric fields on bioactive compounds and antioxidant activity of orange juice in comparison with traditional thermal processing. J. Agric. Food Chem. 53, 4403–4409.

Saravacos, G.D., Maroulis, Z.B., 2011. Food Process Engineering Operations. CRC Press, Boca Raton, FL.

Saveyn, H., Pauwels, G., Timmerman, R., Van der Meeren, P., 2005. Effect of polyelectrolyte conditioning on the enhanced dewatering of activated sludge by application of an electric field during the expression phase. Water Res. 39 (13), 3012–3020.

Scholtz, V., Julák, J., Kříha, V., 2010. The microbicidal effect of low-temperature plasma generated by corona discharge: comparison of various microorganisms on an agar surface or in aqueous suspension. Plasma Process. Polym. 7 (3–4), 237–243.

Seo, Y.C., Choi, W.Y., Kim, J.S., Yoon, C.S., Lim, H.W., Cho, J.S., Ahn, J.H., Lee, H.Y., 2011. Effect of ultra high pressure processing on immuno-modulatory activities of the fruits of *Rubus coreanus* Miquel. Innov. Food Sci. Emerg. Technol. 12, 207–215.

Sharada, S., 2013. Studies on effect of various operating parameters & foaming agents – drying of fruits and vegetables. Int. J. Mod. Eng. Res. 3 (3), 1512–1519.

Sun, D.-W. (Ed.), 2005. Emerging Technologies for Food Processing. Academic Press, Elsevier, London, UK.

Sunderland, J.G., 1987. Electrokinetic dewatering and thickening. I. Introduction and historical review of electrokinetic applications. J. Appl. Electrochem. 17, 889–898.

Takai, E., Kitano, K., Kuwabara, J., Shiraki, K., 2012. Protein inactivation by low-temperature atmospheric pressure plasma in aqueous solution. Plasma Process. Polym. 9 (1), 77–82.

Thuwapanichayanan, R., Prachayawarakorn, S., Soponronnarit, S., 2008. Drying characteristics and quality of banana foam mat. J. Food Eng. 86, 573–583.

Thuwapanichayanan, R., Prachayawarakorn, S., Soponronnarit, S., 2012. Effects of foaming agents and foam density on drying characteristics and textural property of banana foams. LWT − Food Sci. Technol. 47, 348–357.

Tuan, P.-A., Jurate, V., Mika, S., 2008. Electro-dewatering of sludge under pressure and non-pressure conditions. Environ. Technol. 29 (10), 1075–1084.

Tuan, P.-A., Mika, S., 2010. Effect of freeze/thaw conditions, polyelectrolyte addition, and sludge loading on sludge electro-dewatering process. Chem. Eng. J. 164 (1), 85–91.

Tuan, P.-A., Mika, S., Pirjo, I., 2012. Sewage sludge electro-dewatering treatment – a review. Drying Technol. 30 (7), 691–706.

Valentas, K.J., Rotstein, E., Singh, R.P., 1997. Handbook of Food Engineering Practice. CRC Press, Boca Raton, FL.

Van Eylen, D., Bellostas, N., Strobel, B.W., Oey, I., Hendrickx, M., Van Loey, A., Sørensen, H., Sørensen, J.C., 2009. Influence of pressure/temperature treatments on glucosinolate conversion in broccoli (*Brassica oleraceae* L. cv Italica) heads. Food Chem. 112 (3), 646–653.

Velis, C.A., Longhurst, P.J., Drew, G.H., Smith, R., Pollard, S.J.T., 2009. Biodrying for mechanical-biological treatment of wastes: a review of process science and engineering. Bioresource Technol. 100 (11), 2747–2761.

Viji, A.K., 2004. Electrochemical effects in biological materials: electro-osmotic drying of cancerous tissue as the mechanistic proposal for the electrochemical treatment of tumors. J. Mater. Sci.: Mater. Med. 10 (7), 419–423.

Wang, S., Tank, J., Johnson, J.A., Mitcham, E., Hansen, J.D., Hallman, G., Drake, S.R., Wang, Y., 2003. Dielectric properties of fruits and insects as related to radio-frequency and microwave treatments. Biosystems Eng. 85, 201–212.

Widyastuti, T.E.W., Srianta, I., 2011. Development of functional drink based on foam-mat dried papaya (*Carica papaya* L.): optimisation of foam-mat drying process and its formulation. Int. J. Food Nutr. Public Health 4 (2), 167–176.

Xi, J., 2006. Application of high hydrostatic pressure processing of food to extracting lycopene from tomato paste waste. High Pressure Res. 26 (1), 33–41.

Xia, B., Sun, D.-W., Li, L.-T., Li, X.-Q., Tatsumi, E., 2003. Effect of electro-osmotic drying on the quality of tofu sheet. Drying Technol. 21 (1), 129–145.

Xingmin, X., Yaxi, L., Guanjun, Z., Yue, M.A., Xianjun, S., 2011. Experimental study on inactivation of bacterial endotoxin by using dielectric barrier discharge. Plasma Sci. Technol. 13 (6), 651–655.

Yang, G.C.C., Chen, M.-C., Yeh, C.-F., 2011. Dewatering of a biological industrial sludge by electrokinetics-assisted filter press. Separ. Purif. Technol. 79 (2), 177–182.

Yoshida, H., 1993. Practical aspects of dewatering enhanced by electro-osmosis. Drying Technol. 11 (4), 787–814.

Yoshida, H., Fujimoto, T., Hishamudi, H., 2004. Electro-osmotic drying under electric fields with combination of constant voltage and constant current. Kagaku Kogaku Ronbun. 30 (5), 633–635.

Zhao, Y., Flugstad, B., Kolbe, E., Park, J.W., Wells, J.H., 2000. Using capacitive (radio frequency) dielectric heating in food processing and preservation – a review. J. Food Process. Eng. 23 (1), 25–55.

Ziuzina, D., Patil, S., Cullen, P.J., Keener, K.M., Bourke, P., 2013. Atmospheric cold plasma inactivation of *Escherichia coli* in liquid media inside a sealed package. J. Appl. Microbiol. 114 (3), 778–787.

EMERGING MACRO- AND MICROMOLECULES SEPARATION

10

Krishnamurthy Nagendra Prasad*, Giorgia Spigno**,
Paula Jauregi†, N.N. Misra‡, Patrick J. Cullen§

*Chemical Engineering Discipline, School of Engineering, Monash University of Malaysia, Malaysia;
**Institute of Oenology and Agro-Food Engineering, Università Cattolica del Sacro Cuore, Piacenza, Italy;
†Department of Food and Nutritional Sciences, University of Reading, Reading, UK; ‡School of Food Science and
Environmental Health, Dublin Institute of Technology, Dublin, Ireland; §University of New South Wales, Sydney, Australia

10.1 INTRODUCTION

The word separation means, "being moved apart". It is a technique to convert a mixture of substances into two or more distinct product mixtures. Generally, separation is carried out based on the chemical or physical (size, shape, mass, density) properties of the molecules (Wilson et al., 2000). Macromolecules include polysaccharides, dietary fibers, lipids, and proteins, while micromolecules include polyphenols, simple sugars, acids, minerals, and other compounds (Galanakis et al., 2010a, c, 2012a, 2015a; Galanakis, 2011, 2015). Traditional technologies for macro- and micromolecules separation are discussed in detail in Chapter 5, while this chapter covers emerging technologies including colloidal gas aphrons, ultrasound assisted crystallization, and pressurized microwave assisted extraction (Galanakis, 2012, 2013; Galanakis and Schieber, 2014).

10.2 COLLOIDAL GAS APHRONS (CGA)

10.2.1 GENERAL

CGA, also called microfoams, are surfactant-stabilized microbubbles. They differ from regular foams since the bubbles are encapsulated in a multilayered shell consisting of surfactant and liquid, instead of a monolayer of surfactant molecules. The most important characteristics of CGA are:

1. their high stability compared with conventional foams,
2. their high interfacial area due to their small size,
3. sufficient stability to allow them to be pumped from the generation point to the point of use without loss of their original structure,
4. they can be easily separated from the bulk liquid without mechanical aid, as opposed to conventional liquid/liquid extraction methods and the aqueous two-phase separations that need centrifugation for phase separation.

In case of using biodegradable and nontoxic surfactants, this methodology could result in an environment friendly process, while final products could also be safe for human consumption. Due to these

Food Waste Recovery. http://dx.doi.org/10.1016/B978-0-12-800351-0.00010-9

properties, CGA are highly suitable for separation and mass-transfer applications. For example, CGA have found numerous applications for the recovery of a wide variety of valuable materials including bioproducts (e.g. proteins, bacterial cells, enzymes, carotenoids, dyes) and antioxidants (Dermiki et al., 2010; Spigno et al., 2010, 2014), and for pollution remediation (Hashim et al., 2011). Despite these promising applications, in practice CGA have not been utilized to their full potential.

10.2.2 STRUCTURE OF CGA

CGA are surfactant-stabilized microbubbles (closely packed spherical bubbles between 10–100 μm) generated by intense stirring of a surfactant solution at high speeds (>8000 rpm). CGA are composed of multilayers of surfactant molecules (Fig. 10.1) where surfactant molecules adsorb at the interface and their hydrophilic heads are oriented towards the aqueous phase and the hydrophobic tails towards the gas phase (Sebba, 1987). In an attempt to confirm this structure Jauregi et al. (2000) investigated the drainage kinetics of CGA and compared measured drainage rates with those obtained by applying predictive models for foams. Interestingly, the model giving the best prediction was the one that took into account differences in structural features between foams and CGA. In addition, for the first time Jauregi and coworkers used small angle X-ray diffraction in an attempt to determine the thickness of the surfactant film and the number of surfactant layers of CGA generated by the anionic surfactant sodium bis-2-ethylhexyl sulfosuccinate (AOT). Samples containing aphrons gave similar scattering regardless of the surfactant concentration and this corresponded to 5.4 nm (equivalent to seven layers of surfactant assuming the full length of the surfactant molecule arranged in layers and vertically at the interface), while the same surfactant solutions with no aphrons gave a different scattering signal. For example, at concentrations around the critical micellar concentration (cmc), the scattering corresponded to a bilayer and subsequently to micelles. Above the cmc, the scattering corresponded to a multilayer of 3–5 molecules confirming in this way the hypothesis that AOT forms lamellar structures (Jauregi et al., 2000).

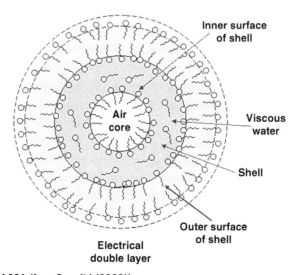

FIGURE 10.1 Structure of CGA (from Dermiki (2009))

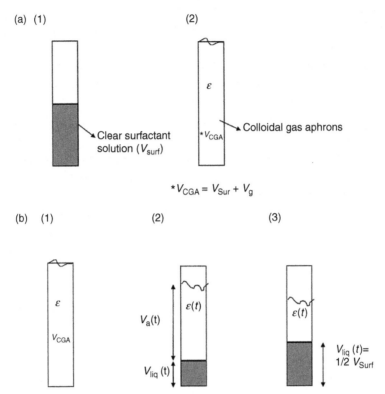

FIGURE 10.2 Schematic of CGA Generation (a) and Separation into Two Phases, an Aphron Phase V_a and a Liquid Phase V_{liq} (b)

(a) (1) Before stirring, (2) after stirring. (b) (1) Stirring ends ($t - 0$), (2) after stirring ends at ($t - t$), (3) at ($t - t$).

10.2.3 CHARACTERISTICS OF CGA

Due to their unique structure CGA possess the following important properties:

1. Higher stability compared with conventional foams. Due to the multilayer structure CGA exhibit high stability. Stability of CGA is characterized in terms of gas hold-up (ε), which is defined as the ratio between the gas volume (V_g) incorporated into the surfactant solution (V_{Sur}) and the dispersion final volume after stopping stirring and at time = 0 of drainage (V_{CGA}) (Fig. 10.2):

$$\varepsilon = \frac{V_g}{V_{CGA}} = \frac{V_{CGA} - V_{Sur}}{V_{CGA}} \tag{10.1}$$

2. The buoyancy of the encapsulated gas leads to easy separation of the aphron phase from the bulk liquid phase, without mechanical aid (Fig. 10.2). On the other hand, creaming can be avoided (when necessary) by stirring CGA at such a rate that the lateral movement conveyed to the bubbles is greater than the upward buoyancy.

3. Adsorption of particles and molecules to the surface of CGA. Depending on the surfactant used to generate the CGA, the outer surface of the bubble can be negatively (anionic surfactant), positively (cationic surfactant), or noncharged (nonionic surfactant). Consequently, oppositely charged molecules will adsorb. Thus, the selectivity of adsorption can be modified by changing the type of surfactant (Jauregi and Dermiki, 2010). Furthermore, small size bubbles provide a large interfacial area per unit volume and subsequently high adsorption capacity. Bubble size depends on the concentration and type of surfactant, ionic strength, and the presence of other molecules or particles.

4. Low viscosity of the system. Flow properties are similar to those of water, as long as the gas hold-up does not exceed 65% (Roy et al., 1995), regardless of the surfactant type used to generate them. Consequently, CGA can be pumped easily from the generation point to the point where they are going to be used. The characteristics of conventional foams change during pumping due to the elastic nature of the bubble. This is advantageous when CGA are pumped into a flotation column, as in flotation for the removal/recovery of products.

10.2.4 GENERATION OF CGA

The formation of CGA requires a horizontal disc that rotates at very high speeds. Baffles are also recommended in order to achieve the required mixing regime and produce smaller bubbles. The minimum power to generate CGA is dependent on the surfactant concentration that affects the volume of air incorporated. For surfactant concentrations below and above the cmc the minimum power requirement to generate CGA was 45 kWm^{-3} (Jauregi et al., 1997). A four-bladed impeller with a high shear screen (Fig. 10.3) was found to generate CGA dispersions with small bubble sizes (average diameters ranged between 35–70 μm) (Jauregi, 1997). The use of sonication leads to CGA with higher gas hold-up, smaller bubble size, higher number of bubbles, and larger interfacial area than mechanical agitation (Xu et al., 2008). However, sonication is an expensive method that cannot be easily scaled up.

A wide range of surfactants can be used for the generation of CGA:

1. anionic, e.g. sodium dodecyl sulfate,
2. cationic, e.g. cetyltrimethylammonium bromide (CTAB),
3. nonionic, e.g. Tween 20.

Characteristics of CGA generated with these surfactants are summarized in Table 10.1. Apart from the surfactant type, the main operating parameters affecting the characteristics of the CGA (stability, gas hold- up, and bubble size) at constant stirring speed are:

1. surfactant concentration,
2. salt concentration (ionic strength),
3. pH value,
4. stirring time, and
5. temperature.

At constant values of surfactant concentration, ionic strength, and shearing speed, CGA of constant interfacial area are produced.

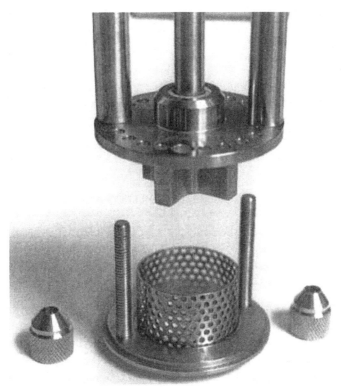

FIGURE 10.3 Image of the Four-Bladed Impeller with a High Shear Screen Used to Generate CGA by Jauregi (1997)

Table 10.1 Characteristics of CGA Generated by Ionic (CTAB Cationic and AOT Anionic) and Nonionic (Tween 20 and Tween 60) Surfactants (Values Found at Optimum Conditions)

Surfactant	Concentration (mM)	Gas Hold-Up (%)	Half Life (s)	References
CTAB cmc = 0.9 mM (in H_2O)	1 2	62.96 67.00	493 509	Dermiki (2009)
Tween 20 cmc = 0.08 mM	1 10 20	42.30 60.00 62.95	166 407 582	Dermiki (2009)
Tween 60 cmc = 0.023 mM	20	62.00	410	Dermiki et al. (2009)
AOT cmc = 0.84 mM	0.1 2.5 34	25.00 56.00 58.00	60 310 758	Jauregi et al. (1997)

10.2.5 CGA-BASED SEPARATION PROCESSES

As explained earlier, CGA offer a large interfacial area and are relatively stable. This is important if CGA are to be used in separations. Moreover, after contacting and adsorbing the target compound to the aphrons, CGA separate easily (e.g. in seconds) into a bulk liquid phase (bottom phase) and a CGA phase (top phase) (Fig. 10.2). This is likely to offer an advantage in terms of separation economics because it does not need an additional centrifugation step.

A conventional flotation process is carried out by sparging air into a solution that contains the compound to be removed (the colligend). The compound adsorbs to the gas bubbles and is removed by flotation. Surfactants are often added to the solution of the colligend to improve its removal. When surfactants contact with the colligend, they act as collectors causing in some cases precipitation. The precipitate is removed by flotation. The latter process is called precipitate flotation (Pinfold, 1972). In these processes, small bubble sizes are desirable so that the interfacial area is maximized.

As a further improvement to conventional flotation techniques, CGA can be sparged in place of air bubbles to remove/recover products. For example, Subramaniam et al. (1990) used CGA to clarify a palm oil mill effluent, whereas Hashim and Sengupta (1998) used CGA for the flotation of cellulose fiber from paper mill wastewater. Palm fruit wastes contain typically fibers, plant cell debris, polyphenols, and flavonoids (Subramaniam et al., 1990; Heng et al., 2015). Subramaniam et al. (1990) claimed to remove 96% of solids successfully using an anionic surfactant. Similarly, CGA have been used for the recovery of microbial (Save and Pangarkar, 1995; Hashim et al., 1995a, b, 1998). A typical CGA-based separation process occurs in four steps as illustrated in Fig. 10.4. In Step 1, CGA are generated as indicated in Section 10.2.4. Based on the gas hold-up (Eq. 10.1) and the surfactant concentration of the solution used to generate CGA ($[Sur]_{Sol}$), the surfactant concentration in the CGA ($[Sur]_{CGA}$), can be calculated as:

$$[Sur]_{CGA} = (1-\varepsilon)[Sur]_{Sol} \qquad (10.2)$$

In this step, it is necessary to select the surfactant and its concentration. Surfactant selection depends mainly on the expected use of the colligend and its chemical structure. For example, if the target compound is intended for food application, a food-grade surfactant should be preferably used in order to avoid Step 5. If charged compounds are separated, a suitable cationic or anionic surfactant will be most suitable in order to exploit electrostatic interactions between the compound and the surfactant. In the case of colligends mixtures with different chemical structures, nonionic surfactants could also be used, as separation is also driven by hydrophobic interactions (see example of polyphenols recovery below).

In Step 2, the CGA are pumped typically into a flotation column where they come in contact with the colligend solution. In view of industrial implementation, rheological properties of CGA and their stability during flowing should then be properly investigated (Larmignat et al., 2008; Arabloo and Shahri, 2014). If the flotation column is operated in batch, the colligend sample (feed) is loaded into the column and then the CGA are introduced, generally from the bottom. Flow rate of CGA combined with the column and feed volume determine the mixing time, according to Eq. 10.3:

$$\frac{V_{CGA}}{V_{feed}} = \frac{\text{Flow rate mixing time}}{V_{feed}} \qquad (10.3)$$

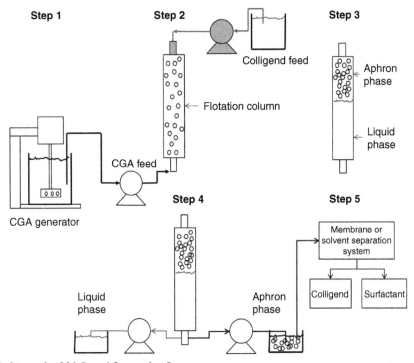

FIGURE 10.4 Steps of a CGA-Based Separation Process

Step 1, CGA generation; Step 2, CGA are mixed with the colligend feed into a flotation column; Step 3, the mixture is left separating into an aphron and a liquid phase; Step 4, collection of aphron phase; Step 5, surfactant removal from the collected aphron phase.

After CGA are introduced and flow (turbulence) stops, the separation of phases occurs as liquid starts draining from the foam. The time allowed for the phase separation is called separation time (Step 3). Total contact time (mixing time plus separation) can influence the process performance as detailed below. Other important operating parameters related to Steps 2 and 3 are the volumetric and molar ratio. The volumetric ratio (Eq. 10.4) is the ratio between the CGA volume and the feed solution:

$$\text{Volumetric ratio} = \frac{V_{CGA}}{V_{feed}} \tag{10.4}$$

The molar ratio (Eq. 10.5) also takes into account the concentration of the CGA and the feed solution:

$$\text{Molar ratio} = \frac{V_{CGA} \cdot [\text{Sur}]_{CGA}}{V_{feed} \cdot [\text{Colligend}]_{feed}} \tag{10.5}$$

This is the ratio between the moles of surfactant and the moles of colligend. The influence of volumetric and molar ratio on the separation process strongly depends on the type of surfactant–colligend interactions (see Chapter 3, Section 3.1). Generally, the separation is enhanced, increasing the molar ratio (see examples given in later sections).

In Step 3, drainage of liquid phase occurs and an aphron phase separates on the top. Ideally, all the colligend molecules present in the feed remain in this phase giving a 100% recovery yield. Process recovery (RE%) is, in fact, defined by Eq. 10.6 as the ratio between the amount of colligend partitioned into the aphron phase and the total amount in the feed.

$$RE(\%) = \frac{[\text{Colligend}]_{\text{Aphron phase}}}{V_{\text{feed}} \cdot [\text{Colligend}]_{\text{feed}}} \qquad (10.6)$$

Higher recovery can be obtained by decreasing the contact time and thus reducing the drained liquid phase. However, at the same time, it will result in a low concentration in the aphron phase. Besides RE%, two other important parameters have to be considered in order to evaluate the efficiency of the separation process: the separation and the enrichment factor. The separation factor (SF) is similar to the partition coefficient and is defined as the ratio of the colligend concentration in the aphron phase to that in the liquid phase, as described in Eq. 10.7.

$$SF = \frac{[\text{Colligend}]_{\text{Aphron phase}}}{[\text{Colligend}]_{\text{Drained liquid phase}}} \qquad (10.7)$$

SF is therefore a measure of the affinity of the colligend for the aphron phase. The enrichment factor (EF) is calculated according to Eq. 10.8 and it is a measure of how much the colligend concentration is increased in the aphron phase in relation to the feed.

$$EF = \frac{[\text{Colligend}]_{\text{Aphron phase}}}{[\text{Colligend}]_{\text{feed}}} \qquad (10.8)$$

Selectivity can be measured as the ratio of SF of the target compound and the SF of another by-product or contaminant. Thus, an efficient separation should result in high recovery. However, in some cases, high recovery may be achieved in combination with low EF and SF values. This process would still be considered a successful separation.

In Step 4, the separated aphron phase is simply collected by pumping the CGA phase out and allowing it to collapse. At this point, the removal of the surfactant may be required (Step 5). This will inevitably increase the cost of the process due to the use of membranes or the addition of specific solvents (Dermiki, 2009). Table 10.2 resumes the basic steps of a CGA-based separation process and the main operating parameters that can influence the process efficiency.

10.2.6 SEPARATION OF PROTEINS BY CGA

The separation of proteins from whey using CGA generated with an anionic surfactant AOT has been investigated (Fuda, 2004). In particular, selective separation of lactoferrin and lactoperoxidase could be achieved using conditions that promote strong electrostatic interactions between the proteins and surfactant (pH = 4). Interestingly, high ionic strength led to higher purity, as at these conditions drainage of CGA is favored. The latter process results in a higher number of contaminant proteins partitioning

Table 10.2 Main Operating Parameters of a CGA-Based Separation Process and Their Influence on Process Performance

Process Step	Operating Parameter	Influence on
CGA generation	Surfactant type Surfactant concentration pH and ionic strength	CGA stability and colligend interaction
CGA mixing with sample feed (in a flotation column)	Volumetric ratio Molar ratio Feed concentration Mixing time (flow rate and column size) Batch or continuous operation	CGA stability Separation efficiency and selectivity
Separation of aphron and liquid phase	Separation time	Separation efficiency and selectivity
Aphron phase collection	This step does not have any influence on the process performance	
Surfactant removal	Use of membranes or specific solvents	Process recovery

into the liquid phase. Moreover, an increase in whey volume resulted in lower separation, too. This could be partly explained in terms of CGA capacity, but also in terms of protein competition, i.e. at high protein load, an increase in protein/protein interaction reduces selectivity.

Further investigations into the mechanism of the separation with CTAB generated CGA and whey (Fuda and Jauregi, 2006) confirmed that CGA when generated with ionic surfactants can act as ion exchangers and hence selectivity can be manipulated by the type of surfactant (cationic or anionic), pH, and ionic strength of the solution. First contact between proteins and surfactant molecules in CGA occurs by electrostatic interactions as promoting hydrophobic interactions (e.g. by increasing the ionic strength of the solution) resulted in poor recoveries. Strength of interaction between protein and surfactant molecules is dependent on the conformational features of the protein and the extent to which these are affected upon interaction with the surfactant. Selectivity is enhanced by the formation of aggregates and their subsequent flotation into the aphron phase.

Other studies have used nonionic surfactants for the recovery of proteins. Jarudilokkul et al. (2004) showed that hydrophobic interactions are important when nonionic surfactants are used for the recovery of proteins such as lysozyme and β-casein. Noble et al. (1998) applied CGA generated with Triton X-100 to a range of proteins and found that highest recoveries (74%) were obtained with the most hydrophobic protein (thaumatin). Similarly, Fernandes et al. (2002) found that the enzyme with a hydrophobic fusion tag was recovered with CGA generated with the nonionic detergent Triton X-114 at higher yield than the wild-type cutinase.

10.2.7 APPLICATION OF CGA TO THE RECOVERY OF POLYPHENOLS FROM WINE-MAKING WASTE

CGA can also be applied for the purification of phenolic extracts in order to separate phenolic compounds from nonphenolic compounds (such as sugars and minerals), but also to fractionate different classes of phenolic compounds. Besides, CGA could be applied for the direct recovery of phenolic compounds from wastewaters, i.e. those generated in the boiling cork process (Benitez et al., 2005).

To further support this application, some literature works have been reported about the interactions between surfactants and phenolic compounds. For example, specific surfactants in micellar form (e.g. Tween 20, Tween 80, Citrem, sodium cholate) were able to:

1. solubilize or precipitate specific phenolic compounds (Löf et al., 2011),
2. alter their partitioning in oil-in-water emulsions (Richards et al., 2002; Sørensen et al., 2008),
3. enable phenolic compounds analytical determination exploiting different affinities (Wang et al., 2007),
4. protect phenolic compounds from oxidation (Lin et al., 2007),
5. solubilize aromatic compounds (Yoshida and Moroi, 2000; Wei et al., 2012),
6. improve phenolic compounds efficiency in topical formulations (Scognamiglio et al., 2013; Yutani et al., 2012).

The surfactant to be used for the generation of CGA should be able to adsorb and interact with different phenolic compounds. Thereby, its selection is driven by molecular structure and polarity. For instance, CGA generated from a cationic surfactant (CTAB) was used to recover gallic acid from aqueous solutions (Spigno et al., 2010), obtaining maximum recovery (close to 80%) at conditions that promote electrostatic interactions between the cationic surfactant and the anionic form of gallic acid, that is at pH > pK_a = 3.4. However, in order to preserve the antioxidant capacity of the phenolic acid, pH should not be over 6. An increased ionic strength, which could be the case of real extracts containing natural salts, hindered electrostatic interactions, which led to decreased RE%.

CGA have also been applied to real crude phenolic extracts obtained with aqueous ethanol solvent extraction from waste red grape pomace and skins (Dahmoune et al., 2013; Spigno et al., 2014). In these trials, carried out in a batch flotation column, the influence of surfactant type (CTAB and the food-grad nonionic Tween 20) and feed concentration was investigated. In view of industrial implementation, the crude extract was preconcentrated to recover the solvent (ethanol), and then diluted with water to reach a predefined total phenols concentration before mixing with CGA or a surfactant solution.

When CTAB was used, the natural extract showed a stabilizing effect on CGA (reduced drainage rate), leading to higher recovery. Anthocyanins revealed more affinity for the CGA than other phenolics, probably due to the fact that they are polarized because of their high electron density in the aromatic ring. When the separation was carried out at pH 2, only a slight reduction in anthocyanins recovery (but not in that of other phenolics) was observed. This confirmed that separation was driven by electrostatic and hydrophobic interactions.

When Tween 20 was used, recovery decreased for both anthocyanins and other phenolics, confirming that electrostatic interactions enhanced the partitioning. Maximum recovery (78% total phenols and 76% total anthocyanins) and SF (4 and 3 for total phenols and anthocyanins, respectively) were obtained at the highest volumetric ratio (22) and highest feed concentration (5 g/L of gallic acid equivalents). However, a too concentrated feed led to the formation of insoluble aggregates in the aphron phase. Tween 20, interestingly, led to a lower loss of antioxidant capacity than CTAB.

10.2.8 GENERAL REMARKS

In general, the use of CGA as a flotation technique is advantageous when compared with conventional precipitate flotation and sublation. CGA are more efficient because of the large interfacial area that results in smaller operating times and higher removal yields. CGA can selectively separate compounds from a mixture as shown in the examples of protein recovery. Indeed, selectivity is enhanced when the separation is driven by electrostatic interactions between ionized colligend and oppositely charged surfactant. Hydrophobic interactions are also an important driving force for the partitioning of compounds in the CGA phase; however, they lead to high yield, but less specific interactions as referred in the case of polyphenols from wine waste extracts.

Another advantage of CGA could be the lower operating cost compared with other conventional methods. An economical evaluation of the astaxanthin separation from a fermentation broth (Dermiki, 2009) has shown that the operating cost for CGA was lower compared with supercritical carbon dioxide and solvent extraction. Nevertheless, the purchase cost was higher for CGA if a surfactant removal step was added. Finally, another advantage of CGA compared with adsorption methods (e.g. ion exchange adsorption) is that it can handle particulate feedstock. This fact enables the integration of the solids removal and recovery steps leading to process simplification and reduction of overall cost.

10.3 **ULTRASOUND-ASSISTED CRYSTALLIZATION**

10.3.1 **INTRODUCTION**

The possibility of growing pure crystals of a controlled size has made crystallization one of the most important purification and separation techniques in the food industries, particularly in the sugar and dairy sectors. Waste streams from the food industry are often subjected to microbial fermentation or other (e.g. membrane or emerging) processes in order to isolate metabolites or biomolecules of interest (Galanakis et al., 2010b, e, 2012b, 2013b, 2014, 2015b; Patsioura et al., 2011; Rahmanian et al., 2014; Deng et al., 2015; Roselló-Soto et al., 2015). Crystallization is an operation used for separation purposes or the improvement of productivity during fermentation processes. To quote an example, calcium lactate could be efficiently produced during lactic acid fermentation in broken rice and then recovered by crystallization (Nakano et al., 2012). The latter operation can be carried out either as an ex situ or in situ technique, whereas the second can prevent inhibition of microbes in some cases of fermentation (Xu and Xu, 2014). On the other hand, it is a complex process that is difficult to control (e.g. to obtain uniform crystal sizes) due to nonuniform nucleation, inefficient cooling caused by surface encrustation of cooling coils, and nonuniform crystal growth due to uneven mixing (Deora et al., 2013). Therefore, it is important to control the nucleation phenomena and progress crystallization in a repeatable and predictable manner.

It is well known that ultrasound can enhance the reaction rate and product yield via facilitating mass transfer and reactant diffusion. The main mechanism of action in power ultrasound is cavitation, a phenomenon that can be either stable or transient. Stable cavitation is associated with small bubbles dissolved in a liquid, while transient cavitation occurs when the bubble size changes quickly, and collapses and locally produces very high pressure (100 MPa) and temperature (5000 K) (Deora et al., 2013). Ultrasound has been successfully used to enhance the rate of crystal growth as well as to induce more nuclei through cavitation and acoustic streaming (Chen and Huang, 2008; Mason, 2007). The application of ultrasound energy to control crystallization is referred as sonocrystallization. The key benefits of ultrasound application during crystallization and its mode of action are summarized in Fig. 10.5.

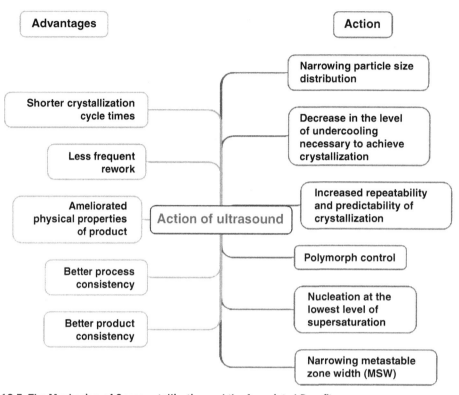

FIGURE 10.5 The Mechanism of Sonocrystallization and the Associated Benefits

Cavitation is effective to induce nucleation and there is evidence of dramatic improvements in reproducibility obtained through sononucleation. In particular, ultrasound decreases the apparent order of the primary nucleation rate and increases the rate of appearance of the solid. Furthermore, it allows the size distribution of the crystals to be decreased. In conventional crystallization, the conditions that enable control over the crystallization process and crystal properties (i.e. operating at low supersaturation levels inside the metastable zone width) cannot offer the maximum process performance (Zamanipoor and Mancera, 2014). Ultrasound provides a means of reducing the induction time and metastable zone width and thus it increases dramatically the nucleation rate.

10.3.2 SONOCRYSTALLIZATION OF LACTOSE FROM WHEY

Whey, the liquid fraction obtained from the cheese processing industry and a rich source of lactose, has drawn considerable interest from the food industries and from pharmaceutical sectors. At industrial-scale operation, lactose is mainly manufactured by cooling crystallization of whey permeate, as the lactose solubility/temperature curve is not flat and its solubility at the ambient temperature is low (Vu et al., 2006). However, this process is time consuming and slow. Fortunately, recent studies show that

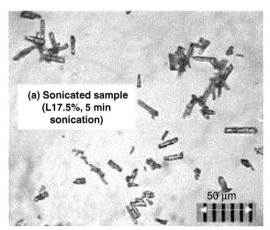 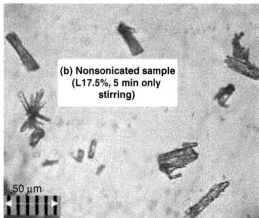

(a) Sonicated sample (L17.5%, 5 min sonication)

(b) Nonsonicated sample (L17.5%, 5 min only stirring)

FIGURE 10.6 Lactose Crystals Observed Under the Microscope (40× Magnification)

(a) Sonicated sample (L17.5%, 5 min sonication); (b) nonsonicated sample (L17.5%, 5 min only stirring).

Adapted from Bund and Pandit (2007c) with permission

ultrasound has a potential application in this area since it is able to improve productivity and quality of lactose crystallized from cheese whey.

Bund and Pandit (2007b) employed sonocrystallization to speed-up lactose recovery from heat-induced deproteinated and concentrated cottage cheese whey using ethanol as an antisolvent. These authors reported a lactose recovery of 91.5% after 5 min of sonication time, from a reconstituted lactose solution (17.5% w/v, pH 4.2, no protein). This is much higher and orders of magnitude faster compared with conventional methods. The latter involve high concentrations of antisolvent with stirring or heating for 1–12 h (Bund and Pandit, 2007a). Interestingly, even shorter sonication time (e.g. 1 min) was found to be effective enough to produce significant changes in the nucleation rate and yield (Zamanipoor et al., 2013).

A direct consequence of the rapid crystallization is a better uniformity and smaller crystal size distribution of lactose samples (see Fig. 10.6) (Bund and Pandit, 2007c). A narrow crystal size distribution during crystallization in the presence of ultrasound, compared with *in absentia* continuous, is confirmed by Zisu et al. (2014). The generation of a more symmetrical cubic-structured crystal of sodium chloride in the presence of ultrasound has also been observed in a recent study (Lee et al., 2014). It is worth mentioning that obtaining a desired uniform crystal size distribution is a challenge due to the variable nucleation and crystal growth rate inherent to conventional crystallization approaches, e.g. using high shear mixers.

10.3.3 SONOCRYSTALLIZATION OF AMINO ACIDS

Ultrasound application towards crystallization of amino acids and proteins has also been explored. Crystallization improvements have been reported for aspartame, adipic acid, L-leucine, L-phenylalanine, and L-histidine (Mccausland, 2003). For example, the application of ultrasound decreases the metastable zone of L-leucine from around 22.5 K to around 9.8 K, after sonication at

20 kHz in short bursts during the cooling stage. As another advantage McCausland and Cains (2004b) reported that ultrasound application increased the crystal size. For example, the 90th percentile particle size for L-histidine increases from 166.3 μm to 370.8 μm, following ultrasound application. The key advantage of an increased particle size in this case is the ease of crystal separation. Such improved results for sonocrystallization of amino acids could be utilized to obtain amino acids from many food waste sources. For example, the discarded sheep visceral mass (including stomach and large and small intestines) is a rich source of essential amino acids (Bhaskar et al., 2007). Ultrasound may interact with the protein macromolecule and alter the primary, secondary, tertiary, or quaternary structures of proteins, which could either increase or decrease protein solubility. Thus, the application of ultrasound towards crystallization of proteins is much more challenging as it is difficult to overcome activity loss (McCausland and Cains, 2004a).

10.3.4 SCALE-UP ASPECTS

The most successful studies on ultrasound applications have been conducted in batch food processes. The ability to achieve desired effects in continuous operation needs to be confirmed in many cases. Fortunately, sonocrystallization of lactose in concentrated whey has been shown to be scalable up to 12 L/min flow rates and applied energy densities of ≥3 J/mL (Zisu et al., 2014). A two-stage application has been recommended in this case to stimulate nuclei formation towards the metastable limit. Luque de Castro and Priego-Capote (2007) identified that the effectiveness of sonocrystallization is influenced by the nature and location of the ultrasonic source as well as the properties of the medium to which it is applied. Thus, depending on the nature of material of interest, the process parameters for ultrasound application should be explored and adopted. In addition, extensive research to examine the complex mechanisms involved in sonocrystallization and the parameters governing scale-up need focus (Deora et al., 2013). Numerical modeling and molecular dynamics simulations are likely to provide insights into the process dynamics and thus facilitate better control. Finally, the possibility of integrating ultrasound-assisted extraction and crystallization in a single process line and this should be further explored.

10.4 PRESSURIZED MICROWAVE EXTRACTION
10.4.1 INTRODUCTION

Microwave-assisted extraction (MAE) has drawn significant attention in the extraction of phytochemicals because it is cheap, fast, and economical with minimal solvent usage (Camel, 2000; Michel et al., 2011). Microwave is a noncontact heat source, which can make heating more selective and effective, and help to accelerate energy transfer. This technique has a promising future and has the potential to be developed further for the extraction of essential oils, polyphenols, flavonoids, pigments, and aromas from plant resources (Chan et al., 2011; Li et al., 2013). MAE is efficient for the separation of hydrophilic or polar compounds from macromolecules, as heating depends on microwave absorbing capacity of the solvent, which is high for water and alcohol. The application of microwave irradiation can be performed either at atmospheric pressure (open MAE) or under controlled pressure (pressurized MAE) during sample extraction (Camel, 2000). If extraction is carried

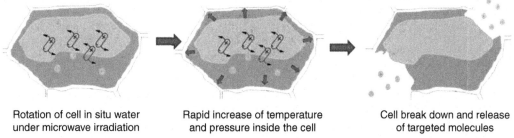

| Rotation of cell in situ water under microwave irradiation | Rapid increase of temperature and pressure inside the cell | Cell break down and release of targeted molecules |

FIGURE 10.7 Extraction Mechanism of Bioactive Molecules by Pressurized Microwave Assisted Extraction

Picture obtained with permission from Li et al. (2013)

out without any solvent, PMAE is referred to as pressurized solvent-free microwave extraction (PSFME). Extraction is carried out in a sealed vessel under controlled pressure and temperature, which allows fast and efficient extraction. Inside the extraction vessel, the process is controlled in such a way that it would not exceed the working pressure of the vessel, while the temperature can be regulated above the normal boiling point of the extraction solvent. The increase in temperature and pressure accelerates MAE due to the ability of extraction solvent to absorb microwave energy (Michel et al., 2011; Routray and Orsat, 2012).

Microwaves are nonionizing electromagnetic waves, which consist of electric and magnetic fields, which oscillate perpendicularly to each other at frequencies ranging from 0.3 to 300 GHz. During PSFME, plant materials are kept inside a closed vessel with little or no solvent and then subjected to microwave treatment. A typical solvent for polyphenols recovery could be hydroethanolic mixtures (Galanakis et al., 2010d, 2013a; Tsakona et al., 2012). Microwaves in combination with increased pressure can penetrate plant materials and interact fast with water to produce heat. The latter acts directly on the molecules by ionic conduction and dipole rotation. The in situ water of fresh plant material heats up instantly beyond the boiling point and consecutively the pressure inside the plant cells increases, too. This fact leads to cell disruption and release of the target molecules (Chan et al., 2011; Li et al., 2013) (Figs 10.7 and 10.8).

10.4.2 APPLICATIONS, SAFETY, AND ENVIRONMENTAL IMPACT

PSFME has several advantages in the form of reduced extraction time, no solvent usage, higher extraction yield, and higher bioactivity with minimal energy requirement. For instance, Li et al. (2013) described that PSFME needs 0.5 kWh of energy, whereas conventional methods need 4.5 kWh. Regarding the carbon footprints in order to obtain 1 kWh of electricity from coal or fuel, 800 g of CO_2 is emitted. Hence, SFME uses up 200 g of CO_2/g of essential oil, while conventional methods use 3600 g of CO_2/g essential oil. PSFME has certain limitations, like, it strongly depends on the moisture content of the plant materials and hence, fresh plant material with moisture content of 70–80% is required during extraction. There is also a potential explosion risk as a result of using a pressurized closed vessel. In addition, it will be difficult to scale up for industrial applications. Besides, the operation of the instrument requires skilled workers with sufficient training.

PSFME has increased the recovery yield of pectin from orange albedo more 10-fold compared with conventional extractions, while pressure in combination with superheated ethanol has been reported

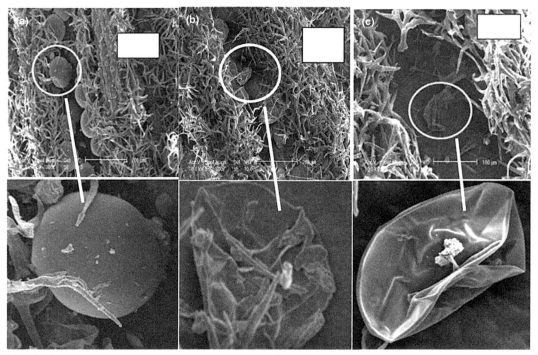

FIGURE 10.8 The Structure of Lavender Flowers Observed by SEM

(a) Before extraction, (b) after microwave extraction, and (c) after conventional extraction.

Picture obtained with permission from Li et al. (2013)

to extract phenols very efficiently from olive mill wastewater, too (Galanakis, 2012). Table 10.3 provides a comparative analysis and advantages for using PSFME for bioactive compounds compared with other extraction methods. From the table, it is clear that PSFME provides higher yield compared with conventional extraction methods. It should also be noted that the time taken for the extraction is significantly reduced. For instance, gensonosides obtained by PSFME are 43.3 mg/g with an extraction time of 10 min, whereas for other extraction methods, namely Soxhlet, ultrasound, and heat reflux, the yield and extraction time were 37.1, 35.8, and 37.8 mg/g and 2, 1, and 2 h, respectively. Also, the antioxidant activity of the extract obtained by PSFME is significantly higher (90%) compared with other extraction methods (25–75%) (Wang et al., 2008). The essential oils obtained by SFME are more valuable than oils obtained by HD because of higher oxygenated compounds and high antioxidant activity.

PSFME has been successfully used to extract essential oils, antioxidants, and bioactive compounds in concentrated form. This is a promising method, which requires no solvent since residual water in the plant materials is sufficient to be used as an extraction solvent. The final products are free of solvent residue and impurities and could be obtained in a short span of time. Very little data on the safety of microwaves has been reported and this aspect needs to be investigated further. In conclusion, the PSFME system is considered a promising technique for the extraction of bioactive compounds from plant materials.

Table 10.3 Comparison of Pressurized Solvent-Free Microwave Extraction of Bioactive Components with Other Extraction Methods

Raw Material	Bioactivity	Extraction Method	Extraction Time	Results	References
Panax ginseng root	DPPH radical scavenging activity (%)	PSFME	250 s	90	
		Pressing	NR	70	
		Maceration	10 min	75	
		PLE	25 min	25	
	Gensenosides (mg/g DW)	PSFME	10 min	43.3	Wang et al. (2008)
		Soxhlet	2 h	37.1	
		UAE	1 h	35.8	
		Heat reflux	2 h	37.8	
Pine saw dust	Total phenolic content (mg GAE/g)	PSFME	40 m	74.6	Meullemiestre et al. (2014)
		HD	120 min	54.1	
		UAE	120 min	58.6	
	DPPH radical scavenging activities (IC_{50} values μg/mL)	HD	120 min	123	
		UAE	120 min	59.8	
		PSFME	40 min	15.4	
Sweet cherries	Cyanidin-3-O-glucoside (mg/100 g FW)	PSFME	180 s	30	Grigoras et al. (2012)
	Cyanidin-3-O-rutinoside (mg/100 g FW)		180 s	60	
Ajowan	Thymol (g/kg)	HD	1 h	35.4	Lucchesi et al. (2004a)
Mint	Carvone (%)	PSFME	30 min	64.9	Lucchesi et al. (2004b)
		HD	270 min	52.3	
		PSFME	8 h	60.3	
Schisandra chinensis	DPPH radical scavenging activities (IC_{50} values μg/mL)	SFME	30 min	31.9	Chun-hui Ma et al. (2012)
		HD	3 h	49.5	
		SD	4 h	85.8	
Dryopetris fragrans	Sesquiterpene hydrocarbons (relative peak area %)	PSFME	34 min	24.4	Li et al. (2012)
		HD	5 h	20.1	
Oregano	Oxygenated compounds (mg/mL)	PSFME	5 min	979.1	Bayramoglu et al. (2008)
		SD	46 min	678.3	
Rosemary	Oxygenated compounds (%)	HD	90 min	33.9	Filly et al. (2014)
		PSFME	30 min	36.2	

PSFME, pressurized solvent-free microwave extraction; UAE, ultrasonic-assisted extraction; PLE, pressurized liquid extraction; HD, hydrodistillation; SD, steam distillation; NA, not applicable; NR, not reported; FW, fresh weight.

REFERENCES

Arabloo, M., Shahri, M.P., 2014. Experimental studies on stability and viscoplastic modelling of colloidal gas aphrons (CGA) based drilling fluids. J. Petrol. Sci. Eng. 113, 8–22.

Bayramoglu, B., Sahin, S., Sumnu, G., 2008. Solvent-free microwave extraction of essential oil from oregano. J. Food Eng. 88, 535–540.

Benitez, F.J., Real, F.J., Acero, J.L., Leal, A.I., Cotilla, S., 2005. Oxidation of acetovanillone by photochemical processes and hydroxyl radicals. J. Environ. Sci. Health A 40 (12), 2153–2169.

Bhaskar, N., Modi, V.K., Govindaraju, K., Radha, C., Lalitha, R.G., 2007. Utilization of meat industry by products: protein hydrolysate from sheep visceral mass. Bioresour. Technol. 98 (2), 388–394.

Bund, R.K., Pandit, A.B., 2007a. Rapid lactose recovery from buffalo whey by use of anti-solvent, ethanol, HYPERLINK "http://www-scopus-com.ezproxy.lib.monash.edu.au/source/sourceInfo.url?sourceId=20586&origin=resultslist" J. Food Eng. 82 (3), 333–341.

Bund, R.K., Pandit, A.B., 2007b. Rapid lactose recovery from paneer whey using sonocrystallisation: a process optimization. Chem. Eng. Process. 46 (9), 846–850.

Bund, R.K., Pandit, A.B., 2007c. Sonocrystallisation: effect on lactose recovery and crystal habit. Ultrason. Sonochem. 14 (2), 143–152.

Camel, V., 2000. Microwave-assisted solvent extraction of environmental samples. Trends Anal. Chem. 19, 229–248.

Chan, C., Yusoff, R., Ngoh, G., Kung, F.B., 2011. Microwave-assisted extractions of active ingredients from plants. J. Chromatogr. A 1218, 6213–6225.

Chen, W.-S., Huang, G.-C., 2008. Ultrasound-assisted crystallisation of high purity of 2,4-dinitrotoluene from spent acid. Ultrason. Sonochem. 15 (5), 909–915.

Chun-hui Ma, Yang, L., Zu, Y., Liu, T., 2012. Optimization of conditions of solvent-free microwave extraction and study on antioxidant capacity of essential oil from *Schisandra chinensis* (Turcz.) Baill. Food Chem. 134, 2532–2539.

Dahmoune, F., Madani, K., Jauregi, P., De Faveri, D.M., Spigno, G., 2013. Fractionation of a red grape marc extract by colloidal gas aphrons. Chem. Eng. Trans. 32, 1903–1908.

Deng, Q., Zinoviadou, K.G., Galanakis, C.M., Orlien, V., Grimi, N., Vorobiev, E., Lebovka, N., Barba, F.J., 2015. The effects of conventional and non-conventional processing on glucosinolates and its derived forms, isothiocyanates: extraction, degradation and applications. Food Eng. Rev. DOI: 10.1007/s12393-014-9104-9.

Deora, N.S., Misra, N.N., Deswal, A., Mishra, H.N., Cullen, P.J., Tiwari, B.K., 2013. Ultrasound for improved crystallisation in food processing. Food Eng. Rev. 5 (1), 36–44.

Dermiki, M., 2009. Recovery of astaxanthin using colloidal gas aphrons. PhD thesis, The University of Reading, Reading, UK.

Dermiki, M., Gordon, M.H., Jauregi, P., 2009. Recovery of astaxanthin using colloidal gas aphrons (CGA): a mechanistic study. Sep. Purif. Technol. 65, 54–64.

Dermiki, M., Bourquin, A., Jauregi, P., 2010. Separation of astaxanthin from cells of *Phaffia rhodozyma* using colloidal gas aphrons (CGA) in a flotation column. Biotechnol. Prog. 26 (2), 477–487.

Fernandes, S., Hatti-Kaul, R., Mattiasson, B., 2002. Selective recovery of lactate dehydrogenase using affinity foam. Biotechnol. Bioeng. 79, 472–480.

Filly, A., Fernandez, X., Minuti, M., Visinoni, F., Cravotto, G., Chemat, F., 2014. Solvent-free microwave extraction of essential oil from aromatic herbs: from laboratory to pilot and industrial scale. Food Chem. 150, 193–198.

Fuda, E., Jauregi, P., 2006. An insight into the mechanism of protein separation by colloidal gas aphrons (CGA) generated from ionic surfactants. J. Chromatogr. B 843, 317–326.

Fuda, E., 2004. Selective separation of whey proteins using colloidal gas aphrons (CGA) generated with ionic surfactants. PhD thesis, The University of Reading, Reading, UK.

Galanakis, C.M., 2011. Olive fruit and dietary fibers: components, recovery and applications. Trends Food Sci. Technol. 22, 175–184.

Galanakis, C.M., 2012. Recovery of high added value components from food wastes: conventional, emerging technologies and commercialized applications. Trends Food Sci. Technol. 26, 68–87.

Galanakis, C.M., 2013. Emerging technologies for the production of nutraceuticals from agricultural by-products: a viewpoint of opportunities and challenges. Food Bioprod. Process. 91, 575–579.

Galanakis, C.M., 2015. Separation of functional macromolecules and micromolecules: from ultrafiltration to the border of nanofiltration. Trends Food Sci. Technol. 42, 44–63.

Galanakis, C.M., Chasiotis, S., Botsaris, G., Gekas, V., 2014. Separation and recovery of proteins and sugars from Halloumi cheese whey. Food Res. Int. 65, 477–483.

Galanakis, C.M., Fountoulis, G., Gekas, V., 2012a. Nanofiltration of brackish groundwater by using a polypiperazine membrane. Desalination 286, 277–284.

Galanakis, C.M., Goulas, V., Tsakona, S., Manganaris, G.A., Gekas, V., 2013a. A knowledge base for the recovery of natural phenols with different solvents. Int. J. Food Prop. 16, 382–396.

Galanakis, C.M., Kanellaki, M., Koutinas, A.A., Bekatorou, A., Lycourghiotis, A., Kordoulis, C.H., 2012b. Effect of pressure and temperature on alcoholic fermentation by *Saccharomyces cerevisiae* immobilized on γ-alumina pellets. Bioresour. Technol. 114, 492–498.

Galanakis, C.M., Kotanidis, A., Dianellou, M., Gekas, V., 2015a. Phenolic content and antioxidant capacity of Cypriot wines. Czech J. Food Sci. 33, 126–136.

Galanakis, C.M., Markouli, E., Gekas, V., 2013b. Fractionation and recovery of different phenolic classes from winery sludge via membrane filtration. Sep. Purif. Technol. 107, 245–251.

Galanakis, C.M., Patsioura, A., Gekas, V., 2015b. Enzyme kinetics modeling as a tool to optimize food biotechnology applications: a pragmatic approach based on amylolytic enzymes. Crit. Rev. Food Sci. Technol. 55, 1758–1770.

Galanakis, C.M., Schieber, A., 2014. Editorial. Special issue on recovery and utilization of valuable compounds from food processing by-products. Food Res. Int. 65, 230–299.

Galanakis, C.M., Tornberg, E., Gekas, V., 2010a. A study of the recovery of the dietary fibres from olive mill wastewater and the gelling ability of the soluble fibre fraction. LWT – Food Sci. Technol. 43, 1009–1017.

Galanakis, C.M., Tornberg, E., Gekas, V., 2010b. Clarification of high-added value products from olive mill wastewater. J. Food Eng. 99, 190–197.

Galanakis, C.M., Tornberg, E., Gekas, V., 2010c. Dietary fiber suspensions from olive mill wastewater as potential fat replacements in meatballs. LWT – Food Sci. Technol. 43, 1018–1025.

Galanakis, C.M., Tornberg, E., Gekas, V., 2010d. Recovery and preservation of phenols from olive waste in ethanolic extracts. J. Chem. Technol. Biotechnol. 85, 1148–1155.

Galanakis, C.M., Tornberg, E., Gekas, V., 2010e. The effect of heat processing on the functional properties of pectin contained in olive mill wastewater. LWT – Food Sci. Technol. 43, 1001–1008.

Grigoras, C.G., Destandau, E., Zubrzycki, S., Elfakir, C., 2012. Sweet cherries anthocyanins: an environmental friendly extraction and purification method. Sep. Purif. Technol. 100, 51–58.

Hashim, M.A., Sengupta, B., 1998. The application of colloidal gas aphrons in the recovery of fine cellulose fibres from paper mill wastewater. Bioresour. Technol. 64, 199–204.

Hashim, M.A., Sengupta, B., Kumar, S.V., 1995a. Clarification of yeast by colloidal gas aphrons. Biotechnol. Tech. 9, 403–408.

Hashim, M.A., Sengupta, B., Subramaniam, M.B., 1995b. Investigations on the flotation of yeast cells by colloidal gas aphron(CGA) dispersions. Bioseparation 5 (3), 167–173.

Hashim, M.A., Sengupta, B., Kumar, S.V., Lim, R., Lim, S.E., Tan, C.C., 1998. Effect of air to solid ratio in the clarification of yeast by colloidal gas aphrons. J. Chem. Technol. Biotechnol. 71, 335–339.

Hashim, M.A., Mukhopadhyay, S., Sengupta, B., Sahu, J.N., 2011. Application of colloidal gas aphrons for pollution remediation. J. Chem. Technol. Biotechnol. 87 (3), 305–324.

Heng, W.W., Xiong, L.W., Ramanan, R.N., Hong, T.L., Kong, K.W., Galanakis, C.M., Prasad, K.N., 2015. Two level factorial design for the optimization of phenolics and flavonoids recovery from palm kernel by-product. Ind. Crop. Prod. 63, 238–248.

Jarudilokkul, S., Rungphetcharat, K., Boonamnuayvitaya, V., 2004. Protein separation by colloidal gas aphrons using nonionic surfactant. Sep. Purif. Technol. 35, 23–29.

Jauregi, P., 1997. Colloidal gas aphrons (CGA): a novel approach to protein recovery. PhD thesis, The University of Reading, Reading, UK.

Jauregi, P., Dermiki, M., 2010. Separation of value-added bioproducts by colloidal gas aphrons. In: Rizvi, S.S.H. (Ed.), Separation, Extraction, and Concentration Processes in the Food, Beverage and Nutraceutical Industries. Woodhead Publishing Ltd, Cambridge, pp. 284–313.

Jauregi, P., Gilmour, S., Varley, J., 1997. Characterisation of colloidal gas aphrons for subsequent use for protein recovery. Chem. Eng. J. 65, 1–11.

Jauregi, P., Mitchell, G.R., Varley, J., 2000. Colloidal gas aphrons (CGA): dispersion and structural features. AIChE J. 46, 24–36.

Larmignat, S., Vanderpool, D., Lai, H.K., Pilon, L., 2008. Rheology of colloidal gas aphrons (microfoams). Colloid. Surface. A 322, 199–210.

Lee, J., Ashokkumar, M., Kentish, S.E., 2014. Influence of mixing and ultrasound frequency on antisolvent crystallisation of sodium chloride. Ultrason. Sonochem. 21 (1), 60–68.

Li Y., Fabiano-Tixier A.S., Vian M.A., Chemat, F., 2013. Solvent-free microwave extraction of bioactive compounds provides a tool for green analytical chemistry. Trend Anal. Chem. 47, 1–11.

Li, X.J., Wang, M., Luo, M., Li, C.Y., Zu, Y.G., Mu, P.S., Fu, Y.J., 2012. Solvent-free microwave extraction of essential oil from *Dryopteris fragrans* and evaluation of antioxidant activity. Food Chem. 133, 437–444.

Lin, Q.L., Wang, J., Qin, D., Björn, B., 2007. Influence of amphiphilic structures on the stability of polyphenols with different hydrophobicity. Sci. China Ser. B 50, 121–126.

Lucchesi, M.E., Chemat, F., Smadj, J., 2004a. An original solvent free microwave extraction of essential oils from spices. Flav. Fragr. J. 19, 134–138, 2004.

Lucchesi, M.E., Chemat, F., Smadja, J., 2004b. Solvent-free microwave extraction of essential oil from aromatic herbs: comparison with conventional hydro-distillation. J. Chromatogr. A 1043, 323–327.

Löf, D., Schillén, K., Nilsson, L., 2011. Flavonoids: precipitation kinetics and interaction with surfactant micelles. J. Food Sci. 76 (3), N35–N39.

Luque de Castro, M.D., Priego-Capote, F., 2007. Ultrasound-assisted crystallisation (sonocrystallisation). Ultrason. Sonochem. 14 (6), 717–724.

Mason, T.J., 2007. Sonochemistry and the environment: providing a "green" link between chemistry, physics and engineering. Ultrason. Sonochem. 14 (4), 476–483.

Mccausland, L.J., 2003. Production of crystalline materials by using high intensity ultrasound. Espacenet. Accentus PLC.

McCausland, L.J., Cains, P.W., 2004a. Power ultrasound: a means to promote and control crystallisation in biotechnology. Biotechnol. Gen. Eng. Rev. 21 (1), 3–10.

McCausland, L.J., Cains, P.W., 2004b. Power ultrasound: a means to promote and control crystallisation in biotechnology. Biotechnol. Gen. Eng. Rev. 21, 3–10.

Meullemiestre, A., Kamal, I., Maache-Rezzoug, Z., Chemat, F., Rezzoug, R.A., 2014. Antioxidant activity and total phenolic content of oils extracted from *Pinus pinaster* sawdust waste. Screening of different innovative isolation techniques. Waste Biomass Valor. 5, 283–292.

Michel, T., Destandau, E., Elfakir, C, 2011. Evaluation of a simple and promising method for extraction of antioxidants from sea buckthorn (*Hippophaë rhamnoides* L.) berries: pressurised solvent-free microwave assisted extraction. Food Chem. 126, 1380–1386.

Nakano, S., Ugwu, C.U., Tokiwa, Y., 2012. Efficient production of D-(−)-lactic acid from broken rice by *Lactobacillus delbrueckii* using Ca(OH)$_2$ as a neutralizing agent. Bioresour. Technol. 104, 791–794.

Noble, M., Brown, A., Jauregi, P., Kaul, A., Varley, J., 1998. Protein recovery using gas–liquid dispersions. J. Chromatogr. B 711, 31–43.

Patsioura, A., Galanakis, C.M., Gekas, V., 2011. Ultrafiltration optimization for the recovery of β-glucan from oat mill waste. J. Membr. Sci. 373, 53–63.

Pinfold, T.A., 1972. Precipitate flotation. In: Lemlich, R. (Ed.), Adsorptive Bubble Separation Techniques. Academic Press, New York, pp. 74–90.

Rahmanian, N., Jafari, S.M., Galanakis, C.M., 2014. Recovery and removal of phenolic compounds from olive mill wastewater. J. Am. Oil Chem. Soc. 91, 1–18.

Richards, M.P., Chaiyasit, W., McClements, D.J., Decker, E.A., 2002. Ability of surfactant micelles to alter the partitioning of phenolic antioxidants in oil-in-water emulsions. J. Agric. Food Chem. 50, 1254–1259.

Roselló-Soto, E., Barba, F.J., Parniakov, O., Galanakis, C.M., Grimi, N., Lebovka, N., Vorobiev, E., 2015. High voltage electrical discharges, pulsed electric field and ultrasounds assisted extraction of protein and phenolic compounds from olive kernel. Food Bioprocess Technol. 8, 885–894.

Routray, W., Orsat, V., 2012. Microwave-assisted extraction of flavonoids: a review. Food Bioprocess Technol. 5, 409–424.

Roy, D., Kommalapati, R.R., Valsaraj, K.T., Constant, W.D., 1995. Soil flushing of residual transmission fluid: application of colloidal gas aphron suspensions and conventional surfactant solutions. Water Res. 29, 589–595.

Save, S.V., Pangarkar, V.G., 1995. Harvesting of *Saccharomyces cerevisiae* using colloidal gas aphrons. J. Chem. Technol. Biotechnol. 62, 192–199.

Scognamiglio, I., De Stefano, D., Campani, V., Mayol, L., Carnuccio, R., Fabbroccini, G., Ayala, F., La Rotonda, M.I., DeRosa, G., 2013. Nanocarriers for topical administration of resveratrol: a comparative study. Int. J. Pharm. 440, 179–187.

Sebba, F., 1987. Foams and Biliquid Foams – Aphrons. John Wiley & Sons Ltd, Chichester.

Sørensen, A.-D.M., Haahr, A.-M., Becker, E.M., Skibsted, L.H., Bergenståhl, B., Nilsson, L., Jacobsen, C., 2008. Interactions between iron, phenolic compounds, emulsifiers, and pH inomega-3-enriched oil-in-water emulsions. J. Agric. Food Chem. 56, 1740–1750.

Spigno, G., Amendola, D., Dahmoune, F., Jauregi, P., 2014. Colloidal gas aphrons based separation process for the purification and fractionation of natural phenolic extracts. Food Bioprod. Process. Available from: http://dx.doi.org/10.1016/j.fbp.2014.06.002.>

Spigno, G., Dermiki, M., Pastori, C., Casanova, F., Jauregi, P., 2010. Recovery of gallic acid with colloidal gas aphrons generated from a cationic surfactant. Sep. Purif. Technol. 71, 56–62.

Subramaniam, M.B., Blakebrough, N., Hashim, M.A., 1990. Clarification of suspensions by colloidal gas aphrons. J. Chem. Technol. Biotechnol. 48, 41–60.

Tsakona, S., Galanakis, C.M., Gekas, V., 2012. Hydro-ethanolic mixtures for the recovery of phenols from Mediterranean plant materials. Food Bioprocess Technol. 5, 1384–1393.

Vu, T.T.L., Durham, R.J., Hourigan, J.A., Sleigh, R.W., 2006. Dynamic modelling optimisation and control of lactose crystallisations: comparison of process alternatives. Sep. Purif. Technol. 48 (2), 159–166.

Wang, X.-K., He, Y.-Z., Qian, L.-L., 2007. Determination of polyphenol components in herbal medicines by micellar electrokinetic capillary chromatography with Tween 20. Talanta 74, 1–6.

Wang, Y., You, J., Yu, Y., Qu, C., Zhang, H., Ding, L., Zhang, H., Li, H, 2008. Analysis of ginsenosides in *Panax ginseng* in high pressure microwave-assisted extraction. Food Chem. 110, 161–167.

Wei, J., Huang, G., Zhu, L., Zhao, S., An, C., Fan, Y., 2012. Enhanced aqueous solubility of naphthalene and pyrene by binary and ternary Gemini cationic and conventional non-ionic surfactants. Chemosphere 89, 1347–1353.

Wilson, I.D., Adlard, E.D., Cooke, M., Poole, C.F., 2000. Encyclopedia of Separation Science. Academic Press, San Diego.

Xu, K., Xu, P., 2014. Efficient calcium lactate production by fermentation coupled with crystallisation-based *in situ* product removal. Bioresour. Technol. 163C, 33–39.

Yoshida, N., Moroi, Y., 2000. Solubilization of polycyclic aromatic compounds into *n*-decyltrimethylammonium perfluorocarboxylate micelles. J. Colloid Interface Sci. 232, 33–38.

Yutani, R., Morita, S., Teraoka, R., Kitagawa, S., 2012. Distribution of polyphenols and a surfactant component in skin during aerosol OT microemulsion-enhanced intradermal delivery. Chem. Pharm. Bull. 60, 989–994.

Zamanipoor, M.H., Dincer, T.D., Zisu, B., Jayasena, V., 2013. Nucleation and growth rates of lactose as affected by ultrasound in aqueous solutions. Dairy Sci. Technol. 93 (6), 595–604.

Xu, Q., Nakajima, M., Ichikawa, S., Nakamura, N., Shiina, T., 2008. A comparative study of microbubble generation by mechanical agitation and sonication. Innov. Food Sci. Emerg. Technol. 9, 489–494.

Zamanipoor, M.H., Mancera, R.L., 2014. The emerging application of ultrasound in lactose crystallisation. Trends Food Sci. Technol.

Zisu, B., Sciberras, M., Jayasena, V., Weeks, M., Palmer, M., Dincer, T.D., 2014. Sonocrystallisation of lactose in concentrated whey. Ultrason. Sonochem.

EMERGING EXTRACTION

Francisco J. Barba*, Eduardo Puértolas**, Mladen Brnčić†, Ivan Nedelchev Panchev‡,
Dimitar Angelov Dimitrov§, Violaine Athès-Dutour¶, Marwen Moussa#, Isabelle Souchon#

*Department of Nutrition and Food Chemistry, Faculty of Pharmacy, Universitat de València, Valencia, Spain;
**Food Research Division, AZTI-Tecnalia, Bizkaia, Spain; †Department of Process Engineering, Faculty of Food Technology
and Biotechnology, University of Zagreb, Zagreb, Croatia; ‡Department of Physics, University of Food Technologies,
Plovdiv, Bulgaria; §LADEC Ltd., Plovdiv, Bulgaria; ¶Génie et Microbiologie des Procédés Alimentaires,
INRA/AgroParisTech, Paris, France; #Génie et Microbiologie des Procédés Alimentaires, INRA/AgroParisTech, Paris, France

11.1 INTRODUCTION

At present, food manufacturing is considered increasingly as an integrated biorefinery where food products are part of a diverse array of coproducts. Within this context and the framework of bioeconomy development, there is an increasing drive toward the diversification of product outputs and minimization of waste and residues.

Moreover, following the demands of final consumers for both safe and high quality foods as well as the expectation of manufacturers for environmentally sustainable food production practices, new techniques based on nonthermal principles are developed. These technologies include, for example, high hydrostatic pressure, ultrasound, pulsed electric fields (PEF), radio-frequencies, high voltage electric discharge, cold plasma, and pulsed lights (Knorr, 2004; Barbosa-Canovas et al., 2005; Toepfl and Knorr, 2006; Fernandes Fabiano et al., 2008; Chemat et al., 2011; Liu et al., 2011; Mújica-Paz et al., 2011; Rastogi, 2011; Puértolas et al., 2012; Galanakis, 2012, 2013; Galanakis and Schieber, 2014; Roselló-Soto et al., 2015; Deng et al., 2015). Each of the mentioned technologies is capable for the treatment of various foods and by-products.

Over the past few years, laser ablation has also attracted the interest of food researchers for the extraction of valuable substances from the fish processing and canning industries. Moreover, interest in the use of sustainable processes such as integrated membranes is growing since it allows better performance in terms of product quality, plant compactness, flexibility, and moderate energy needs. In this respect, new membrane processes such as pervaporation (PV)/vapor permeation (VP) and membrane contactors are becoming key technologies for recovery purposes. The following chapter provides a detailed description of these emerging extraction processes, including the fundamental principles and applications.

11.2 ULTRASOUND-ASSISTED EXTRACTION (UAE)
11.2.1 MECHANISM AND PRINCIPLES

UAE is an innovative, rapid, reproducible, cost/benefit, and clean technique. It is a technology based on the creation of cavitations that lead to disruption within the applied system, known as transient cavitation. Subsequently, disruption allows the diffusion of inner cell material without any significant increase of temperature, and thus without degrading thermally the contained valuable compounds. Disruption of cell walls could be caused by pressures up to 50 MPa and temperatures up to 5000 K (Brnčić et al., 2010). The advantages of ultrasound (US) are less power and energy usage, shortening of processing time, elimination of the addition of chemicals, significantly increased product yield, and simplified maintenance, due to the nonmoving parts of the devices and simple handling of the US systems. In order to increase the yield of the extracted compounds, US requires optimization in the following parameters: US frequency, propagation cycle (continuous or discontinuous), nominal power of the device, amplitude of work, type, and geometry of the system (e.g. length and diameter of the probe). In addition, input power and US frequency are crucial for the choice of the UAE system.

The high power ultrasound systems (HPUS) in liquids are based on low frequencies within the US range and high power inputs. In this case, the achieved intensity levels during treatment could be between 10 and 1000 W/cm^2, causing irreversible changes throughout the applied food system (Pingret et al., 2012; Dujmić et al., 2013). Cavitation bubbles rise and collapse violently under the conditions of low US frequency and high power inputs. Such activity also creates strong acoustic streaming movement/mixing of the fluids making UAE more efficient (Vilkhu et al., 2008). For instance, when lower US frequency (20 kHz) and higher power are applied, more aggressive changes occur. Therefore, an optimal coupling of US parts with solvent and plant properties is required. On the other hand, when frequency reaches 2 MHz, cavitations are impossible to create (Cvjetko Bubalo et al., 2013). During UAE treatment, molecules of solids and liquids are intensively accelerated by the forces generated by the variation of sound pressure. The consequence of this activity is rapid movement of fluids with compression and rarefaction cycles.

11.2.2 APPLICATIONS

Numerous authors have demonstrated the significantly improved extraction of bioactive compounds (e.g. dietary fibers, phenolics, sugars, starches, minerals, vitamins, pigments, essential oils, and organic acids) from fruits, vegetables, and other plants using UAE (Fernandes Fabiano et al., 2008; Virot et al., 2010; Chemat et al., 2011; Da Porto et al., 2013).

Among other applications, both food researchers and the food industry have shown an increased interest in polyphenol recovery from plant food materials. For instance, Virot et al. (2010) stated that apple pomace prolonged treatment time and higher temperatures increased polyphenol extraction yield from apple pomace. A combination of heat treatment at 70°C and ultrasonics at 35 kHz (so-called thermosonication) has also been used to increase polyphenols extraction from apple pomace by 50% compared with conventional extraction (Corrales et al., 2008). On the other hand, Da Porto et al. (2013) compared Soxhlet and US extraction of polyphenols from grape seeds. Both techniques showed similar extraction yield (14 g/100 g), but UAE took only 0.5 h compared with 6 h of Soxhlet extraction. Pingret et al. (2012) evaluated and compared various conventional and nonconventional extraction techniques to recover polyphenols from apple pomace extracts and found that UAE increased yield between 30%–50% compared with the classic techniques.

UAE is also able to increase the extraction of lipophilic antioxidants. Eh and Teoh (2012) indicated that UAE increased lycopene extraction yield from tomato waste by 26% compared with conventional extraction. Indeed, UAE was optimized with 37 kHz, 140 W of power for 23.2–56.8 min. Hossain et al. (2014) investigated the influence of UAE on the recovery of steroidal alkaloids from potato peel waste. UAE was carried out in a continuous flow cell system with immersed probe (1500 W, 20 kHz, cycle 50%, 3–17 min). Results showed that 17 min of US provides a yield of 1102 μg total steroidal alkaloids/g dried potato peel compared with 710.51 μg/g obtained with solid/liquid extraction. Moreover, UAE also increased the yield of α-solanine, α-chaconine, solanidine, and demissidine recovery (273, 542.7, 231, and 55.3 μg /g, respectively) compared with solid/liquid extraction (180.3, 337.6, 160.2, and 32.4 μg/g, respectively).

Waste generated in coffee manufacturing was the subject of investigation conducted by Abdullah and Bulent Koc (2013). The authors enhanced the two-phase extraction procedure of waste coffee grounds using UAE. The waste occurred in the coffee industry after brewing of coffee grounds containing up to 16% oil (dry weight basis, dwb).

In another study, the impact of UAE on the recovery of valuable compounds, especially polyphenols, from carob pulp kibbles, which are by-products of carob gum manufacturing, was investigated (Roseiro et al., 2013). Moreover, the effects of UAE on polyphenol recovery from carob pulp kibbles were compared with those obtained after using supercritical fluid extraction (SFE) and conventional solid/liquid extraction. It was found that UAE combined with SFE could be a potential tool to improve the recovery of polyphenols from this source. In addition, the ability of ultrasound as a potential pretreatment technique for disruption of wheat-dried distillers' grain cell walls, thus improving mass transfer and favoring the recovery of polyphenols, was established (Izadifar, 2013).

On the other hand, Widyasari and Rawdkuen (2014) compared the effects of UAE and acid extraction of gelatine from chicken feet. It was found that both acid and ultrasonic treatments showed a rather low recovery yield of gelatine (4.1% wet weight basis and 12.64% dwb, respectively). Similar research was conducted by Ming (2013) who optimized UAE of gelatine from tropic carp fish scale (*Cyprinus caprio haematopterus*). Bi et al. (2010) extracted astaxanthin from shrimp waste using ionic liquids UAE. The yield was almost twofold higher than the conventional extraction.

UAE offers a cleaner approach of regarding its mechanism of work, less usage of solvents and significantly increased yields of bioactive compounds. Moreover, it leads to shortening of processing time with meaningful savings in energy consumption.

11.3 LASER ABLATION

11.3.1 THEORETICAL RATIONALE

Laser ablation is the process of focusing a pulsed laser beam with sufficient energy onto a sample in order to remove a small amount of its mass. The term laser ablation refers to the explosive interaction of the laser beam and the target material resulting in the release of particles from the matter. The power density and the thermo-optical properties of the exposed material are critical parameters that influence the effectiveness of the process. When the laser pulses have a duration of several microseconds (or longer) and energy density less than 10^6 W/cm^2, evaporation of the material occurs. The optical coefficients of absorption $A (\lambda)$ and reflection $R (\lambda)$ of the material determine how much of the laser beam irradiation will penetrate the material and at what depth using a power density higher than 10^6 W/cm^2 with nanosecond and shorter pulses an explosive separation of the particles from the material occurs (laser ablation). Power density in the range of 10^6–10^9 W/cm^2 can cause both evaporation and ablation.

The physical theory of laser ablation has been elaborated in a huge number of publications, whereas some of the issues dealing with the practical applications have been discussed in the papers of Anisimov and Luk'yanchuk (2002) and Kautek et al. (2005). Laser ablation in analytical chemical practice has been described in detail in the literature (Russo, 1995; Russo et al., 2002, 2011, 2013; Arai et al., 2007). Applying a laser beam to a material can produce three types of interaction mechanisms: First, thermal effects through the propagation of heat in the material. In this case, the temperature of the material rises quickly and may cause melting or vaporization, depending on the power density involved. Thermal effects are the results of three distinct phenomena: (i) conversion of light into heat, (ii) heat transfer, and (iii) a tissue reaction linked with heating. This interaction generally leads to the denaturation or destruction of a tissue volume. Second, photochemical effects: in this case the links between the atoms are disturbed, which can alter the surface link energy. Third, mechanical effects: these occur when the pulse durations are of the order of 10 ns and the power densities are at least 0.1 MW/cm^2. An intense plasma is then emitted, which expands and exerts the surface pressure, which is liable to deform the material.

The power density required to initiate ablation of biological tissue with nanosecond laser pulses is 10-fold less than that required for vaporization (Irving et al., 1995). When the laser pulse duration is shorter than a certain time period, the material is "inertially confined". This means that it does not have time to expand and heating takes place at a constant time. Most materials are weaker in tension than in compression and they will fail wherever the induced tensile stresses exceed the tensile strength. When a biological object absorbs laser energy, the resulting nonuniform temperature distribution causes internal forces leading to thermomechanical transients and thermoelastic deformation. These changes are the driving force of all laser ablation processes that are not photochemically mediated. In the absence of photomechanical or phase transition processes, the power absorbed by the tissue in response to pulsed laser radiation is entirely converted to a temperature rise.

Almost any laser treatment of the material causes evaporation whose intensity varies between laser pulses of milli and nanoseconds (Fig. 11.1). Short high-power pulses are characterized by low evaporation and small amount of removed substances, while pulses of long duration and low power cause the removal of a large amount of material, which involves the formation of deep craters. The release of steam from the irradiated surface takes place at an almost sonic speed in the form of a plume directed from the surface. In addition to steam, this plume contains drops of liquid phase as well as solid

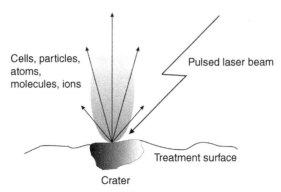

FIGURE 11.1 Laser Treatment of Materials

particles removed from the bottom and sides of the crater. As a result of the recoil pulse (related to the scattering of erosion products) and the high temperature gradient, a wave of pressure and thermomechanical stress is formed, leading to mechanical destruction of the material. Usually evaporation starts at the beginning of the front edge of the pulse. The higher the intensity of radiation, the higher the density of the steam flame. The steam begins to absorb the subsequent irradiation at a higher intensity and overheats, causing ionization of the flame. The formed plasma additionally absorbs, reflects, and diffuses the subsequent laser irradiation, thus shielding the surface. At the end of the pulse the intensity of the laser irradiation falls and the plasma density is significantly reduced due to its scattering, as laser irradiation impacts directly the surface of the material. The shielding phenomenon determines the optimal intensity at which maximum material is removed.

11.3.2 PRACTICAL APPLICATIONS

The unique properties of laser radiation (monochromaticity, time, and space coherence, high focus of the laser beam, brightness, etc.) have been manifested during the irradiation of different substances. Laser ablation intensifies heat and mass-exchange processes, accelerates chemical reactions, modifies macromolecules, and has an antimicrobial effect. Initial applications of low power lasers in food technology were oriented to the use of lasers as a reference tool, retaining or accelerating factors in laser-chemical reactions and biotechnological processes (e.g. accumulation of biomass) or controlling pests in preserved foods. The emergence of new powerful lasers working in the UV and IR bands of the electromagnetic spectrum aroused the interest of scientists concerning their use as intensifying means in food technology.

The photodynamic effect of laser irradiation has been assayed to increase the amount and improve the quality of biologically active substances (anthocyanins, enzymes, polysaccharides, proteins, microelements, waxes, and aromatic substances, etc.) extracted from biological entities such as peels of water melons, melons, and citrus fruits. The latter compounds are nowadays recovered with conventional techniques such as solvent extraction and membrane separations (Patsioura et al., 2011; Tsakona et al., 2012; Galanakis et al., 2010d, 2014; Galanakis, 2013; Heng et al., 2015) with advantages and disadvantages (see Chapter 6).

Laser ablation has also been applied in onion, tomato, potato, pepper, and other vegetable peelings, as well as for the treatment of peanuts, cocoa, and wheat grains whose husks – waste materials – contain valuable substances. Although these studies are still at the laboratory level, the results obtained are encouraging for future industrial applications of laser ablation as an alternative to traditional methods of processing food waste.

11.3.3 EXTRACTION OF USEFUL SUBSTANCES FROM WASTE PRODUCTS IN FOOD TECHNOLOGIES

The possibility of using laser ablation in food technologies has been investigated by Panchev et al. (2011). Laser ablation was able to peel fruit and vegetables efficiently, while preserving the organoleptic properties, such as freshness, naturalness, and texture. It was shown that laser ablation had important advantages compared with the classical thermal peeling (e.g. red long fleshy peppers, tomatoes, onions, potatoes, and other vegetables). For its efficient use, it was necessary to design an automated contraption for feeding the item to the laser beam to provide access to all parts of the treated item. After laser treatment of citrus fruit (oranges and lemons), the fruits were peeled and pectin was extracted from the peels via a classical acid extraction with a solution of hydrochloric acid. Laser ablation of the materials led to an increase in pectin yield, gel strength, and purity for all the samples. Laser ablation of fruits and vegetables has also been used as an alternative method to produce edible films. Finally, interesting results are expected from the use of laser ablation for the extraction of vegetable waxes, aroma, and other useful substances from food waste materials. For instance, during the preliminary laser ablation of the fruits, an intensive release of aroma substances was conducted. This fact indicates the need to carry out separate experiments aimed at extracting aroma substances from different materials, where traditional extraction processes are not efficient enough.

The waste from fish and seafood can constitute a source of precursor material for different applications in the field of biomedicine as bioceramic coatings to improve the osteointegration of metallic implants. In addition, more than a third of fish catches turns into waste product. For instance, an Nd:YAG laser with an average power of 500 W has been used to extract biocompatible coating from fish processing wastes. Preliminary experiments were designed to keep mean laser power, powder mass flow and feed rate as processing parameters to identify the optimum combination to obtain coatings with strong adhesion (Lusquinos et al., 2006).

11.4 PULSED ELECTRIC FIELD (PEF)

11.4.1 BASIC PRINCIPLES AND TECHNICAL ASPECTS

In conventional extraction, the cell membranes of food tissues act as physical barriers, hindering the extraction process. The pretreatment of foods by PEF provokes the formation of pores on these cell structures based on an increase of the transmembrane potential (Vorobiev and Lebovka, 2011). This electroporation (also called electropermeabilization) assists the release of the intracellular compounds, without any significant increase of temperature (Soliva-Fortuny et al., 2009). As action is mainly localized on a microscopic scale, PEF gentle treatment maintains the basic structure of food and thus the achieved extracts are purer than those obtained by other more aggressive technologies (Puértolas et al., 2012).

Typical PEF system (Fig. 11.2) is composed of:

1. a pulse generator, which is in turn basically composed of a DC high voltage generator, a pulse forming unit, and, if it is needed, a pulse transformer to increase voltage,
2. a treatment reactor, basically comprising electrodes (high voltage and ground) separated by insulating material,
3. a suitable product handling system, and
4. a set of monitoring and controlling devices.

Based on this system configuration, PEF-assisted extraction involves essentially the application of direct current electric pulses of high voltage (up to 40 kV) and short duration (less than 10 ms) at a determined repetition rate (Hertz). The target material is placed in a static reactor or flows in a continuous one. These electric pulses generate PEF (typically up to 10 kV/cm), depending on the delivered voltage, the electrode geometry, and their disposition on the reactor. Besides electric field strength, pulse width, and repetition rate, the rest of the process parameters characterizing PEF treatment includes: the number of pulses applied, the treatment time (μs or ms; number of pulses multiplied by pulse width), the

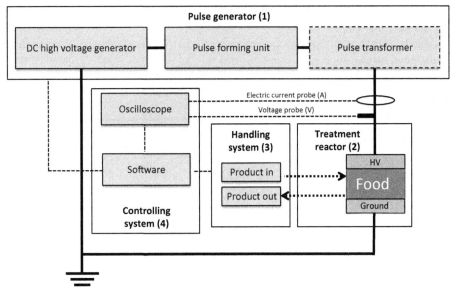

FIGURE 11.2 General Scheme of a Pulsed Electric Field (PEF) System

pulse shape, which depends on pulse forming unit (mainly exponential decay and square waveform), and the total specific energy (kJ/kg) (Puértolas et al., 2012). PEF-assisted extraction typically involves total specific energies lower than 20 kJ/kg. This parameter depends on the voltage applied, treatment time, and resistance of the reactor, which varies according to its geometry and the electric conductivity of the treated food (Heinz et al., 2001). Efficiency of the PEF-assisted extraction not only depends on the PEF process parameters, but also on extraction variables (temperature, product size, solvent nature, and concentration), physicochemical characteristics of the treated matrix (pH, electric conductivity), and the characteristics of the cells that form the matrix (size, shape, membrane, envelope structure) (Puértolas et al., 2012; Vorobiev and Lebovka, 2011). Furthermore, the potential effectiveness of the eletropermeabilization procedure depends on the nature of the molecule and its location within the cell (cytoplasm or vacuoles) (Soliva-Fortuny et al., 2009).

11.4.2 EFFECT OF PEF ON THE RECOVERY OF HIGH VALUE COMPONENTS FROM FOOD

In order to demonstrate the potential of PEF, most of the published work has been based on minimally processed foods (e.g. peeled, sliced) instead of food waste. Regarding polyphenols, important advances have been published on extraction yield and extraction rate for a wide range of vegetable tissues, including fruits, leaves, tubers, and vegetable seeds (Donsi et al., 2010; Puértolas et al., 2012, 2013). For instance, PEF pretreatment increased the aqueous extraction of anthocyanins (red polyphenol) from pretreated red cabbage (2.5 kV/cm; 16.63 kJ/kg) by 115% (Gaschovska et al., 2010). PEF pretreatment could also be a useful tool for lessening the intensity of the parameters that define conventional extraction. Puértolas et al. (2013) studied the influence of solvent concentration and temperature on PEF-assisted extraction of anthocyanins from purple-fleshed potato. For a fixed anthocyanin yield (around 60 mg/100 g fresh weight (fw)), PEF pretreatment (3.4 kV/cm; 105 ms; 8.92 kJ/kg) allowed substitution of ethanol solvent with water (both at 40°C) or reduction of the extraction temperature up to 15°C, maintaining the same solvent.

PEF has also been extensively studied for enhancing the extraction of other bioactive molecules, obtaining similar good results, such as chlorophylls and carotenoids from algae, carotenoids from carrot, betalains from red beetroot or tocopherols (vitamin E), and phytosterols from olive (Donsi et al., 2010; Puértolas et al., 2012; Puértolas and Martínez de Marañón, 2015; Soliva-Fortuny et al., 2009).

11.4.3 EFFECT OF PEF ON THE RECOVERY OF HIGH VALUE COMPONENTS FROM FOOD WASTE

Concerning PEF-assisted extraction from real food waste, only a small number of works have been published (Boussetta et al., 2012, , 2014; Corrales et al., 2008; Luengo et al., 2013). For instance, Corrales et al. (2008) reported an increase in polyphenol yield (100%) when red grape skins from wine making were subjected to a PEF treatment (3 kV/cm, 30 pulses, 10 kJ/kg). The antioxidant activity of the recovered extracts was fourfold higher than control samples. More recently, Luengo et al. (2013) studied the effect of PEF (1–7 kV/cm) on posterior pressing extraction of polyphenols from orange peel by-product, obtaining increments in extraction yield and antioxidant activity up to 159 and 192%, respectively. Specifically, a PEF treatment of 5 kV/cm increased the quantity of naringin and hesperidin in the extract from 1 to 3.1 mg/100 g fw and from 1.3 to 4.6 mg/100 g fw, respectively.

Besides soft tissues, PEF technology has been recently proposed to increase polyphenol extraction from hard vegetable by-products like grape seeds and flaxseed hulls (Boussetta et al., 2012, 2014). The best results were observed when product was partially rehydrated. For example, the highest polyphenol increase from flaxseed hulls (around 37%) was obtained after 40 min of rehydration (Boussetta et al., 2014). For this specific application, more intense PEF treatment conditions are needed than those implemented in soft tissues (10–20 kV/cm; 0–20 ms). Since a powerful PEF generator and higher energy consumption are required, the investment and treatment costs would be higher.

Another potential benefit of PEF treatment is the substitution or grinding reduction. The latter is conventional pretreatment to increase the surface area to solvent volume ratio. Grinding results in extracts with higher turbidity than those obtained using PEF. As a consequence, PEF facilitates a subsequent purification procedure (Boussetta et al., 2014). Moreover, if conditions are optimized PEF energy cost is lower than that required for grinding, maintaining higher extraction yield. For example, in order to improve polyphenol extraction from flaxseed hulls, Boussetta et al. (2014) proposed a PEF treatment of 300 kJ/kg to get a good compromise between reduced energy input and high level of polyphenol content (2.3 times higher than control).

11.4.4 ADVANTAGES AND INDUSTRIAL FEASIBILITY

According to results reported over the last 10 years, the principal benefits of PEF against conventional processes are: (i) improved extraction yield, (ii) decreased extraction time, (iii) decreased intensity of the conventional extraction parameters (e.g. extraction temperature, solvent concentration), (iv) facilitation of purified extract (e.g. reducing grinding), and (v) reduced energy cost and environmental impact compared with conventional extraction (e.g. reduce temperature, grinding).

Concerning industrialization, PEF is technically ready to be implemented. In fact, different industrial prototypes have been developed over the past few years, reaching continuous product flow rates up to 10,000 kg/h (Puértolas et al., 2012). One of the most important advantages of PEF compared with other emerging technologies, such as high hydrostatic pressure, is the possibility of applying it in a continuous mode. Moreover, since a reactor could be designed basically as a pipe, it could be easily adapted to preexisting facilities, decreasing set-up costs. In any case, the industrialization of PEF for a determined application would always require a previous study to optimize the process parameters and scale the needed PEF unit, including continuous reactor design.

11.5 HIGH VOLTAGE ELECTRICAL DISCHARGE
11.5.1 BASIC PRINCIPLES AND TECHNICAL ASPECTS

This technology is based on both chemical reactions and physical processes. When high voltage electrical discharge (HVED) is produced directly in water, it injects energy directly into an aqueous solution through a plasma channel formed by a high current/high voltage electrical discharge between two submersed electrodes (Boussetta and Vorobiev, 2014). Over the past few years, a considerable interest was accorded to the study of HVED. Typical an HVED system is composed of:

1. a high voltage pulsed power generator, a treatment chamber (basically comprising electrodes of needle-plate geometry separated by insulating material),
2. a suitable product handling system, and
3. a set of monitoring and controlling devices.

In Fig. 11.3, a typical treatment chamber for HVED treatments is shown (adapted with permission from Parniakov et al. (2014) and Boussetta et al. (2012b)).

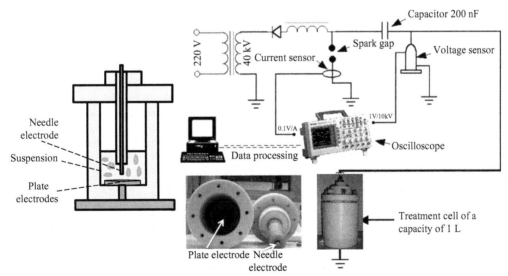

FIGURE 11.3 Typical Scheme of a High Voltage Electrical Discharge (HVED) Unit

Adapted with permission from Parniakov et al. (2014) and Boussetta et al. (2012b)

HVED technology has attracted much interest from the scientific community due to its potential in several applications such as water cleaning of organic chemical impurities, insulators in high voltage pulsed power systems, acoustic sources in medical or sonar, and selective separation of solids and plasma blasting mining applications. In particular, the technology of HVED has been recently studied for enhancing extraction of bioactive compounds from different raw materials. HVED leads to the generation of hot, localized plasmas that strongly emit high-intensity UV light, produce shock waves, and generate hydroxyl radicals during water photodissociation (Boussetta and Vorobiev, 2014).

11.5.2 HVED-ASSISTED RECOVERY OF HIGH ADDED-VALUE COMPOUNDS FROM PLANT MATERIALS, FOOD WASTES, AND BY-PRODUCTS

Over the past few years, several studies have been developed to evaluate the impact of HVED on high added-value compounds recovery from different plant food materials, food wastes, and by-products. One of the most important factors in HVED-assisted extraction is the treatment energy input. Moreover, the efficiency of the HVED-assisted extraction not only depends on the HVED energy inputs, but also on the distance gap of the electrodes, the liquid-to-solid ratio, the extraction temperature, and duration solvent composition (Boussetta and Vorobiev, 2014).

For instance, the effects of HVED on phenolic compounds and protein recovery from winery by-products were evaluated and compared with other emerging technologies and conventional extraction techniques (Boussetta et al., 2009a, b, 2011, 2012a, 2013a; Rajha et al., 2014). It was demonstrated that total polyphenols content (1.37 ± 0.11 g GAE/100 g dry matter, DM), the content of four main polyphenols from grape pomace (catechin, epicatechin, quercetin-3-O-glucoside, and kaempferol-3-O-glucoside), and antioxidant activity (23.02 ± 3.06 g TEAC/kg DM) of grape pomace suspensions were augmented by increasing HVED energy input up to 80 kJ/kg. However, a significant decrease in polyphenols content was found when energy input was augmented up to 800 kJ/kg (Boussetta et al., 2011). This fact can be attributed to the formation of chemical products of electrolysis and free reactive radicals during HVED treatment, which can reduce the nutritional quality of antioxidant compounds.

In addition, polyphenol recovery from grape pomace, seeds, skins, and stems was significantly improved at the optimized energy input conditions (Boussetta et al., 2012a, b).

Another study demonstrated the effectiveness of HVED treatment to recover polyphenols and proteins from vine shoots (Rajha et al., 2014). HVED treatment provided the higher polyphenol yield (34.5 mg of gallic acid equivalent/g DM) and the high purity (89%) in comparison to PEF and ultrasound treatments. Moreover, HVED had the lowest energy consumption.

On the other hand, the impact of HVED at different temperatures (20–60°C) and ethanol concentration (0–25%) on the recovery of lignans from whole and crushed flaxseed cake was investigated (Boussetta et al., 2013b). In this work, a significant improvement in the recovery of total polyphenols using HVED was found. Moreover, the addition of ethanol had a synergistic effect and increased the extraction efficiency.

In addition, the effects of HVED (40 kV/cm, 10 kA, 20–1000 kJ/kg, 0.5 Hz, 20°C) on oil recovery from linseed meal were evaluated (Li et al., 2009). It was observed that HVED of 380 kJ/kg at 15°C, pH 7, and with a liquid-to-solid ratio equal to 6 was the optimum treatment and allowed a higher yield of oil extraction (68%).

In another study, the efficiency of HVED (40 kV/cm, 10 kA, 90 kJ/kg, 0.5 Hz) was demonstrated for the extraction of mucilage from whole linseeds (Barbara variety) immersed in 500 mL of demineralized water (20°C) (Gros et al., 2003).

More recently, the effects of HVED-assisted extraction of phenolic compounds and proteins from papaya peels were investigated (Parniakov et al., 2014). It was found that HVED exhibited the best extraction efficiency of proteins, total phenolics, and carbohydrates compared with PEF treatments and conventional thermal extraction, at different temperatures (20–60°C) and pH values (2.5–11).

11.5.3 ADVANTAGES AND DISADVANTAGES OF HVED-ASSISTED EXTRACTION

HVED treatment is a green extraction technique that allows the extraction of high added-value compounds at low energy input compared with other emerging technologies such as PEF, US, and microwave-assisted extraction, among others. Indeed, it can reduce the required diffusion temperature and time as well as the amount of solvent. However, the application of HVED can be less selective (as compared with PEF), producing small particles that lead to a more difficult solid-to-liquid separation. The energy input of the process depends on the conductivity of the medium. In addition, the feasibility of the application of HVED at pilot or industrial scales is still unknown.

11.6 EMERGING MEMBRANE EXTRACTION
11.6.1 MEMBRANE PERVAPORATION AND VAPOR PERMEATION
11.6.1.1 Process principle
Pervaporation (PV) and VP are separation processes that employ dense nonporous membranes. In PV, a binary or multicomponent liquid stream undergoes a selective and partial evaporation through the membrane: "per" (across) and "vaporation" (vapor formation). The liquid bulk (feed) is in direct contact with the upstream side of the membrane while a solute-enriched vapor phase (permeate) is removed from the opposite side. The remaining components of the liquid are called retentate. In principle, VP is similar to PV, whereas the only difference is that the feed phase is vapor (Neel et al., 1985). Accordingly, PV operates at milder temperature levels than VP. The basic principle of both processes is illustrated in Fig. 11.4.

The driving force is created by the gradient in chemical potential μ_i (or partial pressure p_i) of a given solute i. This gradient is generally maintained owing to the vacuum (Fig. 11.4), which reduces the total pressure on the permeate side of the membrane and the partial pressure p_i. The latter can be reduced by sweeping an inert gas on the permeate side (Fig. 11.5a). A less common variant is called thermopervaporation, where a thermal gradient is used as the driving force (Franken et al., 1990). In this case, the permeate side of the dense membrane is coated with a porous membrane, which is brought into contact with a cold permeate absorbing liquid. Recently, Volkov and Borisov (2012) presented a modified thermopervaporation system where the permeate vapor is condensed on a cooled plate without direct contact with the cooling liquid and the membrane (Fig. 11.5b).

11.6.1.2 Membrane typology
Different types of membranes are used in PV/VP. They include organic, inorganic, and hybrid membranes consisting of polymeric and inorganic materials. Polymeric membranes remain the most widely used materials at the industrial scale. These membranes are attractive because of their economical and fabrication advantages (Wee et al., 2008). Polymers such as polydimethylsiloxane (PDMS), polyvinylalcohol, and polyimide are the examples. Although some polymeric membranes are commercially used for PV, their limited thermal and solvent stability, which add to unfavorable swelling and fouling

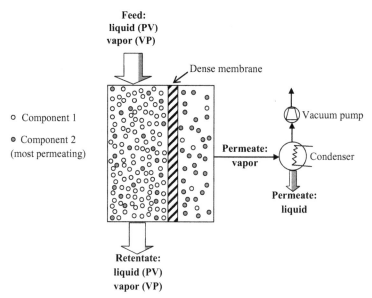

FIGURE 11.4 Schematic Diagram of PV and VP Principles where a Vacuum Pump is Operating on the Permeate Side

behaviors, still remain the major drawbacks (Chapman et al., 2008; He et al., 2012). Over the past few years, considerable efforts have been put into the development of inorganic and hybrid membranes, as they are free of swelling and show better chemical, hydrothermal, and mechanical stability (Wee et al., 2008). Most of these membranes are commercially available. Inorganic membranes include zeolite, silica, and carbon membranes (Tin et al., 2011). Hybrid membranes combine organic and inorganic membranes such as the mixed matrixes (Amnuaypanich et al., 2009), the ceramic-supported organic membranes (Peters et al., 2008), and the so-called HybSi® membranes (Van Veen et al., 2011).

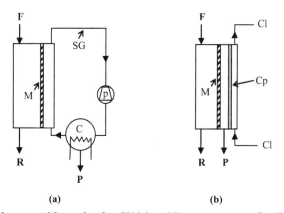

(a)   (b)

FIGURE 11.5 Schematic Diagrams of Sweeping Gas PV (a) and Thermopervaporation (b)

F, feed; R, retentate; P, permeate; M, membrane; C, condenser; p, pump; SG, sweeping gas circuit; Cl, cooling liquid; Cp, cooled plate.

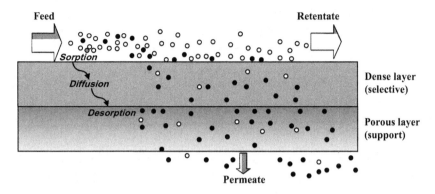

FIGURE 11.6 Sorption/Diffusion Transport Mechanism through a PV Membrane where a Selective Dense Layer is Supported by a Porous Layer

11.6.1.3 Sorption/diffusion transport mechanism

There are many different views on the separation mechanism of PV at the molecular level. The sorption/diffusion phenomenological model is commonly agreed (Bettens et al., 2010; Schafer and Crespo, 2005; Shao and Huang, 2007). In this model, the transport of a component from the feed side to the permeate side consists of the following steps (Fig. 11.6): (i) sorption to the membrane, (ii) diffusion through the membrane, and (iii) desorption of the permeate to the vapor phase.

11.6.1.4 Evaluation of membrane performance

The separation performance is based on the capability of the membrane to separate components from each other. This can be characterized using the following indicators (Baker et al., 2010; Schafer and Crespo, 2005; Wang et al., 2013):

1. Total flux:

$$J = \frac{Q}{t \times A} \tag{11.1}$$

where Q is the total permeate mass transferred during the operation time t and A is the membrane area.

2. Partial flux:

$$J_i = W_i \times J = \frac{S_i \times D_i}{z} \times \left(p_{if} - p_{ip}\right) = \frac{K_i}{z} \times \left(p_{if} - p_{ip}\right) = P_i \times \left(p_{if} - p_{ip}\right) \tag{11.2}$$

where W_i is the weight fraction of the solute i, S_i is its sorption coefficient, D_i is its diffusivity, p_{if} is its partial pressure in the feed side, p_{ip} its partial pressure in the permeate side, and z is the membrane thickness. K_f and P_i are the membrane permeability and permeance, respectively.

3. Membrane selectivity, which is defined as the ratio of the permeabilities or permeances of two different components i and j through the membrane:

$$\alpha_{i/j} = \frac{k_i}{k_j} \, or = \frac{P_i}{P_j} \tag{11.3}$$

11.6.1.5 Applications in food waste recovery

Numerous sectors of the food industries (fat and oil production, distilleries, seafood production, etc.) produce large amounts of effluents containing volatile molecules, which in general are responsible for unpleasant odors and are difficult to eliminate from effluents. The odorous volatile compounds contained in these effluents can be considered valuable compounds due to their natural origin. The treatment of these effluents should be conducted with a nondestructive process, permitting on the one hand the deodorization and on the other hand the recovery of a "natural" aromatic fraction destined to total or partial compensation. In addition, certain constraints must be taken into account when choosing the process to recover aroma compounds from liquid effluents. For instance, it is necessary to respect the molecule integrity (according to the low thermostability of aroma compounds) to have an efficient process that is easy to install and maintain.

This is well illustrated in the case of blanching waters (Souchon et al., 2002a, b), seafood cooking juices (Bourseau et al., 2014), evaporation condensates (Bengtsson et al., 1992), and whey (Rajagopalan et al., 1994), which contain numerous volatile organic compounds. Those molecules have typical odors (e.g. fermented cabbage, crude onion, Munster-like cheese, etc.) and have a particularly low detection threshold (about a few micrograms per cubic meter). Thus, even at very low concentrations, their presence in blanching waters confers a smelly aspect. On the other hand, these aroma compounds are the key components of many food aromas. Accordingly, blanching waters and cooking juices may constitute, with selective treatment, a source of molecules of great industrial interest.

In PV, membrane perm-selectivity is the key performance that determines the relevance of using this technique for the recovery of a given molecule. Hydrophobic membranes such as PDMS and polyetherblockamide (PEBA) can be used for the separation of organics from aqueous streams. There are also hydrophilic membranes (e.g. polyvinylalcohol) for the separation of water and methanol from organics. In addition, the hydrophobicity (philicity) and the volatility of the target molecule under PV conditions should be taken into account. For example, aroma compounds like vanillin have a very low saturation vapor pressure (0.29 Pa at 298.15 K) and are often subjected to concentration polarization on the permeate side of the membrane (Schafer and Crespo, 2005).

Some studies have reported the use of PV for the recovery of aroma compounds from food waste streams. In their patent, Blume and Baker (1990) claimed the recovery of aromatics from orange juice evaporator condensates, using a silicon rubber PV membrane. They reported a total permeate flux of about 0.3 kg/h/m² in which several alcohols, esters, and aldehydes were enriched up to 20-fold. Bengtsson et al. (1992) studied PV as a process alternative for the recovery of natural aroma compounds from an evaporation condensate obtained during the concentration of apple juice. These authors used a PDMS membrane and reported the enrichment of 12 aroma compounds (2 alcohols, 3 aldehydes, and 5 esters) in the PV permeate.

On the other hand, Rajagopalan et al. (1994) performed a systematic study on the performance of the PDMS PV membrane for the recovery of diacetyl from whey permeate. They reported a decrease in total flux with downstream pressure and an increase with temperature and diacetyl concentration in the feed. No effect on whey permeate components (lactose and proteins) was reported. The partial flux of diacetyl was less than 1 g/h/m^2.

Souchon et al. (2002a) studied the pervaporative recovery of three sulfuric compounds identified as typical compounds of cauliflower odor using industrial blanching waters: dimethyl disulfide, dimethyl trisulfide, and S-methyl thiobutyrate. Two types of PV membranes were used: PDMS and PEBA. The permeation fluxes observed in this study were very low (less than 1 g/h/m^2), proportionally to the very low concentrations of the studied compounds in the raw feed, but resulted in a significant deodorization of the PV retentate. Moreover, due to the membrane selectivity, both the permeate composition and odor were found to be very different from the initial industrial effluent and closer to the cheese one, showing a real valorization.

Based upon the aforementioned examples, the applicability of VP for the recovery of aroma compounds from food waste effluents can be addressed (Ribeiro et al., 2005). In this case, the volatility of targeted molecules brings a supplemental selectivity to the separation process in comparison with PV.

11.6.1.6 General comments

PV and VP are looked over as effective techniques with a real potential to replace conventional energy-intensive methods for selective and efficient separation of specific compounds, allowing reliable options for sustainable applications in the agro-food industry. However, these techniques are still poorly adopted at the industrial scale. An obstacle lies in the lack of knowledge about their capability, even if they showed highly interesting results at the laboratory and pilot scales. Amongst other drawbacks, one should mention the high membrane cost and unguaranteed reliability/life span (Jonquières et al., 2002). In this respect, membrane fouling under some operating conditions is still an unsolved problem that leads to a decrease in membrane efficiency and durability.

11.6.2 MEMBRANE CONTACTORS

11.6.2.1 Fundamentals and basic principles

The use of membrane contactors for a membrane-supported extraction (also called perstraction) is an interesting alternative to established extraction processes. Membrane contactors are porous membranes whose only role is to stabilize an interface between two or more fluids. Different separation principles can be implemented within membrane contactors, mainly liquid/liquid, liquid/gas, or liquid/gas/liquid separation. The structural properties of the membrane are an essential factor, offering the greatest porosity (and thus a large surface for exchange) while guaranteeing immobility of the interface between fluids. The stabilization and control of the interface between the feed and the stripping phase inside the membrane contactor is a key issue. Many advantages over conventional dispersive solvent extraction appear. For example, since mixing of phases is avoided, problems of flooding, loading, or downstream phase separation do not occur. The interface is stable and constant under variable feed and stripping phase flow rates, which can be varied independently one from the other. Using a hydrophobic membrane, its pores are wet by the organic solvent and the interface between the two phases is maintained at the pores' mouths, applying a slight overpressure to the aqueous phase. Modularity and compaction are other important aspects of this technique. The compactness is achieved by using a hollow-fiber membrane contactor, for which the exchange area is 200 larger compared with a packed column (Gabelman and Hwang, 1999).

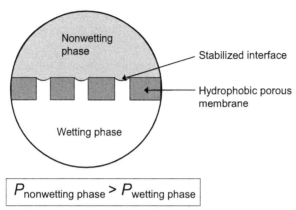

FIGURE 11.7 Principle of Membrane-Based Solvent Extraction with the Interface between a Wetting Fluid Phase and a Nonwetting Solvent Phase in a Hydrophobic Membrane

P refers to the pressure of the considered liquid.

Adapted with permission from Dupuy et al. (2011a)

The main drawback is the additional mass-transfer resistance caused by the presence of the membrane, which hinders diffusion from one phase to another and thus slows down the separation. In most cases, the large surface area per volume offered by hollow-fiber modules overcomes this disadvantage.

Figure 11.7 shows the principle of membrane-based solvent extraction with the interface between a wetting fluid phase and a nonwetting solvent phase in a hydrophobic membrane.

Hydrophilic or hydrophobic membranes can be employed. By using a hydrophilic membrane, the pores are filled with the aqueous phase. By using a hydrophobic one, they are filled with the organic phase or gas, depending on the principle of extraction. Table 11.1 presents an overview of the different separation techniques using membrane contactors.

Membranc-based solvent extraction is a concentration-driven operation. The mass flow rate of the solute i ($\dot{m}_i$) from one phase to another is defined by:

$$\dot{m}_i = K_i A(C_{i\,aq} - C^*_{i\,aq})$$

$$(11.4)$$

where K is the overall mass transfer coefficient, A is the surface area of the membrane, C_{aq} is the concentration in the aqueous phase, and C^*_{aq} is the concentration in an aqueous phase in equilibrium with the solvent phase. C^*_{aq} and C_{org} are related by the partition coefficient (P) that expresses the equilibrium state as:

$$P_i = \frac{C_{i\,org}}{C^*_{i\,aq}} = \frac{C^{eq}_{i\,org}}{C^{eq}_{i\,aq}}$$

$$(11.5)$$

This equilibrium parameter is dependent upon temperature. P is one of the main criteria for choosing the solvent. Indeed, the higher the partition coefficient, the more the equilibrium state is displaced towards the solvent phase resulting in a more complete extraction.

Table 11.1 Different Separation Techniques Involving Microporous Membrane Contactors, with Chemical Potential as the Driving Force

Terminology	Principle	References
Membrane-based solvent extraction	Two immiscible liquids in contact through membrane pores, one of the liquids wetting the membrane pores.	Prasad and Sirkar (1992)
Membrane air stripping	Stripping volatile compounds from a liquid feed with a gas. Membrane pores, impermeable to liquid, are filled with stripping gas.	Kreulen et al. (1993)
Membrane distillation	Liquid stream flowing on upstream side of membrane, impermeable to liquid. Vacuum is maintained on downstream side of membrane, allowing vaporization of volatile compounds from feed stream.	Urtiaga et al. (2001)
Osmotic distillation	An aqueous feed and a brine are flowing on either side of a hydrophobic membrane, with pores filled with air. Water is vaporized through the pores and condensed into the brine.	Kunz et al. (1996)
Supported liquid membrane	Ternary system: aqueous feed and extracting aqueous phase flowing on both sides of membrane, and membrane pores filled of an organic solvent. Extraction occurs in two steps: partition between feed and organic solvent, followed by partition between solvent and the aqueous extracting phase.	Breembroek et al. (1998)
Hollow-fiber contained liquid membrane	A variant of supported liquid membrane: feed and extracting phase are flowing in two independent fibers bundles, interlaced. An intermediate solvent is present in the membrane pores and in the shell in which the fibers bundles are immerged.	Basu and Sirkar (1991)

Mass transfer is always described as a sequence of diffusion steps: from the aqueous boundary layer, through the solvent-filled membrane pores, and finally to the solvent phase boundary layer. A resistance-in-series model is used to relate the overall mass-transfer coefficient (K) to the local mass-transfer coefficients (k) (Pierre et al., 2001). When the aqueous phase flows inside the tube of a hydrophobic hollow-fiber membrane, and the organic phase flows outside the tube, the resistance-in-series model is expressed by:

$$\frac{1}{K_{i\,aq}} = \frac{1}{k_{i\,aq}} + \frac{d_{t\,in}}{P_i k_{i\,mb} d_{t_{lm}}} + \frac{d_{t\,in}}{P_i k_{i\,org} d_{t\,out}} \tag{11.6}$$

where k_{aq} and k_{org} are the local mass-transfer coefficients relative to the aqueous and organic boundary layers, respectively, k_{mb} is the local mass transfer coefficient in the membrane, $d_{t\,in}$ and $d_{t\,out}$ are the internal and external fiber diameters, respectively, and $d_{t_{lm}}$ is their logarithmic mean value (Prasad and Sirkar, 1992).

11.6.2.2 Applications in food waste recovery

More than 20 years of studies have shown the high potential of membrane contactors for extracting organic molecules from aqueous feeds. The extracting phase should either be an organic solvent (Pierre et al., 2001), an edible oil (Baudot et al., 2001), air (Gascons-Viladomat et al., 2006), or even a dense gas (Gabelman et al., 2005). As mentioned earlier, typical examples are odorous volatile compounds contained in some effluents of the industry. For such molecules, nondispersive extraction using membrane contactors can be a key choice.

The applicability of membrane-based solvent extraction to the deodorization of a wastewater coupled with the recovery of valuable compounds has been studied by Pierre et al. (2001, 2002) with two solvents. The first one is hexane, for which an easy separation by evaporation enables a quality product, and the second one is miglyol, for which no additional reprocessing and further purification are needed thereby reducing the costs. This approach has also been applied to recover valuable aromatic fractions from an odorous tomato industry aqueous effluent (Souchon et al., 2002b). Vegetable oils could also be used as an alternative organic solvent, but in this case, there is an additional mass-transfer resistance on the solvent side and inside the hydrophilic membrane pores. For example, Baudot et al. (2001) have performed a systematic study of volatile aroma transfer in a liquid/liquid membrane contactor, using sunflower oil as an extractant. Membrane air stripping was also applied to the extraction of highly diluted aroma compounds in aqueous solutions (Gascons-Viladomat et al., 2006). The membrane air stripping coupled with multistep condensation appears to be capable of separating molecular mixtures with a high degree of selectivity and specificity for the highest volatility molecules compared with a membrane-based solvent extraction, without degrading the sensitive aroma compounds (low temperature) (Souchon et al., 2004). Aroma compounds such as vanillin could also be recovered using membrane-based solvent extraction (Sciubba et al., 2009), but the goal is to perform the one-line extraction of the aroma obtained by bioconversion of ferulic acid. From an environmental point of view, membrane contactors under liquid/liquid extraction mode have been applied to remove ammonia from wastewater streams (Ashrafizadeh and Khorasani, 2010; Mandowara and Prashant, 2011), and reactive solvent extraction in membrane contactors can be used to remove organic and inorganic compounds from aqueous wastes (Alex et al., 2009).

The technology of osmotic distillation has been successfully applied in previous studies to fruit juice concentrates (Kunz et al., 1996; Valdés et al., 2009; Alves and Coelhoso, 2006; Soni et al., 2008). It is based on the use of a salty solution as a stripping agent that extracts selectively the water from aqueous solutions under atmospheric pressure and room temperature, causing no thermal degradation of the volatile components. Some authors (Vaillant et al., 2001) have performed pilot-scale trials for passion fruit juice concentrate, without membrane fouling for 28 h. This technology could be favorably applied to valuable compounds in aqueous effluents in the food industry.

11.6.2.3 General comments

Membrane contactors could be applied to various liquid waste effluents, even though in some cases breakthrough pressure needs to be increased in order to stabilize properly the liquid/liquid interface, in particular for a membrane-based solvent extraction in systems with low interfacial tension. In the specific case of oxygenated terpenes recovery from lemon essential oil, Dupuy et al. (2011b) studied the use of polymeric and ceramic porous membranes, which were able to achieve a robust stabilization of the liquid/liquid interface.

A recent state-of-the-art review on hollow-fiber contactor technology and membrane-based extraction processes (Pabby and Sastre, 2013) underlined the perspective of this technology for several pilot-plant and full-scale applications in the fields of chemical and pharmaceutical technology, biotechnology, food processing, and environmental engineering.

REFERENCES

Abdullah, M., Bulent Koc, A., 2013. Oil removal from waste coffee grounds using two-phase solvent extraction enhanced with ultrasonication. Renew. Energ. 50, 965–970.

Alex, S., Biasottoa, F., Aroca, G., 2009. Extraction of organic and inorganic compounds from aqueous solutions using hollow fibre liquid–liquid contactor. Desalination 241, 337–341.

Alves, V.D., Coelhoso, I.M., 2006. Orange juice concentration by osmotic evaporation and membrane distillation: a comparative study. J. Food Eng. 74, 125–133.

Amnuaypanich, S., Patthana, J., Phinyocheep, P., 2009. Mixed matrix membranes prepared from natural rubber/poly(vinyl alcohol) semi-interpenetrating polymer network (NR/PVA semi-IPN) incorporating with zeolite 4A for the pervaporation dehydration of water–ethanol mixtures. Chem. Eng. Sci. 64, 4908–4918.

Anisimov, S.I., Luk'yanchuk, B.S., 2002. Selected problems of laser ablation theory. Uspekhi Fizicheskikh Nauk Russian Acad. Sci. 45, 293–324.

Arai, T., Hirata, T., Takagi, Y., 2007. Application of laser ablation ICPMS to trace the environmental history of chum salmon *Oncorhynchus keta*. Marine Environ. Res. 63, 55–66.

Ashrafizadeh, S.N., Khorasani, Z., 2010. Ammonia removal from aqueous solutions using hollow-fiber membrane contactors. Chem. Eng. J. 162, 242–249.

Baker, R.W., Wijmans, J.G., Huang, Y., 2010. Permeability, permeance and selectivity: a preferred way of reporting pervaporation performance data. J. Membr. Sci. 348, 346–352.

Barbosa-Canovas, G.V., Tapia, M.S., Cano, M.P., 2005. Novel Food Processing Technologies. CRC Press, Boca Raton.

Basu, R., Sirkar, K.K., 1991. Hollow fiber contained liquid membrane separation of citric acid. AIChE J. 37, 383–393.

Baudot, A., Floury, J., Smorenburg, H.E., 2001. Liquid–liquid extraction of aroma compounds with hollow fiber contactor. AIChE J. 47, 1780–1793.

Bengtsson, E., Tragardh, G., Hallstrom, B., 1992. Concentration of apple juice aroma from evaporator condensate using pervaporation. Lebensmittel – Wissenschaft Technologie 25, 29–34.

Bettens, B., Verhoef, A., Van Veen, H.M., Vandecasteele, C., Degrève, J., van der Bruggen, B., 2010. Pervaporation of binary water-alcohol and methanol–alcohol mixtures through microporous methylated silica membranes: Maxwell-Stefan modeling. Comp. Chem. Eng. 34, 1775–1788.

Bi, W., Tian, M., Zhou, J., Row, K.H., 2010. Task-specific ionic liquid-assisted extraction and separation of astaxanthin from shrimp waste. J. Chromatogr. B 878, 2243–2248.

Blume, I., Baker, R.W., 1990. Treatment of evaporator condensates by pervaporation. Membrane Technology & Research, Inc., Menlo Park, CA, USA. Patent No. 4,952,751.

Bourseau, P., Masse, A., Cros, S., Vandanjon, L., Jaouen, P., 2014. Recovery of aroma compounds from seafood cooking juices by membrane processes. J. Food Eng. 128, 157–166.

Boussetta, N., De Ferron, A., Reess, T., Pecastaing, L., Lanoiselle, J.L., Vorobiev, E., 2009a. Improvement of polyphenols extraction from grape pomace using pulsed arc electro-hydraulic discharges, in: PPC2009 – Seventeenth IEEE International Pulsed Power Conference, pp. 1088–1093.

Boussetta, N., Lebovka, N., Vorobiev, E., Adenier, H., Bedel-Cloutour, C., Lanoiselle, J.-L., 2009b. Electrically assisted extraction of soluble matter from chardonnay grape skins for polyphenol recovery. J. Agr. Food Chem. 57, 1491–1497.

Boussetta, N., Vorobiev, E., Deloison, V., Pochez, F., Falcimaigne-Cordin, A., Lanoiselle, J.-L., 2011. Valorisation of grape pomace by the extraction of phenolic antioxidants: application of high voltage electrical discharge. Food Chem. 128, 364–370.

Boussetta, N., Vorobiev, E., Le, L.H., Cordin-Falcimaigne, A., Lanoiselle, J.L., 2012. Application of electrical treatments in alcoholic solvent for polyphenols extraction from grape seeds. LWT – Food Sci. Technol. 46, 127–134.

Boussetta, N., Vorobiev, E., Le, L.H., Cordin-Falcimaigne, A., Lanoiselle, J.-L., 2012a. Application of electrical treatments in alcoholic solvent for polyphenols extraction from grape seeds. LWT – Food Sci. Technol. 46, 127–134.

Boussetta, N., Vorobiev, E., Reess, T., De Ferron, A., Pecastaing, L., Ruscassie, R., Lanoiselle, J.-L., 2012b. Scale-up of high voltage electrical discharge for polyphenols extraction from grape pomace: effect of the dynamic shock waves. Innov. Food Sci. Emerg. Technol. 16, 129–136.

Boussetta, N., Lesaint, O., Vorobiev, E., 2013a. A study of mechanisms involved during the extraction of polyphenols from grape seeds by pulsed electrical discharges. Innov. Food Sci. Emerg. Technol. 19, 124–132.

Boussetta, N., Turk, M., De Taeye, C., Larondelle, Y., Lanoiselle, J.L., Vorobiev, E., 2013b. Effect of high voltage electrical discharge, heating and ethanol concentration on the extraction of total polyphenols and lignans from flaxseed cake. Ind. Crop. Prod. 49, 690–696.

Boussetta, N., Vorobiev, E., 2014. Extraction of valuable biocompounds assisted by high voltage electrical discharge: a review. C.R. Chim. 17, 197–203.

Boussetta, N., Soichi, E., Lanoisellé, J.L., Vorobiev, E., 2014. Valorization of oilseed residues: extraction of polyphenols from flaxseed hulls by pulsed electric fields. Ind. Crop. Prod. 52, 347–353.

Breembroek, G.R.M., van Straalen, A., Witkamp, G.J., van Rosmalen, G.M., 1998. Extraction of cadmium and copper using hollow fiber supported liquid membranes. J. Membr. Sci. 146, 185–195.

Brnčić, M., Karlović, S., Rimac Brnčić, S., Penava, A., Bosiljkov, T., Ježek, D., Tripalo, B., 2010. Textural properties of infra red dried apple slices as affected by high power ultrasound pre-treatment. Afr. J. Biotechnol. 9, 6907–6915.

Chapman, P.D., Oliveira, T., Livingston, A.G., Li, K., 2008. Membranes for the dehydration of solvents by pervaporation. J. Membr. Sci. 318, 5–37.

Chemat, F., Huma, Z., Kamran Khan, M., 2011. Applications of ultrasound in food technology: processing, preservation and extraction. Ultrason. Sonochem. 18, 813–835.

Corrales, M., Toepfl, S., Butz, P., Knorr, D., Tauscher, B., 2008. Extraction of anthocyanins from grape by-products assisted by ultrasonics, high hydrostatic pressure or pulsed electric fields: a comparison. Innov. Food Sci. Emerg. Technol. 9, 85–91.

Cvjetko Bubalo, M., Sabotin, I., Radoš, I., Valentinčič, J., Bosiljkov, T., Brnčić, M., Žnidaršič-Plazl, P., 2013. A comparative study of ultrasound, microwave and microreactor-assisted imidazolium-based ionic liquid synthesis. Green Process. Synth. 2, 579–590.

Da Porto, C., Porretto, E., Decorti, D., 2013. Comparison of ultrasound-assisted extraction with conventional extraction methods of oil and polyphenols from grape (Vitis vinifera L.) seeds. Ultrason. Sonochem. 20, 1076–1080.

Deng, Q., Zinoviadou, K.G., Galanakis, C.M., Orlien, V., Grimi, N., Vorobiev, E., Lebovka, N., Barba, F.J., 2015. The effects of conventional and non-conventional processing on glucosinolates and its derived forms, isothiocyanates: extraction, degradation and applications. Food Eng. Rev. DOI: 10.1007/s12393-014-9104-9.

Donsi, F., Ferrari, G., Pataro, G., 2010. Applications of pulsed electric field treatments for enhancing of mass transfer from vegetable tissues. Food Eng. Rev. 2, 109–130.

Dujmić, F., Brnčić, M., Karlović, S., Bosiljkov, T., Ježek, D., Tripalo, B., Mofardin, I., 2013. Ultrasound-assisted infrared drying of pear slices: textural issues. J. Food Process. Eng. 36, 397–406.

Dupuy, A., Athès, V., Schenk, J., Jenelten, U., Souchon, I., 2011a. Solvent extraction of highly valuable oxygenated terpenes from lemon essential oil using a polypropylene membrane contactor: potential and limitations. Flavour Frag. J. 26, 192–203.

Dupuy, A., Athès, V., Schenk, J., Jenelten, U., Souchon, I., 2011b. Experimental and theoretical considerations on breakthrough pressure in membrane-based solvent extraction: focus on citrus essential oil/hydroalcoholic solvent systems with low interfacial tension. J. Membr. Sci. 378, 203–213.

Eh, A.L.-S., Teoh, S.-G., 2012. Novel modified ultrasonication technique for the extraction of lycopene from tomatoes. Ultrason. Sonochem. 19, 151–159.

Fernandes Fabiano, A.N., Linhares, Jr., Francisco, E., Rodrigues, S., 2008. Ultrasound as pre-treatment for drying of pineapple. Ultrason. Sonochem. 15, 1049–1054.

Franken, A.C.M., Mulder, M.H.V., Smolders, C.A., 1990. Pervaporation process using a thermal gradient as the driving force. J. Membr. Sci. 53, 127–141.

Gabelman, A., Hwang, S.-T., 1999. Hollow fiber membrane contactors. J. Membr. Sci. 159, 61–106.

Gabelman, A., Hwang, S.T., Krantz, W.B., 2005. Dense gas extraction using a hollow fiber membrane contactor: experimental results versus model predictions. J. Membr. Sci. 257, 11–36.

Galanakis, C.M., 2012. Recovery of high added-value components from food wastes: conventional, emerging technologies and commercialized applications. Trends Food Sci. Technol. 26, 68–87.

Galanakis, C.M., 2013. Emerging technologies for the production of nutraceuticals from agricultural by-products: a viewpoint of opportunities and challenges. Food Bioprod. Process. 91, 575–579.

Galanakis, C.M., Chasiotis, S., Botsaris, G., Gekas, V., 2014. Separation and recovery of proteins and sugars from Halloumi cheese whey. Food Res. Int. 65, 477–483.

Galanakis, C.M., Schieber, A., 2014. Editorial. Special issue on recovery and utilization of valuable compounds from food processing by-products. Food Res. Int. 65, 230–299.

Galanakis, C.M., Tornberg, E., Gekas, V., 2010d. Recovery and preservation of phenols from olive waste in ethanolic extracts. J. Chem. Technol. Biotechnol. 85, 1148–1155.

Gaschovska, T., Cassada, D., Subbiah, J., Hanna, M., Thippareddi, H., Snow, D., 2010. Enhanced anthocyanin extraction from red cabbage using pulsed electric field processing. J. Food Sci. 75, 23–29.

Gascons-Viladomat, F., Souchon, I., Athès, V., Marin, M., 2006. Membrane air-stripping of aroma compounds. J. Membr. Sci. 277, 129–136.

Gros, C., Lanoiselle, J.-L., Vorobiev, E., 2003. Towards an alternative extraction process for linseed oil. Chem. Eng. Res. Des. 81, 1059–1065.

He, Y., Bagley, D.M., Leung, K.T., Liss, S.N., Liao, B.Q., 2012. Recent advances in membrane technologies for biorefining and bioenergy production. Biotechnol. Adv. 30, 817–858.

Heng, W.W., Xiong, L.W., Ramanan, R.N., Hong, T.L., Kong, K.W., Galanakis, C.M., Prasad, K.N., 2015. Two level factorial design for the optimization of phenolics and flavonoids recovery from palm kernel by-product. Ind. Crop. Prod. 63, 238–248.

Heinz, V., Álvarez, I., Angersbach, A., Knorr, D., 2001. Preservation of liquid foods by high intensity pulsed electric fields: basic concepts for process design. Trends Food Sci. Technol. 12, 03–11.

Hossain, M.B., Tiwari, B.K., Gangopadhyay, N., O'Donnell, C.P., Brunton, N.P., Rai, D.K., 2014. Ultrasonic extraction of steroidal alkaloids from potato peel waste. Ultrason. Sonochem. 21, 1470–1476.

Irving, I., Albagli, D., Dark, M., Perelman, L., Rosenbery, C., Feld, M., 1995. The thermoelastic basis of short pulsed laser ablation of biological tissue. Biophysics 92, 1960–1964.

Izadifar, Z., 2013. Ultrasound pretreatment of wheat dried distiller's grain (DDG) for extraction of phenolic compounds. Ultrason. Sonochem. 20, 1359–1369.

Jonquières, A., Clément, R., Lochon, P., Néel, J., Dresch, M., Chrétien, B., 2002. Industrial state-of-the-art of pervaporation and vapour permeation in the western countries. J. Membr. Sci. 206, 87–117.

Kautek, W., P., Rudolph, P., Daminelli, G., Kruger, J., 2005. Physico-chemical aspects of femtosecond-pulse-laser-induced surface nanostructures. Appl. Phys. A 81, 65–70.

Knorr, D., 2004. Applications and potential of ultrasonics in food processing. Trends Food Sci. Technol. 15, 261–266.

Kreulen, H., Smolders, C.A., Versteeg, G.F., Swaaij, W. P. M. v., 1993. Microporous hollow fibre membrane modules as gas–liquid contactors. Part 1. Physical mass transfer processes. J. Membr. Sci. 78, 197–216.

Kunz, W., Benhabiles, A., Ben-Aïm, R., 1996. Osmotic evaporation through macroporous hydrophobic membranes: a survey of current research and applications. J. Membr. Sci. 121, 25–36.

Li, L., Lanoisellé, J.L., Ding, L., Clausse, D., 2009. Aqueous extraction process to recover oil from press-cakes. In: Proceedings of WCCE8. Montréal, Canada.

Liu, D., Vorobiev, E., Savoire, R., Lanoiselle, J.L., 2011. Intensification of polyphenols extraction from grape seeds by high voltage electrical discharge and extract concentration by dead-end ultrafiltration. Separ. Purif. Technol. 81, 134–140.

Luengo, E., Álvarez, I., Raso, J., 2013. Improving the pressing extraction of polyphenols of orange peel by pulsed electric fields. Innov. Food Sci. Emerg. Technol. 17, 79–84.

Lusquinos, F., Comesana, R., Riveiro, A., Quintero, F., Pou, J., 2006. Laser cladding from marine wastes for biomedical applications. 545ICALEO 2006 – Twenty-fifth International Congress on Application of Laser and Electro-Optics. Congress Proceedings.

Mandowara, A., Prashant, K.B., 2011. Simulation studies of ammonia removal from water in a membrane contactor under liquid–liquid extraction mode. J. Environ. Manage. 92, 121–130.

Ming, X., 2013. Optimization of ultrasonic assisted hot water extraction of gelatin from *Cyprinus caprio haematopterus* scale and establishment of its mathematical model. Food Sci. 34, 101–105.

Mújica-Paz, H., Valdez-Fragoso, A., Samson, C.T., Welti-Chanes, J., Torres, J.A., 2011. High-pressure processing technologies for the pasteurization and sterilization of foods. Food Bioprocess Technol. 4, 969–985.

Neel, J., Aptel, P., Clement, R., 1985. Basic aspects of pervaporation. Desalination 53, 297–326.

Pabby, A.K., Sastre, A.M., 2013. State-of-the-art review on hollow fibre contactor technology and membrane-based extraction processes. J. Membr. Sci. 430, 263–303.

Panchev, I.N., Dimitrov, D.A., Kirchev, N.A., 2011. Possibility for application of laser ablation in food technologies. Innov. Food Sci. Emerg. Technol. 12, 369–374.

Parniakov, O., Barba, F.J., Grimi, G., Lebovka, N., Vorobiev, E., 2014. Impact of pulsed electric fields and high voltage electrical discharge on extraction of high-added value compounds from papaya peels. Food Res. Int. 65, 337–343.

Patsioura, A., Galanakis, C.M., Gekas, V., 2011. Ultrafiltration optimization for the recovery of β-glucan from oat mill waste. J. Membr. Sci. 373, 53–63.

Peters, T.A., Benes, N.E., Keurentjes, J.T.F., 2008. Hybrid ceramic-supported thin PVA pervaporation membranes: long-term performance and thermal stability in the dehydration of alcohols. J. Membr. Sci. 311, 7–11.

Pierre, F.X., Souchon, I., Marin, M., 2001. Recovery of sulfur aroma compounds using membrane-based solvent extraction. J. Membr. Sci. 187, 239–253.

Pierre, F.X., Souchon, I., Athès-Dutour, V., Marin, M., 2002. Membrane-based solvent extraction of sulfur aroma compounds: influence of operating conditions on mass transfer coefficients in a hollow fiber contactor. Desalination 148, 199–204.

Pingret, D., Fabiano-Tixier, A.S., Le Bourvellec, C., Renard, M.G.C.C., 2012. Lab and pilot scale ultrasound-assisted water extraction of polyphenols from apple pomace. J. Food Eng. 111, 73–81.

Prasad, R., Sirkar, K.K., 1992. Membrane-based solvent extraction. In: Ho, W.S.W., Sirkar, K.K. (Eds.), Membrane Handbook. Chapman & Hall, New York, pp. 727–763.

Puértolas, E., Cregenzán, O., Luengo, E., Álvarez, I., Raso, J., 2013. Pulsed-electric-field-assisted extraction of anthocyanins from purple-fleshed potato. Food Chem. 136, 1330–1336.

Puértolas, E., Martínez de Marañón, I., 2015. Olive oil pilot-production assisted by pulsed electric field: impact on extraction yield, chemical parameters and sensory properties. Food Chem. 167, 497–502.

Puértolas, E., Luengo, E., Álvarez, I., Raso, J., 2012. Improving mass transfer to soften tissues by pulsed electric fields. Fundamentals and applications. Annu. Rev. Food Sci. Technol. 3, 263–282.

Rajagopalan, N., Cheryan, M., Matsuura, T., 1994. Recovery of diacetyl by pervaporation. Biotechnol. Tech. 8, 869–872.

Rajha, H.N., Boussetta, N., Louka, N., Maroun, R.G., Vorobiev, E., 2014. A comparative study of physical pretreatments for the extraction of polyphenols and proteins from vine shoots. Food Res. Int. 65, 462–468.

Rastogi, N.K., 2011. Opportunities and challenges in application of ultrasound in food processing. Crit. Rev. Food Sci. 51, 705–722.

Ribeiro, C.P.J., Borges, P.C., Lage, P.L.C., 2005. A new route combining direct-contact evaporation and vapor permeation for obtaining high-quality fruit juice concentrates. Part I. Experimental analysis. Ind. Eng. Chem. Res. 44, 6888–6902.

Roseiro, L.B., Duarte, L.C., Oliveira, D.L., Roque, R., Bernardo-Gil, M.G., Martins, A.I., Sepúlveda, C., Almeida, J., Meireles, M., Gírio, F.M., Rauter, A.P., 2013. Supercritical, ultrasound and conventional extracts from carob (*Ceratonia siliqua* L) biomass: effect on the phenolic profile and antiproliferative activity. Ind. Crop. Prod. 47, 132–138.

Roselló-Soto, E., Galanakis, C.M., Brncic, M., Orlien, V., Trujillo, F.J., Mawson, R., Knoerzer, K., Tiwari, B.K., Barba, F.J., 2015. Clean recovery of antioxidant compounds from plant foods, by-products and algae assisted by ultrasounds processing. Modeling approaches to optimize processing conditions. Trends Food Sci. Technol. 42 (2), 134–149.

Russo, R.E., 1995. Laser ablation. Appl. Spectrosc. 49, 14A–28A.

Russo, R.E., Mao, X., Mao, S.S., 2002. The physics of laser ablation in microchemical analysis. Anal. Chem. 1, 72A–77A.

Russo, R.E., Bol'shkov, A.A., Mao, X., McKay, C.P., Perry, D.L., Sorkhabi, O., 2011. Laser ablation molecular isotopic spectrometry. Spectrochim. Acta Part B 66, 99–104.

Russo, R.E., Mao, X., Mao, S., Gonzalez, J, J., Zorba, V., Yoo, J., 2013. Laser ablation in analytical chemistry. ACS Pub. 85, 6162–6177.

Schafer, T., Crespo, J.G., 2005. Vapor permeation and pervaporation. In: Afonso, C.A.M., Crespo, J.G. (Eds.), Green Separation Processes Fundamentals and Applications. Wiley-VCH, Lisboa, pp. 271–289.

Sciubba, L., Di Gioia, D., Fava, F., Gostoli, C., 2009. Membrane-based solvent extraction of vanillin in hollow fiber contactors. Desalination 241, 357–364.

Shao, P., Huang, R.Y.M., 2007. Polymeric membrane pervaporation. J. Membr. Sci. 287, 162–179.

Soliva-Fortuny, R., Balasa, A., Knorr, D., Martín-Belloso, O., 2009. Effects of pulsed electric fields on bioactive compounds: a review. Trends Food Sci. Technol. 20, 544–556.

Soni, V., Abildskov, J., Jonsson, G., Gani, R., 2008. Modeling and analysis of vacuum membrane distillation for the recovery of volatile aroma compounds from black currant juice. J. Membr. Sci. 320, 442–455.

Souchon, I., Pierre, F.X., Athès-Dutour, V., Marin, M., 2002a. Pervaporation as a deodorization process applied to food industry effluents: recovery and valorisation of aroma compounds from cauliflower blanching water. Desalination 184, 79–85.

Souchon, I., Pierre, F.X., Samblat, S., Bes, M., Marin, M., 2002b. Recovery of aroma compounds from industrial food aqueous effluent using membrane-based solvent extraction. Desalination 184, 87–92.

Souchon, I., Athès, V., Pierre, F.X., Marin, M., 2004. Liquid–liquid extraction and air stripping in membrane contactor: application to aroma compounds recovery. Desalination 163, 39–46.

Tin, P.S., Lin, H.Y., Ong, R.C., Chung, T.-S., 2011. Carbon molecular sieve membranes for biofuel separation. Carbon 49, 369–375.

Toepfl, S., Knorr, D., 2006. Pulsed electric fields as a pretreatment in drying processes. Stewart Postharvest. Rev. 3, 4–6.

Tsakona, S., Galanakis, C.M., Gekas, V., 2012. Hydro-ethanolic mixtures for the recovery of phenols from Mediterranean plant materials. Food Bioprocess Technol. 5, 1384–1393.

Urtiaga, A.M., Gorri, E.D., Ruiz, G., Ortiz, I., 2001. Parallelism and differences of pervaporation and vacuum membrane distillation in the removal of VOCs from aqueous streams. Separ. Purif. Technol. 22-23, 327–337.

Valdés, H., Romero, J., Saavedra, A., Plaza, A., Bubnovich, V., 2009. Concentration of noni juice by means of osmotic distillation. J. Membr. Sci. 330, 205–213.

Vaillant, F., Jeanton, E., Dornier, M., O'Brien, G.M., Reynes, M., Decloux, M., 2001. Concentration of passion fruit juice on an industrial pilot scale using osmotic evaporation. J. Food Eng. 47, 195–202.

Van Veen, H.M., Rietkerk, M.D.A., Shanahan, D.P., van Tuel, M.M.A., Kreiter, R., Castricum, H.L., ten Elshof, J.E., Vente, J.F., 2011. Pushing membrane stability boundaries with HybSi® pervaporation membranes. J. Membr. Sci. 380, 124–131.

Vilkhu, K., Mawson, R., Simons, L., Bates, D., 2008. Applications and opportunities for ultrasound assisted extraction in the food industry – a review. Innov. Food Sci. Emerg. Technol. 9, 161–169.

Virot, M., Tomao, V., Le Bourvellec, C., Renard, C., Chemat, F., 2010. Towards the industrial production of antioxidants from food processing by-products with ultrasound-assisted extraction. Ultrason. Sonochem. 17, 1066–1074.

Volkov, V.V., Borisov, I.L., 2012. Thermopervaporation membrane bioreactor as a new concept for the low-cost production of biobutanol. Procedia Eng. 44, 278–280.

Vorobiev, E., Lebovka, L., 2011. Pulsed electric field-assisted extraction. In: Lebovka, L., Vorobiev, E., Chemat, F. (Eds.), Enhancing Extraction Processes in the Food Industry. CRC Press, Boca Raton, pp. 25–84.

Wang, Y., Widjojo, N., Sukitpaneenit, P., Chung, T.S., 2013. Membrane pervaporation. In: Ramaswamy, S., Huang, H.J., Ramarao, B.V. (Eds.), Separation and Purification Technology in Biorefineries. Wiley, Lisboa, pp. 259–299.

Wee, S.L., Tye, C.T., Bhatia, S., 2008. Membrane separation process: pervaporation through zeolite membrane. Separ. Purif. Technol. 63, 500–516.

Widyasari, R., Rawdkuen, S., 2014. Extraction and characterization of gelatin from chicken feet by acid and ultrasound assisted extraction. Food Appl. Biosci. J. 2, 85–97.

EMERGING PURIFICATION AND ISOLATION

Arijit Nath*, Ooi Chien Wei**, Sangita Bhattacharjee†, Chiranjib Bhattacharjee*

*Department of Chemical Engineering, Jadavpur University, Kolkata, West Bengal, India;
**Discipline of Chemical Engineering, School of Engineering, Monash University, Malaysia;
†Department of Chemical Engineering, Heritage Institute of Technology, Kolkata, West Bengal, India

12.1 INTRODUCTION

In the twenty-first century, an outstanding proliferation of biotechnology as well as concern about recycling and reuse of waste material led to the development of upstream and downstream processing technology. Moreover, conservation of resources through the green route is also generating a lot of attention (Bhattacharjee et al., 2006).

Isolation and purification is the last recapture step of the *5-Stage Universal Recovery Process,* as the last step concerns product formation, which is not a recovery process. Therefore, isolation should be more compact, low cost, environmental friendly, and high throughput with pure end-product. For more than two decades, conventional technologies, such as adsorption, nanofiltration, chromatography, and electrodialysis, have been widely used for this purpose (Galanakis, 2011, 2012; Galanakis et al., 2012; Rahmanian et al., 2014). However, due to some limitations with respect to purity and yield of target molecules (Galanakis et al., 2010a, b, c, d, e; Galanakis and Schieber, 2014), more advanced processing technologies have recently been proposed (Galanakis et al., 2012; Deng et al., 2015; Roselló-Soto et al., 2015). Nowadays, the emerging isolation techniques include magnetic fishing, aqueous two-phase system, and ion-exchange membrane chromatography (Table 12.1). This chapter discusses the characteristics of these emerging technologies.

12.2 MAGNETIC FISHING
12.2.1 CHARACTERISTICS

Isolation and purification of various types of biomolecules have received a lot of attention in almost all areas of biotechnology. In the field of downstream biotechnology the isolation of nucleic acid, antibodies, proteins, and peptides is usually performed using electrophoresis, a size exclusion-based membrane separation technique, precipitation, and a variety of chromatographies (Ghosh, 2006). Among these methods, affinity ligand technique is considered the most powerful tool in terms of both purity and product yield. Although column affinity chromatography has been successfully applied, the disadvantage of this system is the unfeasible separation of target molecules in the presence of suspended solid and fouling components.

Table 12.1 Characteristics of Conventional Versus Emerging Isolation and Purification Techniques

Conventional	Emerging
Adsorbtion: Depends upon porosity of surface and thus capacity is decreased when the pores are saturated. Deadsorbtion takes place often, too. As the process is not selective, so that it is limited to separate particular one from multi-component mixture. Moreover, reusability of adsorbent is one of the major problems.	Aqueous two-phase system: It is performed by selective distribution of solutes between the two phases which are composed majorly aqueous solutions. The composition of aqueous two-phase system could be made up of compounds like polymers, surfactant, salt, alcohol and ionic liquid. Its separation efficacy is mainly dependent upon on the hydrophobic interaction and electrostatic interactions between the target compound and the phase components. The process is more specific because it is depending on the distribution coefficient of each molecule.
Nanofiltration: A pressure-driven size exclusion-based membrane separation process, where concentration polarization and fouling often take place. Reuse of membrane is also a major problem. Therefore, it is not applicable to separate solutes with similar molecular weights.	Magnetic fishing: Depends on bearing carriers (affinity or hydrophobic ligand, or ion-exchange groups, or magnetic biopolymer particles having affinity to the isolated solute) that are mixed with a sample containing target compounds. This process is usually a very simple and fine-tuned separation technique. Solutes are gently separated and are especially useful in large-scale operations. Due to the magnetic properties of the adsorbents, the solutes can be relatively easily and selectively removed from the sample. This method can purify small magnetic particles (diameter ca. $0.1–1$ μm) in the presence of biological debris and other fouling material of a similar size.
Chromatography: Gel permeation chromatography is a size exclusion-based separation process. This technique cannot separate solutes with similar molecular weight. Electrophoresis separates solute molecules (mainly protein) depending on their charge and mass ratio. Thin layer chromatography separates solutes based on their refractive factor. The major disadvantages of these techniques is the low throughput.	Membrane ion-exchange chromatography: Performance depends on the pH of the solutes (ionic charge) and elution buffer. Therefore, it can satisfactorily separate similar molecular weight. The solid adsorbents are either positively or negatively charged and adsorbed by the charged adsorbents (stationary phase). During elution with selective buffer, individual proteins could be separated. This method provides the high purity of the desired molecules, whereas the membrane could also be reused.
Electrodialysis: Charged solutes are separated by concentration and an ion-exchange membrane. Higher molecular weight, noncharged, and less mobile ionic species are not removed satisfactory. Membrane fouling and concentration polarization are the two major disadvantages. Electrodialysis is not cost effective since it requires high electric demands and feed pretreatment.	

In this case, magnetic fishing techniques (in a batch mode or in the form of stabilized fluidized beds or magnetically modified two-phase systems) have received a lot of attention (Safarik and Safarikova, 2004).

Magnetic fishing is used to isolate active ingredients from crude cell lysates, whole blood, plasma, ascites fluid, urine, cultivation broth, wastes from the food and fermentation industries, milk, whey, and others. Some immobilized affinity, ion exchange, hydrophobic ligand, or magnetic biopolymer particle could possess affinity toward the isolated compounds. Magnetic carriers bearing such particles/ligands are mixed with a sample containing target molecules. Following a particular incubation period where

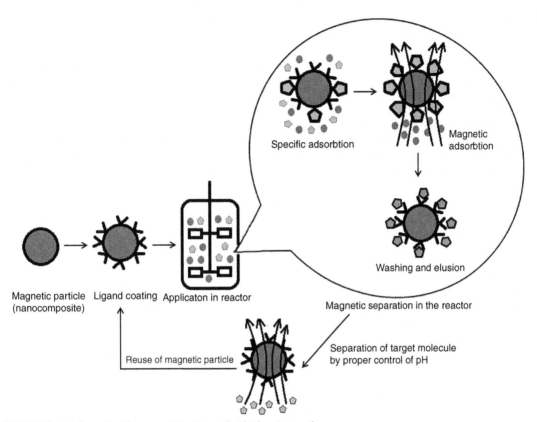

Specific adsorbtion

Magnetic adsorbtion

Washing and elusion

Magnetic particle (nanocomposite)   Ligand coating   Applicaton in reactor

Magnetic separation in the reactor

Separation of target molecule by proper control of pH

Reuse of magnetic particle

FIGURE 12.1 Schematic Diagram of the Magnetic Fishing Separation

the target compound is attached to the magnetic particles, the whole magnetic complex is carefully and rapidly removed from the sample medium using an appropriate magnetic separator. After washing out the contaminants, the isolated target compound is eluted and used for further processing (Safarik and Safarikova, 2004; Schuster et al., 2000; Hofmann et al., 2002; Alche and Dickinson, 1998). A schematic diagram of the magnetic selective bioseparation procedure can be seen in Fig. 12.1.

Magnetic separation techniques have several advantages in comparison with standard isolation procedures, since they are very simple with few handling steps, which can take place in one single test tube or simple solitary vessel. In addition, the separation process can be performed directly in crude samples containing suspended solid materials. As the magnetic separation technique is so selective there is no need to have pretreatment step like precipitation, centrifugation, filtration, etc. In some cases (e.g. isolation of intracellular proteins) it is even possible to integrate the disintegration and separation steps to reduce the total separation time. In fact, magnetic separation is the only feasible method for recovery of small magnetic particles (diameter ca. $0.1–1$ μm) in the presence of biological debris and other fouling material of similar size in large-scale operations. Usually large protein complexes have a tendency to break up by traditional column chromatography techniques, whereas, magnetic separation is very gentle for the separation of target proteins or peptides (Hofmann et al., 2002).

12.2.2 APPLICATIONS

The composition of food waste is usually complex and thus common separation technology cannot provide satisfactory results in terms of product purity and yield. On the other hand, magnetic fishing is very selective, which is why it is used for the recovery of active ingredients from food wastes. Generally, magnetic affinity isolations can be performed in two different ways. In the direct method, magnetic affinity particles are prepared as an appropriate affinity ligand, which is coupled directly to magnetic particles (or biopolymers) that show an affinity toward target compound(s). In the indirect method the free affinity ligand is introduced in the solution or suspension to allow the interaction with the target compounds, and the resulting complex is captured by magnetic particles. For magnetic affinity adsorption, particles with immobilized affinity ligands could be used. For example, magnetic particles with immobilized ligands, such as, streptavidin, antibodies, protein A, and protein G, could also be used as the generic solid phase where native or modified affinity ligands can be immobilized. On the other hand, the free affinity ligands can be biotinylated (covalently attaching biotin to a protein, nucleic acid, or other molecule) and magnetic particles with immobilized streptavidin or avidin are used to incarcerate the formed complexes. In both methods, isolated target compounds are removed by a series of washing steps after changing the pH value of the solution. Although both methods perform generally well, the direct technique is more attractive since it is precise, easy, and controllable. However, the indirect procedure may perform better if ligands have poor affinity for the target compounds. Batch magnetic adsorption is also popular for the separation of larger magnetic particles (with diameters >ca. 1 μm). Usually high-gradient magnetic separators have been used in a batch process. Alternatively magnetically stabilized fluidized beds are used for continuous separation process (Lochmuller et al., 1988; Burns and Graves, 1985; Chetty and Burns, 1991; Safarik and Safarikova, 2004).

12.2.2.1 Purification of proteins from whey

In 2004, Anders Heebøll-Nielsen described a method to isolate lactoferrin and lactoperoxidase from crude bovine whey using superparamagnetic ion exchangers. Initially, crude bovine whey was treated with a superparamagnetic cation exchanger to adsorb the valuable proteins with minimum losses (40% of the maximum lysozyme binding). In the second step, removal of lactoferrin and lactoperoxidase was performed by an anion exchanger with contaminant immunoglobulins. The latter were desorbed from the proteins using a low concentration of sodium chloride ($\leq$0.4 mol/L). Indeed, lactoperoxidase and lactoferrin were co-eluted in a purer form when the concentration of sodium chloride was increased to 0.4–1 molL (Heebøll-Nielsen et al., 2004). Chen et al. (2007) synthesized micron-sized monodisperse superparamagnetic polyglycidyl methacrylate (PGMA) particles coupled with heparin (PGMA-heparin) for the isolation of lactoferrin from bovine whey. In this procedure, adsorption was initially conducted with the above supermagnetic nanoparticles and eluted using the same buffer in a different concentration to recover the target proteins (Chen et al., 2007). Furthermore, Meyer et al. (2005) developed a gradient magnetic fishing methodology for the separation of superoxide dismutase (SOD), an antioxidant protein from whey. In that method, metal chelate supports, charged with copper (II) ions, were considered a suitable ligand and used in a novel high-gradient magnetic separator. Finally, a high-gradient magnetic separator was used for the isolation of SOD from crude sweet whey in a batch reactor. Subsequently, proteins were separated from support by suitable pH of elution buffer. In this technique, the purification fold of SOD was increased 50 times with a yield of >85% at purification factor of approximately 21 (Meyer et al., 2005).

12.2.2.2 Purification of other compounds

A rapid procedure for the large-scale purification of concanavalin A and agglutinin from leguminous extracts has been developed using large-scale high-gradient magnetic fishing. Specifically, three types of magnetic adsorbent were prepared. The first was developed by the direct attachment of maltose or glucose to amine-terminated iron oxide particles, which could bind levels of up to 280 mg concanavalin A/g. The adsorbent's characteristic, coupled tentacular dextran chains have displayed a maximum binding capacity (238 mg/g), and a dissociation constant of 0.13 mM. Adsorbents derivatized with mixed mode or hydrophobic charge induction ligands showed very high capacities for both concanavalin A and *Lens culinaris* agglutinin separation (more than 250 mg/g). Dextran-linked supports have also been employed for the isolation of concanavalin A from the extract of jack beans (Heebøll-Nielsen et al., 2004). In 2010, a simple protocol for the purification of lectin from large-scale potato starch industry wastewater was developed (Safarik et al., 2010). In this case, magnetic chitosan microparticles were used as an affinity adsorbent in a flow separation system, whereas the adsorbed lectin was recovered by elution with glycine/HCl buffer (pH 2.2). The specific activity of purified lectin was increased approximately 27-fold after purification (Safarik et al., 2010). Sabatkova et al. (2008) prepared low-cost magnetic adsorbents from egg ovalbumin using methanol precipitation and subsequently glutaraldehyde cross-linking. The latter process was used to preconcentrate two plant lectins from potato tuber and wheat germ extracts. However, the adsorbed lectins were eluted with diluted hydrochloric acid. Indeed, the specific activities of both purified lectins were increased approximately 30 to 40-fold (Sabatkova et al., 2008). Low-cost and easily prepared magnetic chitosan microparticles cross-linked with glutaraldehyde were also used for the purification of lectin from potato wastewater; the prepared magnetic chitosan microparticles could be reused at least three times with a simple regeneration step (Kateřina et al., 2012). The partial purification of lectin by magnetic composite particles has also been developed by several researchers (Šafaříková and Šafařík, 2000).

12.3 AQUEOUS TWO-PHASE SYSTEM

12.3.1 CHARACTERISTICS

The aqueous two-phase system is a gentle and nondisruptive liquid/liquid partitioning technique, consisting of two types of phase-forming components. When these components are mixed above a critical concentration in an aqueous solution, they display immiscibility and the mixture becomes turbid. Below the critical concentration, both phase-forming components become miscible and revert to a single homogeneous phase (Albertsson, 1986). The interfacial tension between the phases is low (0.0001–0.1 dyne/cm), which is about 400-fold lower as compared with the typical water/organic solvent systems. This fact enables a better and rapid migration process through the interface (Hatti-Kaul, 2000). In addition, small droplets of the dispersed phases are formed upon mixing or stirring. The droplets contain large interfacial areas that can accelerate the partition and the equilibrium mass transfer in the aqueous two-phase system. The aqueous two-phase system offers a better alternative to downstream processing of various biological products such as proteins, nucleic acids, virus particles, microorganisms, and plant and animal cells (Albertsson, 1986; Hatti-Kaul, 2000; Johansson, 1985). Indeed, it integrates the operations of clarification, concentration, and partial purification into a single recovery step, while it can improve the yield and circumvent the shortcoming of product loss in a multistep processing procedure. Besides, dilute products can be concentrated by partitioning the desired substances into smaller volumes of the extraction phase or at the interface of the aqueous two-phase system. Furthermore, polyols such as polyethylene glycol used in the aqueous two-phase system have a stabilizing effect on the proteins, hence maintaining the biological activities and native structure of the protein (Albertsson, 1986). For an aqueous two-phase system composed of polyethylene glycol and dextran, both polymers are nontoxic and classified as safe for the recovery of proteins and industrial use in food processing. The implementation of large-scale aqueous two-phase system purification is feasible due to its easy handling, reliable scale-up, large loading capacity, possibility of continuous steady-state and automated operations, low investment cost, and minimum energy input (Hatti-Kaul, 2000). By recycling the phase-forming components used in the aqueous two-phase system, the problems of chemicals consumption and downstream pollution could be partially eliminated.

12.3.2 APPLICATIONS

The application of the aqueous two-phase system in recovery of biological products has been mainly focused on the purification of proteins, which are produced via microbial fermentation. The aqueous two-phase system can be conveniently used at the early stages of product recovery from the microbial feedstock, containing whole cells, cell debris, or solid particles (Gupta et al., 2004). In addition to microbial sources, food waste is becoming an immerse source of protein as well as other compounds with commercial significance, such as saccharides, antioxidants, and flavonoids. The aqueous two-phase system has been successfully implemented in the recovery of these valuable compounds from food processing wastes such as by-products from cereals (bran, straw, mill waste), fruit crops (peels, skins), animals (skin, bones, shells), and processed food that have passed their expiration date. Table 12.2 lists a number of valuable compounds recovered by using the aqueous two-phase system.

Aqueous two-phase system preparation entails basic steps of equilibration and phase separation (Hustedt et al., 1985). Equilibration is the process of mixing the phase-forming components and dispersing the phases into homogeneity. Gentle shaking or agitation is sufficient for mixing the content

Table 12.2 Valuable Compounds Recovered from Food Waste by Using Aqueous Two-Phase System

Target Molecule	Source	References
Lipase	Rice bran	Wang et al. (2013)
Bromelain	Pineapple peel	Ketnawa et al. (2010); Novaes et al. (2013)
Serine protease	Mango peel	Amid et al. (2012)
Pectinase	Mango peel	Amid et al. (2013)
Vanillin and L-ascorbic acid	Vanilla diet pudding	Reis et al. (2012)
β-Lactoglobulin and α-lactalbumin	Cheese whey	Jara and Pilosof (2011)
Alkaline protease	Chicken intestine, giant catfish viscera	Sarangi et al. (2011); Ketnawa et al. (2013)
Protease	Albacore tuna stomach	Nalinanon et al. (2009)

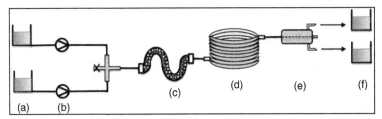

FIGURE 12.2 A Simplified Scheme of the Continuous Separator

(a) Pre-equilibrated phases and sample solution; (b) peristaltic pumps; (c) phase mixer and turbulence generator; (d) continuous tubular separator; (e) phase collector with interface harvesting port; (f) collected top and bottom phases.

Figure adapted from Vázquez-Villegas et al. (2011)

thoroughly. Phase separation can be performed either by settling (under gravitational force) or centrifugation. Settling requires a longer separation time if the discrepancy of density between both phases is low or the viscosity of phases is high (Hustedt et al., 1985). A new separation technique was proposed recently for the continuous operation of the aqueous two-phase system (Vázquez-Villegas et al., 2011). This technique utilizes a tubular reactor and a separator without the need for the centrifugation process (see Fig. 12.2). In large-scale and continuous processes, equilibration and phase separation can be performed by mixer/settler, column contactor, or centrifugal separators (see Fig. 12.3) (Cunha and Aires-Barros, 2000; Espitia-Saloma et al., 2014). In general, the phase-forming components are prepared in stock solution to avoid inaccuracy of pipetting owing to the high viscosity of polymers (e.g. dextran and polyethylene glycol with high molecular weight). Mixture of the phase systems is made up on a weight per weight basis, using the required stock solution, an appropriate buffer, and the feedstock (Hatti-Kaul, 2000).

The distribution of a substance in the aqueous two-phase system is governed by the substance's properties (e.g. size, net charge, and surface properties) and phase-forming components (e.g. type, concentration, or the polymer's molecular weight), together with the surrounding environment, e.g.

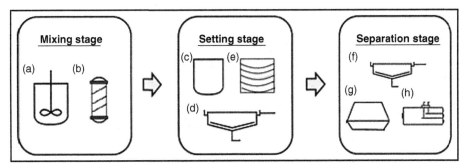

FIGURE 12.3 Devices Employed for Aqueous Two-Phase System Continuous Operation

A mixer tank (a) or a static mixer (b) for the mixing stage, a basic tank (c), an extended tubular settler (d), or a decanter (e) for the settling stage, and finally a decanter (f), a centrifuge (g), or a novel tubular separator (h) for the separating stage.

Figure adapted from Espitia-Saloma et al. (2014)

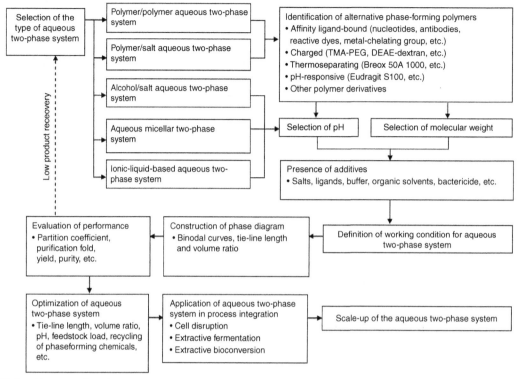

FIGURE 12.4 Practical Strategy for the Process Development of Aqueous Two-Phase System

pH value, temperature, and the presence of additives (e.g. salts or organic compounds) (Albertsson, 1986). In addition, the interaction between phase-forming components and target compounds is through hydrogen bond, van der Waals' forces, electrostatic interactions, steric effects, hydrophobicity, biospecific affinity interactions, and conformational effects (Albertsson, 1986; Albertsson et al., 1990). These system properties can be manipulated to direct the partitioning of the target compounds into one specific phase, while unwanted contaminants such as cell debris can be partitioned into the opposite phase or accumulated at the interface. A practical strategy for the development of the aqueous two-phase system is presented in Fig. 12.4. In the first step of the process design, the selection of phase-forming components is based on the selected type of aqueous two-phase system. In general, there are five types of aqueous two-phase system made up of phase-forming components such as polymers, salts, surfactants, and ionic liquids. A comparison of these five types of aqueous two-phase system is shown in Table 12.3.

The selection of a suitable type of aqueous two-phase system can be facilitated through a preliminary study or characterization of the physicochemical properties (e.g. molecular weight, hydrophobicity, electrochemical charge, and solubility) of target product and typical major contaminants (Benavides and Rito-Palomares, 2008), which have influences on the partition behavior of the desired product. Once the operating condition of the aqueous two-phase system has been defined, a phase diagram should be referred to

Table 12.3 A Comparison of Different Types of Aqueous Two-Phase System

Type	Advantages	Disadvantages
Polymer/polymer aqueous two-phase system	Easily amenable to modification or derivatization	High viscosity
Polymer/salt aqueous two-phase system	Lower viscosity; short separation time	High ionic strength environment
Alcohol/salt aqueous two-phase system	Low cost; low viscosity; short separation time; recycling of alcohol	Denaturation of labile biomolecules; high volatility of alcohol
Aqueous micellar two-phase system	High selectivity of biomolecule solubilization; recycling of surfactant	High cost; denaturation of thermosensitive biomolecules
Ionic liquid (IL)-based aqueous two-phase system	Nonflammability (negligible vapor pressure); strong solubilizing power; low viscosity; high chemical and thermal stability	High cost

or constructed in order to evaluate the influence of the system parameters such as tie-line length, pH, and volume ratio upon the partition behavior of target product (Rito-Palomares, 2004; Rosa et al., 2010). If the initially selected aqueous two-phase system has resulted in unsatisfactory product recovery, other types of aqueous two-phase system should be further examined until an acceptable preliminary partition result is obtained (Benavides and Rito-Palomares, 2008). The process and system parameters of the prototype aqueous two-phase system can then be manipulated to obtain optimum product recovery. Furthermore, the optimized aqueous two-phase system can be integrated to the bioprocesses of either upstream cell culture or downstream processing (Rito-Palomares, 2004). In addition, the aqueous two-phase system holds a great potential in scaling-up for the use in commercial application.

Although the general rules of partitioning in the aqueous two-phase system have been outlined (Huddleston et al., 1991), the main obstacle for aqueous two-phase system implementation is the complex mechanism associated with phase equilibrium and the unique partition behavior of the product (Hatti-Kaul, 2000). The development of the aqueous two-phase system purification strategy is relatively empirical and time-consuming (Hatti-Kaul, 2000). The random screening approach used in the design of the aqueous two-phase system purification strategy results in a less elaborated evaluation of parameters as well as a low predictability related to the potential changes in the performance of the purification process (Rosa et al., 2010). Statistical design of experiments can be alternatively applied for the rapid screening and optimization of the purification process (Hart et al., 1995; Kammoun et al., 2009; Rosa et al., 2007) in order to minimize the number of required experiments for the various conditions of the aqueous two-phase system. Furthermore, a statistical approach could be useful for the study and evaluation of possible interactions between different experimental parameters and their influences on the effective separation of target molecules (Hart et al., 1995; Rosa et al., 2007). The prediction of partition behavior has been attempted by using modeling such as the lattice model, Flory–Huggins solution theory, virial expansions, or the thermodynamic approach (Abbott et al., 1990; Baskir et al., 1989). However, these fundamental models are not sufficient to provide reliable prediction because they depend on a broad range of factors, while the knowledge of the theoretical partitioning mechanisms in the aqueous two-phase system is still limited.

12.4 ION-EXCHANGE MEMBRANE CHROMATOGRAPHY
12.4.1 GENERAL

Isolation of biomolecules is typically based on chemical separation processes. A totally new type of separation technique has been devised under the broad category of chromatography (Ghosh, 2006). Chromatography is the collective term for a set of laboratory techniques used for the separation of mixtures. In the twentieth century, the first known chromatography was traditionally attributed to Russian botanist Mikhail Tswett who used columns of calcium carbonate to separate plant compounds during his research about chlorophyll. Further, in 1952, Archer John Porter Martin and Richard Laurence Millington Synge received a Nobel Prize for their contribution to chromatography. Within the last decade there has been increasing interest in liquid chromatographic processes because of the growing biotechnology industry and the special needs of the pharmaceutical and chemical industries (Smithers, 2008).

In chromatography, the mixture under separation is dissolved in a fluid called the mobile phase, which carries the mixture through another material called the stationary phase. The sample of the mixture to be separated or analyzed is injected as a pulse at the inlet to the column. The individual components are reversibly adsorbed but have different affinities for the adsorbent. The various constituents of the mixture travel at different speeds, causing their isolation. The separation mechanism is based on differential partitioning between the mobile and stationary phases. Due to suitable differences in the partition coefficient of each compound, differential retention on the stationary phase occurs, which leads to separation. Various components emerge from the column at different times and their concentrations in the outlet mobile phase are measured as a function of time by a detector. The response of the detector, which is a measure of the concentration of a component, appears as distinct "peaks" on a plot called a "chromatogram" (Ghosh, 2006).

12.4.2 TECHNICAL ASPECTS OF ION-EXCHANGE MEMBRANE CHROMATOGRAPHY

Chromatographic separations are conventionally carried out in packed beds. Membrane chromatography was developed to overcome the mass-transfer limitations associated with conventional resin-based chromatography. For example, in reversed-phase chromatography (a type of liquid/solid chromatography), the most common packing consists of chemically modified silica gel formed by several particles, with covalently bonded hydrophobic groups (octyl, C_8, or octadecyl, C_{18}). Some of the major disadvantages of using packed beds include:

1. in the case of a liquid chromatography (LC) column, the column diameter is very small and packing particles are very fine, whereas the pressure drop across this type of column is very high (even >50 bar),
2. increase in pressure drop during operation,
3. binding of packed bed by biological macromolecules,
4. solute transport is depended on pore diffusion,
5. difficulty in scaling up.

Although fluidized bed and expanded bed adsorption can solve some of these problems, an alternative approach has been developed by scientists and researchers using stacks of synthetic microporous or macroporous membranes as chromatographic media. In the case of packed bed adsorption using soft porous (or gel-based) media, diffusional resistance governs the transport of solute molecules to the binding sites. Thereby, the process is slow compared with membrane adsorption where the solute transport takes place mainly by fast convection (Ghosh, 2006). The advantages of membrane adsorption/chromatography over packed bed chromatographic operation are:

1. lower pressure drop,
2. low process liquid requirement,
3. possibility of using higher flow rates,
4. easy to set up, and
5. easy scale-up.

Depending on the separation mechanism, membrane chromatography may be categorized as either affinity binding or hydrophobic or ion-exchange interaction. Membrane adsorption processes are carried out in two different pulse- and step-input modes. The pulse input mode is similar to pulse chromatography using packed beds while the step-input mode is similar to conventional adsorption.

The major applications of membrane chromatography include pharmaceutical separation, where the conventional purification techniques are not suitable. In affinity chromatography, the stationary phase has an "affinity ligand" attached to it. The ligand binds selectively the targeted biomolecules, which are afterwards eluted by adjusting the pH and ionic strength of the medium. For biomolecules with a net charge, ion-exchange chromatography is useful since the separation takes place in contact with a stationary phase, where ionic groups are attached. Ion-exchange membrane chromatography (Pedersen et al., 2003; Charcosset, 1998) is an important variation of liquid chromatography. It is a high-resolution separation technique most suited for protein-purification protocols. The separation is based on the reversible electrostatic interaction between a charged protein molecule and the oppositely charged chromatographic membrane (ion exchanger). An ion exchanger consists of an insoluble matrix (stationary phase) such as cellulose, silica, or styrene divinylbenzene where charged groups have been covalently bound. The charged groups are associated with mobile counter-ions. The latter may be reversibly exchanged with other ions of the same charge. The stationary phase surface displays ionic functional groups (R—X) that interact with analyte ions of opposite charge. This type of chromatography is further subdivided into cation and anion exchange chromatography. The ionic compound consisting of the cationic species M^+ and the anionic species B^- can be retained by the stationary phase. Cation exchange chromatography retains positively charged cations because the stationary phase displays a negatively charged functional group:

$$R\text{-}X^-C^+ + M^+B^- \rightleftarrows R\text{-}X^-M^+ + C^+ + B^- \tag{12.1}$$

Anion exchange chromatography retains anions using a positively charged functional group:

$$R\text{-}X^+A^- + M'B^- \rightleftarrows R\text{-}X^+B^- + M^+ + A^- \tag{12.2}$$

12.4.3 APPLICATIONS

Ion-exchange membrane chromatography is widely used for chemical analysis and ion separation. For example, in biochemistry it is widely used to separate charged molecules such as proteins. Another important application is the purification of biologically produced substances (e.g. proteins, amino acids, and DNA/RNA). Other applications include whey protein purification (Bhattacharjee et al., 2006; Goodall et al., 2008). This method provides additional benefits due to the very high association rate between target proteins and functional groups as well as the absence requirements for extreme pH values or heat or chemical pretreatments that could compromise protein structure and functionality (Drioli and Romano, 2001). Amino acids have both acidic and basic properties with a definite role on protein reactivity to ion-exchange media. The protein must displace the counter-ions and therefore bind to the exchanger on the membrane. Before choosing the right column, one must consider whether the target protein would bind as polyanion or polycation. As proteins are usually handled in buffer media their isoelectric points (IEP) are critical. Protein is positively charged below its IEP and thus binds to any cationic-exchange membrane. At a pH higher than the IEP, the target protein will be negatively charged and bind to an anionic-exchange membrane. Desorption of biomolecules from the ion-exchange membranes begins after increasing the ionic strength or changing the pH of the elution buffer.

12.4.4 IMPLEMENTATION IN THE DAIRY INDUSTRY

Casein whey is a waste material of the dairy industries because of its high BOD and COD values. It contains a mixture of different proteins such as immunoglobulins (IgG, IgA, and IgM), bovine serum albumin, lactoperoxidase, and lactoferrin, which have their distinct physicochemical and nutritional values. In this case, ion-exchange membrane chromatography is used for the separation of individual protein molecules and polypeptides. Over the last decade, several experiments were conducted to separate individual protein molecules with high throughput and product purity. The detailed work in this field is reported in Table 12.4.

In many cases, instead of a single operation, ion-exchange membrane chromatography has been coupled with other operations such as gel permeation chromatography, size reduction-based membrane separation and electrodialysis. It is expected that the emerging technology, ion-exchange membrane chromatography, will lead to complete utilization of whey, resulting in implementation of the "zero discharged process".

12.5 CONCLUSIONS

In this chapter, different types of emerging purification technologies such as magnetic fishing, aqueous two-phase separation, and ion-exchange membrane chromatography have been elucidated in the field of food waste upgradation. All these technologies have provided reliable throughput with high product purity. The research focus is now placed on the development of magnetic nanoparticles with suitable carriers for more selective separation of biomolecules at the industrial scale. On the other hand, the development of more stable ionic liquids and liquid membranes will improve the aqueous two-phase separation technique. Finally, design and development of the stationary phase with a uniform pore size and membrane thickness will ion-facilitate the implementation of exchange membrane chromatography for the separation of biomolecules. Although operational cost is high for some cases, these technologies promise high product purity as well as selective bioseparation, and thus should be further explored.

Table 12.4 Separation of Proteins from Dairy Products by Ion-Exchange Membrane Chromatography

Feed Stock	Protein Interest	Specification of Chromatography	References
Whey of lactic casein	Glycomacropeptide	Desaltation, lypholization, ion-exchange chromatography	Tanimoto et al. (1991)
Milk whey	Glycomacropeptide	Heat treatment, ethanol precipitation, centrifugation, and ion exchange chromatography	Saito et al. (1991)
Milk whey	Glycomacropeptide	Membrane chromatography: ion-exchange resins and ultrafiltration	Kawasaki and Dosako (1991)
Acid casein, sodium caseinate, or calcium caseinate	Glycomacropeptide	Membrane chromatography: ion-exchange resin	Dosako et al. (1991)
Single α-lactalbumin, Single bovine serum albumin, binary α-lactalbumin and bovine serum albumin	α-Lactalbumin, bovine serum albumin	Membrane chromatography: MemSep 1010 (Millipore), ligand: SP	Weinbrenner and Etzel (1994)
Whey	β-Globulin, α-lactalbumin, bovine serum albumin	Membrane chromatography: Sartobind MA Q15, MA Q100, MA D15 (Sartorius), ligand: Q, DEAE	Splitt et al. (1996)
Whey	β-Globulin, α-lactalbumin, bovine serum albumin, IgG	Membrane chromatography: Sartobind MA Q15, MA S15 (Sartorius), ligand: Q, SP	Freitag et al. (1996)
Whey	Lactoferrin, lactoperoxidase	Membrane chromatography: Sartobind MA S120 (Sartorius), ligand: SP	Chiu and Etzel (1997)
Whey	β-Globulin, α-lactalbumin, bovine serum albumin	Membrane chromatography: MemSep 1000 (Millipore), ligand: DEAE	Girardet et al. (1998)
Milk-derived products with phenylalanine concentration of 0.5% (w/w)	Glycomacropeptide	Ion exchange chromatography	Ayers et al. (1998)
Whey	Glycomacropeptide	Two step of ion exchange membrane chromatography (anion exchangers of opposite polarity) in series	Etzel (1999)
Nondialyzable fraction of whey	Glycomacropeptide	Anion-exchange chromatography	Nakano and Ozimek (1999)

(Continued)

Table 12.4 Separation of Proteins from Dairy Products by Ion-Exchange Membrane Chromatography *(cont.)*

Feed Stock	Protein Interest	Specification of Chromatography	References
Milk whey	IgG and glycomacropeptide	Anion-exchange chromatography with resin and Amicon YM100 membrane	Xu et al. (2000)
Solution of caseinate hydrolyzed by chymosin	Glycomacropeptide	Exclusion chromatography, and membrane chromatography	Nakano and Ozimek (2000)
Whey	α-Lactalbumin, β-globulin, lactoperoxidase, lactoferrin	Membrane chromatography: 1. SP-Toyopearl™ (Toyosoda), dimension 1.5 cm × 18 cm 2. Quaternary aminoethyl-toyopearl (Toyosoda), dimension 1.5 cm × 18 cm Ligand: anion and cation exchange	Ye et al. (2000)
Whey	Glycomacropeptide	Affinity chromatography	Etzel (2001)
Whey	Lactoferrin	Membrane chromatography: Sartobind S-type cat. # S-10k-15-25 (Sartorius), ligand: SP	Ulber et al. (2001)
Whey protein isolate	α-Lactalbumin	Membrane chromatography: polyhydroxylated methacrylate – TosoHaas AF Chelate 650, ligand: affinity – peptide	Gurgel et al. (2001)
Whey	β-Globulin	Membrane chromatography: calcium biosilicate column, ligand: affinity – all-*trans*-retinal	Vyas et al. (2002)
Whey protein isolate	α-Lactalbumin	Affinity – peptide ligand chromatography	Nakano et al. (2002)
Whey	β-Globulin, α-lactalbumin, bovine serum albumin	Membrane chromatography: 1. Sephadex G-50 (GE Healthcare Technologies), column volume – 131 mL, dimension 1.6 cm × 65 cm 2. DEAE column (GE Healthcare Technologies), column volume – 5 mL Ligand: size exclusion and anion exchange – DEAE	Neyestani et al. (2003)
Whey	β-Globulin	Membrane chromatography, column: Macro-Prep ceramic hydroxyapatite (BioRad), column dimension 12 mm × 88 mm	Schlatterer et al. (2004)
Lactic acid whey	α-Lactalbumin, whey protein isolate	Membrane chromatography, column: SP Sepharose Big Beads, column volume – 80 mL, ligand: cation exchange – SP	Turhan and Etzel (2004)

Table 12.4 Separation of Proteins from Dairy Products by Ion-Exchange Membrane Chromatography *(cont.)*

Feed Stock	Protein Interest	Specification of Chromatography	References
Whey	α-Lactalbumin, whey protein isolate, lactoperoxidase, lactoferrin	Membrane chromatography: SP Sepharose Big Beads, column volume – 80 Ml, ligand: cation exchange – SP	Doultani et al. (2004)
Milk whey	Glycomacropeptide	Ion exchange chromatography (chitosan as anion exchanger)	Nakano et al. (2004)
Whey	β-Globulin, α-lactalbumin, bovine serum albumin, IgG	Membrane chromatography, column: Sephadex™ G-200 (GE Healthcare Technologies); 2.6 cm × 70 cm, ligand: gel filtration	Liang et al. (2006)
Milk	Lactoferrin, lactoperoxidase	Membrane chromatography, column: SP Sepharose Big Beads, column volume – 5 mL, cation exchange – SP	Fee and Chand (2006)
Permeate from two stage ultrafiltration	β-Globulin, α-lactalbumin	Membrane chromatography: Vivapure Q Mini-H (Vivasciences), ligand: Q	Bhattacharjee et al. (2006)
Whey	Lactoferrin, lactoperoxidase, LFcin	Membrane chromatography: Sartobind MA S15, S-type cat. # S-10k-15-25 (Sartorius), ligand: SP	Plate et al. (2006)
Whey, single β-globulin, α-lactalbumin and bovine serum albumin, binary β-globulin and bovine serum albumin	β-Globulin, α-lactalbumin, bovine serum albumin	Membrane chromatography: Sartobind MA D-type and Q- type (Sartorius), ligand: Q, DEAE	Goodall et al. (2008)
Microfiltered whey	β-Lactalbumin	Membrane chromatography, column: HyperCel™ column (Pall BioSepra), column volume – 2.5 mL, 5 mL, 10 mL, ligand: mixed mode – hexyl amine	Brochier et al. (2008)
Whey	Whey protein isolate	Membrane chromatography, column: Mono™ S column (GE Healthcare Technologies), column volume – 2.38 L, 10 cm diameter, ligand: cation exchange – methyl sulfonate	Etzel et al. (2008)
Whey	Whey protein isolate	Membrane chromatography, column: SP Sepharose Big Beads™ (GE Healthcare Technologies), column volume – 5.34 L, 20 cm diameter, 17 cm height, ligand: cation exchange – SP	Etzel et al. (2008)

REFERENCES

Abbott, N.L., Blankschtein, D., Hatton, T.A., 1990. On protein partitioning in two-phase aqueous polymer systems. Bioseparation 1, 191–225.

Albertsson, P.-Å., 1986. Partition of Cell Particles and Macromolecules, third ed. Wiley Publication, New York.

Albertsson, P.-Å., Johansson, G., Tjerneld, F., 1990. Aqueous two-phase separations. In: Asenjo, J.A. (Ed.), Separation Processes in Biotechnology. Marcel Dekker, New York, pp. 287–327.

Alche, J.D., Dickinson, K., 1998. Affinity chromatographic purification of antibodies to a biotinylated fusion protein expressed in *Escherichia coli*. Protein Expr. Purif. 12, 138–143.

Amid, M., Shuhaimi, M., Islam Sarker, M.Z., Abdul Manap, M.Y., 2012. Purification of serine protease from mango (*Mangifera indica* Cv. Chokanan) peel using an alcohol/salt aqueous two phase system. Food Chem. 132, 1382–1386.

Amid, M., Abdul Manap, M.Y., Mustafa, S., 2013. Purification of pectinase from mango (*Mangifera indica* L. cv. Chokanan) waste using an aqueous organic phase system: a potential low cost source of the enzyme. J. Chromatogr. B 931, 17–22.

Baskir, J.N., Hatton, T.A., Suter, U.W., 1989. Protein partitioning in two-phase aqueous polymer systems. Biotechnol. Bioeng. 34, 541–558.

Benavides, J., Rito-Palomares, M., 2008. Practical experiences from the development of aqueous two-phase processes for the recovery of high value biological products. J. Chem. Technol. Biotechnol. 83, 133–142.

Bhattacharjee, S., Bhattacharjee, C., Datta, S., 2006. Studies on the fractionation of β-lactoglobulin from casein whey using ultrafiltration and ion-exchange membrane chromatography. J. Membr. Sci. 275, 141–150.

Brochier, V.B., Schapman, A., Santambien, P., Britsch, L., 2008. Fast purification process optimization using mixed-mode chromatography sorbents in pre-packed mini-columns. J. Chromatogr. A 1177, 226–233.

Bund, T., Allelein, S., Arunkumar, A., Lucey, J.A., Etzel, M.R., 2012. Chromatographic purification and characterization of whey protein-dextran glycation products. J. Chromatogr. A 1244, 98–105.

Burns, M.A., Graves, D.J., 1985. Continuous affinity chromatography using a magnetically stabilized fluidized bed. Biotechnol. Prog. 1, 95–103.

Casal, E., Corzo, N., Moreno, F.J., Olano, A., 2005. Selective recovery of glycosylated casein macropeptide with chitosan. J. Agric. Food Chem. 53, 1201–1204.

Charcosset, C., 1998. Purification of proteins by membrane chromatography. J. Chem. Technol. Biotechnol. 71, 95–110.

Chen, L., Guo, C., Guan, Y., Liu, H., 2007. Isolation of lactoferrin from acid whey by magnetic affinity separation. Sep. Purif. Technol. 56, 168–174.

Chetty, A.S., Burns, M.A., 1991. Continuous protein separations in a magnetically stabilized fluidized bed using nonmagnetic supports. Biotechnol. Bioeng. 38, 963–971.

Chiu, H.C., Lin, C.W., Suen, S.Y., 2007. Isolation of lysozyme from hen egg albumen using glass fiber-based cation-exchange membranes. J. Membr. Sci. 290, 259–266.

Cunha, M.T., Aires-Barros, M.R., 2000. Large-scale extraction of proteins. In: Hatti-Kaul, R. (Ed.), Aqueous Two-Phase Systems: Methods and Protocols. Humana Press, New Jersey, pp. 391–409.

Deng, Q., Zinoviadou, K.G., Galanakis, C.M., Orlien, V., Grimi, N., Vorobiev, E., Lebovka, N., Barba, F.J., 2015. The effects of conventional and non-conventional processing on glucosinolates and its derived forms, isothiocyanates: extraction, degradation and applications. Food Eng. Rev., in press.

Dosako, S., Nishiya, T., Deya, E., 1991. Process for the production of kappa-casein glycomacropeptide. US Patent No. 5,061,622.

Doultani, S., Turhan, K.N., Etzel, M.R., 2003. Whey protein isolate and glycomacropeptide recovery from whey using ion exchange chromatography. J. Food Sci. 68, 1389–1395.

Doultani, S., Turhan, K.N., Etzel, M.R., 2004. Fractionation of proteins from whey using cation exchange chromatography. Process Biochem. 39, 1737–1743.

Drioli, E., Romano, M., 2001. Progress and new perspectives on integrated membrane operations for sustainable industrial growth. Ind. Eng. Chem. 40, 1277–1300.

Espitia-Saloma, E., Vázquez-Villegas, P., Aguilar, O., Rito-Palomares, M., 2014. Continuous aqueous two-phase systems devices for the recovery of biological products. Food Bioprod. Process 92, 101–112.

Etzel, M.R., 1999. Production of k-casein macropeptide for nutraceutical uses. US Patent No. 5,968,586.

Etzel, M.R., 2001. Production of substantially pure kappa casein macropeptide. US Patent No.6,168,823B1.

Etzel, M.R., Helm, T.R., Vyas, H.K., 2008. Methods involving whey protein isolates. US Patent No. 7,378,123.

Fee, C.J., Chand, A., 2006. Capture of lactoferrin and lactoperoxidase from raw whole milk by cation exchange chromatography. Sep. Purif. Technol. 48, 143–149.

Freitag, R., Splitt, H., Reif, O.-W., 1996. Controlled mixed-mode interaction chromatography on membrane adsorbers. J. Chromatogr. A 728, 129–137.

Galanakis, C.M., 2011. Olive fruit dietary fiber: components, recovery and applications. Trends Food Sci. Technol. 22, 175–184.

Galanakis, C.M., 2012. Recovery of high added-value components from food wastes: conventional, emerging technologies and commercialized applications. Trends Food Sci. Technol. 26, 68–87.

Galanakis, C.M., Schieber, A., 2014. Editorial. Special issue on recovery and utilization of valuable compounds from food processing by-products. Food Res. Int. 65, 299–484.

Galanakis, C.M., Tornberg, E., Gekas, V., 2010a. A study of the recovery of the dietary fibres from olive mill wastewater and the gelling ability of the soluble fibre fraction. LWT – Food Sci. Technol. 43, 1009–1017.

Galanakis, C.M., Tornberg, E., Gekas, V., 2010b. Clarification of high-added value products from olive mill wastewater. J. Food Eng. 99, 190–197.

Galanakis, C.M., Tornberg, E., Gekas, V., 2010c. Dietary fiber suspensions from olive mill wastewater as potential fat replacements in meatballs. LWT – Food Sci. Technol. 43, 1018–1025.

Galanakis, C.M., Tornberg, E., Gekas, V., 2010d. Recovery and preservation of phenols from olive waste in ethanolic extracts. J. Chem. Technol. Biotechnol. 85, 1148–1155.

Galanakis, C.M., Tornberg, E., Gekas, V., 2010e. The effect of heat processing on the functional properties of pectin contained in olive mill wastewater. LWT – Food Sci. Technol. 43, 1001–1008.

Galanakis, C.M., Fountoulis, G., Gekas, V., 2012. Nanofiltration of brackish groundwater by using a polypiperazine membrane. Desalination 286, 277–284.

Ghosh, R., 2006. Principles of Bioseparations Engineering. World Scientific Publishing Pvt. Ltd, Singapore.

Girardet, J.M., Saulnier, F., Linden, G., Humbert, G., 1998. Rapid separation of bovine whey proteins by membrane convective liquid chromatography, perfusion chromatography, continuous bed chromatography, and capillary electrophoresis. Lait 78, 391–400.

Goodall, S., Grandison, A.S., Jauregi, P.J., Price, J., 2008. Selective separation of the major whey proteins using ion exchange membranes. J. Dairy Sci. 91, 1–10.

Gupta, R., Gupta, N., Rathi, P., 2004. Bacterial lipases: an overview of production, purification and biochemical properties. Appl. Microbiol. Biotechnol. 64, 763–781.

Hart, R.A., Ogez, J.R., Builder, S.E., 1995. Use of multifactorial analysis to develop aqueous two-phase systems for isolation of non-native IGF-I. Bioseparation 5, 113–121.

Hatti-Kaul, R., 2000. Aqueous two-phase systems: a general overview. In: Hatti-Kaul, R. (Ed.), Aqueous Two-Phase Systems: Methods and Protocols. Humana Press, Totowa, New Jersey, pp. 1–10.

Heebøll-Nielsen, A., Dalkiær, M., Hubbuch, J.J., Thomas, O.R.T., 2004. Superparamagnetic adsorbents for high-gradient magnetic fishing of lectins out of legume extracts. Biotechnol. Bioeng. 87, 311–323.

Hofmann, I., Schnolzer, M., Kaufmann, I., Franke, W.W., 2002. Symplekin, a constitutive protein of karyo- and cytoplasmic particles involved in mRNA biogenesis in *Xenopus laevis* oocytes. Mol. Biol. Cell 13, 1665–1676.

Holst, H.H., Chatterton, D.E. W., 2002. Process for preparing a kappa-caseino glycomacropeptide or a derivative thereof, US Patent No. 6,462,181 B1.

Huddleston, J., Veide, A., Köhler, K., Flanagan, J., Enfors, S.-O., Lyddiatt, A., 1991. The molecular basis of partitioning in aqueous two-phase systems. Trends Biotechnol. 9, 381–388.

Hustedt, H., Kroner, K.H., Kula, M.-R., 1985. Applications of phase partitioning in biotechnology. In: Walter, H., Brooks, D.E., Fisher, D. (Eds.), Partitioning in Aqueous Two-Phase Systems: Theory, Methods, Uses and Application in Biotechnology. Academic Press, Orlando, Florida, pp. 529–587.

Jara, F., Pilosof, A.M.R., 2011. Partitioning of α-lactalbumin and β-lactoglobulin in whey protein concentrate/hydroxypropylmethylcellulose aqueous two-phase systems. Food Hydrocolloid. 25, 374–380.

Johansson, G., 1985. Partition in Aqueous Two-Phase Systems: Theory, Methods, Uses and Applications to Biotechnology. Academic Press, Orlando, Florida.

Kammoun, R., Chouayekh, H., Abid, H., Naili, B., Bejar, S., 2009. Purification of CBS 819.72 α-amylase by aqueous two-phase systems: modelling using response surface methodology. Biochem. Eng. J. 46, 306–312.

Kateřina, H., Ivo, Š., Mirka, Š., 2012. Flow-through magnetic separation of solanum tuberosum lectin from potato starch wastewater, NANOCON 2012, October 23–25, Brno, Czech Republic, EU.

Ketnawa, S., Rawdkuen, S., Chaiwut, P., 2010. Two phase partitioning and collagen hydrolysis of bromelain from pineapple peel *Nang lae* cultivar. Biochem. Eng. J. 52, 205–211.

Ketnawa, S., Benjakul, S., Ling, T.C., Martínez-Alvarez, O., Rawdkuen, S., 2013. Enhanced recovery of alkaline protease from fish viscera by phase partitioning and its application. Chem. Cent. J. 7, 1–9.

Li, E.W.Y., Mine, Y., 2004. Comparison of chromatographic profile of glycomacropeptide from cheese whey isolated using different methods. J. Dairy Sci. 87, 174–177.

Liang, M., Chen, V.Y.T., Chen, H.-L., Chen, W., 2006. A simple and direct isolation of whey components from raw milk by gel filtration chromatography and structural characterization by Fourier transform Raman spectroscopy. Talanta 69, 1269–1277.

Lochmuller, C.H., Ronsick, C.S., Wigman, L.S., 1988. Fluidized-bed separators reviewed: a low pressure drop approach to column chromatography. Prep. Biochem. 1, 93–108.

Meyer, A., Hansen, D.B., Gomes, C.S.G., Hobley, T.J., Thomas, O.R.T., Franzreb, M., 2005. Demonstration of a strategy for product purification by high-gradient magnetic fishing: recovery of superoxide dismutase from unconditioned whey. Biotechnol. Prog. 21, 244–254.

Nakano, T., Ozimek, L., 1999. Purification of glycomacropeptide from non-dialyzable fraction of sweet whey by anion exchange chromatography. J. Biotechnol. Tech. 13, 739–742.

Nakano, T., Ozimek, L., 2000. Purification of glycomacropeptide from caseinate hydrolysate by gel chromatography and treatment with acidic solution. J. Food Sci. 65, 588–590.

Nakano, T., Silva-Hernandez, E.R., Ikawa, N., Ozimek, L., 2002. Purification of k-casein glycomacropeptide from sweet whey with undetectable level of phenylalanine. Biotechnol. Prog. 18, 409–412.

Nalinanon, S., Benjakul, S., Visessanguan, W., Kishimura, H., 2009. Partitioning of protease from stomach of albacore tuna (*Thunnus alalunga*) by aqueous two-phase systems. Process Biochem. 44, 471–476.

Neyestani, T.R., Djalali, M., Pezeshki, M., 2003. Isolation of alpha-lactalbumin, beta-lactoglobulin, and bovine serum albumin from cow's milk using gel filtration and anion-exchange chromatography including evaluation of their antigenicity. Protein Expr. Purif. 29, 202–208.

Novaes, L.C.L., Ebinuma, V.C., Mazzola, P.G., Júnior, A.P., 2013. Polymer-based alternative method to extract bromelain from pineapple peel waste. Biotechnol. Appl. Biochem. 60, 527–535.

Pedersen, L., Mollerup, J., Hansen, E., Jungbauer, A., 2003. Whey proteins as a model system for chromatographic separation of proteins. J. Chromatogr. B 790, 161–173.

Pessela, B.C., Torres, R., Batalla, P., Fuentes, M., Mateo, C., Fernández-Lafuente, R., Guisán, J.M., 2006. Simple purification of immunoglobulins from whey proteins concentrate. Biotechnol. Prog. 22, 590–594.

Plate, K., Beutel, S., Buchholz, H., Demmer, W., Fischer-Fruhholz, S., Reif, O., Ulber, R., Scheper, T., 2006. Isolation of bovine lactoferrin, lactoperoxidase and enzymatically prepared lactoferricin from proteolytic digestion of bovine lactoferrin using adsorptive membrane chromatography. J. Chromatogr. A 1117, 81–86.

Qingdao, L., Amarjeet, B., Jing-Xu (Jesse), Z., Argyrios, M., 2002. Continuous protein recovery from whey using liquid-solid circulating fluidized bed ion-exchange extraction. Biotechnol. Bioeng. 78, 157–163.

Rahmanian, N., Jafari, S.M., Galanakis, C.M., 2014. Recovery and removal of phenolic compounds from olive mill wastewater. J. Am. Oil Chem. Soc. 91, 1–18.

Reis, I.A., Santos, S.B., Santos, L.A., Oliveira, N., Freire, M.G., Pereira, J.F., Ventura, S.P., Coutinho, J.A., Soares, C.M., Lima, Á.S., 2012. Increased significance of food wastes: selective recovery of added-value compounds. Food Chem. 135, 2453–2461.

Rito-Palomares, M., 2004. Practical application of aqueous two-phase partition to process development for the recovery of biological products. J. Chromatogr. B 807, 3–11.

Rojas, E.E.G., dos Reis Coimbra, J.S., Minim, L.A., Zuniga, A.D.G., Saraiva, S.H., Minim, V.P.R., 2004. Size-exclusion chromatography applied to the purification of whey proteins from the polymeric and saline phases of aqueous two-phase systems. Process Biochem. 39, 1751–1759.

Rosa, P.A.J., Azevedo, A.M., Aires-Barros, M.R., 2007. Application of central composite design to the optimisation of aqueous two-phase extraction of human antibodies. J. Chromatogr. A 1141, 50–60.

Rosa, P.A.J., Ferreira, I.F., Azevedo, A.M., Aires-Barros, M.R., 2010. Aqueous two-phase systems: a viable platform in the manufacturing of biopharmaceuticals. J. Chromatogr. A 1217, 2296–2305.

Roselló-Soto, E., Barba, F.J., Parniakov, O., Galanakis, C.M., Grimi, N., Lebovka, N., Vorobiev, E., 2015. High voltage electrical discharges, pulsed electric field and ultrasounds assisted extraction of protein and phenolic compounds from olive kernel. Food Bioprocess Technol., in press.

Sabatkova, Z., Safarikova, M., Safarik, I., 2008. Magnetic ovalbumin and egg white aggregates as affinity adsorbents for lectins separation. Biochem. Eng. J. 40, 542–545.

Safarik, I., Safarikova, M., 2004. Magnetic techniques for the isolation and purification of proteins and peptides. BioMag. Res. Technol. 2 (7), 1–17.

Safarik, I., Horska, K., Martinez, L.M., Safarikova, M., 2010. Large scale magnetic separation of *Solanum tuberosum* tuber lectin from potato starch waste water. In: Häfeli, U., Schütt, W., Zborowski, M. (Eds.), 8th International Conference on the Scientific and Clinical Applications of Magnetic Carriers, American Institute of Physics Conference Proceedings, Rostock, Germany, pp. 146–151.

Šafaříková, M., Šafařík, I., 2000. One-step partial purification of *Solanum tuberosum* tuber lectin using magnetic chitosan particles. Biotechnol. Lett. 22, 941–945.

Saito, T., Yamaji, A., Itoh, T., 1991. A new isolation method of caseinoglycopeptide from sweet cheese whey. J. Dairy Sci. 74, 2831–2837.

Sarangi, B.K., Pattanaik, D.P., Rathinaraj, K., Sachindra, N.M., Madhusudan, M.C., Mahendrakar, N.S., 2011. Purification of alkaline protease from chicken intestine by aqueous two phase system of polyethylene glycol and sodium citrate. J. Food Sci. Technol. 48, 36–44.

Schlatterer, B., Baeker, R., Schlatterer, K., 2004. Improved purification of b-lactoglobulin from acid whey by means of ceramic hydroxyapatite chromatography with sodium fluoride as a displacer. J. Chromatogr. B 807, 223–228.

Schuster, M., Wasserbauer, E., Ortner, C., Graumann, K., Jungbauer, A., Hammerschmid, F., Werner, G., 2000. Short cut of protein purification by integration of cell-disrupture and affinity extraction. Bioseparation 9, 59–67.

Smithers, G.W., 2008. Whey and whey proteins – from "gutter-to-gold". Int. Dairy J. 18, 695–704.

Splitt, H., Mackenstedt, I., Freitag, R., 1996. Preparative membrane adsorber chromatography for the isolation of cow milk components. J. Chromatogr. A 729, 87–97.

Strati, I.F., Oreopoulou, V., 2011. Effect of extraction parameters on the carotenoid recovery from tomato waste. Int. J. Food Sci. Technol. 46, 23–29.

Tanimoto, M., Kawasaki, Y., Shinmoto, H., Dosako, S., Tomizawa, A., 1990. Process for producing kappacasein glycomacropeptide. European Patent Application 0393850A2.

Tellez, C.M., Cole, K.D., 2000. Preparative electrochromatography of proteins in various types of porous media. Electrophoresis 21, 1001–1009.

Turhan, K.N., Etzel, M.R., 2004. Whey protein isolate and α-lactalbumin recovery from lactic acid whey using cation-exchange chromatography. J. Food Sci. 69, 66–70.

Ulber, R., Plate, K., Weiss, T., Demmer, W., Buchholz, H., Scheper, T., 2001. Downstream processing of bovine lactoferrin from sweet whey. Acta Biotechnol. 21, 27–34.

Vázquez-Villegas, P., Aguilar, O., Rito-Palomares, M., 2011. Study of biomolecules partition coefficients on a novel continuous separator using polymer-salt aqueous two-phase systems. Sep. Purif. Technol. 78, 69–75.

Volkov, G.L., Gavriliuk, S.P., Skalka, V.V., Krasnobryzha, Ie.N., Gavryliuk, E.S., Zhukova, A.I., Tseren Bukha, P., 2010. Comparative analysis of ion exchange membrane chromatography and packing bed methods with model whey protein mixture separation. Russ. J. Biopharm. 2, 24–31.

Wang, Q., Lin, Q., Li, X., 2013. Extraction of lipase from rice bran using aqueous two-phase system. Adv. Mater. Res. 651, 384–388.

Wolman, F.J., Maglio, D.G., Grasselli, M., Cascone, O., 2007. One-step lactoferrin purification from bovine whey and colostrum by affinity membrane chromatography. J. Membr. Sci. 288, 132–138.

Xu, Y., Sleigh, R., Hourigan, J., Johnson, R., 2000. Separation of bovine immunoglobulin G and glycomacropeptide from dairy whey. Process Biochem. 36, 393–399.

Ye, X., Yoshida, S., Ng, T.B., 2000. Isolation of lactoperoxidase, lactoferrin, alpha-lactalbumin, beta-lactoglobulin B and beta-lactoglobulin A from bovine rennet whey using ion exchange chromatography. Int. J. Biochem. Cell Biol. 32, 1143–1150.

EMERGING PRODUCT FORMATION

13

Seid Mahdi Jafari*, Milad Fathi**, Ioanna Mandala†

*Department of Food Materials and Process Design Engineering, Faculty of Food Science and Technology, University of Agricultural Sciences and Natural Resources, Gorgan, Iran; **Department of Food Science and Technology, Faculty of Agriculture, Isfahan University of Technology, Isfahan, Iran; †Department of Food Science & Human Nutrition, Agricultural University of Athens, Food Process Engineering Laboratory, Athens, Greece

13.1 INTRODUCTION

Delivery of bioactive compounds to the specific location is an object of interest in food sectors. Direct fortification of nutritionals might lead to significant loss during processing, storage, and consumption. Researchers should manage to achieve scaleup without affecting the functional properties of the target compounds and develop a product that meets the consumers' high quality standards in terms of health, safety, and organoleptic characteristics (Galanakis, 2012). Utilization of conventional encapsulation methods is often restricted by several problems that are difficult to overcome. These include:

1. overheating of the food matrix,
2. high energy consumption and general cost,
3. loss of functionality and poor stability of the final product,
4. accomplishment of increasingly stringent legal requirements on materials safety.

The disadvantages of conventional techniques could be overcome using new trends, the so-called emerging technologies (Galanakis and Schieber, 2014; Roselló-Soto et al., 2015). These modern technologies are today suggested for their application in various processes within the food industry and they could be easily adapted in the product formation of valuable compounds from corresponding wastes (Galanakis, 2012). Nanotechnology and nanoparticles are one of the most popular emerging technologies applied in the broad field of food science. This technology enhances stability, provides moisture- and pH-triggered controlled release, and enhances bioavailability and consecutive delivery of multiple active ingredients.

Emerging product formation techniques like nanoencapsulation and delivery of bioactive components have been well documented in pharmaceutics, cosmetics, and the food sciences. Recent studies have revealed that the formation of nanocarriers could lead to considerable advantages such as more stability, longer release time, and sustained release profile over the conventional product formation systems. However, future trends in nanodelivery systems should focus more on pertaining the physicochemical properties of the nanocarriers and interactions of nanoencapsulated bioactives with food systems. In the following sections, different kinds of nanoencapsulation systems along with their scaleup techniques are introduced and discussed.

Food Waste Recovery. http://dx.doi.org/10.1016/B978-0-12-800351-0.00013-4

13.2 NANOCAPSULES

Nanocapsules are novel delivery systems in nanometer scales for entrapment, protection (from unfavorable conditions such as oxidation, acidic, and enzymatic degradation), and controlled release of bioactives. Nanocapsules provide more surface area and have a potential to enhance solubility, improve bioavailability, and ameliorate controlled release of the encapsulated ingredients compared with micron-sized carriers (Mozafari, 2006; Weiss et al., 2009). Nanocapsules could be produced from synthetic or natural compounds. However, they should be biocompatible and biodegradable for their use in food systems. Agro-waste materials could be considered potential excellent sources for encapsulating of materials. Coating materials for the production of nanocapsules are typically lipid (Fathi et al., 2012), carbohydrate (Fathi et al., 2014), or protein based. These materials could be frequently obtained from food by-products or wastes (Oreopoulou and Russ, 2006).

13.2.1 LIPID-BASED DELIVERY SYSTEMS

Lipid-based delivery systems are composed of biodegradable lipids and might be formed in different structures. The main advantages of lipid carriers include their limited required modification to be encapsulated, and their easy scale-up. However, the most important limitation of lipid nanocarriers is their physical structure. The latter is changed during high-temperature processes. Recently, different lipid delivery systems, i.e. nanoemulsions, nanoliposomes (Fathi et al., 2012), solid lipid nanoparticles (SLN), nanostructure lipid carriers (NLC) (Fathi et al., 2013), lipid nanocapsules (LNCs), and lipid drug conjugates (LDC) with diverse functionalities have been proposed. Table 13.1 shows the production methods, unique properties, advantages, and disadvantages of each LNC system. Almost all of these carriers are favorable for hydrophobic bioactives, whereas LDC could also be used for hydrophilic compounds.

13.2.2 CARBOHYDRATE-BASED DELIVERY SYSTEMS

These systems are suitable for industrial applications since they are biocompatible, biodegradable, and can be modified to achieve the required properties. In contrast to the lipid carriers that are usually favorable to entrap hydrophobic bioactives, carbohydrate-based delivery systems can interact with a wide range of compounds with different hydrophobicity via their functional groups. Likewise, their temperature stability is higher compared with lipid- or protein-based systems (e.g. might be melted or denatured at high temperatures) and thus are more suitable for the encapsulation of thermal labile compounds (Fathi et al., 2014). Different native and modified carbohydrate materials, e.g. starch, cellulose, pectin, guar gum, chitosan, alginate, and dextran, have been applied to nanoencapsulate food bioactives, whereas the first four of them are frequently obtained from agro-waste materials (Galanakis, 2011, 2012). Fung et al. (2011) investigated the potential use of soluble dietary fiber wastes from agro-wastes, okara (soybean solid waste), oil palm trunk, and oil palm frond for the production of nanofibers to encapsulate *Lactobacillus acidophilus*. Produced nanocarriers were able to protect the probiotics against thermal treatments and unfavorable storage conditions. Different carbohydrate-based delivery systems, their production methods, and features are presented in Table 13.2.

Table 13.1 Different Lipid-Based Nanocapsules, Their Production Methods and Features

Lipid-Based Delivery Systems	Production Method	Features, Advantages, and Disadvantages
Nanoemulsion	High-pressure homogenization; ultrasonication; solvent diffusion (Unger et al., 2004); emulsification–evaporation (Cheong et al., 2008)	Can be used for delivery of bioactive compounds and also preparation of stable emulsions; no limitation for using specific lipid compounds; possibility of large-scale production (Khan et al., 2012); rapid release; low stability and possibility of burst release in acidic solutions (Klinkesorn and McClements, 2009); low encapsulation efficiency for hydrophilic compounds.
Nanoliposomes	Ultrasonication (Schroeder et al., 2009); high-pressure homogenization	Ability of simultaneously entrapment of hydrophilic molecules in their interior volume, and hydrophobic compounds in the hydrophobic part of the lipid bilayer (Acosta, 2008); rapid release (Zaru et al., 2009).
SLN	Hot homogenization (Silva et al., 2011) and cold homogenization (Wise, 2000); emulsification–evaporation (Varshosaz et al., 2010); ultrasonication (Silva et al., 2011)	Better encapsulation efficiency than nanoemulsions; high and low encapsulation efficiency for lipophilic and hydrophilic compounds; possibility of burst release and explosion during storage (Fathi et al., 2012).
Nanostructure lipid carriers	Similar to SLN	The crystallinity of lipid structure is decreased by using diverse lipids with different molecular mass, which leads to higher efficiency, better release profile, and lower explosion possibility than SLN during storage (Fathi et al., 2013); low efficiency for hydrophobic compounds.
LNCs	Phase inversion temperature via heat shock treatment: caprylic and capric acid are two main lipids used for the production of LNC in combination with a nonionic hydrophilic surfactant and a lipophilic surfactant. The emulsion is heated to about 90°C. Then, nanoparticles are formed during three cycles of progressive heating and cooling between 90°C and 60°C following by an irreversible shock induced by dilution with cool water (Weyland et al., 2013).	Easy scale-up production methods; there is a limitation for using specific lipid compounds; low encapsulation efficiency for hydrophilic compounds; rapid release.
LDC	Conjugation: an insoluble bioactive (drug)-lipid conjugate bulk is first prepared either by salt formation (e.g. with a fatty acid) or by covalent linking (e.g. to esters or ethers). The obtained LDC is then processed with an aqueous surfactant solution using high-pressure homogenization to form nanoparticles (Jyoti Das et al., 2013).	This carrier is appropriate for hydrophilic bioactive encapsulation (up to 33% encapsulation load).

Table 13.2 Different Carbohydrate-Based Nanocapsules, Their Production Methods and Features

Carbohydrate-Based Delivery System	Production Method	Features, Advantages, and Disadvantages	Applied Encapsulant
Starch	Electrospinning (Kong and Ziegler, 2013); spray drying (Jain et al., 2008); reversed-phase microemulsion (Yang et al., 2014)	Does not need intensive purification procedures to be used for nanoencapsulation; sensitive to acid hydrolysis; natural starch is highly hydrophobic and shows a limitation for entrapment of lipophilic compounds; high potential to be modified (e.g. dialdehyde starch (Yu et al., 2007), propyl starch (Santander-Ortega et al., 2010), PEGylated (Minimol et al., 2013), and octenyl succinic anhydride (Qi and Xu, 1999), which all could be used for entrapment of lipophilic compounds).	Insulin (Jain et al., 2008) and unsaturated fatty acids (Lesmes et al., 2009; Zabar et al., 2009)
Cellulose	Electrospinning or electrospray (Xu et al., 2008); supercritical fluid (Jin et al., 2009)	High potential to be modified (e.g. cellulose esters (Jin et al., 2009)); enzymatic resistance; insoluble in water and most organic solvents.	Flavors (Heitfeld et al., 2008)
Pectin	Coacervation (Zimet and Livney, 2009); ultrasonication (Dutta and Sahu, 2012)	Acid resistant and could be used for colon delivery (Sinha and Kumria, 2001); resistance to enzymatic digestion in the mouth; low encapsulation efficiency and rapid release especially for hydrophilic encapsulants (Sonia and Sharma, 2012).	Nisin (Khaksar et al., 2014)
Chitosan	Coacervation	Cationic biopolymer with antibacterial effect and bioadhesive property to the colon; in its native form could not be used as an acid resistance carrier; high potential to be modified (e.g. chitosan–tripolyphosphate (Hu et al., 2008) to enhance the acid resistance; amphiphilic N-octyl-N-trimethyl chitosan (Zhang et al., 2007) improving the encapsulation efficiency and release properties of a hydrophobic encapsulants; glycol chitosan (Quiñones et al., 2012) to enhance solubility across a broader range of pH).	Catechins (Hu et al., 2008; Zhang et al., 2007)
Alginate	Coacervation and water-in-oil emulsification-gelation (Paques et al., 2014)	An anionic water soluble biopolymer that forms a gel structure in the presence of divalent cations such as calcium and zinc; insoluble in acidic media and could be used for colon delivery or to increase stability of bioactives in acidic foods; rapid dissolution of alginate nanocapsules in present of sodium ions.	Turmeric oil (Lertsutthiwong et al., 2008) and lipase (Liu et al., 2012)

13.2.3 **PROTEIN-BASED DELIVERY SYSTEMS**

The ability of protein networks to interact with a wide range of active compounds via functional groups of their polypeptide primary structure makes them versatile carriers to bind and entrap a variety of hydrophilic and hydrophobic bioactive food molecules (Elzoghby et al., 2011; Pereira et al., 2009).

Applied proteins might be animal-based (e.g. milk protein, gelatin, collagen, albumin, and silk), plant-based (e.g. zein, gliadin, soy protein, and pea protein), or microorganism-based (García-Garibay et al., 2014). Some of these proteins are commonly obtained from food factory by-products. For example, cheese whey proteins are potential coating materials for the nanoencapsulation of hydrophobic molecules, since they have high emulsifying ability and capability to form gastro-resistant hydro-gels (Gunasekaran et al., 2007). In contrast to caseins, which are thermostable, whey proteins start denaturing at temperatures above 70°C (LaClair, 2008; Galanakis et al., 2014).

Gelatin is a hydrophilic thermostable biopolymer having both cationic and anionic groups and swells in cold water but becomes soluble in hot water (Li et al., 1998). These features make it a favorable shell for encapsulation of both hydrophilic and hydrophobic bioactives. Gelatin is derived from collagen and could be obtained as a by-product from meat and fish processing units. Fibroin from the cocoons of the silkworm or spider has recently been explored as a versatile protein biomaterial possessing a number of applications (Lammel et al., 2010).

Table 13.3 indicates the encapsulation methods and features of some protein-based nanocapsules. For obtaining appropriate properties for nanocapsules, it is suggested to apply carbohydrates and proteins together, e.g. whey and zein (Fabra et al., 2014) to enhance emulsification ability and prevent droplet coalescence.

13.3 **NANOENCAPSULATION METHODS AND SCALE-UP**

Different encapsulation methods have been proposed for natural compounds depending on the properties of shell and core. Physical (e.g. ultrasonication, high pressure homogenization, spray drying, electrospinning, electrospray, and supercritical fluid) or chemical (e.g. coacervation, sol-gel, heat treatment, and emulsification) techniques might be used in food-based nanocapsules.

13.3.1 **ULTRASONICATION**

Ultrasound waves induce cavitation of bubbles that create strong shear forces and tear-off lipids (Fathi et al., 2013) or biopolymer structures to nanoscale particles (Tian et al., 2012). This method could be easily scaledup and is applicable for a wide range of encapsulating materials. Additionally, for entrapment of essential oils, extraction, and encapsulation could be performed in one step (Mantegna et al., 2012).

13.3.2 **HIGH-PRESSURE HOMOGENIZATION**

In this method, the coarse solution of shell and bioactive material is pumped under high pressure in the range of 100–1500 bar. The nanoscale particles are obtained due to the high applied shear force. High pressure homogenization could be performed in hot (Silva et al., 2011) or cold (Wise, 2000) modes. The latter mode is more appropriate for thermal labile compounds, while lipid nanoparticles obtained

Table 13.3 Different Protein-Based Nanocapsules, Their Production Methods and Features

Protein-Based Delivery System	Production Method	Features, Advantages, and Disadvantages	Applied Encapsulant
Casein	Self-assembly method to reassemble nanomicelles: by initial formation a noncovalent binding of hydrophobic bioactive and casein, then resemblance of casein nanoparticles by addition of $C_6H_5K_3O_7$, K_2HPO_4, and $CaCl_2$ and finally particle creation by increasing volume of solution by water (Semo et al., 2007).	Possibility of using a self-assembly procedure; high thermal resistance; possibility of formation of nanoparticles using solvent-free procedures; interaction potential with other materials; good surface and gelling properties.	Omega-3 polyunsaturated fatty acids (Zimet et al., 2011)
Whey proteins	Electrospinning (López-Rubio and Lagaron, 2012). pH-induced cold gelation (a solvent-free procedure): consists of two-step process; first, a native whey protein solution is heated at a pH away from its isoelectric point and at low ionic strength leading to formation of soluble aggregates. Second, cooling of solution and reduction of the electrostatic repulsion by gradually lowering the pH toward the isoelectric point of proteins and/or adding salt. In the second step, gelation is induced at ambient temperature through adding salt or lowering the pH. In contrast to heat-induced gelation, this method makes it possible to entrap the heat-sensitive nutraceuticals at ambient temperatures (Alting et al., 2002).	Acid resistant; thermosensible; interaction potential with other materials; better encapsulation efficiency for hydrophobic molecules; possibility of formation of nanoparticles using solvent-free procedures.	β-Carotene (López-Rubio and Lagaron, 2012)
Gelatin	Desolvation/coacervation: a homogeneous solution of charged gelatin undergoes liquid/liquid phase separation, leads to formation of a polymer-rich dense phase at the bottom and a transparent solution above. Addition of salt or alcohol encourages coacervation that results in nanoparticle formation (Singh and Chaudhary, 2010). Emulsion method: preheated gelation solution is mixed with encapsulant solution at above ambient temperature (40°C), the resultant mixture is then added to the oil phase (with or without emulsifier) and then homogenized (Bajpai and Choubey, 2006). Salting out (Hussain and Maji, 2008): based on the separation of a water miscible solvent from aqueous solution via a salting-out effect. Protein and encapsulant are initially dissolved and then emulsified into an aqueous gel containing the salting-out agent (electrolytes, e.g. calcium chloride or nonelectrolytes, e.g. sucrose) and a stabilizer. The obtained oil-in-water emulsion is diluted with a sufficient volume of water to induce diffusion of solvent into the aqueous phase and formation of nanospheres. Electrospinning (Okutan et al., 2014).	High thermal resistance; low encapsulation efficiency of hydrophobic bioactives; applicable for both hydrophilic and hydrophilic compounds.	Catechin (Chen et al., 2010)

Table 13.3 Different Protein-Based Nanocapsules, Their Production Methods and Features *(cont.)*			
Protein-Based Delivery System	**Production Method**	**Features, Advantages, and Disadvantages**	**Applied Encapsulant**
Zein	Liquid/liquid dispersion: zein is dissolved in binary solutions of ethanol-water, followed by shearing zein solutions into deionized water using a high-speed homogenizer. Zein becomes insoluble and precipitates to form nanoparticles when the ethanol concentration in the solution drops to the below the solubilization limit (Wu et al., 2012b). Electrospinning (Li et al., 2009). Electrospray (Gomez-Estaca et al., 2012). Spray drying (Chen and Zhong, 2014). Supercritical fluid (Liu et al., 2012).	High hydrophobicity and possibility of encapsulation and sustained release of water insoluble compounds; acid resistance; needs organic solvent for nanoparticle formation in some methods; low dispersibility due to high amount of nonpolar amino acid residues.	Essential oil (Wu et al., 2012b)
Soy proteins	Liquid/liquid dispersion (Teng et al., 2012). Cold gelation (Zhang et al., 2012).	Balanced amino acid profile that facilitates the protein/nutraceutical interaction.	Vitamin D_3 (Teng et al., 2012)
Silk protein	Salting out (Lammel et al., 2010). Electrospinning (Sheng et al., 2013). Electrospray (Wantanasiri et al., 2014). Gelation (Wu et al., 2012a). Liquid antisolvent precipitation (Subia and Kundu, 2013).	Possibility of site-specific functionalization; mechanical strength; sustained release.	Curcumin (Gupta et al., 2009)

from the former mode show more uniform morphology. High-pressure homogenization is currently applied for large-scale production and could be applied for both lipid and polymeric compounds.

13.3.3 SPRAY DRYING

One of the most commonly applied technologies for encapsulation is spray drying (Jafari et al., 2008a). It is fast, relatively cheap, reproducible, and possesses high potential for scale-up (Yeo et al., 2001). This process arises from dissolving or dispersing the bioactive in biopolymer solution and atomizing in a heated air chamber, which removes the solvent. Since in food systems usually water-based dispersions are used, the biopolymer should have a high water solubility (de Vos et al., 2010). More information on spray drying is provided in Chapter 8. Recently, a nano spray dryer has been developed that could be used to produce nanopowders. This technology utilizes an ultrasonic-driven actuator, which vibrates stainless steel mesh having an array of precise micron-sized holes for fine droplet generation (Fig. 13.1).

13.3.4 ELECTROSPINNING AND ELECTROSPRAY

Electrohydrodynamic processes, namely electrospinning and electrospraying, have gained widespread interest for the fabrication of fibers and particles with diameters down to a few nanometers. In electrospinning, applied high voltage (1–30 kV) generates a charge in the polymer solution, which is pumped through a syringe needle and leads to formation of repulsive interactions between the same charges in

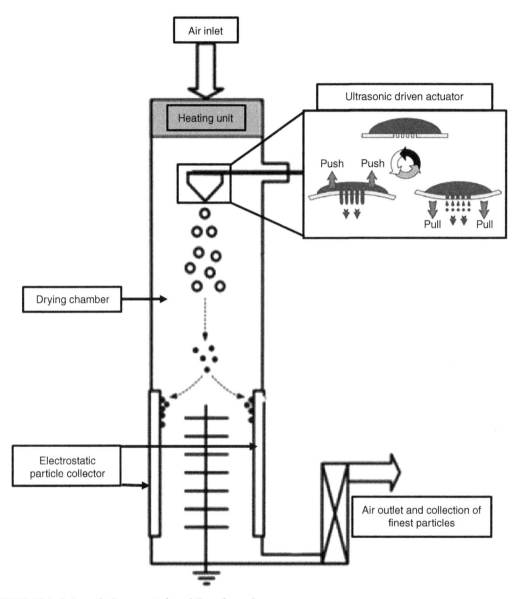

FIGURE 13.1 Schematic Representation of Nano Spray Dryer

Adapted from BÜCHI (BÜCHI Labortechnik, 2009)

the liquid and the attractive forces between the oppositely charged liquid and collector, and consequently elongating and ejecting the pendant drop toward the collector and forming nanofibers or particles (Bhardwaj and Kundu, 2010). Viscosity, concentration, solution conductivity, voltage, needle-collector distance, injection flow rate, temperature, and humidity are among the most important parameters affecting properties of nanofibers (Fathi et al., 2014). Electrospray is a new nanoencapsulation method, which is fundamentally similar to electrospinning. However, instead of nanofiber fabrication, nanoparticles are formed based on biopolymer molecular mass, biopolymer chain entanglement, and solvent evaporation rate. The nanoparticles are formed when the product of solution intrinsic viscosity (η) and biopolymer concentration (c), known as the Berry's number:

$$Be = [\eta]c \tag{13.1}$$

is lower than a certain critical value (Be_{cr}). In the electrospray method, the electrostatic force induced by high voltage atomizes liquid into fine droplets. The most outstanding features of electrospray technique are high encapsulation efficiency and the possibility of production in one step (Hao et al., 2013).

13.3.5 SUPERCRITICAL FLUID

Supercritical fluid (SCF) methods have attracted increasing attention for the encapsulation of heat-sensitive bioactives. Carbon dioxide is one of the most widely used fluids because its supercritical region can be achieved at moderate temperatures and pressures ($T_c = 304.2$ K, $P_c = 7.38$ MPa). Supercritical CO_2 also provides an inert medium suitable for entrapment of easily oxidizable substances such as unsaturated fatty acids (Cocero et al., 2009). Waste materials always contain both bioactives and encapsulating materials (Mattea et al., 2009) and therefore SCF is a technology that can be used for the third and fifth stages of the Universal Recovery Process (see Chapter 3).

13.3.6 COACERVATION

Coacervation is one of the most easily implemented techniques for the production of protein- and/or carbohydrate-based delivery systems. The most common driving force for coacervation in food systems is electrostatic attraction between oppositely charged molecules. Coacervation may be induced between a charged bioactive component and an oppositely charged biopolymer to form an electrostatic complex (simple coacervation). Alternatively, a bioactive may be trapped within a particle formed by electrostatic complexation of positively charged (e.g. chitosan) and negatively charged (e.g. pectin, alginate) biopolymers (complex coacervation). Due to simplicity of the procedure, this technique shows high potential for large-scale production.

13.4 NANOEMULSIONS
13.4.1 DEFINITION AND APPLICATION EXAMPLES

Emulsions consist of two immiscible liquids that create two phases: the dispersed and the continuous phase. They can have very fine droplets of ten to a few hundred nanometers in diameter, namely nanoemulsions. However, they should not be considered similar to microemulsions, which have approximately the same droplet size (e.g. 5–100 nm), and are thermodynamically stable, but are spontaneously created by self-assembly (Jafari et al., 2006).

According to Sivakumar et al. (2014), nanoemulsions are isotropic colloidal systems. Thanks to their nanometer-sized droplets, they have long-term physical stability. In addition, they are resistant to creaming because their Brownian motion is enough to overcome their low gravitational separation force, and resistance to flocculation courtesy their highly efficient steric stabilization (Maswal and Dar, 2014). Therefore, these tiny emulsions of nanodimensions are sometimes referred as "approaching thermodynamic stability" (Tadros et al., 2004).

Nanoemulsions can have several applications. One of the most popular investigated applications in many different systems is as drug or bioactive ingredient delivery carriers with enhanced bioactive solubility and availability, controlled drug release, and protection of environmental stresses. Besides, innovations in material chemistry and nanotechnology have synergistically fueled the development of novel drug delivery systems and nanocarriers, which are biodegradable, biocompatible, targeting, and stimulus responsive (Zhang et al., 2013).

13.4.2 FORMULATION TECHNIQUES

To produce emulsion droplets in the nano-size range, various techniques have been employed. In general, the basic processes used to fabricate the nanoemulsions are classified as (i) low energy emulsification methods, which include spontaneous emulsification, solvent diffusion, and phase inversion temperature, and (ii) high energy emulsification methods, which require large mechanical energy generated by a high pressure homogenizer, ultrasonicator, and Microfluidizer (Sivakumar et al., 2014).

13.4.3 HIGH ENERGY EMULSIFICATION METHODS

High pressure valve homogenization and high pressure impinging jet devices (e.g. Microfluidizer) are the most common high-pressure devices used (Lee et al., 2013). High-pressure (valve) homogenizers (HPH) (Fig. 13.2) have a piston pump to force under high pressure a premixed coarse emulsion through a specially designed valve containing a gap of 10–100 μm and velocities of hundreds of meters per second (Håkansson et al., 2011; Innings and Trägårdh, 2007; Lee et al., 2013). The viscosity change does not affect the final droplet size indicating that the flow is laminar elongational rather than turbulent (Lee et al., 2013). However, it is believed that the flow is transformed into turbulent at the exit of the valve gap (Tcholakova et al., 2004).

The disadvantages of these homogenizers are that several passes through the machine may be required in order to reach small droplet sizes, and sometimes, after several passes, droplet coalescence occurs due to temperature increase.

A microfluidizer is a patented interaction chamber with microchannels of fixed geometry (F), where a coarse emulsion is pumped under high pressure (Fig. 13.3). It splits the coarse emulsion flow into two and redirects it to the chamber, which allows the streams to impinge at approximately 180° (Cook and Lagace, 1985; Jafari et al., 2007). The impact of the two high-velocity and high-pressure streams creates a region of high turbulence and shear for droplet disruption. At the exit of the chamber, the fluid is subjected to elongational flow as in high-pressure homogenizers. Due to this elongation, emulsifier adsorption at the newly formed interfaces is facilitated, resulting in coalescence minimization (Henry et al., 2010).

In general, a microfluidizer and HPH have a similar emulsification efficiency in water-in-oil emulsions giving small droplets of about 60 nm in diameter at the same pressure of 50 MPa. Multiple passes are needed to reduce droplet size regardless of the device used (Lee et al., 2013).

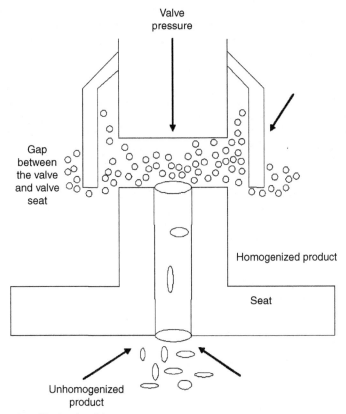

Valve
pressure

Gap
between
the valve
and valve
seat

Homogenized product

Seat

Unhomogenized
product

FIGURE 13.2 Homogenizer Mechanism Diagram

Modified by Spence et al. (2010)

In order to reduce coalescence of broken-up droplets (due to the establishment of a turbulent mixing zone), application of multiple axial flow nozzles (or orifices) in one microsystem has been proposed. A single passage is adequate to produce small droplets. It was interesting to find that a double orifice proved more efficient and virtually independent of the formulation properties (i.e. emulsifier adsorption kinetics, continuous phase viscosity) (Finke et al., 2014).

Apart from high-pressure devices, an ultrasound generator was patented since 1944, and considered an effective way of producing nanoemulsions. High intensity ultrasonication (HIUS) with a frequency range between 16–100 kHz, and power of 10 and 1000 W/cm^2 results in the production of oil droplets in the nanoscale (up to 100 nm), with improved stability over time and the need for small amounts of emulsifying agents (Chemat et al., 2011; Chendke and Fogler, 1975). Microturbulent implosions of the produced cavitation bubbles provide enormous forces that deform and finally break off the droplets in nanometer scale, provided the Laplace pressure is overcome (Sivakumar et al., 2014). Ultrasonication is considered more practicable with respect to production costs as well as equipment contamination compared with microfluidization (Karbstein and Schubert, 1995; Freitas et al., 2006; Jafari et al., 2008b).

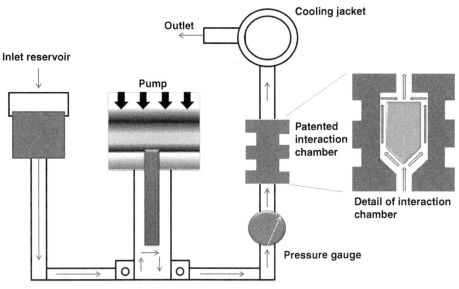

FIGURE 13.3 Microfluidizer Mechanism Diagram

Modified from Nguyen (2013) and Spence et al. (2010)

13.4.4 LOW-ENERGY EMULSIFICATION METHODS

Emulsions can be formed almost spontaneously by taking advantage of the physicochemical proper-ties of the system. Nanoemulsions can then be formed through catastrophic (phase) inversion. This is a process where water-in-oil emulsions suddenly become oil-in-water emulsions and vice versa. The exploitation of this sudden phase transition is interesting in scaling-up low energy methods. Phase in-version also takes place in multiple emulsions where droplets are formed inside droplets.

Emulsion phase inversion (EPI) is a method where an emulsion is formed when water is added to an oil/surfactant mixture. However, there is another method called spontaneous emulsification (SE) in which an emulsion is formed when an oil/surfactant mixture is added to water. A third possibility is to rapidly change the temperature of a surfactant/oil/water mixture below the phase inversion temperature (PIT) under mixing. The first two methods are then isothermal methods whereas the PIT method is an example of a thermal method (McClements, 2013; Komaiko and McClements, 2014).

The main factor that can distinguish SE and EPI from all other preparation methods is that the oil and surfactant phases are intimately mixed together, prior to combining them with water. Furthermore, there must be miscibility between the two mixed components prior to production. The surfactant is also very important because it must be capable of moving from the oil phase into the aqueous phase (Komaiko and McClements, 2014).

The PIT technique has several limitations such as the use of a large quantity of surfactant, careful selection of both surfactant/cosurfactant, and a high polydispersity index. Instability after long-term storage can be observed along with a relatively heavy and tedious workload to identify the system inversion temperature (Sivakumar et al., 2014). However, stability can be controlled by choosing the appropriate storage temperature. It is then the high amount of synthetic surfactants, which may be unsuitable for some applications that should be considered (Komaiko and McClements, 2014).

13.4.5 **RECENT FOOD APPLICATIONS**

One of the major challenges for the food industry is to fabricate food-grade emulsions from label-friendly ingredients. Many food ingredients and products are considered to exist either entirely or partially as emulsions, e.g. beverages, butter, desserts, margarines, dressings, sauces, soups (McClements, 2005; Dolz et al., 2006; Protonotariou et al., 2013; Chung and McClements, 2014; Moore et al., 2012), with the majority of them belonging to the category of oil-in-water emulsions (Chung and McClements, 2014).

Food-grade emulsifiers or stabilizers are increasingly used in the food industry, many of them by utilizing food waste. One recent method of producing emulsions combining food-grade ingredients is heteroaggregation (Chung and McClements, 2014). Droplet flocculation involves fat particle aggregation and can be controlled resulting in microclusters with specific properties. Heteroaggregation is the result of floc creation by mixing two emulsions containing oppositely charged fat droplets. Emulsions with positively charged droplets (protein coated) are mixed with negatively charged droplets (modified starch coated) at low pH (Mao, 2013). Then, a high pressure device is used and the produced emulsions become nanoscale size. Alternatively, two protein-coated emulsions can be mixed with different isoelectric points at a pH where they have opposite charges. Microclusters are formed by mixing an oil-in-water nanoemulsion containing β-lactoglobulin-coated lipid droplets (d32 ~ 0.14 μm) with another oil-in-water nanoemulsion containing lactoferrin-coated lipid droplets (d32 ~ 0.14 μm). By such a method, the viscosity of the final emulsions is controlled resulting in much higher values, which is particularly important for low fat products (Mao and McClements, 2011).

Flavor oil emulsions could be another interesting application of nanoemulsions in the beverage industry. Nanoemulsions based on lemon oil have been produced with a particle size of 81 nm. Food-grade surfactants (e.g. sucrose monoester) have also been applied using a high-pressure device. The addition of cosurfactants and cosolvents increases the stability and inhibits oil particle growth during storage (Rao and McClements, 2013).

Ultrasonication can also be implemented to produce food-grade nanoemulsions. In a recent patent, ultrasonication was used for the production of nanoemulsions containing triglyceride oils (e.g. peanut, flax seed oil). Applications of these emulsions could cover a broad spectrum (Wooster et al., 2010).

In another study, a stable food-grade nanoemulsion with droplet diameter less than 50 nm was formulated by ultrasonic emulsification using basil oil and nonionic surfactant Tween 80. This emulsion was used for food preservation against microbial spoilage as it presented antibacterial properties as well (Ghosh et al., 2013). Additionally, sesame oil was used for the delivery of eugenol, and stable nanoemulsions with antimicrobial activity were fabricated via ultrasonication (Ghosh et al., 2014). Antimicrobial activity was found in D-limonene nanoemulsions including nisin, which was prepared with catastrophic phase inversion (Zhang et al., 2014).

Another example includes the preparation of oil-in-water nanoemulsions by using 10% weight medium chain triglycerides as the oily phase, in which 0.1% citral was dissolved. The fabrication was done using a high-pressure homogenizer (Zhao et al., 2013). Citral stability was maintained by adding an antioxidant at different concentrations. Another method was by coating the oil droplets using cationic biopoylmers (Yang et al., 2012).

The latter technique refers to layer-by-layer deposition. Using this technique, Zhang (2011) attempted to prepare secondary and tertiary beverage clouding emulsions using proteins, such as β-lactoglobulin, to stabilize the primary emulsions and biopolymers, such as ι-carrageenan, gum arabic, pectin, and chitosan to stabilize the secondary and tertiary layers. A combined use of microfluidization and ultrasonication was applied to produce emulsions and increase their stability. Specific combinations resulted in stable emulsions. However, it should be mentioned that final droplet size was around 600–2200 nm; strictly speaking the emulsions produced were out of the nanoscale size.

In the future, nanoemulsions will also be used in food applications. Robust systems, technological advances, and new compounds will be combined and broader applications in many fields of food production will be evident.

13.5 NANOCRYSTALS

Bionanocrystals are crystalline residues with a uniform structure acquired through chemical or enzymatic hydrolysis of natural polysaccharides. Their most important feature is that they can be produced directly from agro-waste materials. Nanocrystals might have needle-like whiskers (e.g. cellulose and chitin) or platelet-like structures (e.g. starch). These bionanocrystals show a high specific strength as well as surface area and therefore contribute as a reinforcing material in composite films. Additionally, the surfaces of polysaccharide nanocrystals are covered with a number of hydroxyl groups, which make it convenient to perform chemical and/or physical conjugation to food bioactives to deliver them to the specific sites (Lin et al., 2011; Zhang et al., 2010).

13.5.1 CELLULOSE

An attractive source of cellulose for industrial uses is agricultural waste. Cellulose nanocrystals (CNC) are a promising hydrophilic material with a broad range of applications such as delivery systems (Rescignano et al., 2014) and an improver of film features (Dehnad et al., 2014a, b; Fortunati et al., 2013). High inherent rigidity (Young's modulus of 167.5 GPa (Tashiro and Kobayashi, 1991)) of CNCs along with their barrier properties make them highly attractive for the preparation of biodegradable films. These hydrophobic, needle-like, and thermostable particles are commonly prepared by acid hydrolysis of plant sources (Azizi Samir et al., 2005). Under controlled conditions, strong acid hydrolysis (usually sulfuric acid) leads to the removal of the amorphous regions. Crystalline parts are acid insoluble and accumulate in the form of crystalline nanoparticles (Peng et al., 2011). The morphology (short or long needle-like) and properties of CNC depend on the source of the original cellulose and preparation conditions.

Table 13.4 shows the properties of CNC obtained from different agro-waste sources. Plant hydrolysis can be performed using hydrochloric acid, while it leads to remaining of the hydroxyl groups on the surface and therefore the surface is weakly charged, which causes low colloidal stability (Araki et al., 1998). Alkaline or enzymatic (mostly pectinase) pretreatment is usually performed to remove excessive amounts of material from the fibers (Galanakis, 2012; Galanakis et al., 2015b).

13.5.2 STARCH

Native starch is found in the form of discrete and partially crystalline granules. Acid treatment (usually hydrochloric acid) of starch below its gelatinization temperature leads to hydrolysis of the amorphous regions allowing the separation of crystalline lamella. Therefore, a water insoluble, highly crystalline,

Table 13.4 Production of Cellulose Nanocrystals Using Agro-Waste Materials

Waste Material Source	Production Method	Features	Applied or Suggested Application	References
Potato peel	Alkali treatment followed by acid hydrolysis	An average length of 410 nm and a diameter of 10 nm (aspect ratio of 41).	Improves thermal, mechanical, and barrier properties of composite.	Chen et al. (2012)
Pineapple leaf	Alkali treatment followed by acid hydrolysis	Degree of crystallinity of 73%, an average length of 249.7 nm and a diameter of 4.45 nm (aspect ratio of around 60).	High potential to be used as surface-active compound due to its high aspect ratio; improve thermal, mechanical, and barrier properties of composite.	dos Santos et al. (2013)
Pristine and carded hemp fibers	Alkaline or pectinase treatment followed by acid hydrolysis	An average length of 100–200 nm and a diameter of 15 nm (aspect ratio of around 6.7–13.3), enzymatically, compared with alkaline pretreated fibers, could be more easily attacked by the acid and consequently would reduce the thermal stability of the CNC.	Improve thermal, mechanical, and barrier properties of composite.	Luzi et al. (2014)
Mango seed	Alkaline or pectinase treatment followed by acid hydrolysis	Degree of crystallinity of 90.6%, an average length of 123.4 nm, and a diameter of 4.59 nm (aspect ratio of around 34.1).	Improves thermal, mechanical, and barrier properties of composite.	Henrique et al. (2013)
Corncob	Alkaline or pectinase treatment followed by acid hydrolysis	Degree of crystallinity of 83.7%, an average length of 210.8 nm, and a diameter of 4.15 nm (aspect ratio of around 53.4).	Improves thermal, mechanical, and barrier properties of composite.	Silvério et al. (2013)

platelet-shaped residue known as starch nanocrystals (SNCs) is obtained (LeCorre et al., 2012). The effect of botanic origin and amylose content on features of SNCs has been investigated, too. For the same amylose content, maize, potato, and wheat starches resulted in a rather similar size and crystallinity of SNCs showing a limited impact on the botanic origin. For the same botanic origin (maize), differences in size were more important, indicating the strong influence of the amylopectin content and molecular structure. Particles tended to show square-like morphology with increasing initial amylopectin content (LeCorre et al., 2011; LeCorre et al., 2012).

SNCs have been used as nanofillers in edible films matrixes to improve their mechanical and/or barrier properties (Kristo and Biliaderis, 2007). In addition, chemical modification of SNCs is provided to enhance their functional properties. For example, surface modification has been performed via different reagents such as stearic acid, chloride, and poly(ethylene glycol) methyl ether (Thielemans et al., 2006) to form better binding sites for delivery of bioactive agents.

13.5.3 CHITIN

Chitin is the second most abundant biopolymer, which can be found both in plant and animal sources. Naturally, chitin is found as ordered crystalline microfibrils, which associate with other materials, such as proteins, lipids, polysaccharides, calcium carbonate, and pigments. Plant chitin is highly crystalline with some disordered or paracrystalline regions that arise from defects. The disordered or paracrystalline regions are preferentially hydrolyzed (using hydrochloric acid) or oxidized under certain conditions, whereas crystalline regions remain intact and rod-like or whiskers of chitin nanocrystals are produced. The sizes of the chitin nanocrystals are generally affected by the origin of the chitin, concentration of the hydrochloric acid solutions, and hydrolysis time, with lengths varying over the range of 150–2200 nm, and widths over the range of 10–50 nm (Wang and Esker, 2014).

In contrast to cellulose and starch, chitin nanocrystals have attracted more attention as they exhibit a unique cationic structure (Li et al., 1997) and could be recovered from marine processing wastes. Chemically modified chitin nanocrystals can be produced by grafting, using different lipophilic compounds (e.g. poly (3-hydroxybutyrate-*co*-3-hydroxyvalerate) and stearic acid chloride) to enhance the hydrophobicity of polysaccharide nanocrystals and thus their dispersion and compatibility with polymer matrices (Thielemans et al., 2006).

13.5.4 POTENTIAL LIMITATIONS

From technological, economic, and environmental points of view it is ideal to convert agro-waste materials to nanocapsules and nanocrystals. In spite of the different applications, there are potential risks that should be studied before usage. If all the components used for fabrication of nanoparticles are digestible, it would be expected to be largely digested in gastrointestinal media. However, if one or more of the components are indigestible (e.g. cellulose- and pectin-based nanoparticles that can resist acidic and enzymatic degradation), they may be able to reach the colon. It should be noted that the colon contains diverse microbes that are capable of breaking down and utilizing cellulose (Azuma et al., 2014). However, if nanoparticles tolerate intestinal and microbial enzymes (such as resistant starch or cellulose-based nanoparticles), they may remain in nanosize and therefore would be absorbed directly into the human body, depending on their size, structure, and surface characteristics (Fathi et al., 2014). There are some reports on the biological fate of metallic nanoparticles (e.g. gold and silver), which show that translocation to the systemic circulation could lead to their excretion or unwanted effects on blood or other organs (e.g. inflammation, oxidative stress, cytotoxicity, fibrosis, and immunologic responses) (Unfried et al., 2007). However, there are very rare reports on biopolymers (McClements, 2013) and more experiments are required for different food-grade nanoparticles on direct absorption to establish their biological fate.

13.6 PULSED FLUIDIZED BED AGGLOMERATION
13.6.1 INTRODUCTION

Food powders are usually characterized by their fine and cohesive particles with low dispersibility in liquids. The agglomeration or granulation is known as an alternative process to improve these product characteristics. Agglomeration of food powders is commonly used to produce porous granules with higher wettability and dispersibility (Schubert, 1987). It is usually carried out by wetting the powder with a liquid that promotes adhesion between a particle and the other by linking bridges, leading to formation of larger particles (Knight, 2001). The agglomeration process promotes particle enlargement as well as reduction of fines and thus results in benefits such as lower rates of particle elutriation and reduction in the dangers of handling or inhalation of powders (Senadeera et al., 2000; Reyes et al., 2007). Food powder agglomeration is commonly used to improve the instant properties of spray-dried products. However, this process is also used when it is desirable to improve the flowability or change the visual properties of the powder due to particle enlargement. The fine particles produced by spray drying present a circular and compact shape, whereas instant particles, or agglomerated products, are characterized by their porous and irregular surface and are easily obtained by a fluid bed agglomeration process. The fluidization behavior of spray-dried particles is commonly characterized by their cracks and channeling formation during their fluidization. Thus, vibration or pulsation systems can be attached to fluid bed equipment to improve bed homogeneity and allow particle fluidization using a smaller fluidizing air flow (Dacanal and Menegalli, 2010).

13.6.2 FLUID BED AGGLOMERATION

Fluid bed agglomeration is an innovative technology used to improve the instant properties of fine and cohesive food powders. This technology is based on fluidization of the bed of a particulate material due to sequential relocation of a fluidizing air stream, which causes bed vibration rather than fluidization (Galanakis, 2012, 2013).

The operational principles of fluidized bed agglomeration consist of particle fluidization by hot air flow and superficial particle wetting caused by the atomization of a solvent or a liquid binder (Fig. 13.4). The collision between wet particles in the fluid bed forms liquid bridges and particle coalescence. On drying, the bridges solidify, resulting in a consolidation of the agglomerates (Dacanal, 2008; Gawrzynski et al., 1999). In the case of the liquid binder, the particle surface becomes more viscous and adherent with sprayed binder drying. At the final drying stage, the particle moisture decreases and glass transition temperature of binder polysaccharides (e.g. maltodextrin) increases, leading to permanent crystallization.

This technology has been used for the encapsulation of biomaterials and nutraceuticals, too, by intensifying as well as accelerating fluid bed dewatering and agglomeration of red pepper slices and instant soy protein isolate, respectively (Dacanal and Menegalli, 2010; Ade-Omowaye et al., 2001). Fluidized bed agglomeration is one of the processes suitable for producing agglomerates with high porosity and good mechanical resistance for handling and packaging (Turchiuli et al., 2005).

13.6.3 PULSED FLUID BED AGGLOMERATION

This emerging technique has some advantages over the conventional fluid bed equipment, including easy fluidization of irregular particles of different sizes (Gawrzynski et al., 1999; Reyes et al., 2008). Pulsed flow may be effective in overcoming defluidization (Turchiuli et al., 2005), especially when applied to cohesive particles. Therefore, since pulsation improves bed fluid dynamics, avoids defluidization, and helps to break liquid binds, pulsed fluidization may reduce channeling and by-passing, so that hard-to-fluidize materials can be fluidized (Iveson et al., 2001).

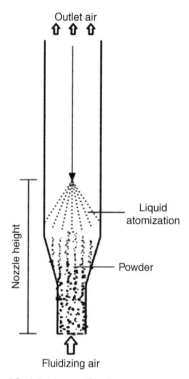

FIGURE 13.4 Schemes of the Fluidized Bed Agglomeration Process

Adapted from Dacanal and Menegalli (2010)

Dacanal and Menegalli (2010) produced soy protein isolate particles with high wettability using agglomeration in a pulsed fluid bed and maltodextrin aqueous solution as a liquid binder. They assessed the influence of various input parameters (product, inlet and outlet air temperature, consumption of liquid binder, granulation liquid-binder spray rate, spray pressure, and drying time) on granulation output properties (granule flow rate, granule size determined using light scattering method and sieve analysis, granule Hausner ratio, porosity, and residual moisture). Compared with raw material, agglomerated particles were more porous and had a more irregular shape, presenting a wetting time decrease, free-flow improvement, and cohesiveness reduction.

REFERENCES

Acosta, E., 2008. Testing the effectiveness of nutrient delivery systems. In: Garti, N. (Ed.), Delivery and Controlled Release of Bioactives in Foods and Nutraceuticals. CRC Press, Boca Raton.

Ade-Omowaye, B.I.O., Angersbach, A., Taiwo, K.A., Knorr, D., 2001. Use of pulsed electric field treatment to improve dehydration characteristics of plant based foods. Trends Food Sci. Technol. 12, 285–295.

Alting, A.C., de Jongh, H.H.J., Visschers, R.W., Simons, J.-W.F.A., 2002. Physical and chemical interactions in cold gelation of food proteins. J. Agric. Food Chem. 50, 4682–4689.

Araki, J., Wada, M., Kuga, S., Okano, T., 1998. Flow properties of microcrystalline cellulose suspension prepared by acid treatment of native cellulose. Colloids Surf. A 142, 75–82.

Azizi Samir, M.A.S., Alloin, F., Dufresne, A., 2005. Review of recent research into cellulosic whiskers, their properties and their application in nanocomposite field. Biomacromolecules 6, 612–626.

Azuma, K., Osaki, T., Ifuku, S., Saimoto, H., Morimoto, M., Takashima, O., Tsuka, T., Imagawa, T., Okamoto, Y., Minami, S., 2014. Anti-inflammatory effects of cellulose nanofiber made from pear in inflammatory bowel disease model. Bioact. Carbohydr. Diet. Fibre 3, 1–10.

Bajpai, A.K., Choubey, J., 2006. *In vitro* release dynamics of an anticancer drug from swellable gelatin nanoparticles. J. Appl. Polym. Sci. 101, 2320–2332.

Bhardwaj, N., Kundu, S.C., 2010. Electrospinning: a fascinating fiber fabrication technique. Biotechnol. Adv. 28, 325–347.

BÜCHI Labortechnik, A., 2009. Switzerland Brochure Nano Spray Dryer B-90.

Chemat, F., Huma, Z., M. Kamran Khan, M., 2011. Applications of ultrasound in food technology: processing, preservation and extraction. Ultrason. Sonochem. 18, 813–835.

Chen, H., Zhong, Q., 2014. Processes improving the dispersibility of spray-dried zein nanoparticles using sodium caseinate. Food Hydrocolloid. 35, 358–366.

Chen, Y.-C., Yu, S.-H., Tsai, G.-J., Tang, D.-W., Mi, F.-L., Peng, Y.-P., 2010. Novel technology for the preparation of self-assembled catechin/gelatin nanoparticles and their characterization. J. Agric. Food Chem. 58, 6728–6734.

Chen, D., Lawton, D., Thompson, M.R., Liu, Q., 2012. Biocomposites reinforced with cellulose nanocrystals derived from potato peel waste. Carbohydr. Polym. 90, 709–716.

Chendke, P.K., Fogler, H.S., 1975. Macrosonics in industry: 4. Ultrasonic innovations in the food industry: from the laboratory to commercial production. Chem. Process. Ultrason. 13, 31–37.

Cheong, J.N., Tan, C.P., Man, Y.B.C., Misran, M., 2008. α-Tocopherol nanodispersions: preparation, characterization and stability evaluation. J. Food Eng. 89, 204–209.

Chung, C., McClements, D.J., 2014. Structure–function relationships in food emulsions: improving food quality and sensory perception. Food Struct. 1 (2), 106–126.

Cocero, M.J., Martín, Á., Mattea, F., Varona, S., 2009. Encapsulation and co-precipitation processes with supercritical fluids: fundamentals and applications. J. Supercrit. Fluids 47, 546–555.

Cook, E.J., Lagace, A.P., 1985. Apparatus for forming emulsions, vol.US4533254. Biotechnology Development Corporation.

Dacanal, G.C., 2008. Experimental study of fluidized bed agglomeration of acerola powder. Braz. J. Chem. Eng. 25, 51–58.

Dacanal, G.C., Menegalli, E., 2010. Selection of operational parameters for the production of instant soy protein isolate by pulsed fluid bed agglomeration. Powder Technol. 203, 565–573.

de Vos, P., Faas, M.M., Spasojevic, M., Sikkema, J., 2010. Encapsulation for preservation of functionality and targeted delivery of bioactive food components. Int. Dairy J. 20, 292–302.

Dehnad, D., Emamjomee, Z., Mirzaei, H.A., Jafari, S.M., Dadashi, S., 2014a. Optimization of physical and mechanical properties for chitosan-nanocellulose biocomposites. Carbohydr. Polym. 105, 222–228.

Dehnad, D., Mirzaei, H.A., Emamjomee, Z., Jafari, S.M., Dadashi, S., 2014b. Thermal and antimicrobial properties of chitosan-nanocellulose films for extending shelf life of ground meat. Carbohydr. Polym. 109, 148–154.

Dolz, M., Hernandez, M.J., Delegido, J., 2006. Oscillatory measurements for salad dressings stabilized with modified starch, xanthan gum, and locust bean gum. J. Appl. Polym. Sci. 102 (1), 897–903.

dos Santos, R.M., Flauzino Neto, W.P., Silvério, H.A., Martins, D.F., Dantas, N.O., Pasquini, D., 2013. Cellulose nanocrystals from pineapple leaf, a new approach for the reuse of this agro-waste. Ind. Crop. Prod. 50, 707–714.

Dutta, R.K., Sahu, S., 2012. Development of a novel probe sonication assisted enhanced loading of 5-FU in SPION encapsulated pectin nanocarriers for magnetic targeted drug delivery system. Eur. J. Pharm. Biopharm. 82, 58–65.

Elzoghby, A.O., Abo El-Fotoh, W.S., Elgindy, N.A., 2011. Casein-based formulations as promising controlled release drug delivery systems. J. Control. Release 153, 206–216.

Fabra, M.J., López-Rubio, A., Lagaron, J.M., 2014. On the use of different hydrocolloids as electrospun adhesive interlayers to enhance the barrier properties of polyhydroxyalkanoates of interest in fully renewable food packaging concepts. Food Hydrocolloid. 39, 77–84.

Fathi, M., Mozafari, M.R., Mohebbi, M., 2012. Nanoencapsulation of food ingredients using lipid based delivery systems. Trends Food Sci. Technol. 23, 13–27.

Fathi, M., Varshosaz, J., Mohebbi, M., Shahidi, F., 2013. Hesperetin-loaded solid lipid nanoparticles and nanostructure lipid carriers for food fortification: preparation, characterization, and modeling. Food Bioprocess Technol. 6, 1464–1475.

Fathi, M., McClements, D.J., Martín, Á., 2014. Nanoencapsulation of food ingredients using carbohydrate based delivery systems. Trends Food Sci. Technol. 39, 18–39.

Finke, J.H., Niemann, S., Richter, C., Gothsch, T., Kwade, A., Buttgenbach, S., Muller-Goymann, C.C., 2014,. Multiple orifices in customized microsystem high-pressure emulsification: the impact of design and counter pressure on homogenization efficiency. HYPERLINK "http://www. sciencedirect. com/science/journal/13504177" \o "Go to Ultrasonics Sonochemistry on ScienceDirect". Chem. Eng. J. 248, 107–121.

Fortunati, E., Peltzer, M., Armentano, I., Jiménez, A., Kenny, J.M., 2013. Combined effects of cellulose nanocrystals and silver nanoparticles on the barrier and migration properties of PLA nano-biocomposites. J. Food Eng. 118, 117–124.

Freitas, S., Hielscher, G., Merkle, H.P., Gander, B., 2006. Continuous contact and contamination free ultrasonic emulsification. Ultrason. Sonochem. 13 (1), 76–85.

Fung, W.-Y., Yuen, K.-H., Liong, M.-T., 2011. Agrowaste-based nanofibers as a probiotic encapsulant: fabrication and characterization. J. Agric. Food Chem. 59, 8140–8147.

Galanakis, C.M., 2011. Olive fruit and dietary fibers: components, recovery and applications. Trends Food Sci. Technol. 22, 175–184.

Galanakis, C.M., 2012. Recovery of high added-value components from food wastes: conventional, emerging technologies and commercialized. Trends Food Sci. Technol. 26, 68–87.

Galanakis, C.M., 2013. Emerging technologies for the production of nutraceuticals from agricultural by-products: a viewpoint of opportunities and challenges. Food Bioprod. Process. 91, 575–579.

Galanakis, C.M., Schieber, A., 2014. Editorial. Special issue on recovery and utilization of valuable compounds from food processing by-products. Food Res. Int. 65, 230–299.

Galanakis, C.M., Chasiotis, S., Botsaris, G., Gekas, V., 2014. Separation and recovery of proteins and sugars from Halloumi cheese whey. Food Res. Int. 65, 477–483.

Galanakis, C.M., Patsioura, A., Gekas, V., 2015b. Enzyme kinetics modeling as a tool to optimize food biotechnology applications: a pragmatic approach based on amylolytic enzymes. Crit. Rev. Food Sci. Technol. 55, 1758–1770.

García-Garibay, M., Gómez-Ruiz, L., Cruz-Guerrero, A.E., 2014. Single cell protein yeasts and bacteria, second ed. Encyclopedia of Food MicrobiologyAcademic Press, Oxford, pp. 431–438.

Gawrzynski, Z., Glaser, R., Kudra, T., 1999. Drying of powdery materials in a pulsed fluid bed dryer. Drying Technol. 17 (7–8), 1523–1532.

Ghosh, V., Mukherjee, A., Chandrasekaran, N., 2013. Ultrasonic emulsification of food-grade nanoemulsion formulation and evaluation of its bactericidal activity. Ultrason. Sonochem. 20, 338–344.

Ghosh, V., Mukherjee, A., Chandrasekaran, N., 2014. Eugenol-loaded antimicrobial nanoemulsions preserves fruit juice against, microbial spoilage. Colloids Surf. B Biointerfaces 114, 392–397.

Gomez-Estaca, J., Balaguer, M.P., Gavara, R., Hernandez-Munoz, P., 2012. Formation of zein nanoparticles by electrohydrodynamic atomization: effect of the main processing variables and suitability for encapsulating the food coloring and active ingredient curcumin. Food Hydrocolloid. 28, 82–91.

Gunasekaran, S., Ko, S., Xiao, L., 2007. Use of whey proteins for encapsulation and controlled delivery applications. J. Food Eng. 83, 31–40.

Gupta, V., Aseh, A., Ríos, C.N., Aggarwal, B.B., Mathur, A.B., 2009. Fabrication and characterization of silk fibroin-derived curcumin nanoparticles for cancer therapy. Int. J. Nanomed. 4, 115–122.

Håkansson, A., Fuchs, L., Innings, F., Revstedt, J., Trägårdh, C., Bergenståhl, B., 2011. High resolution experimental measurement of turbulent flow field in a high pressure homogenizer model and its implications on turbulent drop fragmentation. Chem. Eng. Sci. 66 (8), 1790–1801.

Hao, S., Wang, Y., Wang, B., Deng, J., Liu, X., Liu, J., 2013. Rapid preparation of pH-sensitive polymeric nanoparticle with high loading capacity using electrospray for oral drug delivery. Mater. Sci. Eng. C 33, 4562–4567.

Heitfeld, K.A., Guo, T., Yang, G., Schaefer, D.W., 2008. Temperature responsive hydroxypropyl cellulose for encapsulation. Mater. Sci. Eng. C 28, 374–379.

Henrique, M.A., Silvério, H.A., Flauzino Neto, W.P., Pasquini, D., 2013. Valorization of an agro-industrial waste, mango seed, by the extraction and characterization of its cellulose nanocrystals. J. Environ. Manag. 121, 202–209.

Henry, J.V.L., Fryer, P., Frith, W., Norton, I., 2010. The influence of phospholipids and food proteins on the size and stability of model sub-micron emulsions. Food Hydrocolloid. 24 (1), 66–71.

Hu, B., Pan, C., Sun, Y., Hou, Z., Ye, H., Zeng, X., 2008. Optimization of fabrication parameters to produce chitosan-tripolyphosphate nanoparticles for delivery of tea catechins. J. Agric. Food Chem. 56, 7451–7458.

Hussain, M.R., Maji, T.K., 2008. Preparation of genipin cross-linked chitosan-gelatin microcapsules for encapsulation of *Zanthoxylum* limonella oil (ZLO) using salting-out method. J. Microencapsul. 25, 414–420.

Innings, F., Trägårdh, C., 2007. Analysis of the flow field in a high-pressure homogenizer. Exp. Therm. Fluid Sci. 32 (2), 345–354.

Iveson, S.M., Litster, J.D., Hapgood, K., Ennis, B.J., 2001. Nucleation, growth and breakage phenomena in agitated wet granulation processes: a review. Powder Technol. 117, 3–39.

Jafari, S.M., He, Y., Bhandari, B., 2006. Nano-emulsion production by sonication and microfluidization – a comparison. Int. J. Food Prop. 9, 475–485.

Jafari, S.M., He, Y., Bhandari, B., 2007. Optimization of nano-emulsions production by microfluidization. Eur. Food Res. Technol. 225, 733–741.

Jafari, S.M., Assadpoor, E., He, Y., Bhandari, B., 2008a. Encapsulation efficiency of food flavours and oils during spray drying. Drying Technol. 26 (7), 816–835.

Jafari, S.M., Elham, A., He, Y., Bhandari, B., 2008b. Re-coalescence of emulsion droplets during high-energy emulsification. Food Hydrocolloid. 22 (7), 1191–1202.

Jain, A.K., Khar, R.K., Ahmed, F.J., Diwan, P.V., 2008. Effective insulin delivery using starch nanoparticles as a potential trans-nasal mucoadhesive carrier. Eur. J. Pharm. Biopharm. 69, 426–435.

Jin, H., Xia, F., Jiang, C., Zhao, Y., He, L., 2009. Nanoencapsulation of lutein with hydroxypropylmethyl cellulose phthalate by supercritical antisolvent. Chin. J. Chem. Eng. 17, 672–677.

Jyoti Das, R., Baishaya, K., Pathak, K., 2013. Recent advancement in lipid drug conjugates as nanoparticulate drug delivery system. Int. Res. J. Pharm. 4, 73–78.

Karbstein, H., Schubert, H., 1995. Developments in the continuous mechanical production of oil-in-water macro-emulsions. Chem. Eng. Process.V 34 (3), 205–211.

Khaksar, R., Hosseini, S.M., Hosseini, H., Shojaee-Aliabadi, S., Mohammadifar, M.A., Mortazavian, A.M., khosravi-Darani, K., Haji Seyed Javadi, N., Komeily, R., 2014. Nisin-loaded alginate-high methoxy pectin microparticles: preparation and physicochemical characterisation. Int. J. Food Sci. Technol.

Khan, A.W., Kotta, S., Ansari, S.H., Sharma, R.K., Ali, J., 2012. Potentials and challenges in self-nanoemulsifying drug delivery systems. Expert Opin. Drug Deliv. 9, 1305–1317.

Klinkesorn, U., McClements, D.J., 2009. Influence of chitosan on stability and lipase digestibility of lecithin-stabilized tuna oil-in-water emulsions. Food Chem. 114, 1308–1315.

Knight, P.C., 2001. Structuring agglomerated products for improved performance. Powder Technol. 119, 14–25.

Komaiko, J., McClements, D.J., 2014. Optimization of isothermal low-energy nanoemulsion formation: hydrocarbon oil, non-ionic surfactant and water systems. J. Colloid Interface Sci. 425, 59–66.

Kong, L., Ziegler, G.R., 2013. Quantitative relationship between electrospinning parameters and starch fiber diameter. Carbohydr. Polym. 92, 1416–1422.

Kristo, E., Biliaderis, C.G., 2007. Physical properties of starch nanocrystal-reinforced pullulan films. Carbohydr. Polym. 68, 146–158.

LaClair, C.E., 2008. Purification and use of whey proteins for improved health. University of Wisconsin, Madison.

Lammel, A.S., Hu, X., Park, S.-H., Kaplan, D.L., Scheibel, T.R., 2010. Controlling silk fibroin particle features for drug delivery. Biomaterials 31, 4583–4591.

LeCorre, D.b., Bras, J., Dufresne, A., 2011. Influence of botanic origin and amylose content on the morphology of starch nanocrystals. J. Nanopart. Res. 13, 7193–7208.

LeCorre, D.B., Bras, J., Dufresne, V, 2012. Influence of native starch's properties on starch nanocrystals thermal properties. Carbohydr. Polym. 87, 658–666.

Lee, L., Hancock, R., Noble, I., Norton, I., 2013. Production of water-in-oil nanoemulsions using high pressure homogenisation: a study on droplet break-up. J. Food Eng. 131, 33–37.

Lertsutthiwong, P., Noomun, K., Jongaroonngamsang, N., Rojsitthisak, P., Nimmannit, U., 2008. Preparation of alginate nanocapsules containing turmeric oil. Carbohydr. Polym. 74, 209–214.

Lesmes, U., Cohen, S.H., Shener, Y., Shimoni, E., 2009. Effects of long chain fatty acid unsaturation on the structure and controlled release properties of amylose complexes. Food Hydrocolloid. 23, 667–675.

Li, J., Revol, J.F., Marchessault, R.H., 1997. Effect of degree of deacetylation of chitin on the properties of chitin crystallites. J. Appl. Polym. Sci. 65, 373–380.

Li, J.K., Wang, N., Wu, X.S., 1998. Gelatin nanoencapsulation of protein/peptide drugs using an emulsifier-free emulsion method. J. Microencapsul. 15, 163–172.

Li, Y., Lim, L.T., Kakuda, Y., 2009. Electrospun zein fibers as carriers to stabilize (−)-epigallocatechin gallate. J. Food Sci. 74, C233–C240.

Lin, N., Huang, J., Chang, P.R., Feng, L., Yu, J., 2011. Effect of polysaccharide nanocrystals on structure, properties, and drug release kinetics of alginate-based microspheres. Colloid. Surf. B Biointerfaces 85, 270–279.

Liu, X., Chen, X., Li, Y., Wang, X., Peng, X., Zhu, W., 2012. Preparation of superparamagnetic Fe_3O_4 alginate/chitosan nanospheres for *Candida rugosa* lipase immobilization and utilization of layer-by-layer assembly to enhance the stability of immobilized lipase. ACS Appl. Mater. Interfaces 4, 5169–5178.

López-Rubio, A., Lagaron, J.M., 2012. Whey protein capsules obtained through electrospraying for the encapsulation of bioactives. Innov. Food Sci. Emerg. Technol. 13, 200–206.

Luzi, F., Fortunati, E., Puglia, D., Lavorgna, M., Santulli, C., Kenny, J.M., Torre, L., 2014. Optimized extraction of cellulose nanocrystals from pristine and carded hemp fibres. Ind. Crop. Prod. 56, 175–186.

Mantegna, S., Binello, A., Boffa, L., Giorgis, M., Cena, C., Cravotto, G., 2012. A one-pot ultrasound-assisted water extraction/cyclodextrin encapsulation of resveratrol from *Polygonum cuspidatum*. Food Chem. 130, 746–750.

Mao, Y.Y., 2013. Designing novel emulsion performance by controlled hetero-aggregation of mixed biopolymer systems. Dissertations, paper 826.

Mao, Y.Y., McClements, D.J., 2011. Modulation of bulk physicochemical properties of emulsions by hetero-aggregation of oppositely charged protein-coated lipid droplets. Food Hydrocolloid. 25, 1201–1209.

Maswal, M., Dar, A.A., 2014. Formulation challenges in encapsulation and delivery of citral for improved food quality. Food Hydrocolloid. 33, 182–195.

Mattea, F., Martín, Á., Cocero, M.J., 2009. Carotenoid processing with supercritical fluids. J. Food Eng. 93, 255–265.

McClements, D.J., 2005. Food Emulsions: Principles, Practices, and Techniques. CRC Press, Boca Raton, FL, USA.

McClements, D.J., 2013. Edible lipid nanoparticles: digestion, absorption, and potential toxicity. Prog. Lipid Res. 52, 409–423.

Minimol, P.F., Paul, W., Sharma, C.P., 2013. PEGylated starch acetate nanoparticles and its potential use for oral insulin delivery. Carbohydr. Polym. 95, 1–8.

Moore, R.L., Duncan, S.E., Rasor, A.S., Eigel, W.N., 2012. Oxidative stability of an extended shelf-life dairy-based beverage system designed to contribute to hearth health. J. Dairy Sci. 95, 6242–6251.

Mozafari, M.R., 2006. Nanocarrier Technologies: Frontiers of Nanotherapy. Springer, Dordrecht.

Nguyen, T., 2013. The effects of microfluidization and homogenization on the composition and structure of liposomal aggregates from whey buttermilk and commercial buttermilk. MSc Thesis. California Polytechnic State University.

Okutan, N., Terzi, P.N., Altay, F., 2014. Affecting parameters on electrospinning process and characterization of electrospun gelatin nanofibers. Food Hydrocolloid. 39, 19–26.

Oreopoulou, V., Russ, W., 2006. Utilization of By-Products and Treatment of Waste in the Food Industry. Springer, New York.

Paques, J.P., van der Linden, E., van Rijn, C.J.M., Sagis, L.M.C., 2014. Preparation methods of alginate nanoparticles. Adv. Colloid Interface Sci.

Peng, B.L., Dhar, N., Liu, H.L., Tam, K.C., 2011. Chemistry and applications of nanocrystalline cellulose and its derivatives: a nanotechnology perspective. Can. J. Chem. Eng. 89, 1191–1206.

Pereira, H.V.R., Saraiva, K.P., Carvalho, L.M.J., Andrade, L.R., Pedrosa, C., Pierucci, A.P.T.R., 2009. Legumes seeds protein isolates in the production of ascorbic acid microparticles. Food Res. Int. 42, 115–121.

Protonotariou, S., Evageliou, V., Yanniotis, S., Mandala, I., 2013. The influence of different stabilizers and salt addition on the stability of model emulsions containing olive or sesame oil. J. Food Eng. 117, 124–132.

Qi, Z.H., Xu, A., 1999. Starch-based ingredients for food encapsulation. Cereal Foods World 44, 460–465.

Quiñones, J.P.R., Gothelf, K.V., Kjems, J.R., Caballero, Á.M.H., Schmidt, C., Covas, C.P., 2012. Self-assembled nanoparticles of glycol chitosan-ergocalciferol succinate conjugate, for controlled release. Carbohydr. Polym. 88, 1373–1377.

Rao, J., McClements, D.J., 2013. Optimization of lipid nanoparticle formation for beverage applications: influence of oil type, cosolvents, and cosurfactants on nanoemulsion properties. J. Food Eng. 118, 198–204.

Rescignano, N., Fortunati, E., Montesano, S., Emiliani, C., Kenny, J.M., Martino, S., Armentano, I., 2014. PVA bio-nanocomposites: a new take-off using cellulose nanocrystals and PLGA nanoparticles. Carbohydr. Polym. 99, 47–58.

Reyes, A., Moyano, P., Paz, P., 2007. Drying of potato slices in a pulsed fluidized bed. Drying Technol. 25, 581–590.

Reyes, A., Herrera, N., Vega, R., 2008. Drying suspensions in a pulsed fluidized bed of inert particles. Drying Technol. 26, 122–131.

Roselló-Soto, E., Barba, F.J., Parniakov, O., Galanakis, C.M., Grimi, N., Lebovka, N., Vorobiev, E., 2015. High voltage electrical discharges, pulsed electric field and ultrasounds assisted extraction of protein and phenolic compounds from olive kernel. Food Bioprocess Technol. 8, 885–894.

Santander-Ortega, M.J., Stauner, T., Loretz, B., Ortega-Vinuesa, J.L., Bastos-González, D., Wenz, G., Schaefer, U.F., Lehr, C.M., 2010. Nanoparticles made from novel starch derivatives for transdermal drug delivery. J. Control. Rel. 141, 85–92.

Schroeder, A., Kost, J., Barenholz, Y., 2009. Ultrasound, liposomes, and drug delivery: principles for using ultrasound to control the release of drugs from liposomes. Chem. Phys. Lipids 162, 1–16.

Schubert, H., 1987. Food particle technology. Part I: properties of particles and particulate food systems. J. Food Eng. 6, 1–32.

Semo, E., Kesselman, E., Danino, D., Livney, Y.D., 2007. Casein micelle as a natural nano-capsular vehicle for nutraceuticals. Food Hydrocolloid. 21, 936–942.

Senadeera, W., Bhandari, B., Young, G., Wijesinghe, B., 2000. Methods for effective fluidization of particulate food materials. Dry. Technol. 18, 1537–1557.

Sheng, X., Fan, L., He, C., Zhang, K., Mo, X., Wang, H., 2013. Vitamin E-loaded silk fibroin nanofibrous mats fabricated by green process for skin care application. Int. J. Biol. Macromol. 56, 49–56.

Silva, A.C., González-Mira, E., García, M.L., Egea, M.A., Fonseca, J., Silva, R., Santos, D., Souto, E.B., Ferreira, D., 2011. Preparation, characterization and biocompatibility studies on risperidone-loaded solid lipid nanoparticles (SLN): high pressure homogenization versus ultrasound. Colloid. Surf. B Biointerfaces 86, 158–165.

Silvério, H.A., Flauzino Neto, W.P., Dantas, N.O., Pasquini, D., 2013. Extraction and characterization of cellulose nanocrystals from corncob for application as reinforcing agent in nanocomposites. Ind. Crop. Prod. 44, 427–436.

Singh, V., Chaudhary, A.K., 2010. Development and characterization of rosiglitazone loaded gelatin nanoparticles using two step desolvation method. Int. J. Pharm. Sci. Rev. Res. 5, 100–103.

Sinha, V.R., Kumria, R., 2001. Polysaccharides in colon-specific drug delivery. Int. J. Pharm. 224, 19–38.

Sivakumar, M., Tang, S. Y., Tan W.K., 2014. Cavitation technology – a greener processing technique for the generation of pharmaceutical nanoemulsions. Ultrason. Sonochem. 21, 2069–2083.

Sonia, T.A., Sharma, C.P., 2012. An overview of natural polymers for oral insulin delivery. Drug Discov. Today 17, 784–792.

Spence, K., Venditti, R.A., Rojas, O.J., 2010. Aspects of raw materials and processing conditions on the production and utilization of microfibrillated cellulous, International Conference on Nanotechnology for the Forest Products Industry. 27-29/9/21010, Otaniemi, Espoo, Finland. http://www.tappi.org/content/events/10nano/papers/24.1.pdf.

Subia, B., Kundu, S.C., 2013. Drug loading and release on tumor cells using silk fibroin–albumin nanoparticles as carriers. Nanotechnology 24, 035103.

Tadros, T.F., Izquierdo, P., Esquena, J., Solans, C., 2004. Formation and stability of nano-emulsions. Adv. Colloid Interface Sci., 108–109, 303-318.

Tashiro, K., Kobayashi, M., 1991. Theoretical evaluation of three-dimensional elastic constants of native and regenerated celluloses: role of hydrogen bonds. Polymer 32, 1516–1526.

Tcholakova, S., Denkov, N.D., Danner, T., 2004. Role of surfactant type and concentration for the mean drop size during emulsification in turbulent flow. Langmuir 20, 7444–7458.

Teng, Z., Luo, Y., Wang, Q., 2012. Nanoparticles synthesized from soy protein: preparation, characterization, and application for nutraceutical encapsulation. J. Agric. Food Chem. 60, 2712–2720.

Thielemans, W., Belgacem, M.N., Dufresne, A., 2006. Starch nanocrystals with large chain surface modifications. Langmuir 22, 4804–4810.

Tian, Y., Zhu, Y., Bashari, M., Hu, X., Xu, X., Jin, Z., 2012. Identification and releasing characteristics of high-amylose corn starch-cinnamaldehyde inclusion complex prepared using ultrasound treatment. Carbohydr. Polym. 91, 586–589.

Turchiuli, C., Eloualia, Z., Mansouri, N., Dumoulin, E., 2005. Fluidised bed agglomeration: agglomerates shape and end-use properties. Powder Technol. 157, 168–175.

Unfried, K., Albrecht, C., Klotz, L.-O., Von Mikecz, A., Grether-Beck, S., Schins, R.P.F., 2007. Cellular responses to nanoparticles: target structures and mechanisms. Nanotoxicology 1, 52–71.

Unger, E.C., Porter, T., Culp, W., Labell, R., Matsunaga, T., Zutshi, R., 2004. Therapeutic applications of lipid-coated microbubbles. Adv. Drug Deliv. Rev. 56, 1291–1314.

Varshosaz, J., Minayian, M., Moazen, E., 2010. Enhancement of oral bioavailability of pentoxifylline by solid lipid nanoparticles. J. Liposome Res. 20, 115–123.

Wang, C., Esker, A., 2014. Nanocrystalline chitin thin films. Carbohydr. Polym. 102, 151–158.

Wantanasiri, P., Ratanavaraporn, J., Yamdech, R., Aramwit, P., 2014. Fabrication of silk sericin/alginate microparticles by electrohydrodynamic spraying technique for the controlled release of silk sericin. J. Electrostat. 72, 22–27.

Weiss, J., Gaysinsky, S., Davidson, M., McClements, J., 2009. Nanostructured encapsulation systems: food antimicrobials. In: Gustavo, B.-C., Alan, M., David, L., Walter, S., Ken, B., Paul, C. (Eds.), Global Issues in Food Science and Technology. Academic Press, San Diego, pp. 425–479.

Weyland, M., Griveau, A., Bejaud, J., Benoit, J.P., Coursaget, P., Garcion, E., 2013. Lipid nanocapsule functionalization by lipopeptides derived from human papillomavirus type-16 capsid for nucleic acid delivery into cancer cells. Int. J. Pharm. 454, 756–764.

Wise, D.L., 2000. Handbook of Pharmaceutical Controlled Release Technology. Taylor & Francis, New York.

Wooster, T.J., Andrews, H.F., Sanguansri, P. 2/12/2010. US Patent application 20100305218-Nanoemulsions.

Wu, X., Hou, J., Li, M., Wang, J., Kaplan, D.L., Lu, S., 2012a. Sodium dodecyl sulfate-induced rapid gelation of silk fibroin. Acta Biomater. 8, 2185–2192.

Wu, Y., Luo, Y., Wang, Q., 2012b. Antioxidant and antimicrobial properties of essential oils encapsulated in zein nanoparticles prepared by liquid–liquid dispersion method. LWT – Food Sci. Technol. 48, 283–290.

Xu, S., Zhang, J., He, A., Li, J., Zhang, H., Han, C.C., 2008. Electrospinning of native cellulose from nonvolatile solvent system. Polymer 49, 2911–2917.

Yang, X., Tian, H., Ho, C.T., Huang, Q., 2012. Stability of citral in emulsions coated with cationic biopolymer layers. J. Agric. Food Chem. 60 (1), 402–409, 11.

Yang, J., Huang, Y., Gao, C., Liu, M., Zhang, X., 2014. Fabrication and evaluation of the novel reduction-sensitive starch nanoparticles for controlled drug release. Colloid. Surf. B Biointerfaces 115, 368–376.

Yeo, Y., Baek, N., Park, K., 2001. Microencapsulation methods for delivery of protein drugs. Biotechnol. Bioprocess. Eng. 6, 213–230.

Yu, D., Xiao, S., Tong, C., Chen, L., Liu, X., 2007. Dialdehyde starch nanoparticles: preparation and application in drug carrier. Chin. Sci. Bull. 52, 2913–2918.

Zabar, S., Lesmes, U., Katz, I., Shimoni, E., Bianco-Peled, H., 2009. Studying different dimensions of amylose-long chain fatty acid complexes: molecular, nano and micro level characteristics. Food Hydrocolloid. 23, 1918–1925.

Zaru, M., Manca, M.L., Fadda, A.M., Antimisiaris, S.G., 2009. Chitosan-coated liposomes for delivery to lungs by nebulisation. Colloid. Surf. B Biointerfaces 71, 88–95.

Zhang, J., 2011. Novel emulsion-based delivery systems. Dissertation, University of Minnesota.

Zhang, L., Kosaraju, S.L., 2007. Biopolymeric delivery system for controlled release of polyphenolic antioxidants. Eur. Polym. J. 43, 2956–2966.

Zhang, C., Ding, Y., Yu, L., Ping, Q., 2007. Polymeric micelle systems of hydroxycamptothecin based on amphiphilic N-alkyl-N-trimethyl chitosan derivatives. Colloid. Surf. B Biointerfaces 55, 192–199.

Zhang, X., Huang, J., Chang, P.R., Li, J., Chen, Y., Wang, D., Yu, J., Chen, J., 2010. Structure and properties of polysaccharide nanocrystal-doped supramolecular hydrogels based on cyclodextrin inclusion. Polymer 51, 4398–4407.

Zhang, J., Liang, L., Tian, Z., Chen, L., Subirade, M., 2012. Preparation and *in vitro* evaluation of calcium-induced soy protein isolate nanoparticles and their formation mechanism study. Food Chem. 133, 390–399.

Zhang, Y., Chan, H.F., Leong, K.W., 2013. Advanced materials and processing for drug delivery: the past and the future. Adv. Drug Deliv. Rev. 65 (1), 104–120.

Zhang, Z., Vriesekoop, F., Yuan, Q., Liang, H., 2014. Effects of nisin on the antimicrobial activity of D-limonene and its nanoemulsion. Food Chem. 150, 307–312.

Zhao, Q., Ho, C.T., Huang, Q., 2013. Effect of ubiquinol-10 on citral stability and off-flavor formation in oil-in-water (o/w) nanoemulsions. J. Agric. Food Chem. 61, 7462–7469.

Zimet, P., Livney, Y.D., 2009. Beta-lactoglobulin and its nanocomplexes with pectin as vehicles for w-3 polyunsaturated fatty acids. Food Hydrocolloid. 23, 1120–1126.

Zimet, P., Rosenberg, D., Livney, Y.D., 2011. Re-assembled casein micelles and casein nanoparticles as nano-vehicles for ω-3 polyunsaturated fatty acids. Food Hydrocolloid. 25, 1270–1276.

COMMERCIALIZATION ASPECTS AND APPLICATIONS

IV

COST AND SAFETY ISSUES OF EMERGING TECHNOLOGIES AGAINST CONVENTIONAL TECHNIQUES

Charis M. Galanakis*, Francisco J. Barba**, Krishnamurthy Nagendra Prasad[†]

*Department of Research & Innovation, Galanakis Laboratories, Chania, Greece; **Department of Nutrition and Food Chemistry, Faculty of Pharmacy, Universitat de València, Valencia, Spain; [†]Chemical Engineering Discipline, School of Engineering, Monash University of Malaysia, Malaysia

14.1 INTRODUCTION

Researchers today consider food wastes as a source of valuable compounds such as antioxidants, dietary fibers, flavonoids, polyphenols, glucosinolates, anthocyanins, proteins, and enzymes (Galanakis et al., 2010a, b, c, d, e; Galanakis and Schieber, 2014; Galanakis et al., 2014, 2015;. Rahmanian et al., 2014; Heng et al., 2015; Deng et al., 2015). This ambition is generated by the fact that the current processing methodologies allow the recovery of target compounds and their reutilization inside the food chain as additives, supplements, or food fortification. However, what are the obstacles to making this trend really happen? For instance, the labeled products derived from food wastes are few compared with the plethora of investigations in the field, patented methodologies, and proposed scenarios. The answer could include many aspects, but five of them are more critical:

1. Energy efficiency of conventional techniques is not so high and respectively cost is not low enough to extract compounds that exist in lower concentrations inside by-products compared with whole fruits or vegetables.
2. Traditional thermal processes such as concentration, drum, or spray drying may cause enzymatic or nonenzymatic deterioration of the target compounds, loss of their functionality, and diminution of the final product organoleptic characteristics.
3. Conventional extraction is restricted due to the nonfood-grade nature of many solvents.
4. Target compounds such as antioxidants may not be stable on the shelf due to the inefficient encapsulation of the final product.
5. Consumers are and always will be skeptical about the safety of products derived from by-products or food wastes.

Thereby, not only the yield of the applied technologies, but also scale-up boundaries, product safety, and overall cost govern the industrialization of recovery processing (Galanakis, 2012, 2013). Improving process efficiency in the food-manufacturing environment focuses on minimizing cost. The latter is typically conducted by reducing processing steps, enhancing throughput, restricting overall energy consumption, and finally optimizing plant design. Emerging and typically nonthermal technologies promise to overcome the obstacles of conventional techniques by claiming:

1. increased recovery yield in most cases compared with conventional techniques (Table 14.1),
2. optimized heat and mass transfer that results in lower operation cost,
3. reduced processing time that results in lower operation cost (Table 14.2),
4. gentle treatment of the food waste matrix and control of unwanted Maillard by-products,
5. advanced encapsulation and improved functionality of antioxidants.

Despite their promising advantages, the implementation of emerging technologies in the recovery of downstream processing is still under debate. The main reason is that some techniques (e.g. nano-encapsulation) concern consumers because of their safety, whereas others could be too sophisticated. For example, they require high capital cost and energy overconsumption compared with the additional value induced by the increase in the extraction yield.

In the current chapter, an attempt to clarify cost issues of conventional and emerging technologies is conducted. The analysis of economic efficiency in food processing plants involves two different aspects. First is optimum plant utilization in the short run and the second is the optimum plant size in the long run. In the short run consideration, a company cannot vary its fixed inputs such as major pieces of equipment, available space, and so on. In this case, output is alterable only by changing the usage of variable inputs. In the long run, there are no fixed factors and the company may build a plant of any size to produce at any scale. The total cost of a given output rate is calculated as the sum of the total fixed, variable, and semivariable cost. Total fixed cost includes insurance, taxes, and depreciation. Total semivariable cost includes administrative services, management, supervision, distribution costs, maintenance, and other costs. Total variable costs (raw material, direct labor, utilities, etc.) are directly proportional to the output of food processing plants. Typically, labor costs are calculated on the basis of hourly rates due to the irregularity of supply of raw material and intensive-labor processing (Parin and Zugarramurdi, 1999).

Herein, comparison of cost estimations is performed indirectly, as methodologies found in literature:

1. either investigate specific recovery stages,
2. include different technologies for each stage separately,
3. deal with different food wastes, or
4. have only been assayed inside the laboratory.

In general, the referred values mainly concern variable costs, whereas important semivariable cost (e.g. maintenance) is not determined. This is important for particular cases, e.g. membrane technologies (Patsioura et al., 2011; Galanakis et al., 2012, 2013b) that require cleaning of the membranes. Comparison of safety issues is also conducted indirectly as there are no proven negative effects on consumers, but there is a lack of information about the impact of emerging technologies on them (Galanakis, 2012).

14.2 ASSUMPTIONS AND CALCULATIONS

In order to compare the operational and other costs obtained from individual bibliographic sources (referring to different currencies and assumptions), calculations within the chapter were based on the following assumptions:

1. All costs were given in US dollars ($).
2. Power units were given in W (or kW) and energy units per substrate mass in kJ/kg.

Table 14.1 Approximate Efficacy of Conventional Techniques and Emerging Technologies Applied for the Recovery of Target Compounds from Several Food Wastes (Galanakis, 2012)

Target Compounds	Food Waste Source	Applied Technologies	Recovery Yield		References
			g Compound/100 g Waste Dry Matter	g Compound/100 g Compound Contained in the Waste	
Pectin	Lemon peel, pulp, and pips	Freeze drying, acid-assisted extraction, centrifugation, sequential ethanol precipitation	11.2	51.9	Masmoudi et al. (2008)
	Lemon peel	Drying, acid-assisted extraction	13.0	–	Panchev et al. (2011)
		Laser ablation, drying, acid-assisted extraction	15.2	–	Panchev et al. (2011)
	Orange peel	Drying, acid-assisted extraction	13.1	–	Panchev et al. (2011)
		Laser ablation, drying, acid-assisted extraction	16.5	–	Panchev et al. (2011)
	Orange albedo	Microwave-assisted extraction	0.8	–	Liu et al. (2006)
		Soxhlet extraction	1.7	–	Liu et al. (2006)
		Microwave- and pressure-assisted extraction, filtration, washing and centrifugation	19.6	35.0	Fishman et al. (2000)
Phenols	Olive mill wastewater	Concentration, acid-assisted extraction, ethanol precipitation, concentration, dilution, microfiltration, ultrafiltration	1.0	14.5	Galanakis et al. (2010b)
		Drying, pressurized and superheated ethanol-assisted extraction	4.7 for total phenols, 2.8 for hydroxytyrosol, 1.6 for tyrosol	–	Japón-Luján and Luque de Castro (2007)
	Sea buckthorn berries pomace	Microwave-assisted extraction	1.2 for total phenols, 0.1 for isorhamnetin 3-O-rutinoside	–	Périno-Issartier et al. (2011)

(Continued)

Table 14.1 Approximate Efficacy of Conventional Techniques and Emerging Technologies Applied for the Recovery of Target Compounds from Several Food Wastes (Galanakis, 2012) *(cont.)*

Target Compounds	Food Waste Source	Applied Technologies	Recovery Yield		References
			g Compound/100 g Waste Dry Matter	g Compound/100 g Compound Contained in the Waste	
		Conventional solid/liquid extraction	0.7 for total phenols, 0.2 for isorhamnetin 3-O-rutinoside	—	Périno-Issartier et al. (2011)
	Mango peel	Acid-assisted extraction, resin adsorption, methanol elution, evaporation and freeze drying	0.14 for total phenols, 0.12 for mangiferin	34.9 for total phenols, 70.4 for mangiferin	Berardini et al. (2005)
		Acid-assisted extraction, ethanol precipitation, evaporation, resin adsorption, methanol elution, evaporation and freeze drying	0.10 for total phenols, 0.08 for mangiferin	24.9 for total phenols, 52.9 for mangiferin	Berardini et al. (2005)
	White grape pomace	Water extraction	0.26	—	Boussetta et al. (2009)
		Water extraction and high-voltage electrical discharge	0.44	—	Boussetta et al. (2009)
	Casein whey	Centrifugation, ultrafiltration (three steps) and ion-exchange membrane chromatography	0.4 for total proteins, 0.04 for β-lactoglobulin	33.3 for total proteins, 7.8 for β-lactoglobulin	Bhattacharjee et al. (2006)
Lactose	Paneer whey	Stirring, crystallization and ethanol precipitation	—	14.6	Bund and Pandit (2007)
		Sonocrystallization and ethanol precipitation	—	90.3	Bund and Pandit (2007)

Table 14.2 Comparison of Conventional and Emerging Technologies for Extraction of Essential Components from Plant Materials

Raw Material	Essential Components	Extraction Method	Extraction Time (min)	Results	References
P. ginseng root	Gensenosides (mg/g DW)	PSFME	10	43.3	Wang et al. (2008)
		Soxhlet	120	37.1	
		UAE	60	35.8	
		Heat reflux	120	37.8	
Mint	Carvone (%)	PSFME	30	64.9	Lucchesi et al. (2004)
		HD	270	52.3	
Longan pericarp	Corilagin (mg/g)	Conventional	720	2.35	Prasad et al. (2009b)
		HPE	2.5	9.65	
	Total phenolic content (mg/g)	Conventional	720	14.6	
		HPE	2.5	21.0	
Litchi pericarp	Epicatechin (mg/g)	Conventional	30	0.04	Prasad et al. (2009c)
		HPE	30	0.34	
	Epicatechin gallate (mg/g)	Conventional	30	0.01	
		HPE	30	0.25	
Grape by-product	Anthocyanin (μmol GAE/g)	Conventional	60	210	Corrales et al., 2008
		PEF	60	370	
		Ultrasonic	60	325	
		HPE	60	365	
Pink guava by-product	Lycopene (mg/100 g)	SC-CO$_2$	30	42.9	Kong et al. (2010)
		Solvent	60	33.6	
Grape seeds	Polyphenols (mg GAE/g)	Maceration	720	89.4	Porto et al. (2013)
		Ultrasound	15	105	
	Tannins	Maceration	720	41.7	
		Ultrasound	15	53.3	

PSFME, pressurized solvent free microwave extraction; UAE, ultrasonic-assisted extraction; GAE, gallic acid equivalent; HD, hydrodistillation; SD, steam distillation; HPE, high pressure extraction; PEF, pulsed electric field; SC-CO$_2$, supercritical carbon dioxide; DW, dry weight.

3. 1 kWh costs $0.05.
4. $1 equals €0.76 (currency rates as at 2014).
5. Plant flow rate is given in t/h.

14.3 CONVENTIONAL TECHNIQUES

As stated in Chapters 4–8, several technologies have been developed over the years covering all the recovery steps for the food waste sources up to the encapsulation of the product. Classic processing techniques (e.g. wet milling, mechanical pressing, membrane technologies, isoelectric solubilization, adsorption, microwave, and supercritical fluid extraction (SFE)) are well documented and generally assumed as safe since they have been applied for many years in several sectors of the food industry (Galanakis, 2013). Thermal processes such as concentration or spray drying may be an exception, as they cause deterioration of the food waste matrix and acceleration of Maillard reactions. The latter generate moieties with unknown impact on human health.

Thermal processes like hydrodistillation and steam diffusion induce milder treatment of the waste matrix due to the heat transfer via the vapor phase. These techniques are considered safe, but they possess high energy demands due to the applied temperatures. Besides, they destroy labile compounds such as volatile polyphenols and essential oils. On the other hand, the heat transfer induced by microwaves is cheaper and requires moderate investment in equipment, too. According to a recent study, coupling hydrodistillation with microwaves could reduce the energy cost almost twofold, e.g. microwave-assisted hydrodistillation demands 0.07 kWh/kg (270 kJ/kg, \$3.5/t) compared with hydrostillation that demands up to 0.12 kWh/kg (430 kJ/kg, \$6/t) (Golmakani and Rezaei, 2008).

Recently, economics of microwave-assisted extraction and spray drying were evaluated in a two-step process for the recovery of polyphenols from orange peel extracts. The solvent-to-solid ratio was shown to have an impact on the phenolic compounds of the extract and the powder. According to the results, the two-step process is economically feasible with profits of \$6.1–8.8/kg for solvent-to-solid ratios of 2 and 14, respectively. The uncertain analysis of the economics showed that the selling price, the labor cost, and the orange peel cost are the three most important parameters compared with the costs of electricity, natural gas, and water (Shofinita and Langrish, 2014). Spray drying is considered the cheapest and most effective encapsulation technique since it ensures a continuous operation and delivery of an easy-to-handle powder. The latter can reduce storage and transportation costs. For instance, the processing cost of a plant designed to produce 2.375 t/h of encapsulated butter oil or cream powder (e.g. 50% milk fat) is ~\$0.23/kg, plus the cost of the butter oil or cream, encapsulant, and other ingredients (Holsinger et al., 2000).

Vacuum processes (e.g. concentration or freeze drying) are safer compared with thermal processes, but they require additional energy consumption due to the applied vacuum conditions. The energy cost is by far the largest operating cost in all the pressure-driven membrane processes (nanofiltration, electrodialysis, reverse osmosis, and liquid chromatography). This is mainly due to the high cross-flow velocity and low flux. Reducing the cross-flow velocity will decrease both the frictional pressure drop and the feed flow in a membrane element. Ultrafiltration and diafiltration are pressure-driven processes, too, but they are generally considered cheaper due to the lower values of applied pressures. For instance, the energy requirements for the recovery of kraft lignin from pulping liquors have been calculated. Thereby, the energy demands in a black liquor plant could reach 14 kWh/m^3 (50,400 kJ/m^3 or \$0.7/m^3), whereas in a cooking liquor plant they could reach 6.5 kWh/m^3 (23,400 kJ or \$0.325/m^3). The higher energy requirement for black liquor ultrafiltration is due to the lower flux. In this calculation, it was assumed that neither flux nor retention was influenced by the reduced velocity. In total, lignin could be recovered from a cooking liquor at a cost of ~\$0.078/kg of lignin and from black liquor (withdrawn from the evaporation unit) at \$0.043/kg (Jönsson and Wallberg, 2009). In any case, the cost of membrane and adsorption processes is increasing as a function of material (e.g. membrane sheets, chromatographic columns, adsorption resins) regeneration and discharge.

SFE has both high investment and operational cost. Recently, a technoeconomic and environmental assessment of essential oil extraction from citronella and lemongrass had been conducted by comparing three extraction technologies: supercritical fluid, solvent, and water distillation. In particular, all technologies were evaluated using Aspen Plus, Aspen Process Economic Analyzer, and WAR GUI, carrying out simulation, economic evaluation, and environmental assessment, respectively. For the analysis of the energy consumption impact in each technology, two scenarios with different levels of integration (without or fully) were considered. According to the results, the lowest production cost was obtained using water distillation with full energy integration for both citronella and lemongrass (citronella: $6.48/kg, lemongrass: $7.50/kg) (Moncada et al., 2014).

Another technoeconomical evaluation was conducted for the SFE of spent coffee grounds, which are of interest under the biorefinery concept. Cost of manufacturing and net income calculations were performed for distinct operating conditions and unit arrangements, but the optimal results were obtained for an arrangement of three beds of 1 m^3, extraction time of 2.0 h, 300 bar, and 50°C. Under these conditions, production can reach 454 t/year, whereas the cost of manufacturing could reach $3.16 m and the process net income could reach €74.5 m. A sensitivity analysis varying the unit capacity, extraction time, and precipitation pressure (extract vessel) showed the process economics to remain viable (de Melo et al., 2014).

The capital costs of an SFE plant dealing with the recovery of bioactive compounds and nutraceuticals from flaxseed, microalgae, and ginger have been denoted by Martinez (2008). The estimated equipment costs for processing (24 h/day for 300 days/year) of 3000 Mt flaxseed/year, 50 Mt microalgae/year, and 3000 Mt ginger/year were equal to $2.5 m, $1.5 m, and $4.5 m, respectively. Moreover, the operation costs were equal to $0.23, $5.57, and $0.43/kg feed, for flaxseed, microalgae, and ginger, respectively. Indeed, SFE costs were highly dependent on bioresource, number of vessels, design pressure, size of the vessels, flow rate, automation, and good manufacturing practice. In this case, operation costs included power consumption, CO_2 losses, maintenance, and lab operation. Overall, the capital costs of an SFE plant are higher than those required for a traditional extraction method, while the operation costs are lower. However, when comparing SFE with traditional extraction methods, it is always necessary to take into account different parameters such as additional extraction equipment, building cost, instrumentation, and electrical connections to meet the risk of explosion.

As a rule of thumb, safety of conventional techniques is dominated by the nature of the applied materials. For instance, adsorption and chromatography safety is dependent on the toxicity of the materials involved in the process. On the other hand, acids, alkali, solvents, and supercritical fluids (CO_2) are considered safe when the involved materials exist inherently in foods or are of a food-grade nature (Galanakis, 2012). Nongreen organic solvents such as hexane, diethyl ether, or chloroform are toxic and raise public awareness about their usage in food production processes. These solvents should be totally removed from the recovered extracts prior to their application in food products. Methanol is also toxic and thus not preferred despite the fact that it is very efficient for the extraction of small anti oxidants (Tsakona et al., 2012; Galanakis, 2015; Galanakis et al., 2013a, 2015). Similarly, food-grade acids (e.g. citric or tartaric acid) should be preferred to corrosive acids (e.g. hydrochloric, sulfuric, or nitric acid).

There are three main aspects that dominate the 12 Principles of Green Chemistry: waste, hazard (health, environmental, and safety), and energy. In the food industry, approximately half of the consumed end-use energy is used in processes changing raw materials to products. These processes include conventional techniques such as heating, cooling, refrigeration, mechanical, and electrochemical treatment. Among these, process heating uses approximately 29% of the total energy in the food industry of the United States, while processes like cooling and refrigeration demand about 16% of the total energy inputs (Okos et al., 1998). This fact facilitates the need to explore new nonthermal technologies that could diminish process cost by reducing energy demands.

14.4 **EMERGING TECHNOLOGIES**

Most of the emerging nonthermal technologies require high capital cost, whereas some of them have also increased operational cost that is not necessary related to energy consumption. For instance, low temperature plasma treatment shows low energy consumption ($0.0045/kg), but its overall operational cost is high due to the input feed gas. If the feed gas is helium, the cost can reach $636–9096/h, whereas for nitrogen the cost is much less ($9–72/h) (Niemira, 2012). Besides, toxicological effects of cold plasma on biological matrices have not yet been investigated.

Pulsed electric technologies such as pulsed electric fields (PEF) and high voltage electrical discharges (HVED) have emerged as a potential tool to give an added value to food wastes because of their ability to induce membrane electroporation and therefore increase recovery yield (Roselló-Soto et al., 2015). Considering this, Toepfl (2006) estimated the price for a PEF unit processor for inducing cell disintegration in the range of $197,000, taking into account a flow rate of 10 t/h, an energy input of 10 kJ/kg, and an average output power of 30 kW. This energy input seems to be adequate for the recovery of compounds from numerous substrates such as juices, sugar beets, apples, and chicory (Donsì et al., 2010).

Toepfl also estimated the cost of PEF treatment at $1.52/t, adding a 10% overhead total power of $1.67/t. Moreover, he also reported that the costs for a conventional enzymatic maceration can be estimated at approximately $37.95/t. The cost of orange juice pasteurization using PEF has also been estimated. Total pasteurization cost was estimated to be $0.037/L of juice, whereas capital costs accounted for 54% ($0.02/L), labor costs accounted for 35% ($0.013/L), and utility charges, mainly electricity, accounted for 11% ($0.004/L). The high capital cost led to a total cost of 147% ($0.022/L), higher than that obtained for conventional thermal processing ($0.015/L) (Sampedro et al., 2013).

HVED also requires rather high capital cost. Several studies have been conducted to compare the energy consumption of pulsed electric technologies (HVED and PEF), ultrasounds (US), and conventional techniques such as grinding and thermal treatments (Table 14.3). Moreover, the use of these technologies (PEF, HVED, and US) allows the consumption of toxic solvents to be reduced, which is relevant from a safety point of view.

The effects of PEF, HVED, and grinding at equivalent energy input were evaluated and compared for the case of polyphenol recovery from grape seeds (Boussetta et al., 2012). The maximum yield (9 g gallic acid equivalents/100 g dry matter) was found after applying HVED treatments and subsequent ethanol extraction (30 g ethanol/100 g S). In addition, the feasibility of PEF to obtain extracts with less turbidity was facilitated. This observation is very important to avoid subsequent purification steps and reduce the cost of recovery process.

More recently, it was found that the energy to improve the extraction yield of polyphenols from vine shoots was less when HVED (10 kJ/kg) and PEF (50 kJ/kg) were used in comparison to US (1010 kJ/kg) (Rajha et al., 2014). Indeed, the highest polyphenol and protein recovery and purity corresponded to HVED with the lowest energetic prerequisite.

The effects of PEF and conventional grinding extraction at equivalent energy inputs (720 kJ/kg) have also been compared for the recovery of biocompounds from flaxseed hulls (Boussetta et al., 2014). Polyphenol recovery after grinding and PEF was similar compared with control. However, PEF extracts presented lower turbidity with higher particle size than that obtained after grinding, thus making more difficult the subsequent extraction processes and increasing the costs of the

Table 14.3 Energy Costs of Valuable Compounds Recovery from Food Wastes and By-Products after Pretreatment with Pulsed Electric Technologies

Food Wastes and/or By-Products	Energy Cost Major Findings	References
Vine shoots	HVED had the highest polyphenol and protein extraction yields with the lowest energy input (10 kJ/kg). The same recovery was achieved after PEF at 50 kJ/kg and US at 1010 kJ/kg. Extracts of high polyphenol yield (34.5 mg GAE per gram of dry matter) and high purity (89%) were obtained with HVED.	Rajha et al. (2014)
Flaxseed hulls	Lower turbidity of the extracts obtained after PEF treatment compared with grinding at equivalent energy input (720 kJ/kg). PEF can reduce subsequent steps in purification process, thus reducing cost.	Boussetta et al. (2014)
	For the extraction of oil from hulled and nonhulled rapeseed, maximum permeabilization of the cell membrane was achieved for hulled rapeseed 55% at 7.0 kV/cm and 120 pulses (84 kJ/kg) and 17% at 5 kV/cm and 60 pulses (42 kJ/kg) for nonhulled rapeseed.	Guderjan et al. (2007)
Grape skins	Optimum Z at 0.4–6.7 kJ/kg	Lopez et al. (2008)
Red beetroot	Optimum Z at 2.5 kJ/kg	Lopez et al. (2009b)

Z, cell permeabilization; PEF, pulsed electric fields; HVED, high voltage electrical discharges; GAE, gallic acid equivalent; US, ultrasounds.

recovery. Likewise, PEF treatment allowed the extraction of 80% of polyphenols when applied at 20 kV/cm for 10 ms (operation rate of 3600 kg/h). Despite the increase of recovery yield, the cost of this process was found to be relatively high (720 kJ/kg, which corresponds to $10/t). Experiments were conducted in a batch mode, whereas the operation rate (3600 kg/h) was almost threefold lower compared with that proposed by Toepfl (2006), e.g. energy input of 10 kJ/kg for a plant with a flow rate of 10,000 kg/h. This fact indicates that the extraction and recovery of compounds may need more energy compared with other cell disintegration PEF processes. Moreover, extraction efficiency was smaller using lower PEF electric field strength.

In another study, Guderjan et al. (2007) evaluated oil extraction from hulled and nonhulled rapeseed. The maximum permeabilization of the cell membrane achieved for hulled rapeseed was 55% (at 7.0 kV/cm, 120 pulses, 84 kJ/kg) and 17% for nonhulled rapeseed (at 5 kV/cm, 60 pulses, 42 kJ/kg). Similar energy consumptions for optimum cell permeabilization with PEF treatment have been observed in many experimental works, where energy input was found to be rather low, typically lying within 1–15 kJ/kg (Vorobiev and Lebovka, 2010). For example, it accounted for 6.4–16.2 kJ/kg of potato (Angersbach and Knorr, 1997), 0.4–6.7 kJ/kg of grape skins (Lopez et al., 2008), 2.5 kJ/kg of red beetroot (Lopez et al., 2009a), 3.9 kJ/kg (7 kV/cm) of sugar beet (Lopez et al., 2009b), and 10 kJ/kg (400–600 V/cm) of chicory root (Loginova et al., 2010). According to the conclusion of these studies, PEF technology was ideal to disrupt plant tissues when compared with other methods like mechanical (20–40 kJ/kg), enzymatic (60–100 kJ/kg), and heating or freezing/thawing (100 kJ/kg) (Toepfl, 2006) (Table 14.4).

Regarding high pressure (HP)-assisted extraction, the pressure level is an important parameter to evaluate the extraction achieved with this technique. Most of the published articles reported that the

Table 14.4 Cost Estimation of PEF Treatment for the Extraction of Valuable Compounds from Several Sources

Tissue	Extracted Compound	E (kV/cm)	Energy (kJ/kg)	Cost[a] ($/t)	References
Chicory	Soluble solids	0.5	10	0.14	Loginova et al. (2010)
Grape skin	Polyphenols	5	1.8	0.025	Lopez et al. (2008)
Fennel	Vitamins C and E	0.6	5	0.07	El-Belghiti and Vorobiev (2004)
Flaxseed hulls	Polyphenols	20	720	10	Boussetta et al. (2014)
Rapeseed (hulled)	Oil	7	84	1.17	Guderjan et al. (2007)
Rapeseed (nonhulled)	Oil	5	42	0.58	Guderjan et al. (2007)
Red beetroot	Betanine	1	7	0.097	Fincan et al. (2004)
Red beetroot	Betanine	7	2.5	0.035	Lopez et al. (2009a)
Soybeans	Isoflavonoids	1.3	1.857	0.026	Guderjan et al. (2005)
Sugar beet	Sugar	7	3.9	0.054	Lopez et al. (2009b)
Sugar beet	Sugar	0.5	12	0.167	Jemai and Vorobiev (2003)
Sugar beet	Sugar	0.94	7	0.09	El-Belghiti and Vorobiev (2004)
Sugar beet	Sugar	0.9	6	0.0833	El-Belghiti et al. (2005)
Vine shoots	Polyphenols	13.3	50	0.7	Rajha et al. (2004)

[a]*Calculation of cost was conducted using the following assumptions: (i) 1 kWh = 3600 kJ costs $0.05 (US dollars) (Niemira, 2012), (ii) $1 equals €0.76€ (currency rates as at 2014).*

pressure needed to extract bioactive ingredients from biomaterial was usually in the range of 100 to 600 MPa (Jun, 2013; Prasad et al., 2009a, b, c). Under these conditions, the overall energy consumption (based on the work of compression) was estimated between 5 and 40 kJ/kg (Ardia, 2004). However, it should be noted that the cost of high pressure processing (HPP) will depend on the total cycle time (pressure rise, holding, and loading/unloading times), vessel filling ratio, energy, labor, and capital costs (Mujica-Paz et al., 2011).

Taking into account that the HPP-assisted extraction process is, basically, similar to HPP for food pasteurization, the cost of the process can be estimated approximately to $0.107/L, for processing 16,500,000 L/year (3000 L/h). Capital and labor costs accounted for 59% ($0.063/L) and 37% ($0.04/L), respectively, whereas utility charges (mainly electricity) accounted for 4% ($0.004/L) (Sampedro et al., 2014).

Foam-mat drying shows a higher capital cost compared with conventional drying, as it requires larger surface area to treat large quantities of foamed fruit by-products. However, its energy consumption could be reduced by up to 80% compared with traditional dryers (Rajkumar et al., 2007; Jakubczyk et al., 2011). Radio-frequency drying would only lead to the same energy levels as tunnel oven drying, if the latter consumes more than 1977 kJ/kg. This estimation was based on assumed values for the average productivity increase (40%) and specific primary energy consumption (558 kJ/kg) of radio-frequency drying units. The latter energy consumption was calculated based on an assumed average product moisture content of 7.5% (Lung et al., 2006).

On the other hand, electro-osmotic dewatering requires increased energy consumption compared with conventional dewatering, because of its combination with pressure or vacuum conditions as noted above. This process does not raise safety concerns of the treated material, but similar to any minimal process, it requires safety precautions during handling because of the entrapment of hydrogen gas at the cathodes as well as physical contact of the personnel with electrical apparatus (Jumah et al., 2005; Citeau et al., 2011). The same conclusion could be obtained for laser ablation.

Pervaporation is another emerging technology that requires low investment cost and energy consumption compared with traditional processes such as hydrodistillation. For instance, an economic analysis for the continuous selective fermentation of glucose coupled with pervaporation (to remove ethanol) has been described. Two approaches were studied. According to the results of the study, the variable costs that involve the installed membrane area are the ones that most influence process viability. For a complete plant installation, the highest membrane cost allowed to keep the project feasible is $\sim$\$500/m^2, whereas the adaptation of an existing plant can cost up to \$800/m^2, considering a minimum return rate on investment of 17% (Di Luccio et al., 2002).

US extraction and crystallization as well as pressurized microwave-assisted extraction are considered green technologies, but the investment cost of the second is again much higher because of the presence of pressure conditions. The safety of other low-cost technologies such as colloidal gas aphrons and liquid membranes depends on the nature of the biodegradable surfactants and the applied organic phase. Besides, the instability of liquid membranes may increase the general cost of the process. Magnetic fishing is a more costly technique since magnetic materials are relatively expensive and should be recycled a few times in order to reduce operational cost (Heebool-Nielsen et al., 2004). Aqueous two-phase separation is another technique with increased cost, as the involved polymers (e.g. dextrans) are generally expensive due to their purity and food-grade nature.

On the contrary, product safety is nowadays the main concern about the application of nanoencapsulation. For example, nanoemulsions may alter the route of lipophilic compounds adsorption in the human body, while the impact of nanoparticles on biological cells has not yet been characterized (Frewer et al., 2011). In particular, nanoemulsions consist of a lipophilic core surrounded by a shell of adsorbed material. The lipophilic core may be comprised of triacylglycerols, diacylglycerols, monoacylglycerols, flavor oils, essential oils, mineral oils, fat substitutes, weighting agents, oil soluble vitamins, carotenoids, phytosterols, and coenzyme Q. Some of these lipophilic components are typically digested within the human gastrointestinal tract (e.g. triacylglycerols and diacylglycerols), some normally pass through the gastrointestinal tract without being absorbed (e.g. mineral oils and fat substitutes), and others are absorbed without being digested (e.g. carotenoids and phytosterols). The shell around nanoemulsion droplets may also be composed of proteins or polysaccharides. Phospholipids and proteins may be digested by phospholipases and proteases in the stomach and small intestine, whereas dietary fibers may not be digested until they reach the large intestine (McClements and Rao, 2011; Galanakis, 2011). At present, there is a knowledge gap in understanding how the above compounds' characteristics influence the biological fate of nanoemulsion particles within the gastrointestinal tract, and a lack of research prior to establishing their potential toxicity (Hagens et al., 2007; Bouwmeester et al., 2009). Besides, there is a clear lack of information about the release of nanoparticles in the air and their potential toxicological effects on the environment.

14.5 CONCLUSIONS

Both conventional and emerging technologies have advantages and disadvantages regarding cost and safety issues. This is far more obvious for advanced technologies such as low temperature plasma and nanocapsule formation. The latest techniques induce unknown effects on biological matrices by permeating cell membranes. On the other hand, consumers use price as a signal of quality and this implication is reflected in food technologies, too. Consumers with neutral or positive opinions towards emerging technologies tend to decide based on the price of the final product, whereas those with negative opinions tend to decide based on ideology, which dominates cost considerations (Cox et al., 2007; Costa-Font et al., 2008). In some cases, consumers have negative attitudes towards emerging technologies and claim that lower cost would not convince them to buy irradiated materials, novel products, or foods with an overincreased shelf-life. Nevertheless, consumer demands for personalized products and tailored-made processes would ultimately lead food manufacturers to adapt emerging technologies in the production line. This trend could be appropriate for food waste recovery processes, too. For example, each waste stream consists of several components. Optimum recovery could only be achieved if all compounds of the initial material could be split up into streams that ideally contain one individual compound. This hypothetical scenario can only be performed by minimizing the size of the processes or using technologies that act at the nanoscale (van Loon, 2000). In any case, safety assessments are necessary to ensure the production of healthy food additives using emerging technologies.

ACKNOWLEDGMENTS

F.J. Barba thanks the Valencian Autonomous Government (Consellería d'Educació, Cultura i Esport. Generalitat Valenciana) for the postdoctoral fellowship of the VALi+d program "Programa VALi+d per a investigadors en fase postdoctoral 2013" (APOSTD/2013/092).

REFERENCES

Angersbach, A., Knorr, D., 1997. High intensity electric field pulses as pretreatment for affecting dehydration characteristics and rehydration properties of potato cubes. Nahrung – Food 41, 194–200.

Ardia, A., 2004. Process considerations on the application of high pressure treatment at elevated temperature levels for food preservation. PhD thesis, Technischen Universitat Berlin.

Berardini, N., Knödler, M., Scieber, A., Carle, R., 2005. Utilization of mango peels as a source of pectin and polyphenolics. Innov. Food Sci. Emerg. Technol. 6, 442–452.

Bhattacharjee, S., Bhattacharjee, C., Datta, S., 2006. Studies on the fractionation of beta-lactoglobulin from casein whey using ultrafiltration and ion exchange membrane chromatography. J. Membr. Sci. 275, 141–150.

Boussetta, N., Lanoisellé, J-.L., Bedel-Clotour, C., Vorobiev, E., 2009. Extraction of polyphenols from grape pomace by high voltage electrical discharges: effect of sulphur dioxide, freezing process and temperature. J. Food Eng. 95, 192–198.

Boussetta, N., Vorobiev, E., Le, L.H., Cordin-Falcimaigne, A., Lanoisellé, J.-L., 2012. Application of electrical treatments in alcoholic solvent for polyphenols extraction from grape seeds. LWT – Food Sci. Technol. 46, 127–134.

Boussetta, N., Soichi, E., Lanoiselle, J.-L., Vorobiev, E., 2014. Valorization of oilseed residues: extraction of polyphenols from flaxseed hulls by pulsed electric fields. Ind. Crop. Prod. 52, 347–353.

Bouwmeester, H., Dekkers, S., Noordam, M.Y., Hagens, W.I., Bulder, A.S., de Heer, C., ten Voorde, S.E.C.G., Wijnhoven, S.W.P., Marvin, H.J.P., Sips, A., 2009. Review of health safety aspects of nanotechnologies in food production. Regul. Toxicol. Pharm. 53, 52–62.

Bund, R.K., Pandit, A.B., 2007. Rapid lactose recovery from paneer whey using sonocrystallization: a process optimization. Chem. Eng. Process. 46, 846–850.

Citeau, M., Larue, O., Vorobiev, E., 2011. Effect of electrolytes content on the electro-osmotic dewatering of agro-industrial sludge. In: Taoukis, P.S., Stoforos, N.G., Karathanos, V.T., Saravacos, G.D. (Eds.), Food process engineering in a changing world. Proceedings of the Eleventh International Congress on Engineering and Food, Cosmosware, Athens, pp. 1215–1216.

Corrales, M., Toepfl, S., Butz, P., Knorr, D., Tauscher, B., 2008. Extraction of anthocyanins from grape by-products assisted by ultrasonic, high hydrostatic pressure or pulsed electric fields: a comparison. Innov. Food Sci. Emerg. Technol. 9, 85–91.

Costa-Font, M., Gil, J., Traill, W., 2008. Consumer acceptance, valuation of and attitudes towards genetically modified food: review and implications for food policy. Food Policy 33, 99–111.

Cox, D.N., Evans, G., Lease, H.J., 2007. The influence of information and beliefs about technology on the acceptance of novel food technologies: a conjoint study of farmed prawn concepts. Food Qual. Pref. 18, 813–823.

de Melo, M.M.R., Barbosa, H.M.A., Passos, C.P., Silva, C.M., 2014. Supercritical fluid extraction of spent coffee grounds: measurement of extraction curves, oil characterization and economic analysis. J. Supercrit. Fluid. 86, 150–159.

Deng, Q., Zinoviadou, K.G., Galanakis, C.M., Orlien, V., Grimi, N., Vorobiev, E., Lebovka, N., Barba, F.J., 2015. The effects of conventional and non-conventional processing on glucosinolates and its derived forms, isothiocyanates: extraction, degradation and applications. Food Eng. Rev, in press.

Di Luccio, M., Borges, C.P., Alves, T.L.M., 2002. Economic analysis of ethanol and fructose production by selective fermentation coupled to pervaporation: effect of membrane costs on process economics. Desalination 147, 161–166.

Donsì, F., Ferrari, G., Pataro, G., 2010. Applications of pulsed electric field treatments for the enhancement of mass transfer from vegetable tissue. Food Eng. Rev. 2, 109–130.

El-Belghiti, K., Vorobiev, E., 2004. Mass transfer of sugar from beets enhanced by pulsed electric field. Food Bioprod. Process. 82, 226–230.

El-Belghiti, K., Rabhi, Z., Vorobiev, E., 2005. Effect of centrifugal force on the aqueous extraction of solute from sugar beet tissue pretreated by a pulsed electric field. J. Food Process Eng. 28, 346–358.

Fincan, M., De Vito, F., Dejmek, P., 2004. Pulsed electric field treatment for solid-liquid extraction of red beetroot pigment. J. Food Eng. 64, 381–388.

Fishman, M.L., Chau, H.K., Hoagland, P., Ayyad, K., 2000. Characterization of pectin, flash-extracted from orange albedo by microwave heating, under pressure. Carbohyd. Res. 323, 126–138.

Frewer, L.J., Bergmann, K., Brennan, M., Lion, R., Meertens, R., Rowe, G., Siegrist, M., Vereijken, C., 2011. Consumer response to novel agri-food technologies: implications for predicting consumer acceptance of emerging food technologies. Trends Food Sci. Technol. 22, 442–456.

Galanakis, C.M., 2011. Olive fruit and dietary fibers: components, recovery and applications. Trends Food Sci. Technol. 22, 175–184.

Galanakis, C.M., 2012. Recovery of high added-value components from food wastes: conventional, emerging technologies and commercialized applications. Trends Food Sci. Technol. 26, 68–87.

Galanakis, C.M., 2013. Emerging technologies for the production of nutraceuticals from agricultural by-products: a viewpoint of opportunities and challenges. Food Bioprod. Process. 91, 575–579.

Galanakis, C.M., 2015. Separation of functional macromolecules and micromolecules: from ultrafiltration to the border of nanofiltration. Trends Food Sci. Technol. 42, 44–63.

Galanakis, C.M., Schieber, A., 2014. Editorial. Special issue on recovery and utilization of valuable compounds from food processing by-products. Food Res. Int. 65, 230–299.

Galanakis, C.M., Tornberg, E., Gekas, V., 2010a. A study of the recovery of the dietary fibres from olive mill wastewater and the gelling ability of the soluble fibre fraction. LWT – Food Sci. Technol. 43, 1009–1017.

Galanakis, C.M., Tornberg, E., Gekas, V., 2010b. Clarification of high-added value products from olive mill wastewater. J. Food Eng. 99, 190–197.

Galanakis, C.M., Tornberg, E., Gekas, V., 2010c. Dietary fiber suspensions from olive mill wastewater as potential fat replacements in meatballs. LWT – Food Sci. Technol. 43, 1018–1025.

Galanakis, C.M., Tornberg, E., Gekas, V., 2010d. Recovery and preservation of phenols from olive waste in ethanolic extracts. J. Chem. Technol. Biotechnol. 85, 1148–1155.

Galanakis, C.M., Tornberg, E., Gekas, V., 2010e. The effect of heat processing on the functional properties of pectin contained in olive mill wastewater. LWT – Food Sci. Technol. 43, 1001–1008.

Galanakis, C.M., Fountoulis, G., Gekas, V., 2012. Nanofiltration of brackish groundwater by using a polypiperazine membrane. Desalination 286, 277–284.

Galanakis, C.M., Goulas, V., Tsakona, S., Manganaris, G.A., Gekas, V., 2013a. A knowledge base for the recovery of natural phenols with different solvents. Int. J. Food Prop. 16, 382–396.

Galanakis, C.M., Markouli, E., Gekas, V., 2013a. Fractionation and recovery of different phenolic classes from winery sludge via membrane filtration. Separ. Purif. Technol. 107, 245–251.

Galanakis, C.M., Chasiotis, S., Botsaris, G., Gekas, V., 2014. Separation and recovery of proteins and sugars from Halloumi cheese whey. Food Res. Int. 65, 477–483.

Galanakis, C.M., Kotanidis, A., Dianellou, M., Gekas, V., 2015b. Phenolic content and antioxidant capacity of Cypriot wines. Czech J. Food Sci. 33, 126–136.

Golmakani, M.-T., Rezaei, K., 2008. Comparison of microwave-assisted hydrodistillation with the traditional hydrodistillation method in the extraction of essential oils from *Thymus vulgaris* L. Food Chem. 109, 925–930.

Guderjan, M., Toepfl, S., Angersbach, A., Knorr, D., 2005. Impact of pulsed electric field treatment on the recovery and quality of plant oils. J. Food Eng. 67, 281–287.

Guderjan, M., Elez-Martinez, P., Knorr, D., 2007. Application of pulsed electric fields at oil yield and content of functional food ingredients at the production of rapeseed oil. Innov. Food Sci. Emerg. Technol. 8, 55–62.

Hagens, W.I., Oomen, A.G., de Jong, W.H., Cassee, F.R., Sips, A., 2007. What do we (need to) know about the kinetic properties of nanoparticles in the body? Regul. Toxicol. Pharmacol. 49, 217–229.

Heebool-Nielsen, A., Justesen, S.F.L., Thomas, O.R.T., 2004. Fractionation of whey proteins with high capacity superparamagnetic ion-exchangers. J. Biotechnol. 113, 247–262.

Heng, W.W., Xiong, L.W., Ramanan, R.N., Hong, T.L., Kong, K.W., Galanakis, C.M., Prasad, K.N., 2015. Two level factorial design for the optimization of phenolics and flavonoids recovery from palm kernel by-product. Ind. Crop. Prod. 63, 238–248.

Holsinger, V.H., McAloon, A.J., Onwulata, C.I., Smith, P.W., 2000. A cost analysis of encapsulated spray-dried milk fat. J. Dairy Sci. 83, 2361–2365.

Jakubczyk, E., Gondek, E., Tambor, K., 2011. Characteristics of selected functional properties of apple powders obtained by foam-mat drying method. In: Taoukis, P.S., Stoforos, N.G., Karathanos, V.T., Saravacos, G.D., (Eds.), Food process engineering in a changing world. Proceedings of the Eleventh International Congress on Engineering and Food, Cosmosware, Athens, pp. 1385–1386.

Japón-Luján, R., Luque de Castro, M.D., 2007. Static-dynamic superheated liquid extraction of hydroxytyrosol and other biophenols from Alperujo (a semisolid residue of the olive oil industry). J. Agric. Food Chem. 55, 3629–3634.

Jemai, A.B., Vorobiev, E., 2003. Enhanced leaching from sugar beet cossettes by pulsed electric field. J. Food Eng. 59, 405–412.

Jönsson, A.-S., Wallberg, O., 2009. Cost estimates of kraft lignin recovery by ultrafiltration. Desalination 237, 254–267.

Jumah, R., Al-Asheh, S., Banat, F., Al-Zoubi, K., 2005. Electroosmotic dewatering of tomato paste suspension under AC electric field. Drying Technol. 23, 1465–1475.

Jun, X., 2013. High-pressure processing as emergent technology for the extraction of bioactive ingredients from plant materials. Crit. Rev. Food Sci. Nutr. 53, 837–852.

Kong, K.K., Rajab, N.F., Prasad, K.N., Ismai, A., Markom, M., Tan, C.P., 2010. Lycopene-rich fractions derived from pink guava by-product and their potential activity towards hydrogen peroxide-induced cellular and DNA damage. Food Chem. 123 (4), 1142–1148.

Liu, Y., Shi, J., Langrish, T.A.G., 2006. Water-based extraction of pectin from flavedo and albedo of orange peels. Chem. Eng. J. 120, 203–209.

Loginova, K.V., Shynkaryk, M.V., Lebovka, N.I., Vorobiev, E., 2010. Acceleration of soluble matter extraction from chicory with pulsed electric fields. J. Food Eng. 96, 374–379.

Lopez, N., Puertolas, E., Condon, S., Alvarez, I., Raso, J., 2008. Effects of pulsed electric fields on the extraction of phenolic compounds during the fermentation of must of Tempranillo grapes. Innov. Food Sci. Emerg. Technol. 9, 477–482.

Lopez, N., Puertolas, E., Condon, S., Raso, J., Alvarez, I., 2009a. Enhancement of the extraction of betanine from red beetroot by pulsed electric fields. J. Food Eng. 90, 60–66.

Lopez, N., Puertolas, E., Condon, S., Raso, J., Alvarez, I., 2009b. Enhancement of the solid-liquid extraction of sucrose from sugar beet (*Beta vulgaris*) by pulsed electric fields. LWT – Food Sci. Technol. 42, 1674–1680.

Lucchesi, M.E., Chemat, F., Smadja, J., 2004. Solvent-free microwave extraction of essential oil from aromatic herbs: comparison with conventional hydro-distillation. J. Chromatogr. A 1043, 323–327.

Lung, R.B., Masanet, E., McKane, A., et al., 2006. The role of emerging technologies in improving energy efficiency: examples from the food processing industry. Proceedings of the Twenty-Eighth Industrial Energy Technology Conference, New Orleans, LA, May 9–12, 2006.

Martinez, J.L. (Ed.), 2008. Supercritical Fluid Extraction of Nutraceuticals and Bioactive Compounds. CRC Press, Boca Raton, Florida.

Masmoudi, M., Besbes, S., Chaabouni, M., Robert, C., Paquot, M., Blecker, C., Attia, H., 2008. Optimization of pectin extraction from lemon by-product with acidified date juice using response surface methodology. Carbohyd. Polym. 74, 185–192.

McClements, D.J., Rao, J., 2011. Food-grade nanoemulsions: formulation, fabrication, properties, performance, biological fate, and potential toxicity. Crit. Rev. Food Sci. Nutr. 51, 285–330.

Moncada, J., Tamayo, J.A., Cardona, C.A., 2014. Techno-economic and environmental assessment of essential oil extraction from Citronella (*Cymbopogon winteriana*) and Lemongrass (*Cymbopogon citrus*): a Colombian case to evaluate different extraction technologies. Ind. Crop. Prod. 54, 175–184.

Mujica-Paz, H., Valdez-Fragoso, A., Samson, C.T., Welti-Chanes, J., Torres, A., 2011. High-pressure processing technologies for the pasteurization and sterilization of foods. Food Bioprocess Technol. 4, 969–985.

Niemira, B.A., 2012. Cold plasma decontamination of foods. Annu. Rev. Food Sci. Technol. 3, 125–142.

Okos, M., Drecher, S., Rode, S., Kozak, J., 1998. American Council for an Energy Efficient Economy, Research Report IE981: A Review of Energy Use in the Food Industry, Washington, D.C., http://www.aceee.org/pubs/ie981.htm.

Panchev, I.N., Kirtchev, N.A., Dimitrov, D.D., 2011. Possibilities for application of laser ablation in food technologies. Innov. Food Sci. Emerg. Technol. 12, 369–374.

Parin, M.A., Zugarramurdi, A., 1999. Investment and production costs analysis in food processing plants. Int. J. Prod. Econ. 34, 83–89.

Patsioura, A., Galanakis, C.M., Gekas, V., 2011. Ultrafiltration optimization for the recovery of β-glucan from oat mill waste. J. Membr. Sci. 373, 53–63.

Périno-Issartier, S., Zill-e-Huma, Abert-Vian, M., Chemat, F., 2011. Solvent free microwave-assisted extraction of antioxidants from sea buckthorn (*Hippophae rhamnoides*) food by-products. Food Bioprocess Technol. 4, 1020–1028.

Porto, D.P., Porretto, E., Decorti, D., 2013. Comparison of ultrasound assisted extraction with conventional extraction methods of oil and polyphenols from grape seeds. Ultrason. Sonochem. 20, 1076–1080.

Prasad, N.K., Yang, B., Zhao, M., Wang, B.S., Chen, F., Jiang, Y., 2009a. Effects of high-pressure treatment on the extraction yield, phenolic content and antioxidant activity of litchi (*Litchi chinensis* Sonn.) fruit pericarp. Int. J. Food Sci. Technol. 44, 960–966.

Prasad, N.K., Yang, M., Zhao, X., Wei, Y., Jiang, F., 2009b. High pressure extraction of corilagin from longan (*Dimocarpus longan* Lour.) fruit pericarp. Separ. Purif. Technol. 70, 41–45.

Prasad, N.K., Yang, B., Zhao, M., Ruenroengklin, N., Jiang, Y., 2009c. Application of ultrasonication or high pressure extraction of flavonoids from litchi fruit pericarp. J. Food Process Eng. 32, 828–843.

Rahmanian, N., Jafari, S.M., Galanakis, C.M., 2014. Recovery and removal of phenolic compounds from olive mill wastewater. J. Am. Oil Chem. Soc. 91, 1–18.

Rajha, H.N., Boussetta, N., Louka, N., Maroun, R.G., Vorobiev, E., 2014. A comparative study of physical pretreatments for the extraction of polyphenols and proteins from vine shoots. Food Res. Int. 2014. doi:http://dx.doi.org/10.1016/j.foodres.2014.04.024.

Rajkumar, P., Kailappan, R., Viswanathan, R., Raghavan, G.S.V., Ratti, C., 2007. Foam mat drying of alphonso mango pulp. Drying Technol. 25, 357–365.

Roselló-Soto, E., Barba, F.J., Parniakov, O., Galanakis, C.M., Grimi, N., Lebovka, N., Vorobiev, E., 2015. High voltage electrical discharges, pulsed electric field and ultrasounds assisted extraction of protein and phenolic compounds from olive kernel. Food Bioprocess Technol. 8, 885–894.

Sampedro, F., McAloon, A., Yee, W., Fan, X., Zhang, H.Q., Geveke, D.J., 2013. Cost analysis of commercial pasteurization of orange juice by pulsed electric fields. Innov. Food Sci. Emerg. Technol. 17, 72–78.

Sampedro, F., McAloon, A., Yee, W., Fan, X., Geveke, D.J., 2014. Cost analysis and environmental impact of pulsed electric fields and high pressure processing in comparison with thermal pasteurization. Food Bioprocess Technol. 7, 1928–1937.

Shofinita, D., Langrish, T.A.G., 2014. Spray drying of orange peel extracts: yield, total phenolic content, and economic evaluation. J. Food Eng. 139, 31–42.

Toepfl, S., 2006. Pulsed electric fields (PEF) for permeabilization of cell membranes in food- and bioprocessing-applications, process and equipment design and cost analysis. PhD thesis, Berlin University of Technology, Berlin.

Tsakona, S., Galanakis, C.M., Gekas, V., 2012. Hydro-ethanolic mixtures for the recovery of phenols from Mediterranean plant materials. Food Bioprocess. Technol. 5, 1384–1393.

van Loon, A.J., 2000. From the benefits of micro to the threats of nano for the ore-mining and ore-refining sectors. Earth-Sci. Rev. 58, 233–241.

Vorobiev, E., Lebovka, N., 2010. Enhanced extraction from solid foods and biosuspensions by pulsed electrical energy. Food Eng. Rev. 2, 95–108.

Wang, Y., You, J., Yu, Y., Qu, C., Zhang, H., Ding, L., Zhang, H., Li, H., 2008. Analysis of ginsenosides in *Panax ginseng* in high pressure microwave-assisted extraction. Food Chem. 110, 161–167.

PATENTED AND COMMERCIALIZED APPLICATIONS

Charis M. Galanakis*, Nuria Martinez-Saez**, Maria Dolores del Castillo**, Francisco J. Barba†, Vassiliki S. Mitropoulou‡

*Department of Research & Innovation, Galanakis Laboratories, Chania, Greece; **Food Bioscience Group, Institute of Food Science Research (CSIC-UAM), Madrid, Spain; †Department of Nutrition and Food Chemistry, Faculty of Pharmacy, Universitat de València, Valencia, Spain; ‡Business Consultant, Athens, Greece

15.1 SCALE-UP AND COMMERCIALIZATION PROBLEMS

The industrialization of processes dealing with the recovery of compounds from food wastes includes numerous issues such as laboratory research, transfer to pilot plan and full-scale production, protection of intellectual properties, and development of definite applications. These parameters are necessary in order to ensure the sustainability of the process, the economic benefit for the involved food industry, and the perpetual establishment of the derived products in the market. In addition, a working scenario focused absolutely on the extraction technologies and not on the investigation of tailor-made applications is doomed to fail (Galanakis, 2012, 2013).

A scale-up process should be conducted without diminishing the functional properties (e.g. antioxidant, viscoelastic, etc.) of the target compounds and at the same time a product should be developed that meets consumers' high quality organoleptic standards. This is difficult since compounds recovery development challenges the typical scale-up problems as well as other more complicated procedures related to the particular nature of the recapture procedure (Galanakis and Schieber, 2014).

For instance, scale-up of recovery processes meets the same limitations (e.g. mixing and heating time) as any food manufacture procedure. Transition of batch to continuous processes is usually accompanied with extension of mixing and heating time, heavier handling, increased air incorporation, and higher degree of scrutiny (Galanakis, 2012). All of these parameters generate numerous interactions and loss of product functionality. Subsequently, process cost is increased, as industrially recovered compounds are used in food formulations in higher concentrations compared with laboratory-recovered compounds.

On the other hand, a more specific problem could be waste collection at the source that often requires additional transportation cost and control of microbial growth. Proper management of the collection process, cooling/freezing of the material, and/or addition of chemical preservatives can provide solutions in a particular case. Another complicated problem is the broad variation of target and nontarget compounds from source to source. This fact affects the mass and energy balances as well as the functionality and the organoleptic character of the final products, especially of more crude extracts. The above problem may be monitored by adding a modification pretreatment step. The latter usually includes a selective mixing of by-product streams at the beginning of the process, taking into account basic parameters (e.g. total antioxidants concentration),

Food Waste Recovery. http://dx.doi.org/10.1016/B978-0-12-800351-0.00015-8

or a vacuum concentration of wastewater streams taking into account macroscopic characteristics such as water and solids content.

15.2 PROTECTION OF INTELLECTUAL PROPERTIES

Intellectual property is an important asset of any organization (start-up, spin-off, etc.) involved in the field of valuable compounds recovery from food wastes. Moreover in recent years, recovery of high added-value compounds from food waste streams of special interest often involves development of new methods, industrial processes, and new applications of recovered compounds in food products, for example as food additives, e.g. natural pigments such as recovered lycopene instead of carmine (red), or as food supplements, i.e. antioxidants such as hydroxytyrosol, etc. This new knowledge is a valued asset for the researchers and related organizations and it should be protected and exploited following appropriate scientific and commercial strategies. Intellectual property law is often complex and it is important to know the basic rules regarding the best way to protect intellectual property assets. Nowadays, it is equally important for start-up companies, research organizations, and individual researchers to protect their intellectual property and therefore to be informed as to procedures, filing rules, costs, expected revenues, etc.

Intellectual property has a commercial value since it allows early protection of competitive advantages, applications as a marketing edge, increase of business value, and implementation as a potential revenue stream through licensing or investment attraction. Companies need to take precautionary steps to protect intellectual property by filing patent applications where appropriate, registering trademarks and copyrights as well as taking appropriate steps to protect trade secrets. Protecting intellectual property early during business formation provides safety to investors, builds credibility, and creates a solid foundation that can be capitalized on later. Indeed, it is imperative to be aware of others' intellectual property rights in order to ensure that nobody could prevent or restrict operation of the specific business (e.g. start-up companies). Intellectual property is often created during the earliest stages of company life. In the case of start-up companies, there is a need to develop a specific and robust intellectual property strategy in order to maximize the value of these important assets.

Different types of intellectual property rights exist, meeting the different needs for protection, products, costs, period, etc. Registered intellectual property includes:

1. patents,
2. utility models,
3. designs, and
4. trademarks.

Copyright is intellectual property that is protected without registration (e.g. literature, paintings, films, other forms of artistic or scientific works) and is in force for 70 years after the death of the writer or the originator. Patents and utility models are the most common forms of protection for industrial property, namely inventions of products or processes. A patent is an exclusive right to exploit commercially the invention in the country where the patent is granted. Protection is valid for up to 20 years. A patent forbids others to produce, sell, work, use, import, and possess the invention. However, it does not extend to acts performed for noncommercial purposes, experimental purposes, or other concerning products, which are commercially worked by (or with the consent of) the patentee and individual production of a medicinal product. A patentable invention can be (i) a product, (ii) the apparatus for producing the product, (iii) the process for producing the product, or (iv) the use of the product. A patentable invention must fulfill the following criteria:

1. it must have industrial application,
2. it should be new (e.g. above the state of the art), and
3. it must differ essentially from what is already known (e.g. involve an inventive step).

The invention must have at least one practical purpose, and must be reproducible.

When a new product or process is invented, or when a need must be fulfilled by inventing the appropriate solution, the inventor must adhere to defining a patent/intellectual property rights strategy (at an early stage, prior to filing of a patent or taking any important decisions).

Some integral steps of an intellectual property rights strategy are:

1. to think of the idea as a financial asset,
2. to implement policies and processes for identifying, disclosing, and assigning patentable inventions as an integral part of research and development efforts of the organization,
3. to consider advantages and disadvantages of patent protection and to check alternatives (secrecy, utility models, etc.). Also a cost/benefit analysis is crucial as applying for a patent at the right time without neglecting to evaluate if the invention is more complex than the problem merits. Following a consideration of the patentability requirements and the details of what is patentable in your own country is crucial.

At this point, a prior art search should be conducted in order to verify if the invention is not already claimed by another individual or business, and identify who the competitors or potential partners are and what they do. Moreover, all issues relating to rights over the invention between the organization, its employees, and any other business partner (who may have participated either financially or technically in developing the invention) should be clarified. Problems concerning invalidity and/or noninfringement before launching potentially infringing products must be solved. Finally, the invention should be kept secret until the date of filing to protect novelty.

The decision to file an application for a patent should be cautious since legal systems vary between countries and several options are offered. Patent protection can be achieved by filling an application for a national, international, or regional patent. A national patent is typically used to protect invention within the home country market and as a basis to extend protection to other countries (or regions) by providing "priority". International patent applications can be used to protect invention in many countries and extend the prepublication period up to 30 months. A regional patent protects invention in a number of countries within the same region, at a lower cost. Within one year from filling the first patent application, applications can be filled in other countries, too, with a priority inclusion established in the first country. Thereby, novelty is valid from the application filling date of the first country.

The main international/regional patent systems are based on legal agreements:

- Patent Cooperation Treaty (PCT)
- European Patent Convention (EPC)
- Other regional systems (African Intellectual Property Organization, African Regional Intellectual Property Organization, Eurasian Patent Organization)

The World Intellectual Property Office (WIPO) provides international novelty and patentability search, a single place for filing and final decision for countries based on common rules and postponement for 30 months from the priority date. At the European level, the European Patent Organization (EPO) provides a single place for filing, completion, and patent granting for the 27 EPC member countries at lower costs compared with filing in each country separately.

A most decisive step – or more precisely an ongoing process towards achieving successful market introduction and establishment of new products from recovered compounds – is, as stipulated earlier, the choice of a coherent and flexible patent and general intellectual property rights strategy. Olive waste by products valorization is a new promising field for the recovery of phenolic compounds and it could be used to exemplify the process and important decisions to elect an effective intellectual property rights strategy. A modest – only in terms of financial recourses – strategy could be to apply a national patent in the country of origin of the invention (e.g. in Greece) both for the novel processes used for the recovery of phenolic compounds in this example and for all the possible products that could be marketed, e.g. products having very good antioxidant properties and that can be marketed as functional foods, improving health, etc. It is important to note that patent protection is only provided to what is explicitly stated in the claims section of a patent application. This initial step provides one year priority and protects both processes and possible marketable products with low early investment. The second step could be to apply/extend to a regional patent, e.g. an EPO patent if the anticipated product could be marketed in some or all countries in Europe or world patent through WIPO, or if the market is limited to, e.g. a few countries to apply/extend national patents. Another approach after filing a national patent could be to find an investor or licensee who invests in the early stage of marketing products, providing the patent holder finance to develop new research and/or products. An antioxidant phenolic compound recovered from olive waste by-products is hydroxytyrosol. Several commercial products involving hydroxytyrosol are marketed already in the United States where the Food and Drug Administration (FDA) granted generally recognized as safe (GRAS) status, and the European Food Safety Authority (EFSA) handles market release of most products with health claims, such as hydroxytyrosol products, in a preserved manner. Hydroxytyrosol has received European market release acceptance only for low cholesterol spreads. This significantly different approach of market release between FDA and EFSA should be taken into account at an early stage when electing patent filing strategy and exploitation of research results.

15.3 APPLICATIONS AND MARKET PRODUCTS
15.3.1 VEGETABLE AND PLANT BY-PRODUCTS

A collection of valuable compounds recovered from vegetable and plant by-products at a commercial scale are shown in Table 15.1. Verification of market existing products matching with a patented process was herein conducted using a patent applicant name in each case. However, this matching may not always be correct as companies typically are secretive about their production methods and respective data cannot be found in the literature.

For example, industrial commercialization of citrus peel has been practiced for more than 30 years, whereas the solvent extracted "sugar syrup" contains essential oils, flavonoids, sugar, and pectin (Bonnell, 1983). Sugar syrup provides sweetness and flavor in juices, while it replaces artificial sweeteners (e.g. saccharine or aspartame) in foods. Another commercialized application concerns the recovery of albumin from soy protein wastewater, using membrane technology, flash distillation, and spray drying. The albumin-rich powder is applied as a nutritional supplement similar to whey protein concentrates

Table 15.1 Patented Methodologies Leading to Commercial Applications of Fruit, Vegetable, and Plant By-Products

Source	Patents Application Number	Applicant/ Company	Title/Treatment Steps	Products/Brand Names	Potential/ Commercialized Applications	Inventors/ References
Citrus peel waste	AU1983/0011308D	Tropicana Products Inc. (Florida, USA)	Treatment of citrus fruit peel	Sugar syrup	Food natural sweetener	Bonnell (1983)
Fruit and vegetable residues, unsellable fruits with defects	US2001/6296888	Provalor (Hoofddorp, The Netherlands)	Squeezing/ decantation/ centrifugation	Juices	Health drinks	Nell (2001)
	–	Indulleida S.A., Spain	Drying, solvent extraction	Fibers, sugars, polyphenols, and aromas	Food additives	http://www. indulleida.com/
Tomato waste	PCT/ EP2007/061923	Biolyco SRL (Lecce, Italy)	Process for the extraction of lycopene	Lycopene	Food antioxidant and supplement	Lavecchia and Zuorro (2008)
Soy protein isolate wastewater	CN2008/10238791	ShanDong Wonderful Industry Group Co. Ltd (Shandong, China)	Method for extracting and recycling albumin from whey wastewater from production of soy protein isolate	Soybean albumin	Food additive and supplement	Jishan et al. (2009)
Edible biological materials derived from plants (fruits, flowers, leaves, stems, herbs)	WO/2006/099553	Innovative Foods, Inc., South San Francisco, USA	Methods for preparing freeze-dried foods	Dried fruits and vegetables	Snacks and additives in prepared foods and commercially available meals, confectionary	Hirschberg et al. (2006)
Depectinated apple pomace	CN2008/1139768	Yantai Andre Pectin Co. Ltd (Yantai, China)	Process for extracting nonpectin soluble pomace dietary fibers	Apple dietary fiber granules	Dietary supplement	Anming et al. (2010)
Grape and cranberry seed	JP1998/0075070	Kikkoman Corp. (Chiba, Japan)	Protein food	Proanthocyanidin	Coloring additive in soy sauce	Ariga et al. (1999)

(Continued)

Table 15.1 Patented Methodologies Leading to Commercial Applications of Fruit, Vegetable, and Plant By-Products *(cont.)*

Source	Patents Application Number	Applicant/ Company	Title/Treatment Steps	Products/Brand Names	Potential/ Commercialized Applications	Inventors/ References
Wine grapes pomace and seeds	–	WholeVine, Santa Rosa, California, USA	Milling and oil extraction	Grape skin and seed flour	Additives in gluten-free baked goods	http://m. wholevine.com/
Grape pomace or seed	WO/1999/030724	Pierre Fabre Sante (France)	–	Polyphenols	Food supplements, functional foods, cosmetics, and pharmaceutical applications	Rouanet et al. (1999)
Pomegranate rind and seedcase residues	CN2010/1531940	Xi'an App Chem-Bio(Tech) Co., Ltd. (Xi'an, China)	Method for preparing punicalagin and ellagic acid from pomegranate rind	Ellagic acid (40%) and punicalagin (40%)	Food antioxidants and cosmetics	Guangyu & Xiaoyan (2011)
Conifer and pine bark	US1987/4698360	Societe Civile d'Investigations Pharmacologiques d'Aquitane Horphag Overseas Ltd (Bordeaux, France)	Boiling, filtering, and solvent extraction	Proanthocyanidins	Cardiovascular health, oral and topical skin care, and eye health	Masquelier (1987)

(Jishan et al., 2009). Natural shrimp and crab shells have been used as substrates for the extraction of food-grade chitosan with alkali and chloracetic acid treatment (Shenghui, 1995). This product is sold as thickener in vegetable oils or as an antirancidity agent in meat (Kanatt et al., 2008).

The industrial recovery of water insoluble carotenoids from food wastes is in progress, too. Lycopene is one of the most popular natural pigments (red). Recently, the FDA approved the use of higher levels of tomato lycopene to color-processed meats as an alternative to carmine (USDA, 2014). Moreover, in vitro, in vivo, and ex vivo studies have demonstrated that its addition to foods is inversely associated with cancers and cardiovascular diseases (Kong et al., 2010). Besides, the Food Safety and Inspection Service has established that tomato lycopene extracts and concentrates (GRN 000156) of ≤ 50 and ≤ 100 mg/kg, respectively, could be used as coloring agent in ready-to-eat meat, poultry, and egg products (USDA, 2014). This fact will probably lead to wide-scale applications around the world and could open the door for the commercial recovery of lycopene from food by-products. Today, the extraction of lycopene is under industrial development using sequential extraction with a nonpolar and a polar protic solvent (Lavecchia and Zuorro, 2008).

Figure 15.1 illustrates an example of apple pomace valorization for recovery purposes. Specifically, pretreatment of apple pomace (peel, pulp, and seed mixture) includes a sieving process for the separation of the seeds. The latter are rich in oil and proteins. Separation of macro- and micromolecules could be performed using alcohol precipitation, after malaxation of the remaining mixture with hot water. Ethanol would solubilize a major part of the contained polyphenols (Tsakona et al., 2012; Galanakis et al., 2013a;

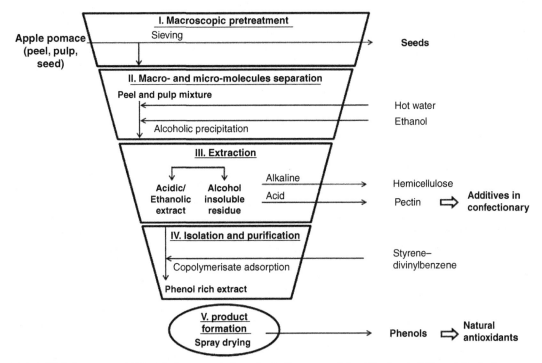

FIGURE 15.1 Recovery of Valuable Compounds from Apple Pomace and Reutilization in Different Products

Heng et al., 2015). Extraction of hemicellulose from the alcohol insoluble residue could be conducted using alkaline solution, whereas the recovery of pectin could be performed using acidic extraction. Water-soluble dietary fibers (e.g. hemicellulose and pectin) could be incorporated into food and confectionary products as cheap noncaloric bulking agents instead of flour, fat, or sugar, due to their ability to retain water, improve emulsions, and oxidative stability. On the other hand, sequential extraction of peel and pulp mixture with inorganic acid and alcohol precipitation can generate an acidic/ethanolic extract used for the isolation of phenolic compounds with adsorption on a hydrophobic styrene–divinylbenzene copolymerisate material (Schieber et al., 2001). Fractionation of phenolic compounds can also be performed using ultrafiltration (Galanakis et al., 2013b, 2015).

15.3.2 OLIVE BY-PRODUCTS

A new tendency in the field of food waste recovery concerns the valorization of olive by-products as a source of phenolic compounds (Galanakis, 2011; Galanakis et al., 2010d, e; Rahmanian et al., 2014; Roselló-Soto et al., 2015) (Table 15.2). For example, commercial hydroxytyrosol has been recovered from olive mill waste in pure form (99.5% per weight) using chromatographic columns filled with two resins: nonactivated ionic and XAD-type nonionic (Fernández-Bolaños et al., 2002). Another popular process is performed using acid treatment of olive mill wastewater, prior to an incubation process that converts oleuropein to hydroxytyrosol. Thereafter, extraction is performed using supercritical fluid extraction and a column operating in the counter-current mode, where a nonselective porous membrane is the barrier interface between the hydroxytyrosol-containing fluid and the dense gas. Encapsulation is conducted using either freeze or spray drying (Crea, 2002a, b). Ultimately, hydroxytyrosol possesses advanced antiradical properties compared with vitamins E and C, and prevents the oxidation of lipids in fish (Fernández-Bolaños et al., 2006). Thereby, it could be used as a functional supplement, a food preservative in bakery products, or as a life prolonging agent (Liu et al., 2008). Hidrox is a commercially available product from CreAgri (Hayward, USA), granting a GRAS status. According to several scientific studies, Hidrox possesses several beneficial (e.g. anti-inflammatory and antimicrobial) properties.

Other commercially available products include: (i) Olnactiv (Glanbia, Milan, Italy), Oleaselect, and Opextan (Indena, Milan, Italy), (ii) Olive Braun Standard 500 (containing 1.0–2.2 g of hydroxytyrosol and 0.2–0.7 g of tyrosol/kg) from Naturex, (iii) olive polyphenols from Albert Isliker (containing 22–23 g hydroxytyrosol and 6.5–8.0 g tyrosol/kg), (iv) Prolivols (containing 35% polyphenols, 2% hydroxytyrosol, and 3% tyrosol) from Seppic Inc., and (v) Olive Polyphenols NLT from Lalilab Inc. (containing 2.0–6% hydroxytyrosol and 0.7–1.1% tyrosol). Another product is Phenolea Complex, which is a natural hydrophilic extract, obtained directly from aqueous olive pulp and olive mill wastewater. This product is produced without any kind of organic solvent, so the biochemical composition of the final extract reflects the original composition of olive fruit but in a more concentrated formulation. The production process comprises the following steps: (i) collection of olive pulp and mill wastewater after the milling process, (ii) pretreatment of the material, (iii) tangential microfiltration with ceramic membranes, and (iv) vacuum evaporation of the permeate (FDA, 2012). Olive pulp extracts (in general) have been approved by FDA with GRAS status (GRN No. 459) for being used as an antioxidant in baked goods, beverages, cereals, sauces and dressings, seasonings, snacks, and functional foods at a level of up to 3000 mg/kg in the final food (FDA, 2014).

Table 15.2 Patented Methodologies Leading to Commercial Applications of Olive Wastes and By-Products

Food Waste Source	Patents Application Number	Applicant/ Company	Title/Treatment Steps	Products/Brand Names	Potential/ Commercialized Applications	Inventors/ References
Olive mill waste	PCT/US2001/027132	CreAgri, Inc. (Hayard, USA)	Method of obtaining a hydroxytyrosol-rich composition from vegetation water	Hydroxytyrosol/ Hidrox	Food supplements and cosmetics	Crea (2002a, b)
	PCT/ES2002/000058	Consejo Superior de Investigaciones Cientificas (Madrid, Spain)/ Genosa I+D S.A. (Malaga, Spain)	Method for obtaining purified hydroxytyrosol from products and by-products derived from olive trees	Hydroxytyrosol (99.5%)/Hytolive	Conserving foods, functional ingredient in bread	Fernández-Bolaños et al. (2002)
	GR2010/1006660	Polyhealth (Larissa, Greece)	Ultrafiltration, ion exchange resin adsorption, solvent elution, spray drying	Medoliva (hydroxytyrosol, tyrosol, caffeic acid and p-coumaric acid)	Food supplements and antioxidants, cosmetics, personal care products	Petrotos et al. (2010)
	PCT/SE2007/001177	Phenoliv AB (Lund, Sweden)	Olive waste recovery	Olive phenols and dietary fibers containing powders	Natural antioxidants in foodstuff and fat replacement in meatballs, respectively	Tornberg and Galanakis (2008)
	US patent 6361803 B1	Usana, Inc. (USA)	Wastewater from olive oil production	Antioxidant compounds	Food supplements and antioxidants, cosmetics, personal care products	Cuomo and Rabovskiy (2004)

(Continued)

Table 15.2 Patented Methodologies Leading to Commercial Applications of Olive Wastes and By-Products *(cont.)*

Food Waste Source	Patents Application Number	Applicant/ Company	Title/Treatment Steps	Products/Brand Names	Potential/ Commercialized Applications	Inventors/ References
Olive leaves extracts	EP 1582512 A1	Cognis IP Management GmbH (Düsseldorf, Germany)	Olive waste recovery	Olive hydroxytyrosol	Natural antioxidants in foodstuff	Beverungen (2005)
Vegetation water from olives	US Patent/ 2005/0103711 A1		Olive waste recovery	Oleuropein aglycon	Natural antioxidant	Emmons and Guttersen (2005)

There are several patents that deal with the extraction of oleuropein and/or hydroxytyrosol from olive and water vegetation (WO/2002/0218310, US/2002/0198415, US 2002/0058078, WO2004/005228, US 6414808, EP-A 1 582 512) (Crea, 2002a, b, 2004; Crea and Caglioti, 2005). In addition, a method for obtaining hydroxytyrosol and/or oleuropein from the vegetation water of depitted olives is disclosed in US 2004/0039066 A1. In most patents, an acidic hydrolysis of the substrate (olive leaves or vegetation water) for 2–12 months is suggested in order to convert at least 90% of the present oleuropein to hydroxytyrosol (Liu et al., 2008). Moreover, a method for the extraction of phenolic compounds from olives, olive pulps, olive oil, and oil mill wastewater has been described by Usana Inc. patents US 6,361,803 and WO01/45514 and in US2002/0004077 (Cuomo and Rabovskiy, 2002, 2004).

Figure 15.2 illustrates an example for the recovery of valuable compounds from olive mill wastewater, which is adapted to the *5-Stage Universal Recovery Process*. In this case, pretreatment includes two processes (centrifugation and skimming) in order to remove remaining fats from three-phase olive mill wastewater. Thereafter, a vacuum concentration process can remove part of the contained water. Treatment of concentrated and defatted olive mill wastewater with acids and ethanol can generate two streams: an alcohol insoluble residue rich in dietary fibers and an ethanolic extract rich in polyphenols. Isolation of the latter compounds can be performed using resin adsorption or chromatography. On the other hand, purification of the water-soluble fraction (mainly pectin) of the residue can be conducted

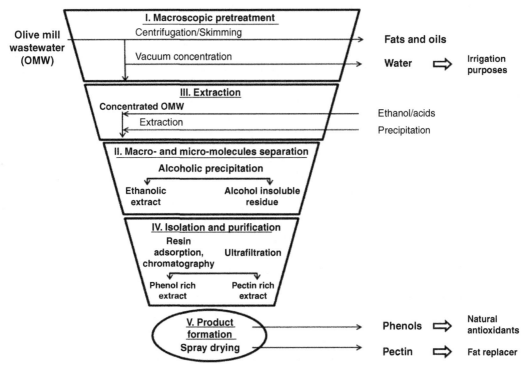

FIGURE 15.2 Recovery of Valuable Compounds from Olive Mill Wastewater and Reutilization in Different Products

using ultrafiltration (Galanakis, 2015). Pectin derived from olive mill wastewater has been proved to restrict oil uptake of low fat meatballs during deep fat frying (Galanakis et al., 2010a, b, c, d, e).

15.3.3 COFFEE BY-PRODUCTS

Coffee is one of the most consumed beverages in the world. Due to the great demand for this product, large amounts of wastes are generated in the coffee industry. Coffee silverskin (CS) and spent coffee grounds (SCG) are the main coffee industry wastes or by-products, obtained during roasting and brewing processes, respectively. CS contains several bioactive compounds such as prebiotic carbohydrates, dietary fiber, and antioxidants (Borrelli et al., 2004; Napolitano et al., 2007; Ballesteros et al., 2014). In agreement, CS has been proposed as a natural source of health promoters or functional food ingredients (Esquivel and Jimenez, 2012; Pourfarzad et al., 2013).

Figure 15.3 illustrates the process for the recovery of bioactive compounds from CS patented by del Castillo et al. (2013). The procedure consists of the extraction of the CS (without prior milling) using subcritical water at moderate temperature (50°C or higher) and high pressure (1500 psi), although unpressurized water at 100°C can be used as well. Under these conditions, extracts with high antioxidant properties are obtained in 10–20 min, with antioxidant activity depending on the extraction conditions: 0.85–3.7 g of equivalents of chlorogenic acid and 150–450 mg of caffeine, both per 100 mg of silverskin, when extracted with only hot water or under subcritical conditions, respectively. The antioxidant properties of the extract resist the in vitro gastrointestinal digestion process. It maintains the antioxidant

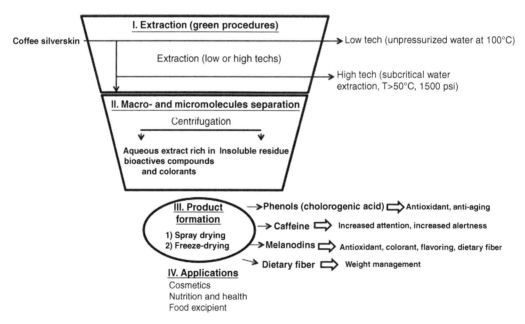

FIGURE 15.3 Recovery of Valuable Compounds from Coffee Silverskin and Reutilization in Different Products

del Castillo et al., 2013

properties for more than 6 months at room temperature and in dry conditions. Therefore, it has been proposed as an antioxidant additive in food and cosmetics manufacture, with potential excipient (preservative, flavoring) functions as well as antioxidant, antiageing, and anticellulite activities. The patented extract also has a potential to reduce body fat accumulation (Martinez-Saez et al., 2014), scavenge dicarbonyls and inhibit the formation of advanced glycation end-products (Mesías et al., 2014). The insoluble residue generated during the extraction process may be recycled as dietary fiber. The patent has been granted and licensed to a Spanish company. Currently, the development of products using the patented extract is being carried out.

Another example is the invention of a CS-containing paper and the production method thereof (Sato and Morikawa, 2011). The inventors of this application have focused on the oil absorbency of silverskin, and they have studied a method for providing oil absorbent paper. The latter shows low water absorbency and it is produced by mixing silverskin with paper pulp. The patented procedure produces useful and highly functional paper that has characteristics of roasting residues such as oil absorbency and low water absorbency. Coffee silverskin discarded during beverage production is effectively used for paper manufacture.

SCG is generated in large amounts, with a worldwide annual generation of 6 million tons (Tokimoto et al., 2005). Consequently, there is an increased interest in new alternatives to add value to this by-product. Information regarding the chemical composition of this food matrix has been recently reviewed by Ballesteros et al. (2014). Some examples of patented applications for SCG are shown in Table 15.3. Mussatto et al. (2013) successfully used SCG for the production of a distilled beverage with coffee aroma. The global process was based on an aqueous extraction of aromatic compounds from SCG, supplementation with sugar, and production of ethanol. This developed spirit showed acceptable flavor, volatile compounds, and different organoleptic character compared with the commercially available spirits. Another patented application for SCG is its use as an ingredient in healthy bakery products (with high level of dietary antioxidant fiber), pastry, confectionary, biscuits, breakfast, cereals, etc. (del Castillo et al., 2014). This patent is under commercialization, whereas national and international applications have been performed. The preparation of this ingredient involves a simple and low-cost method with minimum treatment and direct application of the by-product. The developed formulation employs a coffee by-product as a source of antioxidant fiber in diverse combinations with other basic and/or novel ingredients, e.g. stevia. The resulting formula is rich in insoluble dietary fiber (3–7%) and its content in acrylamide is low. These products might be appropriate for special nutritional needs due to their low glycemic index and low energetic value.

SCG have also been employed for the invention of foods and beverages containing mannooligosaccharides to reduce blood pressure, elevate suppressing effect (Takao et al., 2009), and reduce body fat (especially abdominal) (Asano et al., 2006). The mannooligosaccharides may be produced by the hydrolysis of mannan from coffee materials (especially spent coffee residues and other coffee-containing materials from commercial multistage coffee extraction systems). The object is to provide an economical and simple food or drink, with excellent blood pressure reducing, elevation suppressing, and fat reducing effects, without changing ordinary eating habits. Additionally, Asano et al. (2002) described a nonpatented but important application. In this case, mannooligosaccharides obtained by thermal hydrolysis of SCG resistant to α-amylase, artificial gastric juice, porcine pancreatic enzymes, and enzymes of the intestine of rats were fermented by fecal bacteria in human beings. This suggests that under these hydrolysis conditions, mannooligosaccharides are indigestible for humans. Thus, they could potentially be used as a prebiotic for the probiotic microorganisms present in the large intestine.

Table 15.3 Patented Methodologies Leading to Commercial Applications of Coffee Wastes and By-Products

Source	Patents Application Number	Applicant/ Company	Title/Treatment Steps	Products/Brand Names	Potential/ Commercialized Applications	Inventors/ References
Coffee silverskin	US 7,927,460	Ito En, Ltd. (Tokyo, Japan)	Silverskin-containing paper and method for producing the same	Functional silverskin-containing paper	Paper industry	Sato and Morikawa (2011)
	WO 2013/004873	Consejo Superior de Investigaciones Cientificas/CIAL (Madrid, Spain)	Application of products of coffee silverskin antiageing cosmetics and functional food	Bioactive silversink extract	Cosmetics, nutrition, and health	del Castillo et al. (2013)
Spent coffee grounds	WO 2006/036208	Ajinomoto General Foods, Inc. (Tokyo, Japan), Kraft Foods Global Brands LLC (Northfield, Illinois, USA)	Mannooligosaccharide composition for body fat reduction	Mannooligosaccharides	Functional food ingredient	Asano et al. (2006)
	US2009/0005342 (PCT/JP2006/301025)	Ajinomoto General Foods, Inc. (Japan)	Composition having blood pressure reducing and/ or elevation suppressing effect and food and drink containing the same	Mannologosaccharides	Functional food ingredient	Takao et al. (2009)
	PCT/ES2014/070062	Consejo Superior de Investigaciones Cientificas/CIAL (Madrid, Spain)	Healthy bakery products with high level of dietary antioxidant fiber	Antioxidant insoluble dietary fiber	Functional food ingredient	del Castillo et al. (2013)
	PT 105346	University of Minho. CEB – Centre of Biological Engineering (Braga, Portugal)	Distilled beverage from spent coffee grounds and respective production method	Distilled beverage	Aroma	Mussatto et al. (2013)

Another waste obtained from the coffee bean is the surrounding pulp, known as the coffee cherry, which is discarded during coffee processing. A new interesting ingredient has been developed (Coffee flour, http://www.coffeeflour.com/) which can be used in different formulations such as breads, cookies, muffins, squares, brownies, pastas, sauces, and beverages. This flour has achieved cooking and baking success. It does not taste like coffee, but rather expresses more floral, citrus, and roasted fruit-type notes. In addition, it is gluten free, possesses fivefold more fiber than the whole grain of wheat flour, and has 84% less fat and 42% more fiber than coconut flour. Coffee flour is set for commercial rollout in 2015.

15.3.4 DAIRY, ANIMAL, AND FISHERY BY-PRODUCTS

Whey is the most known food waste source used for recovery and valorization purposes (Table 15.4). It is generated in different forms and compositions depending on the characteristics of cheese manufacturing, whereas protein concentrates, lactose, and respective monosaccharides (glucose and lactose) comprise the target compounds. Figure 15.4 illustrates different technologies adapted in the *5-Stage Universal Recovery Process* as well as commercial applications for the derived products. Initially, skimming is used to remove casein fines and whey cream. Both ingredients are today used in confectionary. Membrane filtration comes next to concentrate proteins from the defatted whey (Galanakis et al., 2014). For instance, Jensen and Larsen (1993) reported a two-step microfiltration process for the sequential concentration of α-lactoalbumin and β-lactoglobulin. Whey protein concentrates and liquid glycoproteins can be further treated using electrodialysis in an alkaline environment. Ion-exchange chromatography has also been industrially employed to deflavor whey protein concentrates and clarify glycoproteins, prior to drying by either spray or freeze drying. Proteins are typically used as nutritional supplements in bodybuilding. Moreover, hydrolyzed whey proteins are known for their ability to reduce total and LDL-cholesterol levels in mammals (Davis et al., 2003) and thus could be used for the production of functional foods. Heat coagulation of the residual deproteinized whey causes precipitation and subsequent removal of glycoproteins from the liquid stream (Davis et al., 2002). Finally, crystallization of the deproteinized whey using seed crystals causes lactose precipitation. Crude lactose is used as a supplement in diet food or as an aroma stabilizer. It can also be hydrolyzed with enzymes to produce galactose- and glucose-rich syrup (Galanakis et al., 2014). The latter can be used as sweetener.

On the other hand, the meat industry generates a large amount of wastes and by-products, which are a good source of nutrients and can be used as food ingredients and additives. Among the different meat by-products, air flotation skimming sludge typically has an important nutritive value. However, it is lost due to microbial degradation. On the other hand, a practical precipitation process of food waste sludge from dissolved air flotation units and sugar by-products has been patented (Lee, 2002). The process is based on the transformation of the skimming sludge and animal blood into a precipitate using centrifugation, screening, or pressing. The precipitate binds most nutrients. Moreover, as water feed is eliminated and product surface area is increased during the process, the cost is reduced. The process can convert the waste skimming sludge into a safe and valuable product for feed and nutritional applications, e.g. it can be used for improving the properties of animal feed block products (Lee, 2002). Moreover, chicken feathers have been used as a source of keratin for packaging (Table 15.4).

The fishery industry generates a large amount of wastes and by-products, rich in fish oils. The latter contains omega-3 and omega-6 fatty acids that could be used in the food and pharmaceutical industries. Several groups have proposed different projects for the valorization of fishery wastes and by-products (Table 15.4). For example, Shenghui (1995) developed a patent for the preparation of a chitosan derivative

Table 15.4 Patented Methodologies Leading to Commercial Applications of Fruit, Dairy, Animal, and Fishery By-Products

Source	Patents Application Number	Applicant/ Company	Title/Treatment Steps	Products/Brand Names	Potential/ Commercialized Applications	Inventors/ References
Cheese whey	PCT/ SE1993/000378	Alfa-Laval Food Engineering AB (Lund, Sweden)	Method for obtaining high-quality protein products from whey	α-Lactoalbumin and β-lactoglobulin containing product	Food supplements and additives	Jensen and Larsen (1993)
	PCT/ US2002/010485	Davisco International Foods Inc. (Le Sueur, USA)	Isolation of glycoproteins from bovine milk	Whey protein isolate/Bipro	Food supplements	Davis et al. (2002)
	US2009/7582326	Kraft Foods Global Brands Llc (Northfield, USA)	Method of deflavoring whey protein using membrane electrodialysis	Deflavored whey proteins	Food supplements	Brown & Crowely C.P. (2009)
Shrimp and crab shell	CN1994/1001978	Qingdao Zhengzhongjiahe Export & Import Co., Ltd (Shandong, China)	Preparation of chitosan derivative fruit and vegetable antistaling agent	Chitosan (≥85%) food grade	Food thickener and fruit antistaling agent	Shenghui (1995)
Salmon viscera, heads, skin, frames, and trimmings	—	Aquaprotein, Chile	Hydrolysis, mechanical extraction, and spray drying	Protein hydrolysates and salmon oil	Petfood, pig weaning, animal breeding, injuries recuperation in animals	http://www. aquaprotein.com/
Chicken feathers	US2012/08182551	Eastern Bioplastics LLC	Washing, grinding, perforated rotating spiral drum, skimming, compounding	Keratin	Polypropylene packaging, as a sorbent of hydrocarbons	Meyerhoeffer and Showalter (2012)

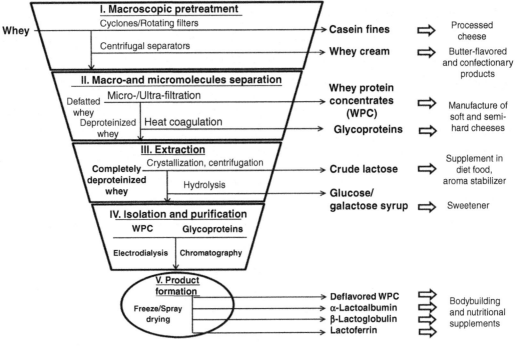

FIGURE 15.4 Recovery of Valuable Compounds from Cheese Whey and Reutilization in Different Products

(Galanakis 2012)

fruit and vegetable antistaling agent from shrimp and crab shell, which is used as a food thickener and fruit antistaling agent. Other commercial applications are based on the recapture of fish protein hydrolysates and fish oil for feed from fish guts, skins, heads, and bones. Finally, fish skin has been used to obtain gelatin. The latter could be used in the preparation of fresh and smoked salmon as well as in salmon oil.

15.4 POTENTIAL USE OF EMERGING TECHNOLOGIES

In recent decades, there has been a considerable increase in the number of commercially available foods processed by high pressure processing (HPP) for food preservation. For example, products have been marketed in Japan since 1990 and in the United States and Europe since 1996 (Zhang et al., 1995; Rizvi and Tong, 1997; Körmendy et al., 1998). Table 15.5 shows fruits, vegetables, and derivatives treated by HPP that are currently in the market. These applications reveal the potential use of HPP for food waste recovery in two different ways: first, reutilization of wastes from food industries, supermarkets, etc., and second, recovery of valuable compounds using HPP as an extraction or pretreatment technique. Other emerging technologies, e.g. ultrasonication, have been used in soy processing in order to enhance protein and sugar yields as well as nisin production (Khanal, 2007). In the last two decades, pulsed electric fields (PEF) and high voltage electrical discharges (HVED) have been

Table 15.5 Recently Marketed Foodstuffs Treated by High Pressure Processing

Country	Products	Pressure (MPa)	Temperature (°C)	Time (min)	Shelf-Life
			Treatment		
Czech Republic	Broccoli and apple or carrot juice (Beskyd Fryčovice)	500	20	10	21 d
France	Juices (Pampryl)	400	20	10	18 d, 4°C
	Orange juice (Ultrafruit)	500	20	5–10	–
Italy	Fruit juices, fruit desserts (Ortogel)	600	17	3–5	1–2 m
Japan	Jams, sauces, jellies (Meidi-Ya Co.)	400	20	10–30	2–3 m, 4°C
	Grape juice (Pokka Co.)	120–400	23	2–20	–
	Mandarin juice (Wakayama Food Industries)	300–400	23	2–20	–
	Sake (Chiyonosono)	400	15	30	6–12 m, 4°C
Lebanon	Fruit juices	500	–	–	1 m
Mexico	Smoothies, citrus juices (Jumex)	500	–	–	–
Portugal	Juices (Frucalba)	450	12	0.33–1.5	28 d
Spain	Prepared vegetable dishes	500	–	–	–
Sweden	Fruit juices (Västerås)	500–600	–	–	–
UK	Orange juice (Orchard House Foods)	500	20	–	–
USA	Avocado purée (Avomex)	700	–	10	–
	Apple juice (Odwalla)		–	–	2–3 times more than fresh
	Avocado products, vegetable salads, orange juice, lemonade, sliced onions				45 d

M, months; D, days.

proposed as a potential tool to recover valuable compounds from different bioresources, including plant and animal food wastes and by-products (Roselló-Soto et al., 2015; Deng et al., 2015). For instance, several patents have been developed by different authors: Kortschack (2007) developed an electroporation method for the processing of meat, fish, and seafood by-products. This trend allows the development of food products based on animal wastes. Moreover, the use of pulsed electric treatments for the recovery of high added-value components from plant food materials has been widely reported recently and some patents have been developed, too (Table 15.6). For instance, a method for extracting liquid from a cellular material using a combination of mechanical pressing and moderated

Table 15.6 Nonconventional Patented Methodologies with Potential to Valorizate Food Wastes and By-Products

Food Waste Source	Patents Application Number	Applicant/ Company	Title/Technology	Potential/ Commercialized Applications	References
Cellular material	PCT/ FR2001/000490	Association Gradient (Compiègne, France)/ Fonderies & Ateliers Lucien Choquenet (Chauny, France)	Treatment of cellular material/PEF	Functional food ingredients	Andre et al. (2001)
Vegetal matrix	PCT/ EP2011/070597	Universite Technologie De Compiègne-UTC (Compiègne, France)	Procede d'extraction de molecules d'interet a partir de tout ou partie d'une matrice vegetale/PEF-HVED	Food supplements and additives	Bousseta et al. (2013)
Plant materials	PCT/ CA2007/001,652/ US Patent 8147879	–	Pulsed electric field enhanced method of extraction/PEF	Food supplements	Gachovska et al. (2008); Ngadi et al. (2012)
Meat, meat by-products, fish, and seafood	CA 2620122	Triton GmbH, Fritz Kortschack	Method for processing raw materials having a cell structure in the meat, meat by-products, fish, and seafood processing industry/PEF	Cosmetics, nutrition, and health	Kortschack (2007)
Vegetable tissues	US Patent App. 13/696040; EP20,110,716,571	Maguin Sas	Method and apparatus for treating vegetable tissues in order to extract therefrom a vegetable substance, in particular a juice/ PEF	Juice obtention	Vidal and Vorobiev (2013a, b)
Plant materials	US Patent 7943190	Minister of Agriculture and Agri-Food, Canada	Extraction of phytochemicals/SCW	Functional food ingredients	Mazza and Cacace (2011)
Capsicum solids	US Patent 6074687	Kalamazoo Holdings, Inc.	High temperature countercurrent solvent extraction of capsicum solids	Cosmetics, nutrition, and health	Todd (2000)

PEF, pulsed electric fields; HVED, high voltage electrical discharges; SCW, subcritical water extraction.

power PEF applied by short and repeated high voltage pulse bursts was developed (Andre et al., 2001). Moreover, this invention has the potential to be used for implementing the extraction of liquid from a cellular material (Andre et al., 2001). PEF can also be used to recover high added-value compounds from vegetable tissue (Gachovska et al., 2008; Ngadi et al., 2012; Vidal and Vorobiev, 2013a, b). For example, some patents have been developed involving a process of compacting the plant tissues and at least one treatment chamber. Subsequently, PEF treatment was applied to the compacted tissues (Vidal and Vorobiev, 2013a, b). In addition, the extraction of a molecule of interest from a plant or similar matrix, especially phenolic compounds from winery by-products such as grape marc and lees by using HVED and/or PEF, has been patented recently (Bousseta et al., 2013).

On the other hand, a patent was developed to extract and concentrate the carotenoid pigments from paprika, red pepper, pungent chili, and other plants of the genus *Capsicum* by using an edible solvent in a series of mixing and high temperature and pressure mechanical pressing steps (Todd, 2000).

Finally, a patent involving the use of a subcritical water extraction process was developed for the extraction of phytochemicals from plant food materials (Mazza and Cacace, 2011). The processing system includes a water supply interconnected with a high pressure pump, diverter valve, a temperature-controllable extraction vessel, a cooler, a pressure relief valve, and a collection apparatus for collecting eluant fractions from the extraction vessel.

15.5 CONCLUSIONS

The key point for commercialization is to develop a recovery strategy that allows flexibility and provides alternative scenarios for each stage of processing. Implementation of nonthermal technologies, addition of green solvents, and safer materials (possessing GRAS status) are strongly recommended. The conducting of integral investigations that include recovery protocols and preservation assays are necessary, too. These parameters will ensure industrial exploitation and sustainability of the final product. Methodologies with fewer recovery steps are cheaper and scale-up is easier, but at the same time they generate cruder products with lower concentrations of target compounds. This fact alters the functional properties of the developed products since some of the target compounds are replaced with coextracted ingredients. Nevertheless, safety assessment and market release permission of purified active compounds is rather demanding. This procedure includes long and sophisticated tests on different species of laboratory animals (similar to synthetic antioxidants). In the case of enriched natural extracts, the criteria are not so strict. This is because natural extracts are considered to exist inherently in foods and thus safety concerns are limited. In any case, the development of tailor-made applications for the recovered products (crude or highly purified) is necessary, as target compounds may not be as beneficial as proposed theoretically, and more importantly, it is difficult to survive competition in the functional foods market. With regard to the recovery stages, the fifth stage (product formation) is the more essential and thus needs deeper investigation. Indeed, encapsulation enhances functionality and extends the shelf-life of the products. Conclusively, researchers will soon deal with the prospect of applying emerging technologies and particularly nanotechniques with an ultimate goal of optimizing overall efficiency of suggested methodologies. This concept will definitely reopen the debate concerning the safety of products recovered from food wastes and the impact (beneficial or not) of recycling them inside the food chain (Galanakis, 2012).

Other problems may arise from the market needs for healthier products. Authorities around the world (especially in Europe via the European Food Safety Authority) have tightened up the way in which companies can advertise health benefits. This policy is driven by the need to protect consumers from false or dubious claims. On the other hand, demonstration of proven health benefits is very costly for companies active in the field. For instance, health claims have only been approved for a small number of compounds (e.g. hydroxytyrosol in olive oil) and products (e.g. cholesterol-reducing yogurts and butters). This fact creates implications for stifling innovation in the field, as the required data are too much and most companies (typically start-ups with low funding) cannot afford them. Besides, the risk of claims rejection by the corresponding authorities is too high. This is not only a problem for the companies involved. A broader discussion is also needed about the implementation of real sustainability in the food industry and the way that research and innovation efforts encourage the development of licensed novel foods. Perhaps the establishment of a new label (similar to organic foods) or establishment of a tax reduction to relevant products could reveal the potentiality of recovering valuable compounds from food waste and reutilizing them in food products.

REFERENCES

Andre, A., Bazhal, M., Bouzrara, H., Vorobiev, E., 2001. Method for extracting liquid from a cellular material and devices therefor. WO Patent App. PCT/FR2001/000,490.

Anming, Z., De, F., Xiaoyan, H., Jianmin, Z., Bing, L., Lei, X., Qiyin, D., 2010. Process for extracting non-pectin soluble pomace dietary fibers. State Intellectual Property Office of the People's Republic of China, CN 101817809 (A).

Ariga, T., Yamazaki, E., Yamashita, K., Sasaki, M., Yamatsugu, N., Ishii, N., 1999. Protein food. Japanese Patent Office, JP 199,980,075,070. http://www.newgemfoods.com/

Asano, I., Hamaguchi, K., Fushii, S., Iino, H., 2002. *In vitro* digestibility and fermentation of manooligosaccharides from coffee mannan. Food Sci. Technol. 9 (1), 62–63.

Asano, I., Fujii, S., Mutoh, K., Takao, I., Ozaki, K., Nakamuro, K., Matsushima, T., 2006. Mannooligosaccharide composition for body fat reduction. Patent WO 2006/036208.

Ballesteros, L.F., Teixeira, J.A., Mussatto, S.L., 2014. Chemical, functional, and structural properties of spent coffee grounds and coffee silverskin. Food Bioprocess Technol. 7, 3493–3503.

Beverungen, C., 2005. Process for obtaining hydroxytyrosol from olive leaves extracts. EP Patent EP 1582512 A1.

Bonnell 1983. Treatment of citrus fruit peel. Australian Government – IP Australia, 1130883 (A).

Borrelli, R.C., Esposito, F., Napolitano, A., Ritieni, A., Fogliano, V., 2004. Characterization of a new potential functional ingredient: coffee silverskin. J. Agric. Food Chem. 52, 1338–1343.

Bousseta, N., Lanoiselle, J.L., Logeat, M., Manteau, S., Vorobiev, E., 2013. Procede d'extraction de molecules d'interet a partir de tout ou partie d'une matrice vegetale. WO Patent App. PCT/EP2011/070,597.

Brown, P.H., Crowely, C.P., 2009. Method of deflavoring whey protein using membrane electrodialysis. US2009/7582326, http://www.google.co.in/patents/US7582326

Coffee flour, http://www.coffeeflour.com.

Crea, R., 2002a. Method of obtaining a hydroxytyrosol-rich composition from vegetation water. World Intellectual Property Organization, WO/2002/0218310.

Crea, R., 2002b. Producing vegetation water from olives; adding acid; incubating the acidified vegetation water for a period of at least two months, until at least 75% of oleoeuropein has been converted to hydroxytyrosol. US Patent 2002/0198415 A1.

Crea, R., 2004. An hydroxytyrosol-rich composition from olive vegetation water and method of use thereof. World Intellectual Property Organization, WO/2004/005228 A1.

Crea, R., Caglioti, L., 2005. Water-soluble extract from olives. US Patent 6936287 B1.

Cuomo, J., Rabovskiy, A.B., 2002. Antioxidant compositions extracted from olives and olive by-products. US Patent 2002/0004077 A1.

Cuomo, J., Rabovskiy, A.B., 2004. Antioxidant compositions extracted from a wastewater from olive oil production. US Patent 6361803 B1.

Davis, M., Su, S., Ming, F., Yang, M., Ichinomiya, A., Melachouris, N., 2002. Isolation of glycoproteins from bovine milk. World Intellectual Property Organization, WO/2002/080961.

Davis, M.E., Nelson, L.A., Keenan, J.M., Pins, J.J., 2003. Reducing cholesterol with hydrolyzed whey protein. World Intellectual Property Organization, WO/2003/063778.

del Castillo, M.D., Ibáñez, E., Amigo, M., Herrero, M., Plaza, M., Ullate, M., 2013. Application of products of coffee silverskin in anti-aging cosmetics and functional food. WO 2013/004873.

del Castillo, M.D., Martinez-Saez, N., Ullate, M., 2014. Healthy bakery products with high level of dietary antioxidant fibre. PCT/ES2014/070062.

Deng, Q., Zinoviadou, K.G., Galanakis, C.M., Orlien, V., Grimi, N., Vorobiev, E., Lebovka, N., Barba, F.J., 2015. The effects of conventional and non-conventional processing on glucosinolates and its derived forms, isothiocyanates: extraction, degradation and applications. Food Eng. Rev., in press.

Emmons, W., Guttersen, C., 2005. Isolation of oleuropein aglycon from olive vegetation water. US Patent/2005/0103711 A1.

Esquivel, P., Jiménez, V.M., 2012. Functional properties of coffee and coffee by-product. Food Res. Int. 46 (2), 488–495.

European Patent Office, www.epo.org.

Food and Drug Administration (FDA), 2012. http://www.fda.gov/ucm/groups/fdagov-public/@fdagov-foods-gen/documents/document/ucm346897.pdf.

Food and Drug Administration (FDA), 2014. http://www.accessdata.fda.gov/scripts/fdcc/?set=GRASNotices&id=459.

Fernández-Bolaños, J., Heredia, A., Rodríguez, G., Rodríguez, R., Guillén, R., Jiménez, A., 2002. Method for obtaining purified hydroxytyrosol from products and by-products derived from the olive tree. World Intellectual Property Organization, WO/2002/064537.

Fernández-Bolaños, J., Rodríguez, G., Rodríguez, R., Guillén, R., Jiménez, A., 2006. Extraction of interesting organic compounds from olive oil waste. Grasas Y Aceites 57, 95–106.

Gachovska, T., Ngadi, M., Raghavan, V., 2008. Pulsed electric field enhanced method of extraction. WO Patent App. PCT/CA2007/001,652.

Galanakis, C.M., 2011. Olive fruit and dietary fibers: components, recovery and applications. Trends Food Sci. Technol. 22, 175–184.

Galanakis, C.M., 2012. Recovery of high added-value components from food wastes: conventional, emerging technologies and commercialized applications. Trends Food Sci. Technol. 26, 68–87.

Galanakis, C.M., 2013. Emerging technologies for the production of nutraceuticals from agricultural by-products: a viewpoint of opportunities and challenges. Food Bioprod. Process. 91, 575–579.

Galanakis, C.M., 2015. Separation of functional macromolecules and micromolecules: from ultrafiltration to the border of nanofiltration. Trends Food Sci. Technol. 42, 44–63.

Galanakis, C.M., Schieber, A., 2014. Editorial. Special issue on recovery and utilization of valuable compounds from food processing by-products. Food Res. Int. 65, 299–484.

Galanakis, C.M., Tornberg, E., Gekas, V., 2010a. A study of the recovery of the dietary fibres from olive mill wastewater and the gelling ability of the soluble fibre fraction. LWT – Food Sci. Technol. 43, 1009–1017.

Galanakis, C.M., Tornberg, E., Gekas, V., 2010b. Clarification of high-added value products from olive mill wastewater. J. Food Eng. 99, 190–197.

Galanakis, C.M., Tornberg, E., Gekas, V., 2010c. Dietary fiber suspensions from olive mill wastewater as potential fat replacements in meatballs. LWT – Food Sci. Technol. 43, 1018–1025.

Galanakis, C.M., Tornberg, E., Gekas, V., 2010d. Recovery and preservation of phenols from olive waste in ethanolic extracts. J. Chem. Technol. Biotechnol. 85, 1148–1155.

Galanakis, C.M., Tornberg, E., Gekas, V., 2010e. The effect of heat processing on the functional properties of pectin contained in olive mill wastewater. LWT – Food Sci. Technol. 43, 1001–1008.

Galanakis, C.M., Goulas, V., Tsakona, S., Manganaris, G.A., Gekas, V., 2013a. A knowledge base for the recovery of natural phenols with different solvents. Int. J. Food Prop. 16, 382–396.

Galanakis, C.M., Markouli, E., Gekas, V., 2013b. Fractionation and recovery of different phenolic classes from winery sludge via membrane filtration. Separ. Purif. Technol. 107, 245–251.

Galanakis, C.M., Chasiotis, S., Botsaris, G., Gekas, V., 2014. Separation and recovery of proteins and sugars from Halloumi cheese whey. Food Res. Int. 65, 477–483.

Galanakis, C.M., Kotanidis, A., Dianellou, M., Gekas, V., 2015. Phenolic content and antioxidant capacity of Cypriot Wines. Czech J. Food Sci. 33, 126–136.

Galanakis, C.M., Patsioura, A., Gekas, V., 2015b. Enzyme kinetics modeling as a tool to optimize food biotechnology applications: a pragmatic approach based on amylolytic enzymes. Crit. Rev. Food Sci. Technol. 55, 1758–1770.

Guangyu, H., Xiaoyan, Z., 2011. Method for preparing punicalagin and ellagic acid from pomegranate rind. State Intellectual Property Office of the People's Republic of China, CN 101974043A.

Heng, W.W., Xiong, L.W., Ramanan, R.N., Hong, T.L., Kong, K.W., Galanakis, C.M., Prasad, K.N., 2015. Two level factorial design for the optimization of phenolics and flavonoids recovery from palm kernel by-product. Ind. Crop. Prod. 63, 238–248.

Hirschberg, E., Pan, Z., McHugh, T.H. 2006. Methods for preparing freeze dried foods. WO/2006/099553.

Jensen, J., Larsen, P.H., 1993. Method for obtaining high-quality protein products from whey. World Intellectual Property Organization, WO/1993/021781.

Jishan, L., Jian, Z., Jianyong, G., Yang, L., Niannian, N., Hongying, G., 2009. Method for extracting and recycling albumin from whey wastewater from production of soy protein isolate. State Intellectual Property Office of the People's Republic of China, CN 101497645.

Kanatt, S.R., Chander, R., Sharma, A., 2008. Chitosan and mint mixture: a new preservative for meat and meat products. Food Chem. 107, 845–852.

Khanal, S.K., 2007. Ultrasonication in soy processing for enhanced protein and sugar yields and subsequent nisin production. US Patent No. 60/914,502.

Kong, K.-W., Khoo, H.-E., Prasad, K.N., Ismail, A., Tan, C.-P., Rajab, N.F., 2010. Revealing the power of the natural red pigment lycopene. Molecules 15, 959–987.

Körmendy, I., Körmendy, L., Ferenczy, A., 1998. Thermal inactivation kinetics of mixed microbial populations. A hypothesis paper. J. Food Eng. 38, 439–453.

Kortschack, F., 2007. Method for processing raw materials having a cell structure in the meat, meat by-products, fish and seafood processing industry. CA Patent App. CA 2,620,122.

Lavecchia, R., Zuorro, A., 2008. Process for the extraction of lycopene. World Intellectual Property Organization, WO/2008/055894. Production, Composition, and Application of Coffee and its Industrial Residues.

Lee, J.H., 2002. Precipitation recovery process for food waste sludge. US Patent 6368657 B1.

Liu, J., Schalch, W., Wang-Schmidt, Y., Wertz, K., 2008. Novel use of hydroxytyrosol and olive extracts/concentrates containing it. WO Patent App. WO/2008/128552.

Martinez-Saez, N., Ullate, M., Martin-Cabrejas, M.A., Martorell, P., Genovés, S., Ramon, D., et al., 2014. A novel antioxidant beverage for body weight control based on coffee silverskin. Food Chem. 150, 227–234.

Masquelier, J., 1987. Plant extract with a proanthocyanidins content as therapeutic agent having radical scavenger effect and use thereof. US Patent 4,698,360.

Mazza, G., Cacace, J.E., 2011. Extraction of phytochemicals. US Patent 7,943,190.

Mesías, M., Navarro, M., Martínez-Saez, N., Ullate, M., del Castillo, M.D., Morales, F.J., 2014. Antiglycative and carbonyl trapping properties of the water soluble fraction of coffee silverskin. Food Res. Int. 62, 1120–1126.

Meyerhoeffer, Jr. C., Showalter, A., 2012. Systems, devices, and/or methods for washing and drying a product. US Patent 12/08,182,551.

Mussatto, S.I., Sampaio, A., Dragone, G.M., Teixeira, J.A., 2013. Bebida destilada a partir da borra de café e respectivo método de produção (Distilled beverage from spent coffee grounds and respective production method). Patent PT105346.

Napolitano, A., Fogliano, V., Tafuri, A., Ritieni, A., 2007. Natural occurrence of ochratoxin A and antioxidant activities of green and roasted coffees and corresponding byproducts. J. Agric. Food Chem. 55, 10499–10504.

Nell, P., 2001. Method for the extraction of vegetable juices from vegetable residue and/or from vegetable remnants residue. US Patent 6,296,888.

Ngadi, M., Raghavan, V., Gachovska, T., 2012. Pulsed electric field enhanced method of extraction. US Patent 8,147,879.

Petrotos, K.B., Goutsidis, P.E., Christodouloulis, K., 2010. Method of total discharge of olive mill vegetation waters with co-production of polyphenol powder and fertilizer. Greek Patent 1006660.

Pourfarzad, A., Mahdavian-Mehr, H., Sedaghat, N., 2013. Coffee silverskin as a source of dietary fiber in bread-making: optimization of chemical treatment using response surface methodology. LWT – Food Sci. Technol. 50, 599–606.

Rahmanian, N., Jafari, S.M., Galanakis, C.M., 2014. Recovery and removal of phenolic compounds from olive mill wastewater. J. Am. Oil Chem. Soc. 91, 1–18.

Rizvi, A.F., Tong, C.H., 1997. Fractional conversion for determining texture degradation kinetics of vegetables. J. Food Sci. 62, 1–7.

Rouanet, M., Potherat, J.-J., Cousse, H., 1999. Polyphenolic composition, useful as food supplement, functional food or cosmetic composition. World Intellectual Property Organization, WO/1999/030724.

Roselló-Soto, E., Barba, F.J., Parniakov, O., Galanakis, C.M., Grimi, N., Lebovka, N., Vorobiev, E., 2015. High voltage electrical discharges, pulsed electric field and ultrasounds assisted extraction of protein and phenolic compounds from olive kernel. Food Bioprocess. Technol. 8, 885–894.

Sato, T., Morikawa, H., 2011. Silver skin-containing paper and method for producing the same. Patent US 7,927,460.

Schieber, A., Stintzing, F.C., Carle, R., 2001. By-products of plant food processing as a source of functional compounds – recent developments. Trends Food Sci. Technol. 12, 401–413.

Shenghui, Z., 1995. Preparation of chitosan derivative fruit and vegetable antistaling agent. State Intellectual Property Office of the People's Republic of China, CN 1106999.

Takao, I., Asano, I., Fujii, S., Kaneko, M., Nielson, J.R., Steffen, D.G., Hatzold, T. 2009. Composition having blood pressure reducing and/or elevation suppressing effect and food and drink containing the same. US Patent 2009/0005342 (PCT/JP2006/301025).

Todd, G.N., 2000. High temperature countercurrent solvent extraction of capsicum solids. US Patent 6,074,687.

Tokimoto, T., Kawasaki, N., Nakamura, T., Akutagawa, J., Tanada, S., 2005. Removal of lead ions in drinking water by coffee grounds as vegetable biomass. J. Colloid Interf. Sci. 281, 56–61.

Tornberg, E., Galanakis, C.M., 2008. Olive Waste Recovery. World Intellectual Property Organization, WO/2008/082343.

Tsakona, S., Galanakis, C.M., Gekas, V., 2012. Hydro-ethanolic mixtures for the recovery of phenols from Mediterranean plant materials. Food Bioprocess. Technol. 5, 1384–1393.

USDA, 2014. http://www.fsis.usda.gov/wps/portal/fsis/topics/regulatory-compliance/new-technologies/new-technology-information-table.

Vidal, O.P., Vorobiev, E., 2013a. Method and apparatus for treating vegetable tissues in order to extract therefrom a vegetable substance, in particular a juice. US Patent 13/696,040.

Vidal, O.P., Vorobiev, E., 2013b. Procede et installation de traitement de tissus vegetaux pour en extraire une substance vegetale, notamment un jus. EP Patent App. EP20,110,716,571.

Zhang, Q., Barbosa-Cánovas, G., Swanson, B., 1995. Engineering aspects of pulsed electric field pasteurization. J. Food Eng. 25 (2), 261–281.

RECOVERY AND APPLICATIONS OF ENZYMES FROM FOOD WASTES

16

Dimitris P. Makris

School of Environment, University of the Aegean, Lemnos, Greece

16.1 INTRODUCTION

Amongst other value-added constituents, food processing by-products and wastes contain an appreciable burden in various biocatalysts, the usefulness of which should not be overlooked. It is irrefutable that most plant enzymes can be used following an extensive, laborious, and costly downstream process, but from the literature available to date it has become clear that several attractive concepts of the low-cost applicability of these substances do exist. A significant emphasis is being given to the utility of oxidative enzymes because of their versatile function, high activity, relative stability, and operational ease.

Plant oxidative enzymes (oxidoreductases), such as peroxidases (PODs) and polyphenol oxidases (PPOs), have been used in several applications, pertaining to biosensors, biocatalysis, immunoassays, organic synthesis, etc. (Hamid and Rehman, 2009; Ryan et al., 2006; Xu, 2005). However, a vast number of studies have been conducted on the use of oxidative enzymes for bioremediation processes, based on concrete evidence that their deployment might be advantageous over conventional treatments, which aim at detoxifying recalcitrant organic pollutants (Alcalde et al., 2006; Demarche et al., 2012; Durán and Esposito, 2000; Karam and Nicell, 1997). The potential advantages of enzymic treatment as compared with conventional treatment include:

1. operation over a wide range of contaminant concentrations,
2. operation over a wide range of pH, temperature, and salinity,
3. absence of problems associated with the acclimatization of biomass,
4. reduction in sludge volume (no biomass generated), and
5. ease and simplicity of controlling the process.

Another significant advantage over conventional chemical treatments is enzyme specificity, which limits undesired side reactions that could increase reactant consumption and raise the cost of treatment.

However, the most intriguing challenge in using enzymes from residual plant materials for bioremediation is that they can be used as crude preparations or extracts, without the necessity for advanced purification, as opposed to other applications, such as organic synthesis, biotechnology, biosensors, etc., where enzyme purity is indispensable. In bioremediation processes involving enzymes, it is salient that the optimal conditions for the enzyme are maintained throughout. This requires enzymes originating from inexpensive sources, with high substrate affinity (K_m in the micromolar range) and low dependency on expensive redox cofactors (e.g. NAD(P)H), which might prove prohibitive in a commercial implementation.

In such a framework, this chapter focuses neither on fungal enzymes, with the exception of edible mushrooms, nor on enzymes from sources other than actual or potential agri-food solid wastes. The sole purpose is to bring out the importance of by-products and wastes generated from the food industry, as a cost-effective and versatile means of recovering highly valuable biocatalysts, as well as to highlight their usefulness by adducing representative fields of applicability.

16.2 ENZYMES FROM PLANT FOOD PROCESSING WASTES

16.2.1 PEROXIDASES (PODS)

PODs are of widespread occurrence in plant material rejected during the processing of edible plant tissues. Thus, this material may be regarded as a low-cost source of PODs, which might have high prospects in several environmental and food treatment applications. Nevertheless, although PODs from several residual sources have been studied with respect to their biochemical properties, including spring cabbage (Belcarz et al., 2008), broccoli (Duarte-Vázquez et al., 2007), asparagus (Jaramillo-Carmona et al., 2013), lentil stubbles (Hidalgo-Cuadrado et al., 2012), and wheat bran (Manu and Prasada Rao, 2009), only PODs from a few sources have been extensively tested for use in various processes.

By far the most studied POD is the enzyme recovered from the roots of horseradish (*Armoracia rusticana*) and, though not a plant processing by-product *sensu stricto*, horseradish roots are the main source of plant POD. Horseradish POD (HRP), along with peanut, soybean (SBP), turnip, tomato, and barley PODs belong to the class III peroxidases (classical secretory plant peroxidases) and catalyse the oxidation of a wide variety of electron donor substrates, such as phenols and aromatic amines, by H_2O_2 (Azavedo et al., 2003).

In heme peroxidases (EC 1.11.1.7) that have a ferriprotoporphyrin IX prosthetic group located at the active site, HRP catalyse the oxidative coupling of phenolic compounds using H_2O_2 as the oxidizing agent. The reaction is a three-step cyclic reaction by which the enzyme is first oxidized by H_2O_2 and then reduced in two sequential one-electron transfer steps from reducing substrates (Fig. 16.1), typically a phenol derivative (Derat and Shaik, 2006; Henriksen et al., 1999).

The greatest effort has been directed on the utilization of HRP for bioremediation purposes. The usefulness of HRP as a detoxifying agent with potential in bioremediation has been pointed out in early studies, which demonstrated the ability of HRP to act on aromatic amines, removing them from aqueous media (Klibanov and Morris, 1981). This pioneering study was followed by numerous investigations, which illustrated the potency of HRP to oxidize a wide range of industrial organic pollutants (Table 16.1), such as phenols, cresols, chlorophenols, dyes, estrogens, polychlorinated biphenyls, and bisphenol A. The vast majority of these studies demonstrated a particularly high efficiency of HRP to remove pollutants through radical formation, polymerization, and consequent sedimentation of the insoluble polymers formed, frequently with the aid of an additive, e.g. polyethylene glycol (PEG), Tweens, etc.

In most instances, the high efficiency of the enzyme, both in its free (soluble) form or immobilized, was expressed in pH values ranging from 5 to 7, although treatments included pH values from 2.5 to 9, depending mainly on the nature of the target molecule and the type of the additive, if used. Likewise, the preferred temperature of the treatments studied was 25°C. This temperature is fairly suitable for the enzyme to express high activity, but also a value close to that encountered under real operating conditions. In spite of the large number of reports on the enzyme efficiency tested in model wastewaters

$$[(Fe(III))Porph^{2-}]^+ + H_2O_2 \rightarrow [(Fe(IV)=O)Porph^{\bullet-}]^{\bullet+} + H_2O$$

Compound I

$$[(Fe(IV)=O)Porph^{\bullet-}]^{\bullet+} + AH \rightarrow [(Fe(IV)=O)Porph^{2-}] + H^+ + A^\bullet$$

Compound II

$$[(Fe(IV)=O)Porph^{2-}] + H^+ + AH \rightarrow [(Fe(III)Porph^{2-}]^+ + H_2O + A^\bullet$$

FIGURE 16.1 Reactions Involved in POD-Catalyzed Oxidation Mechanism of Small Phenolic Substrates

Porph, ferriprotoporphirin IX prosthetic group; AH, substrate.

Based on Henriksen et al. (1999)

(aqueous solutions), early studies demonstrated the potency of free enzymes by treating bleaching plant effluent, where more than 60% removal of industrial dye was achieved (Paice and Jurasek, 1984). The use of minced horseradish as a crude source of POD was also shown to be effective, provoking complete removal of 2,4-dichlorophenol from an industrial effluent (Dec and Bollag, 1994).

In a similar frame, POD from SBP was proven by several investigations to be particularly effective in removing a spectrum of recalcitrant pollutants (Table 16.2) within a pH range of 5.5 to 9. It is noteworthy that in several cases, crude enzyme extracts (Al-Ansari et al., 2009; Wilberg et al., 2002) or even whole soybean seed hulls (Bassi et al., 2004; Flock et al., 1999; Geng et al., 2004), a regularly generated soybean processing waste, were used instead of purified enzymes, providing removal yields of common toxicants (e.g. phenol and chlorophenols) of more than 95%. A similar outcome

Table 16.1 Applications of HRP in Bioremediation

Enzyme Form	Target Compound(s)	Effluent Treated	Conditions	Outcome	References
Immobilized enzyme	Phenol, *p*-chlorophenol	Aqueous solution	pH: 7; *T*: 25–35°C	Nearly complete removal	Bayramoğlou and Arica (2008)
Free enzyme	Industrial dyes	Aqueous solution	pH: 2.5	Over 50% removal	Bhunia et al. (2001)
Immobilized enzyme	4-Chlorophenol	Aqueous solution	pH: 7	Nearly complete removal	Bódalo et al. (2008)
Immobilized enzyme	Phenol	Aqueous solution	pH: 7	Over 90% removal	Cheng et al. (2006)
Immobilized enzyme	4-Chlorophenol	Aqueous solutions	pH: 5.5, *T*: 25°C	Nearly complete removal	Dalal and Gupta (2007)
Partly purified extract	Substituted phenols, chlorophenols	Aqueous solution	pH: 6–7.5; *T*: 15–30°C	35.1–100% removal	Davidenko et al. (2004)
Minced horseradish	2,4-Dichlorophenol	Industrial effluent	pH: 4–7	100% removal	Dec and Bollag (1994)

(Continued)

Table 16.1 Applications of HRP in Bioremediation *(cont.)*

Enzyme Form	Target Compound(s)	Effluent Treated	Conditions	Outcome	References
Free enzyme	Textile dyes	Aqueous solution	pH: 4–5; T: 25°C	52–94% removal	de Souza et al. (2007)
Free enzyme	o-Xylene, naphthalene	Aqueous solution	pH: 6; T: 23°C	Approximately 54% removal	Fang and Barcelona (2003)
Free enzyme	Bisphenol-A	Reverse micelle	pH: 7; T: 40°C	Nearly complete removal	Hong-Mei and Nicell (2008)
Free enzyme	Bisphenol-A	Aqueous solution	pH: 7	Nearly complete removal	Huang and Weber (2005)
Free enzyme	Natural and synthetic estrogens	Aqueous solution	pH: 6.7–7.5; T: 25°C	Approximately 95% removal	Khan and Nicell (2007)
Free enzyme	Pentachlorophenol	Aqueous solution	pH: 6.5; Tween 20, Tween 80	88% removal	Kim et al. (2006)
Free enzyme	Aromatic amines	Aqueous solution	pH: 5.5	98.3–100% removal	Klibanov and Morris (1981)
Free enzyme	Substituted phenols, chlorophenols	Aqueous solution	pH: 7; T: 25°C	Approximately 98% removal	Nazari et al. (2007)
Free enzyme	Dyes	Bleach plant effluent	pH: 5	Over 60% removal	Paice and Jurasek (1984)
Immobilized enzyme	4-Chlorophenol	Aqueous solution	pH: 7.4	Over 80% removal	Siddique et al. (1993)
Free enzyme	Polychlorinated biphenyls	Aqueous solution	pH: 4	60–80% removal	Singh et al. (2000)
Immobilized enzyme	Chlorophenols	Aqueous solution	pH: 7; T: 25°C	Nearly complete removal	Tatsumi et al. (1996)
Minced horseradish	2,4-Dichlorophenol	Aqueous solution	pH: 7; PEG addition	Approximately 90% removal	Tonegawa et al. (2003)
Free enzyme	Phenol and 4-chlorophenol	Aqueous solution	pH: 9; polymer coagulant addition	Nearly complete removal	Tong et al. (1997)
Free enzyme	Phenol, chlorophenols, and cresols	Aqueous solution	pH: 5–8.5; additives	Nearly complete removal	Wu et al. (1997)
Free enzyme	Chlorophenols and cresols	Aqueous solution	pH: 5–7	56.2–100% removal	Yamada et al. (2007)
Free enzyme	Pentachlorophenol	Aqueous solution	pH: 4–5	Nearly complete removal	Zhang and Nicell (2000)
Immobilized enzyme	Substituted phenols, amino-, and chlorophenols	Aqueous solution	pH: 7	34.4–87.6% removal	Zhang et al. (2010)

Table 16.2 Applications of SBP in Bioremediation

Enzyme Form	Target Compound(s)	Effluent Treated	Conditions	Outcome	References
Crude enzyme extract	Phenylenediamines, benzenediols	Aqueous solution	pH: 4.5–8	95% removal	Al-Ansari et al. (2009)
Soybean seed hulls	Phenol, chlorophenols	Aqueous solution	pH: 7–8	80–96% removal	Bassi et al. (2004)
Immobilized enzyme	4-Chlorophenol	Aqueous solution	pH: 7	Over 96% removal	Bódalo et al. (2008)
Free enzyme	Phenol, chlorophenols, crezols, bisphenol A	Aqueous solution	pH: 5.5–8; PEG addition	Over 95% removal	Caza et al. (1999)
Soybean seed hulls	Phenol, 2-chlorophenol	Aqueous solution	pH: 6.5; detergent addition	Over 96% removal	Flock et al. (1999)
Soybean seed hulls	2-Chlorophenol, 2,4-dichlorophenol	Soil slurry	pH: 7	Over 96% removal	Geng et al. (2004)
Free enzyme	2,4-Dichlorophenol	Aqueous solution	pH: 8.2; $T = 22°C$; PEG addition	Over 83% removal	Kennedy et al. (2002)
Free enzyme	Phenol	Aqueous solution	pH: 7, $T = 25 \pm 1°C$; PEG addition	Over 95% removal	Kinsley and Nicell (2000)
Free enzyme	Dinitrotoluenes	Aqueous solution	pH: 5.2	Over 95% removal	Patapas et al. (2007)
Free enzyme	Phenol	Aqueous solution; refinery wastewater	pH: 5.6–8; PEG addition	Over 95% removal	Steevensz et al. (2009)
Crude enzyme extract	Phenol	Aqueous solution	pH: 6	95% removal	Wilberg et al. (2002)
Free enzyme	Phenol, chlorophenols, cresols	Aqueous solution	pH: 6–9	Nearly complete removal	Wright and Nicell (1999)

using minced potato has also been reported (Dec and Bollag, 1994). This finding is of prime importance, considering that processes involving enzyme purification would inevitably be proven far more costly. In a recent comparative assessment, SBP was deemed a more suitable biocatalyst for 4-chlorophenol removal than both HRP and artichoke POD (Máximo et al., 2012), exhibiting also the highest thermostability.

PODs from a few other sources appeared promising means of bioremediation (Table 16.3) and in some cases this ability was well substantiated by treating industrial wastewaters. A crude extract from onion solid wastes was shown to decrease polyphenol concentration in olive mill wastewater (OMW) by more than 50% (Barakat et al., 2010). Reduction of 2,4-dichlorophenol and other phenolics by more

Table 16.3 Applications of PODs from Various Agri-Food Waste Sources in Bioremediation

Enzyme Form	Target Compound(s)	Effluent Treated	Conditions	Outcome	References
Immobilized bitter gourd enzyme	Phenol, chlorophenols	Aqueous solution	pH: 5–6; T: 40°C	96% removal	Akhtar and Husain (2006)
Immobilized bitter gourd enzyme	Dyes	Aqueous solution	pH: 3–4; T: 40°C	Over 80% removal	Akhtar et al. (2005)
Crude onion extract	Phenolics	OMW	pH: 2.76	Over 50% removal	Barakat et al. (2010)
Minced potato	2,4-Dichlorophenol	Industrial effluent	pH: 5	87.9% removal	Dec and Bollag (1994)
Crude artichoke extract	4-Chlorophenol	Aqueous solution	pH: 7	35% removal	López-Molina et al. (2003)
Immobilized bitter gourd enzyme	Benzidine	Aqueous solution	pH: 5; T: 40°C	58% removal	Karim and Husain (2011)
Crude turnip extract	Cresols, chlorophenols, bisphenol A	Aqueous solution	pH: 4–8; T: 25°C	Over 85% removal	Duarte-Vázquez et al. (2002)
Immobilized tomato enzyme	Direct dyes	Aqueous solution	pH: 6; T: 40°C	93% removal	Matto and Husain (2008)
Immobilized turnip enzyme	Phenolics	Industrial dye effluent	pH: 7.2; PEG addition	Over 95% removal	Quintanilla-Guerrero et al. (2008)
Partly purified bitter gourd enzyme	Dyes	Aqueous solution	pH: 3; T: 40°C	90% removal	Satar and Husain (2009)

than 88% was also observed in industrial effluents treated with minced potato (Dec and Bollag, 1994) and immobilized turnip POD (Quintanilla-Guerrero et al., 2008), respectively. Crude enzyme preparations from artichoke (López-Molina et al., 2003), turnip (Duarte-Vázquez et al., 2002), and bitter gourd (Satar and Husain, 2009) displayed variable efficacies, reaching 96% removal, when applied to model effluents.

Despite the extended literature on the utilization of PODs from waste plant sources for bioremediation, the evidence accumulated suggested a significant perspective of these enzymes in the generation of various bio-based compounds as well. The use of POD from bitter gourd, a vegetable widely consumed in China, has been shown to produce a ferulic acid dehydrodimer with strong anti-inflammatory activity (Ou et al., 2003), while further studies on ferulic acid oxidation by this enzyme demonstrated the generation of two other dehydrodimers (Liu et al., 2005). Treatment of a ferulic acid/resveratrol mixture with a partly purified bitter gourd POD extract resulted in the formation of dimers of both ferulic acid and resveratrol, as well as in a heterocoupling oligomer (Yu et al., 2007), exerting strong radical scavenging and cardioprotective effects. Likewise, the oxidation of sinapic acid afforded six oxidation products, mostly dimers, with increased antioxidant potency, compared with the parent molecule (Liu et al., 2007).

Similar evidence has also been drawn on the use of POD from onion solid wastes. The apical trims of the onion bulb, as well as the outer dry and semidry layers that are discarded during onion processing, were used as the source of crude POD preparations. A first detailed study, which used the physiological substrate quercetin (Osman et al., 2008), suggested a rather unusual mechanism of oxidative cleavage, implying a dioxigenase-type mechanism of oxidation. The detection of some quercetin oxidation products in onion solid wastes possessing POD activity further confirmed this assumption (Khiari and Makris, 2012), while findings from investigations on structurally similar flavonols, such as fisetin (Osman and Makris, 2010) and morin (Osman and Makris, 2011), consisted of an additional verification for the oxidation pathways proposed. The pH optima were between 3 and 7, contrasting previous findings, which indicated that quercetin oxidation by a crude onion peroxidase extract was favored at pH 8 (Takahama and Hirota, 2000).

Apart from the above-mentioned flavonoids, onion POD from crude waste preparations was extensively studied with respect to oxidation of simpler polyphenols, including the structurally related hydroxycinnamates hydrocaffeic acid (El Agha et al., 2008a), ferulic acid (El Agha et al., 2008b), caffeic acid (El Agha et al., 2009), p-coumaric acid (El Agha and Makris, 2012), and chlorogenic acid (Osman et al., 2012). In all these cases, it was suggested that POD-mediated oxidation resulted in the formation radicals, followed by combinations thereof to yield dimers up to tetramers, but the generation of dehydrodimers, alone or as adducts with dimers, was also deduced for hydrocaffeic, caffeic, and p-coumaric acids.

A higher value option for crude onion waste POD has been substantiated by experiments pertaining to the synthesis of specific bio-based polyphenols. Treatment of 2',3,4,4',6'-pentahydroxy-chalcone with a crude preparation resulted in the formation of aureusidin, an aurone occurring mainly in the flowers of certain plant species, with peculiar biological properties such as anticancer activity (Moussouni et al., 2010). This class of flavonoids is biosynthesized in plant tissues upon the activity of a PPO-like oxidase on similar substrates. Furthermore, the oxidation of methyl esters of p-coumaric acid, caffeic acid, and ferulic acid with the same crude extract yielded dihydrobenzofuran lignans, which were isolated and their structure was fully elucidated (Moussouni et al., 2011). In a reducing power test, some of these derivatives were proven to be superior antioxidants compared with the parent molecules. The synthetic potential of this crude enzyme was also highlighted by the catalytic formation of coumestans and benzofuroquinolinones, which are heterocycles with an important pharmacological interest (Angeleska et al., 2013).

16.2.2 POLYPHENOL OXIDASES (PPOS)

PPOs, also known as tyrosinases (EC 1.14.18.1), are copper-containing oxidoreductases, widely distributed through the phylogenetic scale, from bacteria to mammals, and even with different biochemical features in various organs of the same organism, such as in the leaves, fruits, and roots of higher plants (Sánchez-Ferrer et al., 1995; Yoruk and Marshall, 2003). PPOs catalyze two distinct reactions: the hydroxylation of a monophenol to produce an o-diphenol (monophenolase activity) and the oxidation of the o-diphenol yielding the corresponding o-quinone (diphenolase activity). Both reactions require molecular oxygen (Fig. 16.2). o-Quinones being highly reactive molecules may spontaneously

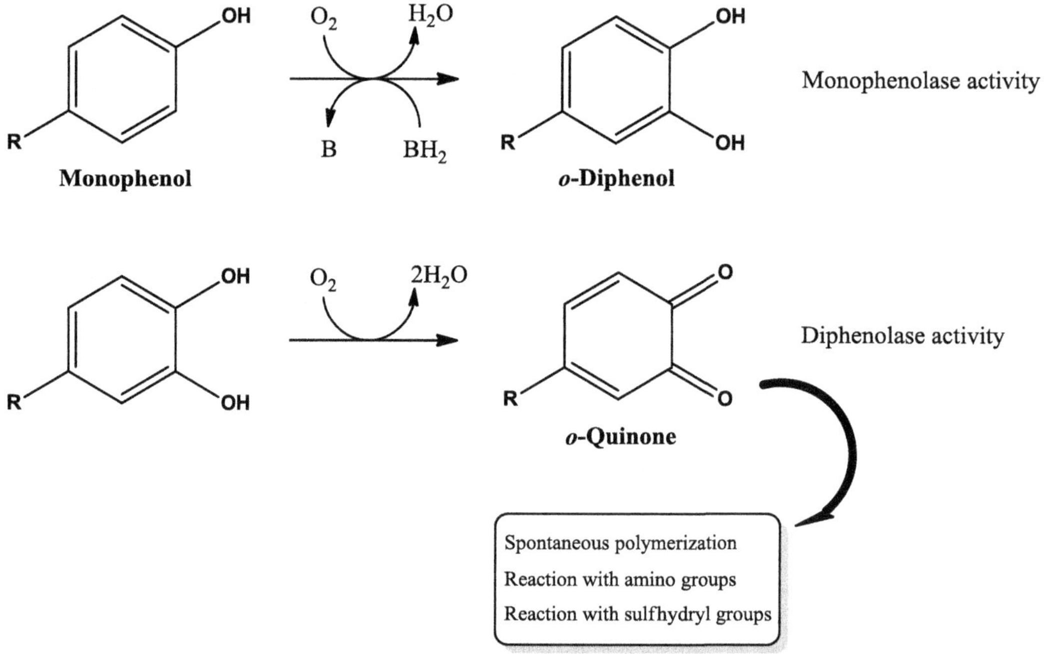

FIGURE 16.2 Cascade of Reactions Involved in PPO-Catalyzed Oxidation of Phenolic Compounds

BH_2, hydrogen donor.

Based on Yoruk and Marshall (2003)

polymerize with no enzyme intervention into large molecular weight, insoluble brown substances, or react with nucleophiles, such as amines and sulfhydryl-containing substances.

PPOs may be considered another enzyme class that could have a potential in various low-cost environmental processes. It is to be mentioned that a plethora of fungi and other microorganisms have been broadly studied as a means of bioremediation for several organic pollutants (Chiacchierini et al., 2004; Mukherjee et al., 2013), due to their potency in the biosynthesis of oxidative enzymes with a wide range of potential substrates, such as laccases (EC 1.10.3.1). Nonetheless, PPOs recovered from plant food processing residues are gaining attention, as witnessed by several recent studies (Table 16.4).

However, unlike PODs, the spectrum of plant waste sources for PPO recovery is much narrower, although a variety of fruit and vegetable PPO-active homogenates have been tested for their ability to oxidize pollutants, including bisphenol A (Imanaka et al., 2005; Yoshida et al., 2002), polyaromatic hydrocarbons (Lau et al., 2003), OMW phenolics (Greco et al., 1999; Toscano et al., 2003), and indigo carmine (Solís et al., 2013), and producing bio-based substances such as theaflavins (Tanaka et al., 2002). In fact, only PPOs from potato and mushrooms have been tested in depth as bioremediation agents, whereas data on PPO from other sources are particularly limited in this regard.

Mushroom PPO, originating from *Agaricus bisporus* and *Pleurotus* spp. processing residues (trims), was the subject of earlier studies, due to its abundance and the relatively wide spectrum of substrates.

Table 16.4 Applications of Potato PPO in Bioremediation

Enzyme Form	Target Compound(s)	Effluent Treated	Conditions	Outcome	References
Crude enzyme extract	Hydrocaffeic acid	Aqueous solution	pH: 3	Over 99% removal	Demian and Makris (2013a)
Crude enzyme extract	Phenolics	OMW	pH: 4; PEG addition	54% removal	Demian and Makris (2013b)
Partly purified enzyme	Pentachlorophenol	Aqueous solution	pH: 5; T: 25°C	Over 70% removal	Hou et al. (2011)
Free and immobilized enzyme	Dyes	Industrial dying effluent	pH: 3–4; T: 37°C	0–96% removal	Khan and Husain (2007a)
Partly purified enzyme	Dyes	Aqueous solution	pH: 3; T: 37°C	18–97% removal	Khan and Husain (2007b)
Partly purified enzyme	Benzo[a]pyrene	Aqueous solution	pH: 6.75	Over 60% removal	Kirso et al. (1981)
Immobilized enzyme	Halogenated phenols	Aqueous solution	pH: 8–9	55% removal	Lončar et al. (2011)
Partly purified enzyme	Dyes	Industrial dying effluent	pH: 3	93–100% removal	Lončar et al. (2012)
Immobilized enzyme	Halogenated phenols	Aqueous solution	pH: 7	Over 90% removal	Lončar and Vujčičs (2011)
Immobilized enzyme	Phenol	Aqueous solution	pH: 7	Over 86% removal	Shao et al. (2007)
Immobilized enzyme	Phenol	Aqueous solution	pH: 7	Over 90% removal	Shao et al. (2009)
Crude enzyme extract	Bisphenol A	Aqueous solution	pH: 8; T: 40–45°C	Over 95% removal	Xuan et al. (2002)

The treatment of frequently encountered pollutants, such as phenol and chlorophenols, both in model and industrially generated effluents, showed a very satisfactory ability for this enzyme, applied under various forms (free, immobilized, crude extract, etc.). Although the pH range used varied from 5 to 7.8, in most cases high removal efficacy was observed at pH 7 (Table 16.5).

On the other hand, potato PPO has for a long time been known to act on substances such as benzo[*a*] pyrene (Kirso et al., 1981), a toxic environmental pollutant, but only recently has there been a renewed interest in the use of this enzyme obtained from potato peels, the most abundant potato-processing waste. Efficient activity has been proven for a variety of recalcitrant effluent constituents, such as industrial dyes (Khan and Husain, 2007a, b; Lončar et al., 2012), halogenated phenols (Hou et al., 2011; Lončar et al., 2011; Lončar and Vujčić, 2011), etc., giving in many cases removal yields over 90%, but the treatment of real wastewaters, such as OMW (Demian and Makris, 2013b), has also had a promising outcome, with a 54% removal of phenolics.

Table 16.5 Applications of Mushroom PPO in Bioremediation

Enzyme Form	Target Compound(s)	Effluent Treated	Conditions	Outcome	References
Partly purified enzyme	Phenols	Industrial effluent	pH: 7.8	100% removal	Atlow et al. (1984)
Immobilized enzyme	Phenol and cresol	Industrial effluent	pH: 5–6.8	Nearly complete removal	Edwards et al. (1999)
Immobilized enzyme	Phenolics and chlorophenols	Aqueous solution	pH: 7	10–92% removal	Grecchio et al. (1995)
Free enzyme	Phenol and chlorophenols	Aqueous solution	pH: 7	25–100% removal	Ikehata and Nicell (2000)
Crude enzyme extract	Phenol	Aqueous solution	No pH control; T: 30°C	90% removal	Kameda et al. (2006)
Crude enzyme extract	Aniline and phenolics	Aqueous solution	pH: 7	21.5–100% removal	Trejo-Hernandez et al. (2001)
Immobilized enzyme	Phenols, cresols, chlorophenols	Aqueous solution	pH: 7; T: 25°C	100% removal	Wada et al. (1993)
Cross-linked aggregate	Phenol, p-cresol, 4-chlorophenol, bisphenol A	Aqueous solution	pH: 6; T: 30°C	100% removal	Xu and Yang (2013)
Free enzyme	p-Alkyphenols	Aqueous solution	pH: 7; T: 45°C	97–100% removal	Yamada et al. (2006)

16.2.3 OTHER ENZYMES

References pertaining to the recovery of enzymes other than oxidoreductases from plant food wastes are particularly confined, apart from those deriving from microorganisms using residual materials as substrates (Dhillon et al., 2013). Essentially, there are reports only on carbohydrate-degrading enzymes, such as a pectinesterase from apple wastes (King, 1991) and amylases deriving from *Opuntia ficus-indica* seeds (Ennouri et al., 2013), *Citrus sinensis* peels (Mohamed et al., 2010), and spent mushroom (Phan and Sabaratnam, 2012). The examination of enzymes from barley, wheat (Jin et al., 2013), and rice bran (Wang et al., 2010), capable of transforming glutamate to γ-aminobutyric acid, has also been reported, but no practical applications embracing the use of these sources has been developed.

16.3 FISH AND SEAFOOD PROCESSING WASTES

It is nowadays a consolidated concept that discarded materials from aquatic organisms may have a variety of possible valorization options in several industrial sectors, including food products for human consumption. By-products may constitute as much as 70% of fish and shellfish after industrial processing and much focus has been on converting these into commercial products

(Olsen et al., 2014). According to the Food and Agricultural Organization, in 2008 around 27 million tonnes of marine biomass were used for nonfood purposes, including fish meal, fish oil, bait, or high added-value compounds production by pharmaceutical or cosmetic industries (Ordóñez-Del Pazo et al., 2014).

However, despite extensive research and development, only a limited number of high value-added products based on fish processing wastes have become established on the market. Major reasons for this are probably overestimation of market demand, small amounts of high-quality by-products available on a regular basis, high costs of isolating specific components often present in small amounts in the by-products, and the availability of alternative, more cost-effective sources (Olsen et al., 2014). The processing residual materials from marine organisms include mainly viscera and heads, and apart from the recovery of functional ingredients, such as ω-fatty acid-enriched fish oil, peptides, collagen (Rustad et al., 2011), proteins (Sanmartín et al., 2012), and chitin (Beaulieu et al., 2009), these materials are also rich in commercially valuable biocatalysts, possessing lipolytic (Sovik and Rustad, 2005), collagenolytic, proteolytic, and caseinolytic activities (Salamone et al., 2012).

The most important digestive proteases of fish viscera are acid stomach enzymes and alkaline intestine enzymes. The main alkaline enzymes are trypsin, chymotrypsin, and elastase, all belonging to the serine protease family (EC 3.4.21) and characterized by serine, histidine, and aspartic residues at the active site. Trypsin is one of the major digestive enzymes, belonging to the serine protease family of enzymes (EC 3.4.21.4) and it acts by cleaving the ester and peptide bonds involving the carboxyl groups of arginine or lysine. Proteases constitute the most important group of industrial enzymes used in the world today, accounting for about 50% of the total industrial enzyme market (Bougatef, 2013), since they play a very significant role in various industrial applications. For example, pepsin and trypsin are widely used in food technology for a number of purposes, such as protein coagulation and hydrolysis, selective tissue degradation, meat tenderization, etc. (Ferraro et al., 2013). Enzymic hydrolysis of proteins allows preparation of bioactive peptides and these can be obtained by in vitro hydrolysis of protein sources using appropriate proteolytic enzymes.

Although microbial proteases dominate the enzyme market worldwide, alternative sources have been investigated and evaluated. The internal organs of fish are a rich source of enzymes, many of which exhibit high catalytic activities at relatively low concentrations. The enzymes available in fish may include pepsin, trysin, chymotrypsin, and collagenase, but also cathepsin B (Sovik and Rustad, 2006). These enzymes are commercially extracted from the fish viscera on a large scale and they may display higher efficiency at lower temperatures, reduced sensitivity to substrate concentrations, and higher stability in a wide range of pH (Ghaly et al., 2013).

Besides viscera, which is regularly rejected following fish gutting (Ferraro et al., 2013) and constitutes the major pool of proteolytic enzymes, other sources, such as surimi wash water (DeWitt and Morrissey, 2002) and shrimp heads (Ganugula et al., 2008) have also been reported as a means of recovering a range of different proteases. Although proteolytic enzymes from various marine organisms have been investigated (Bougatef, 2013; Freitas-Júnior et al., 2012), the most thoroughly studied and abundant fish and seafood by-products and wastes are those deriving from sardine processing (Ben Khaled et al., 2008; Ferraro et al., 2013; Salazar-Leyva et al., 2013). Recently, it was demonstrated that proteolytic preparations from sardine viscera, exhibiting trypsin, chymotrypsin, aminopeptidase, and pepsin activities, were more effective than commercially available proteases, yielding threefold higher hydrolysis (Castro-Ceseña et al., 2012).

16.4 FUTURE PROSPECTS

It is beyond reasonable doubt that the waste streams generated from the food industry are a rich and low-cost source of a variety of valuable natural products. The literature available to date provides sufficient data to spotlight the importance of utilizing food-processing residues for the production of high value-added substances, yet the methodologies proposed for recovery and application in many instances suffer serious shortcomings. This is because the techniques implemented may yield excellent results on a laboratory scale, but do not meet fundamental criteria for scaling-up and industrial production. The high cost, the ideal conditions employed, and the generation of further wastes might render laboratory techniques completely incompatible with a realistic perspective. Thus, a critical evaluation of the data presented in this chapter leads to the following criticism:

1. An issue of significance pertaining to studies related with the use of enzymes for bioremediation is the large-scale applicability of the methodologies proposed, as most of the investigations reported results using purified PODs and PPOs, and model effluents containing specific pollutants. Although the use of pure biocatalysts and media with defined composition of pollutants is of undisputed importance in providing fundamental kinetic and mechanistic data, the lack of testing enzyme potency in real wastewaters would reasonably raise serious concerns for the implementation of such technologies in practice. This uncertainty should be by no means perceived as being interdictory to deploying biocatalysts recovered from plant food residues in wastewater treatment, but the necessity for further and more thorough examination is to be stressed.

2. The use of fish and seafood as sources of industrial enzymes implicates some limitations associated with seasonability, variations in the content and/or activity of enzymes, and the highly perishable nature of the raw material. Although the utilization of these enzymes is desirable in many food applications, economic viability of the process and products must be assessed on a realistic basis. This is mainly because the cost incurred in recovering these enzymes from their natural sources is a limitation for their widespread use. More extensive research to identify the most specific and promising enzymes and to determine optimal conditions for their use is of utmost importance. Research towards this direction would provide the incentive for commercial developments leading to large-scale production of enzymes at a much lower cost.

REFERENCES

Akhtar, S., Husain, Q., 2006. Potential applications of immobilized bitter gourd (*Momordica charantia*) peroxidase in the removal of phenols from polluted water. Chemosphere 65, 1228–1235.

Akhtar, S., Khan, A.A., Husain, Q., 2005. Potential of immobilized bitter gourd (*Momordica charantia*) peroxidases in the decolorization and removal of textile dyes from polluted wastewater and dyeing effluent. Chemosphere 60, 291–301.

Al-Ansari, M.M., Steevensz, A., Al-Aasm, N., Taylor, K.E., Bewtra, J.K., Biswas, N., 2009. Soybean peroxidase-catalyzed removal of phenylenediamines and benzenediols from water. Enzyme Microb. Technol. 45, 253–260.

Alcalde, M., Ferrer, M., Plou, F.J., Ballesteros, A., 2006. Environmental biocatalysis: from remediation with enzymes to novel green processes. Trends Biotechnol. 24, 281–287.

Angeleska, S., Kefalas, P., Detsi, A., 2013. Crude peroxidase from onion solid waste as a tool for organic synthesis. Part III: synthesis of tetracyclic heterocycles (coumestans and benzofuroquinolinones). Tetrahydron Lett. 54, 2325–2328.

Atlow, S.C., Bonadonna-Aparo, L., Klibanov, A.M., 1984. Dephenolization of industrial wastewaters catalysed by polyphenol oxidase. Biotechnol. Bioeng. 26, 599–603.

Azavedo, A.M., Martins, V.C., Prazeres, D.M.F., Vojinović, V., Cabral, J.M.S., Fonseca, L.P., 2003. Horseradish peroxidase: a valuable tool in biotechnology. Biotechnol. Annu. Rev. 9, 199–247.

Barakat, N., Makris, D.P., Kefalas, P., Psillakis, E., 2010. Removal of olive mill waste water phenolics using a crude peroxidase extract from onion by-products. Environ. Chem. Lett. 8, 271–275.

Bassi, A., Geng, Z., Gijzen, M., 2004. Enzymatic removal of phenol and chlorophenols using soybean seed hulls. Eng. Life Sci. 4, 125–130.

Bayramoğlou, G., Arica, M.Y., 2008. Enzymatic removal of phenol and p-chlorophenol in enzyme reactor: horseradish peroxidase immobilized on magnetic beads. J. Hazard. Mater. 156, 148–155.

Beaulieu, L., Thibodeau, J., Bryl, P., Carbonneau, M.-E., 2009. Characterization of enzymatic hydrolyzed snow crab (*Chionoecetes opilio*) by-product fractions: a source of high-valued biomolecules. Bioresource Technol. 100, 3332–3342.

Belcarz, A., Ginalska, G., Kowalewska, B., Kulesza, P., 2008. Spring cabbage peroxidases – potential tool in biocatalysis and bioelectrocatalysis. Phytochemistry 69, 627–636.

Ben Khaled, H., Bougatef, A., Balti, R., Triki-Ellouz, Y., Souissi, N., Nasri, M., 2008. Isolation and characterisation of trypsin from sardinelle (*Sardinella aurita*) viscera. J. Sci. Food Agr. 88, 2654–2662.

Bhunia, A., Durani, S., Wangikar, P.P., 2001. Horseradish peroxidase catalyzed degradation of industrially important dyes. Biotechnol. Bioeng. 72, 562–567.

Bódalo, A., Bastida, J., Máximo, M.F., Montiel, M.C., Murcia, M.D., 2008. A comparative study of free and immobilized soybean and horseradish peroxidases for 4-chlorophenol removal: protective effects of immobilization. Bioprocess Biosyst. Eng. 31, 587–593.

Bougatef, A., 2013. Trypsins from fish processing waste: characteristics and biotechnological applications – comprehensive review. J. Clean. Prod. 57, 257–265.

Castro-Ceseña, A.B., del Pilar Sánchez-Saavedra, M., Márquez-Rocha, F.J., 2012. Characterisation and partial purification of proteolytic enzymes from sardine by-products to obtain concentrated hydrolysates. Food Chem. 135, 583–589.

Caza, N., Bewtra, J.K., Biswas, N., Taylor, K.E., 1999. Removal of phenolic compounds from synthetic wastewater using soybean peroxidase. Water Res. 33, 3012–3018.

Cheng, J., Yu, S.M., Zuo, P., 2006. Horseradish peroxidase immobilized on aluminium-pillared interlayered clay for the catalytic oxidation of phenolic wastewater. Water Res. 40, 283–290.

Chiacchierini, E., Restuccia, D., Vinci, G., 2004. Bioremediation of food industry effluents: recent applications of free and immobilised polyphenoloxidases. Food Sci. Technol. Int. 10, 373–382.

Dalal, S., Gupta, N., 2007. Treatment of phenolic wastewater by horseradish peroxidase immobilized by bioaffinity layering. Chemosphere 67, 741–747.

Davidenko, T.I., Oseychuk, O.V., Sevastyanov, O.V., Romanovskaya, I.I., 2004. Peroxidase oxidation of phenols. Appl. Biochem. Microbiol. 40, 542–546.

de Souza, S.M.A.G.U., Forgiarini, E., de Souza, A.A.U., 2007. Toxicity of textile dyes and their degradation by the enzyme horseradish peroxidase (HRP). J. Hazard. Mater. 147, 1073–1078.

Dec, J., Bollag, J.-M., 1994. Use of plant material for the decontamination of water polluted with phenols. Biotechnol. Bioeng. 44, 1132–1139.

Demarche, P., Junghanns, C., Nair, R.R., Agathos, S.N., 2012. Harnessing the power of enzymes for environmental stewardship. Biotechnol. Adv. 30, 933–953.

Demian, F.D., Makris, D.P., 2013a. Factorial design optimisation of hydrocaffeic acid removal from an aqueous matrix by the use of a crude potato polyphenoloxidase. Biocatal. Agr. Biotechnol. 2, 305–310.

Demian, F.D., Makris, D.P., 2013b. Removal of olive mill wastewater phenolics with the use of a polyphenol oxidase homogenate from potato peel waste. J. Waste Manage., article ID 630209.

Derat, E., Shaik, S., 2006. An efficient proton-coupled electron-transfer process during oxidation of ferulic acid by horseradish peroxidase: coming full cycle. J. Am. Chem. Soc. 128, 13940–13949.

DeWitt, C.A.M., Morrissey, M.T., 2002. Pilot plant recovery of catheptic proteases from surimi wash water. Bioresource Technol. 82, 295–301.

Dhillon, G.S., Kaur, S., Brar, S.K., 2013. Perspective of apple processing wastes as low-cost substrates for bioproduction of high value products: a review. Renew. Sust. Energ. Rev. 27, 789–805.

Duarte-Vázquez, M.A., Ortega-Tovar, M.A., García-Almendarez, B.E., Regalado, C., 2002. Removal of aqueous phenolic compounds from a model system by oxidative polymerization with turnip (*Brassica napus* L. var purple top white globe) peroxidase. J. Chem. Technol. Biotechnol. 78, 42–47.

Duarte-Vázquez, M.A., García-Padilla, S., García-Almendarez, B.E., Whitaker, J.R., Regalado, C., 2007. Broccoli processing wastes as a source of peroxidase. J. Agr. Food Chem. 55, 10396–10404.

Durán, N., Esposito, E., 2000. Potential applications of oxidative enzymes and phenoloxidase-like compounds in wastewater and soil treatment: a review. Appl. Catal. B: Environ. 28, 83–99.

Edwards, W., Bownes, R., Leukes, W.D., Jacobs, E.P., Sanderson, R., Rose, P.D., Burton, S.G., 1999. A capillary membrane bioreactor using immobilized polyphenol oxidase for the removal of phenols from industrial effluents. Enzyme Microb. Technol. 24, 209–217.

El Agha, A., Makris, D.P., 2012. Biocatalytic characteristics, product formation and putative pathway of *p*-coumaric acid oxidation by a crude peroxidase from onion. Acta Aliment. 41, 304–315.

El Agha, A., Makris, D.P., Kefalas, P., 2008a. Hydrocaffeic acid oxidation by a peroxidase homogenate from onion solid wastes. European Food Res. Technol. 227, 1379–1386.

El Agha, A., Makris, D.P., Kefalas, P., 2008b. Peroxidase-active cell free extract from onion solid wastes: biocatalytic properties and putative pathway of ferulic acid oxidation. J. Biosci. Biotechnol. 106, 279–285.

El Agha, A., Abbeddou, S., Makris, D.P., Kefalas, P., 2009. Biocatalytic properties of a peroxidase-active cell-free extract from onion solid wastes: caffeic acid oxidation. Biodegradation 20, 143–153.

Ennouri, M., Khemakhem, B., Hassen, H.B., Ammar, I., Belghith, K., Attia, H., 2013. Purification and characterization of an amylase from *Opuntia ficus-indica* seeds. J. Sci. Food Agr. 93, 61–66.

Fang, J., Barcelona, M.J., 2003. Coupled oxidation of aromatic hydrocarbons by horseradish peroxidase and hydrogen peroxide. Chemosphere 50, 105–109.

Ferraro, V., Carvalho, A.P., Piccirillo, C., Santos, M.M., Castro, P.M.L., Pintado, M.E., 2013. Extraction of high added value biological compounds from sardine, sardine-type fish and mackerel canning residues — a review. Mater. Sci. Eng. C 33, 3111–3120.

Flock, C., Bassi, A., Gijzen, M., 1999. Removal of aqueous phenol and 2-chlorophenol with purified soybean peroxidase and raw soybean hulls. J. Chem. Technol. Biotechnol. 74, 303–309.

Freitas-Júnior, A., Costa, H.M.S., Icimoto, M.Y., Hirata, I.Y., Marcondes, M., Carvalho, Jr., L.B., Oliveira, V., Bezerra, R.S., 2012. Giant Amazonian fish pirarucu (*Arapaima gigas*): its viscera as a source of thermostable trypsin. Food Chem. 133, 1596–1602.

Ganugula, R., Chakrabarti, R., Rao, K.R.S.S., 2008. Distribution of proteolytic activity in the different protein fractions of tropical shrimp head waste. Food Biotechnol. 22, 18–30.

Geng, Z., Bassi, A.S., Gijzen, M., 2004. Enzymatic treatment of soils contaminated with phenol and chlorophenols using soybean seed hulls. Water Air Soil Poll. 154, 151–166.

Ghaly, A.E., Ramakrishnan, V.V., Brooks, M.S., Budge, S.M., Dave, D., 2013. Fish processing wastes as a potential source of proteins, amino acids and oils: a critical review. J. Microbiol. Biochem. Technol. 5, 4.

Grecchio, C., Ruggiero, P., Pizzigallo, M.D.R., 1995. Polyphenoloxidases immobilized in organic gels: properties and applications in the detoxification of aromatic compounds. Biotechnol. Bioeng. 48, 585–591.

Greco, Jr., G., Toscano, G., Cioffi, M., Gianfreda, L., Sannino, F., 1999. Dephenolisation of olive mill waste-waters by olive husk. Water Res. 33, 3046–3050.

Hamid, M., Rehman, K.-U., 2009. Potential applications of peroxidases. Food Chem. 115, 1177–1186.

Henriksen, A., Smith, A.T., Gajhede, M., 1999. The structures of the horseradish peroxidase C-ferulic acid complex and the ternary complex with cyanide suggest how peroxidases oxidize small phenolic substrates. J. Biol. Chem. 274, 35005–35011.

Hidalgo-Cuadrado, N., Peérez-Galende, P., Manzano, T., De Maria, C.G., Shnyrov, V.L., Roig, M.G., 2012. Screening of postharvest agricultural wastes as alternative sources of peroxidases: characterization and kinetics of a novel peroxidase from lentil (*Lens culinaris* L.) stubble. J. Agr. Food Chem. 60, 4765–4772.

Hong-Mei, L., Nicell, J.A., 2008. Biocatalytic oxidation of bisphenol A in a reverse micelle system using horseradish peroxidase. Bioresource Technol. 99, 4428–4437.

Hou, M.-F., Tang, X.-Y., Zhang, W.-D., Liao, L., Wan, H.-F., 2011. Degradation of pentachlorophenol by potato polyphenol oxidase. J. Agric. Food Chem. 59, 11456–11460.

Huang, Q., Weber, Jr., W.J., 2005. Transformation and removal of bisphenol A from aqueous phase via peroxidase-mediated oxidative coupling reactions: efficacy, products, and pathways. Environ. Sci. Technol. 39, 6029–6036.

Ikehata, K., Nicell, J.A., 2000. Characterization of tyrosinase for the treatment of aqueous phenols. Bioresource Technol. 74, 191–199.

Imanaka, M., Yamabe, S.-I., Yamamoto, J., Koezuka, K., Take, S., Sato, A., Sasaki, K., 2005. Oxidative degradation of bisphenol A by fruit homogenates. J. Food Sci. 70, 529–533.

Jaramillo-Carmona, S., Lopez, S., Vazquez-Castilla, S., Rodriguez-Arcos, R., Jimenez-Araujo, A., Guillen-Bejarano, R., 2013. Asparagus byproducts as a new source of peroxidases. J. Agric. Food Chem. 61, 6167–6174.

Jin, W.-J., Kim, M.-J., Kim, K.-S., 2013. Utilization of barley or wheat bran to bioconvert glutamate to γ-aminobutyric acid (GABA). J. Food Sci. 78, 1376–1382.

Kameda, E., Langone, M.A.P., Coelho, M.A.Z., 2006. Tyrosinase extract from *Agaricus bisporus* mushroom and its *in natura* tissue for specific phenol removal. Environ. Technol. 27, 1209–1215.

Karam, J., Nicell, J.A., 1997. Potential applications of enzymes in waste treatment. J. Chem. Technol. Biotechnol. 69, 141–153.

Karim, Z., Husain, Q., 2011. Removal of benzidine from polluted water by soluble and immobilized peroxidase in batch processes and continuous horizontal bed reactor. Environ. Technol. 32, 83–91.

Kennedy, K., Alemany, K., Warith, M., 2002. Optimisation of soybean peroxidase treatment of 2,4-dichlorophenol. Water SA 28, 149–158.

Khan, A.A., Husain, Q., 2007a. Decolorization and removal of textile and non-textile dyes from polluted wastewater and dyeing effluent by using potato (*Solanum tuberosum*) soluble and immobilized polyphenol oxidase. Bioresource Technol. 98, 1012–1019.

Khan, A.A., Husain, Q., 2007b. Potential of plant polyphenol oxidases in the decolorization and removal of textile and non-textile dyes. J. Environ. Sci. 19, 396–402.

Khan, U., Nicell, J.A., 2007. Horseradish peroxidase-catalysed oxidation of aqueous natural and synthetic oestrogens. J. Chem. Technol. Biotechnol. 82, 818–830.

Khiari, Z., Makris, D.P., 2012. Stability and transformation of major flavonols in onion (*Allium cepa*) solid wastes. J. Food Sci. Technol. 49, 489–494.

Kim, E.Y., Choi, Y.J., Chae, H.J., Chu, K.H., 2006. Removal of aqueous pentachlorophenol by horseradish peroxidase in the presence of surfactants. Biotechnol. Bioprocess Eng. 11, 462–465.

King, K., 1991. Characteristics of pectinesterase isolated from Bramley apple waste. J. Sci. Food Agr. 57, 43–48.

Kinsley, C., Nicell, J.A., 2000. Treatment of aqueous phenol with soybean peroxidase in the presence of polyethylene glycol. Bioresource Technol. 73, 139–146.

Kirso, U., Belykh, I., Stom, D., 1981. Oxidation of benzo[a]pyrene catalysed by phenol oxidases of alga *Nitella* sp. and potato tubers. Acta Hydrochim. Hydrobiol. 9, 401–406.

Klibanov, A.M., Morris, E.D., 1981. Horseradish peroxidase for the removal of carcinogenic aromatic amines from water. Enzyme Microb. Technol. 3, 119–122.

Lau, K.L., Tsang, Y.Y., Chiu, S.W., 2003. Use of spent mushroom compost to bioremediate PAH-contaminated samples. Chemosphere 52, 1539–1546.

Liu, H.-L., Kong, L.-Y., Takaya, Y., Niwa, M., 2005. Biotransformation of ferulic acid into two new dihydrotrimers by *Momordica charantia* peroxidase. Chem. Pharm. Bull. 53, 816–819.

Liu, H.-L., Wan, X., Huang, X.-F., Kong, L.-Y., 2007. Biotransformation of sinapic acid catalysed by *Momordica charantia* peroxidase. J. Agric. Food Chem. 55, 1003–1008.

Lončar, N., Vujčič, Z., 2011. Tentacle carrier for immobilization of potato phenoloxidase and its application for halogenophenols removal from aqueous solutions. J. Hazard. Mater. 196, 73–78.

Lončar, N., Božić, N., Anđelković, I., Milovanović, A., Dojnov, B., Vujčić, M., Roglić, G., Vujčić, Z., 2011. Removal of aqueous phenol and phenol derivatives by immobilized potato polyphenol oxidase. J. Serb. Chem. Soc. 76, 513–522.

Lončar, N., Janović, B., Vujčić, M., Vujčić, Z., 2012. Decolorization of textile dyes and effluents using potato (*Solanum tuberosum*) phenoloxidase. Int. Biodeter. Biodegr. 72, 42–45.

López-Molina, D., Hiner, A.N.P., Tudela, J., García-Cánovas, F., Rodríguez-López, H.N., 2003. Enzymatic removal of phenols from aqueous solution by artichoke (*Cynara scolymus* L.) extracts. Enzyme Microb. Technol. 33, 738–742.

Manu, B.T., Prasada Rao, U.J.S., 2009. Calcium modulated activity enhancement and thermal stability study of a cationic peroxidase purified from wheat bran. Food Chem. 114, 66–71.

Matto, M., Husain, Q., 2008. Redox-mediated decolorization of Direct Red 23 and Direct Blue 80 catalyzed by bioaffinity-based immobilized tomato (*Lycopersicon esculentum*) peroxidase. Biotechnol. J. 3, 1224–1231.

Máximo, M.F., Gómez, M., Murcia, M.D., Ortega, S., Barbosa, D.S., Vayá, G., 2012. Screening of three commercial plant peroxidases for the removal of phenolic compounds in membrane bioreactors. Environ. Technol. 33, 1071–1079.

Mohamed, S.A., Drees, E.A., El-Badry, M.O., Fahmy, A.S., 2010. Biochemical properties of α-amylase from peel of *Citrus sinensis* cv. Abosora. Appl. Biochem. Biotechnol. 160, 2054–2065.

Moussouni, S., Detsi, A., Majdalani, M., Makris, D.P., Kefalas, P., 2010. Crude peroxidase from onion solid waste as a tool for organic synthesis. Part I: cyclization of 2′,3,4,4′,6′-pentahydroxy-chalcone into aureusidin. Tetrahydron Lett. 51, 4076–4078.

Moussouni, S., Saru, M.-L., Ioannou, E., Mansour, M., Detsi, A., Roussis, V., Kefalas, P., 2011. Crude peroxidase from onion solid waste as a tool for organic synthesis. Part II: oxidative dimerization–cyclization of methyl *p*-coumarate, methyl caffeate and methyl ferulate. Tetrahydron Lett. 52, 1165–1168.

Mukherjee, S., Basak, B., Bhunia, B., Dey, A., Mondal, B., 2013. Potential use of polyphenol oxidases (PPO) in the bioremediation of phenolic contaminants containing industrial wastewater. Rev. Environ. Sci. Biotechnol. 12, 61–73.

Nazari, K., Esmaeili, N., Mahmoudib, A., Rahimi, H., Moosavi-Movahedi, A.A., 2007. Peroxidative phenol removal from aqueous solutions using activated peroxidase biocatalyst. Enzyme Microb. Technol. 41, 226–233.

Olsen, R.L., Toppe, J., Karunasagar, I., 2014. Challenges and realistic opportunities in the use of by-products from processing of fish and shellfish. Trends Food Sci. Technol. 36, 144–151.

Ordóñez-Del Pazo, T., Antelo, L.T., Franco-Uría, A., Pérez-Martín, R.I., Sotelo, C.G., Alonso, A.A., 2014. Fish discards management in selected Spanish and Portuguese métiers: identification and potential valorisation. Trends Food Sci. Technol. 36, 29–43.

Osman, A., Makris, D.P., 2010. Comparison of fisetin and quercetin oxidation with a cell-free extract of onion trimmings and peel, plant waste, containing peroxidase enzyme: a further insight into flavonol degradation mechanism. Int. J. Food Sci. Technol. 45, 2265–2271.

Osman, A., Makris, D.P., 2011. Oxidation of morin (2′,3,4′,5,7-pentahydroxyflavone) with a peroxidase homogenate from onion. Int. Food Res. J. 18, 1039–1043.

Osman, A., Makris, D.P., Kefalas, P., 2008. Investigation on biocatalytic properties of a peroxidase-active homogenate from onion solid wastes: an insight into quercetin oxidation mechanism. Process Biochem. 43, 861–867.

Osman, A., El-Agha, A., Makris, D.P., Kefalas, P., 2012. Chlorogenic acid oxidation by a crude peroxidase preparation: biocatalytic characteristics and oxidation products. Food Bioprocess Technol. 5, 243–251.

Ou, L., Kong, L.-Y., Zhang, X.-M., Niwa, M., 2003. Oxidation of ferulic acid by *Momordica charantia* peroxidase and related anti-inflammation activity changes. Biol. Pharm. Bull. 26, 1511–1516.

Paice, M.G., Jurasek, L., 1984. Peroxidase-catalyzed color removal from bleach plant effluent. Biotechnol. Bioeng. 26, 477–480.

Patapas, J., Al-Ansari, M.M., Taylor, K.E., Bewtra, J.K., Biswas, N., 2007. Removal of dinitrotoluenes from water via reduction with iron and peroxidase-catalyzed oxidative polymerization: a comparison between *Arthromyces ramosus* peroxidase and soybean peroxidase. Chemosphere 67, 1485–1491.

Phan, C.-W., Sabaratnam, V., 2012. Potential uses of spent mushroom substrate and its associated lignocellulosic enzymes. Appl. Microbiol. Biotechnol. 96, 863–873.

Quintanilla-Guerrero, F., Duarte-Vázquez, M.A., García-Almendarez, B.E., Tinoco, R., Vazquez-Dulhalt, R., Regalado, C., 2008. Polyethylene glycol improves phenol removal by immobilized turnip peroxidase. Bioresource Technol. 99, 8605–8611.

Rustad, T., Storrø, I., Slizyte, R., 2011. Possibilities for the utilisation of marine by-products. Int. J. Food Sci. Technol. 46, 2001–2014.

Ryan, B.J., Carolan, N., Ó'Fágáin, C., 2006. Horseradish and soybean peroxidases: comparable tools for alternative niches? Trends Biotechnol. 24, 355–363.

Salamone, M., Cuttitta, A., Seidita, G., Mazzola, S., Bertuzzi, F., Ricordi, C., Ghersi, G., 2012. Characterization of collagenolytic/proteolytic marine enzymes. Chem. Eng. Trans. 27, 1–6.

Salazar-Leyva, J.A., Lizardi-Mendoza, J., Ramirez-Suarez, J.C., Valenzuela-Soto, E.M., Ezquerra-Brauer, J.M., Castillo-Yañez, F.J., Pacheco-Aguilar, R., 2013. Acidic proteases from Monterey sardine (*Sardinops sagax caerulea*) immobilized on shrimp waste chitin and chitosan supports: searching for a by-product catalytic system. Appl. Biochem. Biotechnol. 171, 795–805.

Sánchez-Ferrer, Á., Rodríguez-López, J.N., García-Cánovas, F., García-Carmona, F., 1995. Tyrosinase: a comprehensive review of its mechanism. Biochim. Biophys. Acta 1247, 1–11.

Sanmartín, E., Arboleya, J.C., Iloro, I., Escuredo, K., Elortza, F., Moreno, F.J., 2012. Proteomic analysis of processing by-products from canned and fresh tuna: identification of potentially functional food proteins. Food Chem. 134, 1211–1219.

Satar, R., Husain, Q., 2009. Use of bitter gourd (*Momordica charantia*) peroxidase together with redox mediators to decolorize disperse dyes. Biotechnol. Bioprocess Eng. 14, 213–219.

Shao, J., Ge, H., Yang, Y., 2007. Immobilization of polyphenol oxidase on chitosan–SiO$_2$ gel for removal of aqueous phenol. Biotechnol. Lett. 29, 901–905.

Shao, J., Huang, L.-L., Yang, Y.-M., 2009. Immobilization of polyphenol oxidase on alginate–SiO$_2$ hybrid gel: stability and preliminary applications in the removal of aqueous phenol. J. Chem. Technol. Biotechnol. 84, 633–635.

Siddique, M.H., St Pierre, C.C., Biswas, N., Bewtra, J.K., Taylor, K.E., 1993. Immobilized enzyme catalyzed removal of 4-chlorophenol from aqueous solution. Water Res. 27, 883–890.

Singh, A., Billingsley, K.A., Ward, O.P., 2000. Transformation of polychlorinated biphenyls with oxidative enzymes. Bioprocess Eng. 23, 421–425.

Solís, A., Perea, F., Solís, M., Manjarrez, N., Pérez, H.I., Cassani, J., 2013. Discoloration of indigo carmine using aqueous extracts from vegetables and vegetable residues as enzyme sources. BioMed Res. Int., article ID 250305.

Sovik, S.L., Rustad, T., 2005. Effect of season and fishing ground on the activity of lipases in byproducts from cod (Gadus morhua). LWT – Food Sci. Technol. 38, 867–876.

Sovik, S.L., Rustad, T., 2006. Effect of season and fishing ground on the activity of cathepsin B and collagenase in by-products from cod species. LWT – Food Sci. Technol. 39, 43–53.

Steevensz, A., Mousa Al-Ansari, M., Taylor, K.E., Bewtra, J.K., Biswas, N., 2009. Comparison of soybean peroxidase with laccase in the removal of phenol from synthetic and refinery wastewater samples. J. Chem. Technol. Biotechnol. 84, 761–769.

Takahama, U., Hirota, S., 2000. Deglucosidation of quercetin glucosides to the aglycone and formation of antifungal agents by peroxidase-dependent oxidation of quercetin on browning of onion scales. Plant Cell Physiol. 41, 1021–1029.

Tanaka, T., Mine, C., Inoue, K., Matsuda, M., Kouno, I., 2002. Synthesis of theaflavin from epicatechin and epigallocatechin by plant homogenates and role of epicatechin quinone in the synthesis and degradation of theaflavins. J. Agric. Food Chem. 50, 2142–2148.

Tatsumi, K., Wada, S., lchikawa, H., 1996. Removal of chlorophenols from wastewater by immobilized horseradish peroxidase. Biotechnol. Bioeng. 51, 126–130.

Tonegawa, M., Dec, J., Bollag, J.-M., 2003. Use of additives to enhance the removal of phenols from water treated with horseradish and hydrogen peroxide. J. Environ. Qual. 32, 1222–1227.

Tong, Z., Qingxiang, Z., Hui, H., Qin, L., Yi, Z., 1997. Removal of toxic phenol and 4-chlorophenol from waste water by horseradish peroxidase. Chemosphere 34, 893–903.

Toscano, G., Colarieti, M.T., Greco, Jr., G., 2003. Oxidative polymerisation of phenols by a phenol oxidase from green olives. Enzyme Microb. Technol. 33, 47–54.

Trejo-Hernandez, M.R., Lopez-Munguia, A., Quintero Ramirez, R., 2001. Residual compost of *Agaricus bisporus* as a source of crude laccase for enzymic oxidation of phenolic compounds. Process Biochem. 36, 635–639.

Wada, S., Ichikawa, H., Tatsumi, K., 1993. Removal of phenols from wastewater by soluble and immobilized tyrosinase. Biotechnol. Bioeng. 42, 854–858.

Wang, L., Xu, D.X., Lv, Y.G., Zhang, H., 2010. Purification and biochemical characterisation of a novel glutamate decarboxylase from rice bran. J. Sci. Food Agr. 90, 1027–1033.

Wilberg, K., Assenhaimer, C., Rubio, J., 2002. Removal of aqueous phenol catalysed by a low purity soybean peroxidase. J. Chem. Technol. Biotechnol. 77, 851–857.

Wright, H., Nicell, J.A., 1999. Characterization of soybean peroxidase for the treatment of aqueous phenols. Bioresource Technol. 70, 69–79.

Wu, Y., Taylor, K.E., Biswas, N., Bewtra, J.K., 1997. Comparison of additives in the removal of phenolic compounds by peroxidase-catalyzed polymerization. Water Res. 31, 2699–2704.

Xu, F., 2005. Applications of oxidoreductases: recent progress. Ind. Biotechnol. 1, 38–50.

Xu, D.-Y., Yang, Z., 2013. Cross-linked tyrosinase aggregates for elimination of phenolic compounds from wastewater. Chemosphere 92, 391–398.

Xuan, Y.J., Endo, Y., Fujimoto, K., 2002. Oxidative degradation of bisphenol A by crude enzyme prepared from potato. J. Agric. Food Chem. 50, 6575–6578.

Yamada, K., Inoue, T., Akiba, Y., Kashiwada, A., Matsuda, K., Hirata, M., 2006. Removal of *p*-alkylphenols from aqueous solutions by combined use of mushroom tyrosinase and chitosan beads. Biosci. Biotechnol. Biochem. 70, 2467–2475.

Yamada, K., Shibuya, T., Noda, M., Uchiyama, N., Kashiwada, A., Matsuda, K., Hirata, M., 2007. Influence of position of substituent groups on removal of chlorophenols and cresols by horseradish peroxidase and determination of optimum conditions. Biosci. Biotechnol. Biochem. 71, 2503–2510.

Yoruk, R., Marshall, M.R., 2003. Physicochemical properties and function of plant polyphenol oxidase: a review. J. Food Biochem. 27, 361–422.

Yoshida, M., Ono, H., Mori, Y., Chuda, Y., Mori, M., 2002. Oxygenation of bisphenol A to quinones by polyphenol oxidase in vegetables. J. Agric. Food Chem. 50, 4377–4381.

Yu, B.-B., Han, X.-Z., Lou, H.-X., 2007. Oligomers of resveratrol and ferulic acid prepared by peroxidase-catalyzed oxidation and their protective effects on cardiac injury. J. Agric. Food Chem. 55, 7753–7757.

Zhang, G., Nicell, J.A., 2000. Treatment of aqueous pentachlorophenol by horseradish peroxidase and hydrogen peroxide. Water Res. 34, 1629–1637.

Zhang, F., Zheng, B., Zhang, J., Huang, X., Liu, H., Guo, S., Zhang, J., 2010. Horseradish peroxidase immobilized on graphene oxide: physical properties and applications in phenolic compound removal. J. Phys. Chem. C 114, 8469–8473.

Subject Index

Printed in the United States
By Bookmasters